INTERNATIONAL REVIEWS IN

AERΟSOL PHYSICS AND CHEMISTRY

VOLUME 1

THE DYNAMICS OF AEROCOLLOIDAL SYSTEMS

THE DYNAMICS OF AEROCOLLOIDAL SYSTEMS

BY

G. M. HIDY
Science Center, North American Rockwell Corporation,
Thousand Oaks, California 91360

AND

J. R. BROCK
University of Texas, Austin, Texas

PERGAMON PRESS
Oxford · New York · Toronto
Sydney · Braunschweig

Pergamon Press Ltd., Headington Hill Hall, Oxford
Pergamon Press Inc., Maxwell House, Fairview Park, Elmsford, New York 10523
Pergamon of Canada Ltd., 207 Queen's Quay West, Toronto 1
Pergamon Press (Aust.) Pty. Ltd., 19A Boundary Street, Rushcutters Bay, N.S.W. 2011, Australia
Vieweg & Sohn GmbH, Burgplatz 1, Braunschweig

First edition 1970

Library of Congress Catalog Card No. 70–104120

PRINTED IN GREAT BRITAIN BY PAGE BROS. (NORWICH) LTD.

08 006613 5

CONTENTS

PREFACE

DESPITE an ever widening technological interest in aerocolloidal systems, the fundamental science of aerosols, on which practical considerations must rely, largely has been set aside until recently in deference to other problems. However, in the last decade, a kind of renaissance in aerosol research has taken place in which much of the classical work of the early twentieth century is being extended. New sophisticated theoretical and experimental techniques are being developed, and are being applied to better understand the behavior of aerosol systems.

As scientists in many diverse fields expand their activity in aerosol research, the results of investigations appear in a wide variety of journals that reach entirely different small groups of workers. The problems of communication of these scientific results are complicated further by the worldwide character of aerosol science. To help focus attention on the variety of important aerosol research presently being published, and to open a new channel for international communication between workers in this field, we have organized a new series entitled, *The International Reviews in Aerosol Physics and Chemistry*. This work will consist of a collection of monographs of book length, and companion volumes of selected review articles dealing with several aspects of aerosol science, and its relationship to the study of the so-called "particulate state of matter." The scope of the *International Reviews* will be limited to results which contribute significantly to the state of fundamental knowledge of aerosol behavior.

Because *The International Reviews in Aerosol Physics and Chemistry* is designed to concentrate on the fundamental aspects of aerosol science, it should have considerable usefulness to both practicing scientists and to graduate students in such widely diverse fields as physics, physical chemistry, meteorology, geophysics, astronomy, chemical engineering, mechanical engineering, aerospace engineering, and medicine.

We have chosen to introduce the *International Reviews* by publishing a monograph on the mechanics of aerocolloidal systems. This first volume contains a brief outline of various aspects of aerosol dynamics including topics ranging from transport processes to single particles to the generation and behavior of collections of particles, clouds. In this book, emphasis is placed on a truly distinguishing feature of particulate matter in the aerosol class. That is, the dynamics of particles in an environment requires consideration of fluid dynamics from the continuum regime to the free molecule regime.

Though some of the theory discussed here involves classical hydrodynamics, the bulk of the topics presented stem from rather recent studies being published at an alarming rate in the last twenty five years.

We anticipate that this book will be useful not only to the scholars but to scientists more concerned with practical aspects of aerosols. It is hoped that this volume also will be helpful to those who would consider teaching aerosol mechanics as part of a curriculum in the atmospheric sciences, or in other applied sciences including applied physical chemistry, and engineering.

It is a pleasure to acknowledge the support during the writing of this book of the National Science Foundation through its contract with the National Center for Atmospheric Research, and of the Public Health Service through its grant No. AP00479-01 to J. R. Brock. Thanks go to the many people who helped us with preparation and review of this manuscript. We are also grateful to those who kindly made the figures, plates and other data available to us for illustrating the many facets of aerosol dynamics.

G.M.H.
J.R.B.

PRINCIPAL NOMENCLATURE

Subscripts

i; 1, 2, 3, ...	Index for particles of species. Indexing refers to conservation of mass, or for unit density spheres $R_i = i^{1/3}R_1$, where species 1 is the unit mass.
α; A, B, ...	Index for gaseous components.
G	Gas.
∞	Property a large distance from a surface.
s	Conditions at a sphere surface.

Superscripts

$\oplus$	Dimensionless quantities.
$*$	Free molecule regime.
c	Continuum regime.
s	Slip flow regime.
$\star$	Critical conditions in a supersaturated vapor.

Italic letters

$\mathbf{A}$, $\mathbf{A}_\alpha$	Vector parameter in particular solution to the Boltzmann equation. Vector symbols in the illustrations are presented in the style **A** for the sake of clarity.
$\tilde{A}$	Correction parameter in the Stokes–Cunningham correction to the mobility of particles for Knudsen number, Kn_i.
a	Radius or length dimension of a body in a gas stream as a particle collector.
B_i	Particle mobility.
$\bar{\bar{B}}_\alpha$, $\bar{\bar{B}}$	Tensor parameter in particular solution to the Boltzmann equation.
b, b_{ij}	Collision parameter in the kinetic equations for coagulation.
C_v, C_p	Specific heat or heat capacity at constant volume or pressure.
$C_{\alpha\beta}$	Maxwell's coefficient of diffusional resistance.
c	Speed of light.
c_α	Mass density ratio, or concentration. ρ_α/ρ_G.
c_α^0	Molar concentration.
c_m	Isothermal slip coefficient.
c_t	Temperature jump coefficient.
c_v	Heat capacity at constant volume per molecule.
$c_{d\alpha}$	Concentration jump coefficient.
c_{tm}	Thermal creep coefficient.
D_i	Brownian diffusivity of particles.
$\mathbf{D}_\alpha$	Vector parameter in generalized flux equation.

$D_{\alpha\beta}$	Diffusivity of gas molecules in gas mixture.
d_α, d_o	Effective molecular diameters of gaseous species.
$\mathbf{d}_\alpha^a$	Parameter in generalized diffusion flux vector.
$\mathbf{d}_{\alpha\beta}$	Vector parameter in particular solution to Boltzmann's equation.
E	Kinetic or total energy or energy density.
$\tilde{e}$	Unit of electrical charge.
$\mathbf{F}, F$	Force acting on a particle (F' force per unit mass).
$\tilde{F}$	Ratio of particle velocity induced by electrical forces at a surface to gas flow velocity.
$\hat{F}$	Helmholtz free energy.
F_{ij}	Any function, $F_{ij} = F_i + F_j$.
f	Boltzmann distribution function.
f_w	Wind constant, $Nu/2$.
G_α, G_k	Chemical or collision "reaction" rate.
$\hat{G}$	Gibbs free energy.
$\tilde{G}$	Ratio of particle velocity induced by an electrical field to free stream velocity U_∞.
G_q	Velocity gradient.
$\mathbf{g}, g$	Molecular collisional parameter, relative molecular velocity.
H	Total rate of heat transfer to a surface.
$\hat{H}$	Specific enthalpy.
$\hat{h}$	Heat transfer coefficient.
$\mathbf{h}$	Heat flux vector.
I	Steady state nucleation rates.
$\tilde{I}$	Dynamical parameter, $v_{si}\rho_i/\rho_G\mathscr{A}$.
$\hat{I}$	Impaction parameter.
J	Collision function in Boltzmann equation.
$\mathbf{j}_\alpha, \mathbf{j}_i$	Mass flux to a surface.
K_k	Equilibrium constant for kth embryo.
$\hat{K}$	Interception parameter R_i/a or R_i/R_j.
k	Boltzmann's constant.
$\hat{k}$	Mass transfer coefficient.
k_G	Thermal conductivity of gas.
k_p, k_{pi}	Thermal conductivity of particle.
L	Characteristic aerodynamic length.
L_{st}	Stopping distance, $\mathscr{A}^{-1}v_s$.
L_{ij}	Binary, collision rate density.
M_α	Molecular weight of species α.
M_{ij}	Symmetry number.
m_i, m_α	Molecular or particle mass.
N	Total number of particles per unit volume.
N_o	Number of particles per unit volume at time $t = 0$.
$n(\tilde{R}), n(\tilde{v})$	Spectrum function for continuous size distribution.
$\mathbf{n}$	Vector normal to a surface.
n_α, n_G	Number density of gas molecules.
n_i	Number density of particles of species i.
$P, \tilde{p}$	Probability function.

p	Gas or vapor pressure.
p_o	Equilibrium vapor pressure over a flat surface.
p_α	Vapor pressure of species α.
Q_i	Electrostatic charge.
$\tilde{Q}_k$	Partition function of kth species.
$\mathbf{q}$, $\mathbf{q}_G$	Mass average gas velocity based on center of mass of the gas, relative to fixed coordinate system.
$\mathbf{q}_i$	Average drift velocity of particle i relative to fixed coordinate system.
$\mathbf{q}_s$	Stefan velocity.
R	Gas constant.
$\tilde{R}$	Continuous spectral variable for particle radius.
R_i	Particle radius.
$R_{11}(\tau)$	Correlation function for gas elements.
$R'_{11}(\tau)$	Correlation function for gas elements near a particle.
$R_p(\tau)$	Correlation function for particles.
$\mathbf{r}$	Spatial coordinate vector (also $\mathbf{x}$).
S	Surface area.
S_p	Supersaturation ratio, p/p_0.
$\hat{S}$	Entropy.
T	Temperature.
T_r	Lagrangian integral time scale for particle motion.
T_{11}	Lagrangian integral time scale for fluid motion.
t	Time.
t_d	Characteristic diffusion time, R_i^2/D_i.
U_∞, $\mathbf{U}_\infty$	Free stream velocity; also q_∞, $\mathbf{q}_\infty$.
$\hat{U}$	Internal energy.
V	Volume.
$\bar{v}_i$, $\bar{v}_G$	Mean thermal speed.
$\mathbf{v}_L$	Gas velocity in Lagrangian frame of reference.
v_ε	Terminal velocity in a steady electrical field.
v_s	Terminal settling velocity under gravity.
$\tilde{v}$	Continuous volume variable for size spectrum.
$\tilde{W}$	Probability function.
$\mathbf{w}$	Molar average velocity.
$\mathbf{x}$	Position vector with cartesian components x, y, z (also $\mathbf{r}$).
x_α	Mole fraction of species α.
Z	Non-equilibrium factor in nucleation theory.
z_i	Number of unit charges on a particle of species α.

Script letters

$\mathscr{A}$	Reciprocal relaxation time, $(B_i m_i)^{-1}$.
$\tilde{a}$	Empirical parameter in evaporation rate equation, Knudsen transition regime.
$\mathscr{B}$	Coefficient of pressure perturbation term in Eq. (3.28), $= 1{\cdot}5\ \rho_G/\rho_i$.
$\tilde{\mathscr{B}}$	Factor in heat transfer equation for slip flow.
$\mathscr{b}$	Molecular interaction parameter.
$\mathscr{C}$	Coefficient of Basset term in eqn. (3.28) $= 9\sqrt{(\rho_G\mu_G/\pi)}/2\rho_i R_i$.

c	Collision sticking coefficient.
$\tilde{\mathscr{C}}$	Coefficient for evaporation equation in Knudsen transition regime.
$\mathscr{D}, \mathscr{D}$	Drag coefficient.
$\tilde{\mathscr{D}}$	Coefficient for heat transfer equation in Knudsen transition regime.
$\hat{\mathscr{D}}$	Generalized diffusivity, D_{AB}, or $k_G/\rho_G C_p$.
$\mathscr{E}_o$	Electrical field intensity.
$\tilde{\mathscr{E}}$	Coefficient for isothermal drag in Knudsen transition regime.
$\mathscr{E}$	External force in Boltzmann equation.
e_G, e_i	Eddy diffusivity for gas and for particles.
$\mathscr{F}$	Stochastic "force" associated with thermal agitation (see 5.40 or 7.29).
$\mathscr{g}$	Acceleration of gravity.
$\mathscr{H}$	Hamiltonian.
$\mathscr{I}_{\mu is}$, $\mathscr{I}_{\mu Gs}$	Interaction parameter.
$\tilde{\mathscr{I}}$	Integral function, eqn. (5.45).
ℓ_L	Lagrangian integral length scale.
$\mathscr{m}$	Shape parameter in theory for nucleation on solid substrates.
$\mathscr{P}_m$, $\mathscr{P}_t$	Constants in theory of slip flow regime, eqns. (6.3) and (6.7).

Greek letters

α_{cv}	Evaporation coefficient.
α_{mv}	Accommodation coefficient for momentum.
α_{tv}	Thermal accommodation coefficient.
β_α	Mass to thermal energy ratio, $m_\alpha/2kT$.
β	Ratio of collision rate density to Smoluchowski's value for coagulation in a continuum.
Δ	Fuchs' jump length.
δ	Boundary layer thickness.
$\tilde{\delta}$	Perpendicular distance to critical particle trajectory from stagnation streamline.
$\hat{\delta}$	Constant in BGK approximate solution to Boltzmann's equation.
ϵ	Dielectric constant.
ϵ_d	Energy dissipated per unit mass.
ϵ_r	Emissivity.
ϵ_t	Turbulent energy dissipation per unit mass.
ζ	Generalized scalar property, for example, T or ρ_α.
η	Collection efficiency.
$\tilde{\eta}$	Size scaling function for self-preserving distributions.
Θ	Correction parameter for Chapman–Enskog solution to Boltzmann equation.
κ	Debye reciprocal length.
$\tilde{\kappa}$	Wave number.
λ_G, λ_i	Mean free path for gas molecules or for particles.
λ_v	Latent heat of vaporization.
$\underline{\mu}_\alpha$	Reduced mass.
μ_G	Dynamic gas viscosity.
$\hat{\mu}_\alpha$	Chemical potential for species.
ν_G	Kinematic gas viscosity.

ν_i	Dimensionless number density, n_i/N_0.
$\bar{\bar{\Pi}}$	Pressure tensor.
π	Dimensionless constant, 3.141.
$\boldsymbol{\rho}$	Dimensionless trajectory, $\mathbf{r}/R_i$.
ρ_G, ρ_i	Gas or particle density.
ρ_α	Partial density of species α.
σ	Surface tension or energy.
$\underline{\sigma}$	Collision cross-section.
σ_{st}	Stefan–Boltzmann radiation constant.
σ_{AB}	Semi-empirical coefficient for diffusiophoretic force in the slip flow regime.
$\tilde{\tau}$	Dimensionless time, $2\pi D_i R_i N_0 t$.
$\hat{\tau}$	Scale factor in thermal force theory.
τ_H, τ_s	Time scales for spinning disc atomization.
$\Phi(\mathbf{x})$	Force potential or interaction potential.
$\tilde{\Phi}_m$	Rate of transfer of mass to a surface of area S.
$\tilde{\Phi}$	Rate of transfer of molecules to a surface of area S.
ϕ	Volume fraction.
$\tilde{\phi}$	Dimensionless size distribution.
χ	Psychrometric ratio.
$\tilde{\psi}$	Self-preserving number density for aerosol size spectrum.
ω, $\boldsymbol{\omega}$	Angular velocity or rotational speed.

Dimensionless numbers

Bo_{I}	Boltzmann number I, R_G/λ_G.
Bo_{II}	Boltzmann number II, R_G/L_G.
Br	Brown number, $\bar{v}_i/\bar{v}_G$.
Co	Coulomb number, $z_i z_j \tilde{e}^2 R_{ij} kT$.
De	Debye number, $z_i z_j \tilde{e}^2 \kappa/kT$.
Kn	Knudsen number, λ_G/R_i.
La	Langmuir number, $\mu_L q_i/\sigma$.
Ma	Mach number, $\lvert \mathbf{q}_i - \mathbf{q}_G \rvert/\bar{v}_G$.
Nu	Nusselt number, $2\hat{h}R_i/k_G$.
Pe	Peclét number, $U_\infty L/D_v$.
Re	Reynolds number, $U_\infty L/\nu_G$.
Sc	Schmidt number, $R_i^2 n_G \lambda_G$; or ν_G/D_v.
Sh	Sherwood number $2\hat{k}R_i/D_{AB}$.
Stk	Stokes number, $U_\infty/\mathscr{A}a$.
We	Weber number, $\rho_G(q_G - q_i)R_i/\sigma$.

CHAPTER 1

1. INTRODUCTION

DISPERSED systems composed of particles suspended in gases are frequently observed in nature, in the laboratory, and in industrial surroundings. Depending on the characteristic length and the time scales of the cloud of particles, definitions of such systems can encompass everything from molecular mixtures to the universe. Historically, however, common usage dictates that aerial dispersions refer to suspended particulate material of microscopic or submicroscopic size, but larger than molecular dimensions. These classes of particulate clouds are known by loose terms like *aerosols*, or *aerocolloidal systems*. Clouds of suspended matter ranging from dusts and smokes to mists, smogs, or sprays fall into this general category.

Aerial dispersions vary widely in physical and chemical properties depending on the nature of the suspended particles, their concentration in the gas, their size, and their spatial homogeneity of dispersion. Both liquid and solid material can become suspended in a gas through a variety of different mechanisms. Near sources like fires or smoke generators, the aerocolloids may have marked uniformity in composition and properties that can be studied relatively easily by known techniques in the laboratory. However in nature, the aged and mixed aerosol suspensions quickly lose the fingerprints of their origins, and they appear as combinations of suspended matter which are highly heterogeneous in composition and particle size. These natural aerosols are much more difficult to describe and study in any quantitative way. Typical samples of natural aerosol particles are shown in Plates 1 and 2. This material, collected from an urban atmosphere (Plate 1), and from a zone in the tropics and the Antarctic (Plate 2), relatively free of human modification, show a wide variety of physical properties.

The behavior of suspended microscopic or submicroscopic particles in gases, especially in cases where the ratio of the mean free path of the suspending gas λ_G to the particle size R_i becomes large, displays certain unique features that are unlike "simple" atomic and molecular systems, or phenomenologically, "continuum" systems containing two or more phases. Aside from the natural scientific interest in the unusual properties of aerial suspensions, workers have been concerned with the substantial practical importance of these systems since the dawn of civilization. Theories for the formation of clouds in the atmosphere to produce rain have appeared in the earliest writings of natural philosophers. Smoke from fires has been known as an annoyance for centuries. Although pollution of the atmosphere has certainly occurred since men discovered fire, widespread regard for this problem has only appeared relatively recently as men have begun to realize how important their environment is to future life on the earth.

Widespread practical concern for application of the special properties of aerial suspensions has truly come to pass in the last 50 years. In our modern world, household sprays for dispensing anything from deodorants to paints are used extensively. Pesticides in the form of sprays and dusts are spread over fields of crops and over areas like swamps or marshes. Fuel injection systems for combustion engines involve the generation of aerosol clouds of fuel. The spurious formation of particles in rocket engines causing unwanted erosion of the rocket nozzles has been examined. On the opposite side of the coin, the formation of particles in rockets may also have potential usefulness in exotic devices for space propulsion (e.g. Norgren and Goldin[1]). Physicians have found that aerial suspensions may have therapeutic effects for treatment of certain respiratory diseases. However, other dispersoids, especially air pollutants, can just as well have detrimental effects on biological systems.

With the ever increasing awareness of dispersed systems in relation to environmental control and military, industrial, and household application, aerosol science began to evolve as a definite part of physical chemistry by the beginning of the twentieth century. With the surge of discoveries in atomic and molecular physics after World War I, the active interests in aerocolloidal systems waned somewhat. However, the post-World War II years have seen a widespread renewal of research on aerosols within the investigations of a variety of disciplines, ranging from geophysics to medicine.

A. Preliminary Definitions and Classification

Because of the indefinite and often poorly understood character of many particulate clouds, the science of such systems has displayed a tradition of a fragmented, rather disorganized, and unsystematic development. A typical illustration of an unnatural division of efforts is the separation between aerosol physics and cloud physics. Most of the underlying principles of behavior of aerosol clouds and clouds of water droplets in the atmosphere are the same. However, workers in these disciplines often have failed to communicate with each other and have tended to work on divergent paths instead of attempting to unify their ideas for describing the general theory of particulate suspensions. Even the definitions of aerial dispersed systems are vague and lie somewhere in between rigorous scientific and popular terminology. As an example, one of the most confusing terms to the layman certainly must be that of an *aerosol*, which can refer to a household spray can, or more precisely to the aerocolloidal suspension, mechanically generated by means of the spray device.

Though no set of rules for classifying aerial dispersions has been commonly accepted, certain kinds of suspensions have been named largely by everyday usage of language. This descriptive nomenclature is so widely established and so well understood, at least for natural clouds,† that it does not seem useful to introduce radically different terminology at this time.

Existing divisions for aerocolloids are distinguished largely in terms of their mechanical origins. Some suspensions may be generated by mechanical dispersion of material, while others form as a result of condensation from supersaturated vapor. The historical classification of aerocolloidal suspensions is grouped as follows.

Dusts

These consist of solid particles dispersed in a gas by the mechanical disintegration of material. Some examples include (a) clouds that result from breakup of rocks or minerals

† As applied in this book, a cloud refers to any aerial dispersion having a concentration of particles sufficiently high to give it definite size and form (dust cloud, rain cloud, smoke puff (cloud), etc.). This definition is used in the sense of Fuchs.[2]

during crushing, grinding, or blasting, (b) subjecting powders to blasts of air. Dusts frequently may appear naturally by winds lifting small particles on the Earth's surface into the atmosphere. In general, dusts are very heterogeneous in composition and have rather poor colloidal stability. They are made up of larger particles than smokes or mists. However, the lower range of the size distribution of dusts typically may extend well into the region of sub-microscopic sizes.

Smokes

In contrast to dusts, smokes cover a wide variety of aerial dispersions and are formed of part by condensation from supersaturated vapors. Such clouds consist of very small particles, composed of material of low vapor pressure, in relatively high concentrations, which fall out slowly under the action of an external force like gravity. Examples include particulate suspensions from combustion processes, from the chemical reaction between gaseous ammonia and hydrogen chloride, the oxidation in a metallic arc, or the photochemical decomposition of materials like iron carbonyl. One example is shown in Plate 3. The principal criterion for defining smokes is one of particle size: the distribution of size ranges downward from about 10 μ radius to submicroscopic sizes less than 0.1 μ radius.

When smoke formation is accompanied by the appearances of a gas or a vapor, it may be called a *fume*—for example, when metallic oxide smokes develop with sulphur in the melting process. The term, fume, may be used more generally, in a confusing way, to refer to a particulate cloud being generated over a substance by chemical reaction at a surface or in the vapor phase.

Mists

Suspensions of liquid droplets formed by condensation of vapor, but sometimes by atomization, are called *mists.* These clouds consist of fairly large particles, exceeding 10 μ radius, and their particulate concentration is relatively low. The size of particles is the principal property that distinguishes mists from smokes. If the dispersoid has sufficiently high particulate concentration to obscure visibility, it may be designated a *fog. Hazes* in the atmosphere may consist of a combination of water droplets, pollutants, and dust. Similarly, the name *smog* (smoke mixed with fog), refers to a particulate system occurring over urban areas, where pollution from automobile exhausts and other combustion devices, combined with other materials forms a noxious, sometimes toxic cloud in the low levels of the atmosphere.

Various ranges of particle size can be used to subdivide classes of natural aerosols. A frequently used table of Podzimek[(3)] illustrates the various groupings for natural aerosols and their place in atmospheric phenomena (Fig. 1.1). A similar diagram, shown in Fig. 1.2, may be constructed for other typical classes of particulate suspensions, including those from industrial origins. Shown also in this diagram are the ranges of operations of several devices for removal of particulates from gas streams.

Aerocolloidal Systems (Aerosols)

The generic term, *aerosol,* evidently was coined by F. G. Donnan near the end of World War I (see also Green and Lane[(4)]). This name was used to describe clouds of microscopic and submicroscopic particles in air, as exemplified by chemical smokes being developed at that time for military purposes. An aerosol was regarded as a direct analogy to a liquid

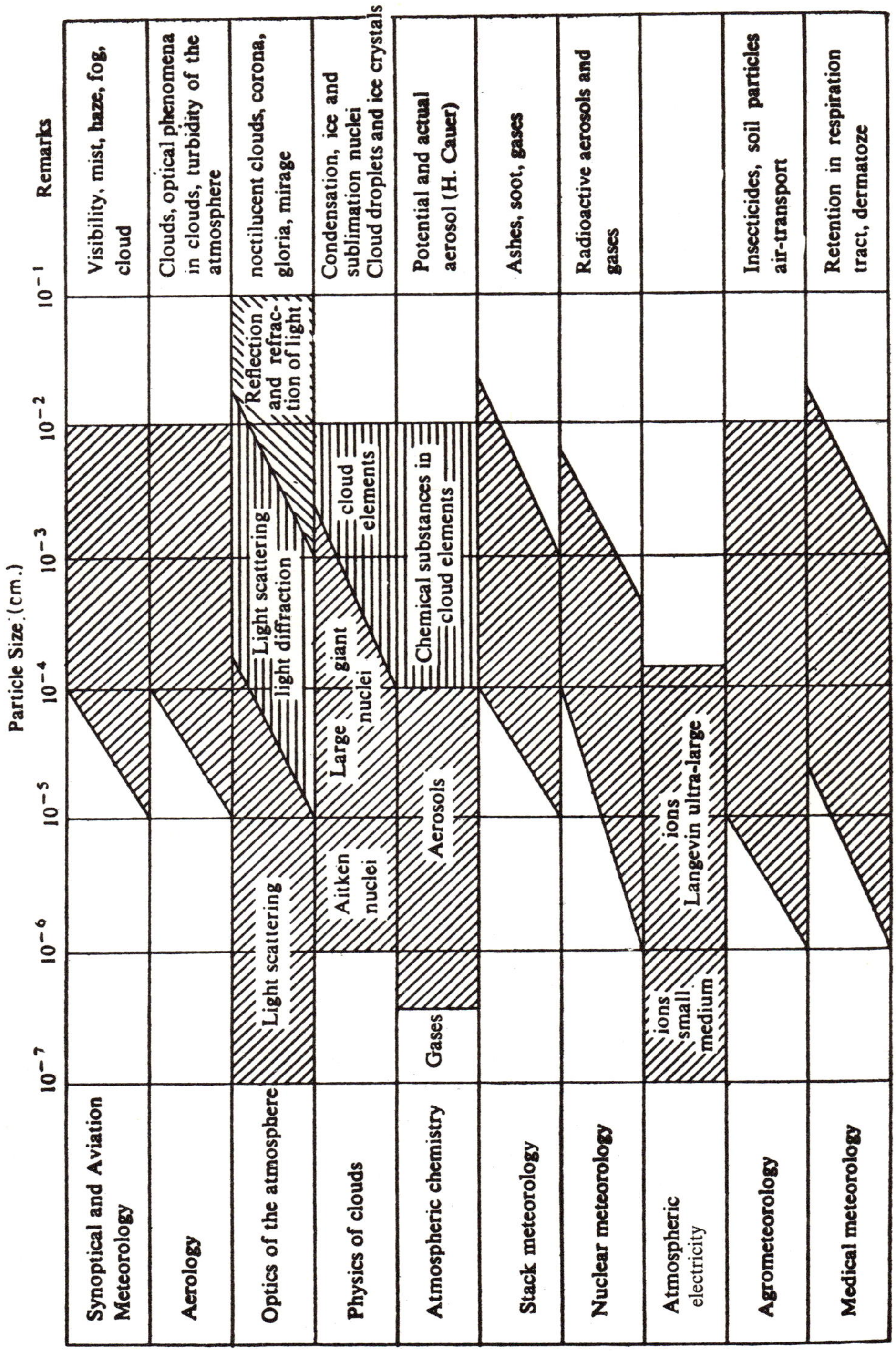
Particle Size (cm.)
10^{-7}
10^{-6}
10^{-5}
10^{-4}
10^{-3}
10^{-2}
10^{-1}
Remarks
Synoptical and Aviation Meteorology
Visibility, mist, haze, fog, cloud
Aerology
Clouds, optical phenomena in clouds, turbidity of the atmosphere
Optics of the atmosphere
Light scattering
Light scattering light diffraction
Reflection and refraction of light
noctilucent clouds, corona, gloria, mirage
Physics of clouds
Aitken nuclei
Large nuclei
giant nuclei
cloud elements
Condensation, ice and sublimation nuclei
Cloud droplets and ice crystals
Atmospheric chemistry
Gases
Aerosols
Chemical substances in cloud elements
Potential and actual aerosol (H. Cauer)
Stack meteorology
Ashes, soot, gases
Nuclear meteorology
Radioactive aerosols and gases
Atmospheric electricity
ions small medium
ions Langevin ultra-large
Agrometeorology
Insecticides, soil particles air-transport
Medical meteorology
Retention in respiration tract, dermatoze

FIG. 1.1

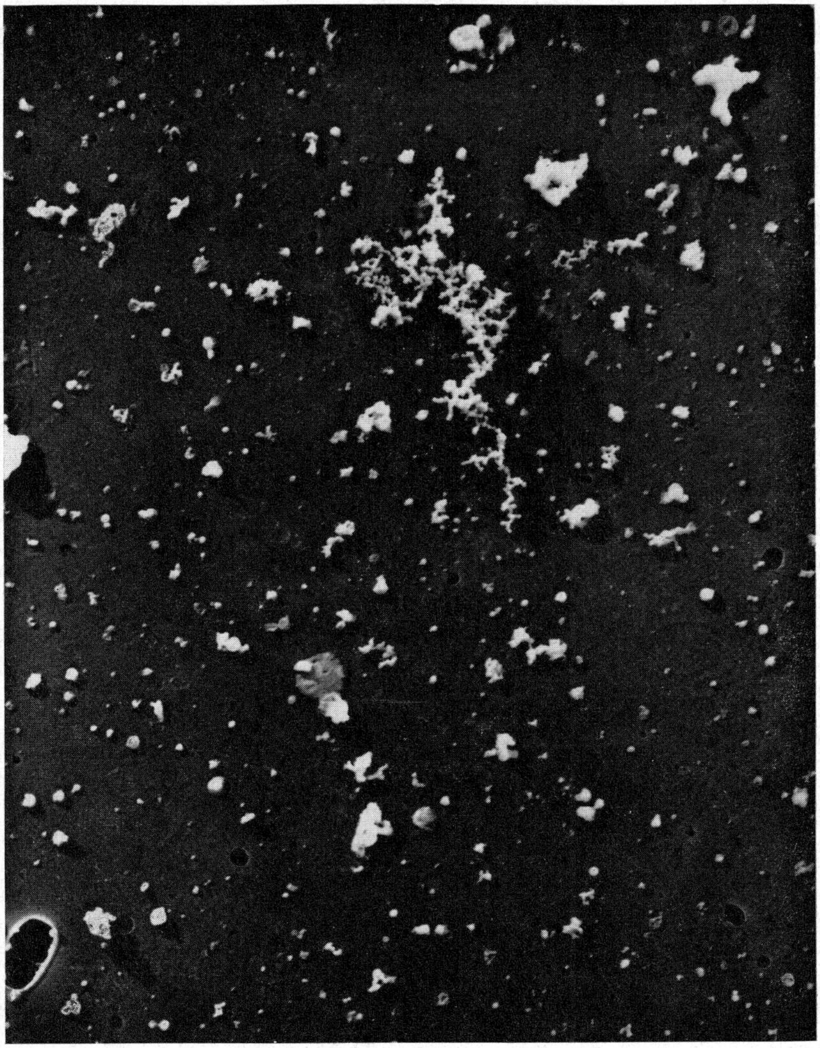

PLATE 1. A photomicrograph of a sample of urban aerosol taken in London during a period of heavy pollution in 1960. Magnification 11,000×. (Photo by R. Waller; with British Medical Research Council Air Pollution Unit, London.)

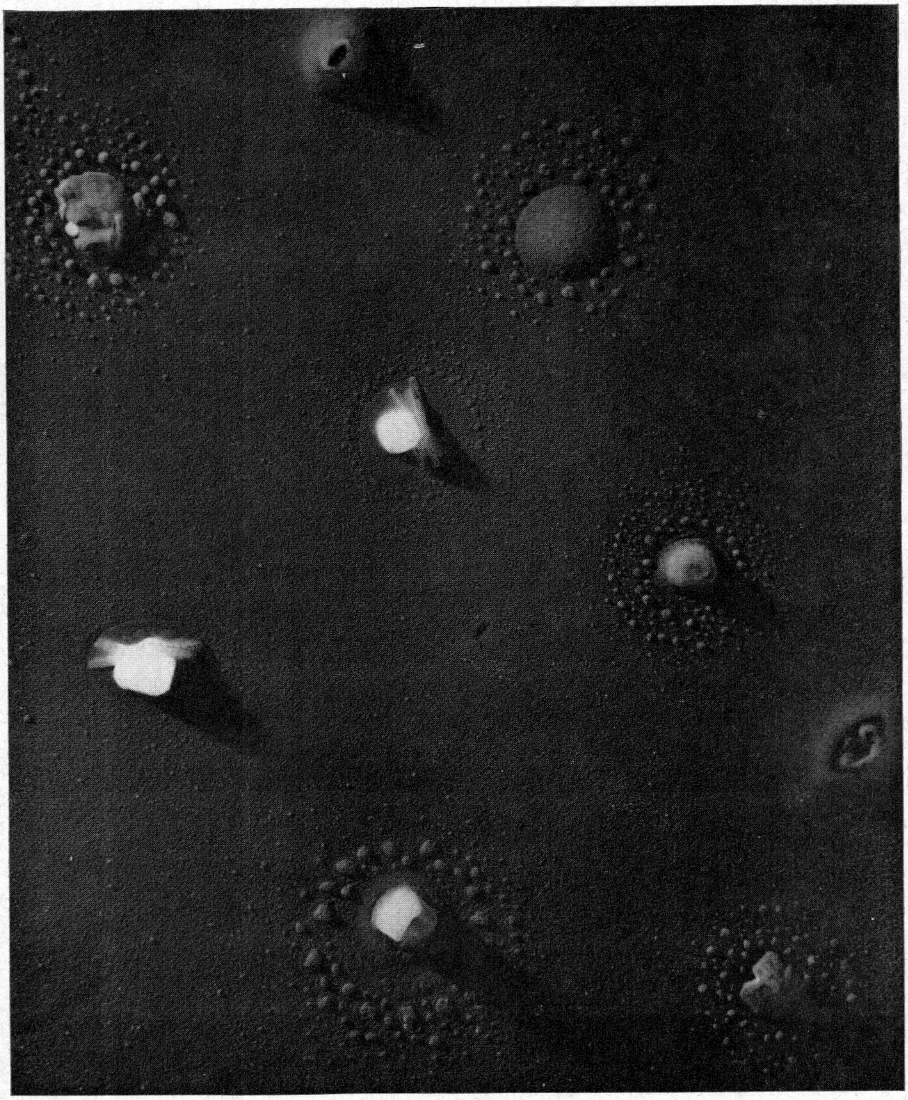

PLATE 2A. A photomicrograph of natural aerosols at 37,300× showing sulfuric acid droplets, characterized by cores surrounded by satellite rings and sodium chloride particles (white). Sample taken at Fort Sherman on the Atlantic side of Panama. (Photo by E. Frank and J. P. Lodge, Jr.)

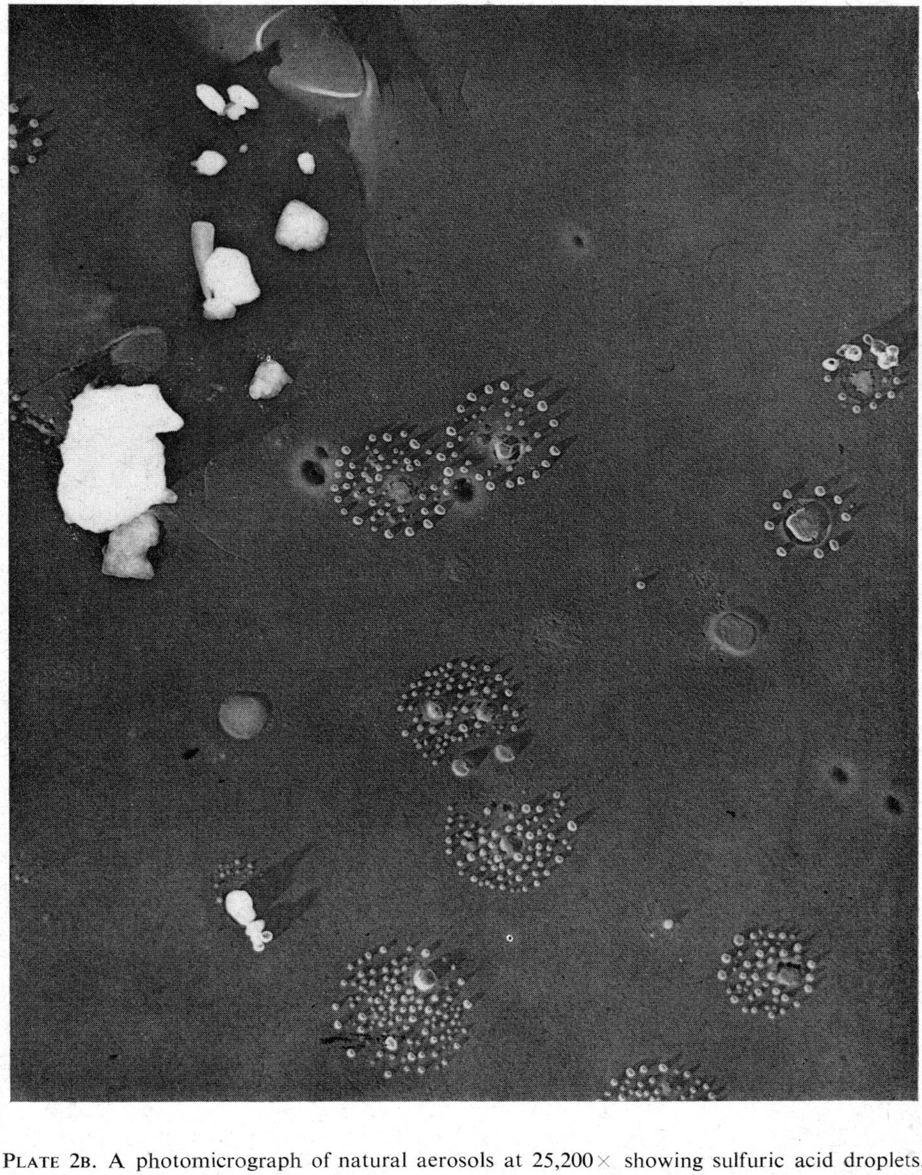

PLATE 2B. A photomicrograph of natural aerosols at 25,200× showing sulfuric acid droplets, ammonium sulfate particles and volcanic dust. Sample from South Polar region of the Antarctic. (Photo by E. Frank.)

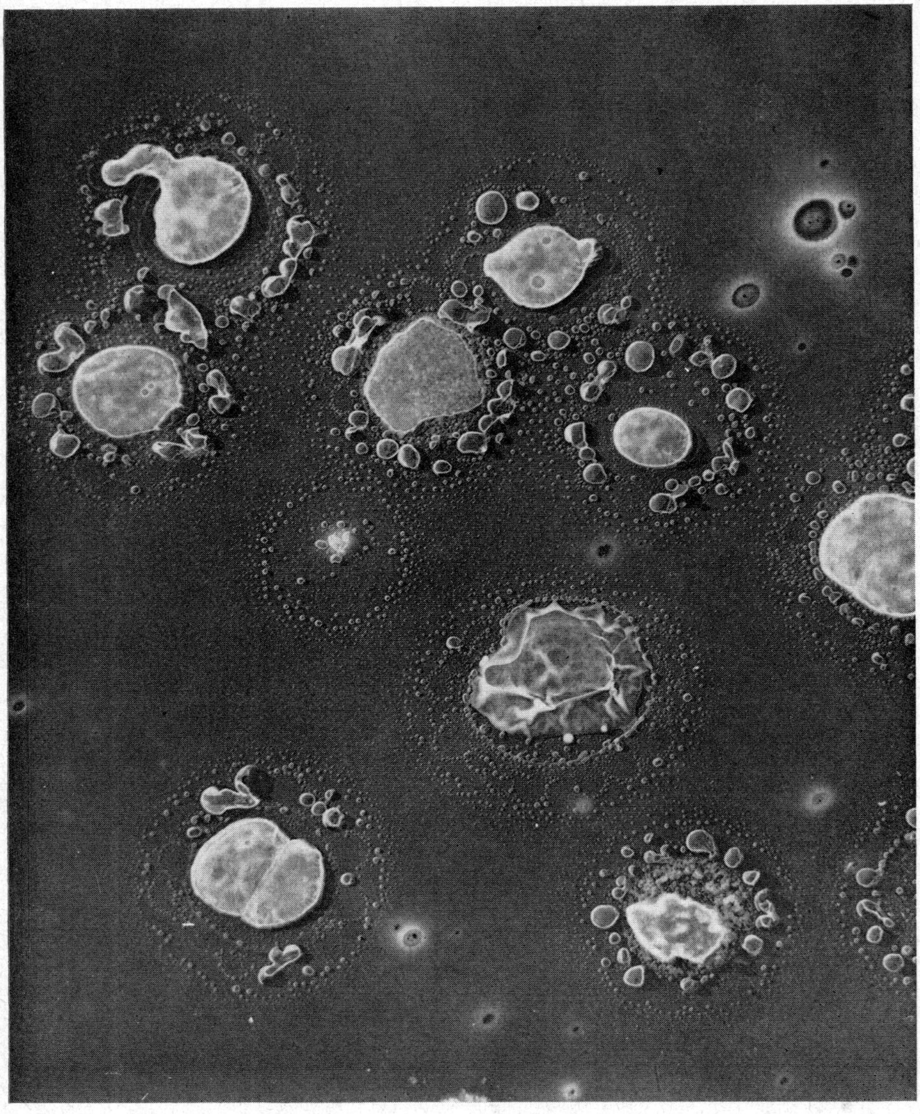

PLATE 2C. A photomicrograph of natural aerosols at 12,100×. Sample of volcanic fume (Halemaumau volcano, Hawaii) showing sulfuric acid droplets. (Photo by E. Frank and J. P. Lodge, Jr.)

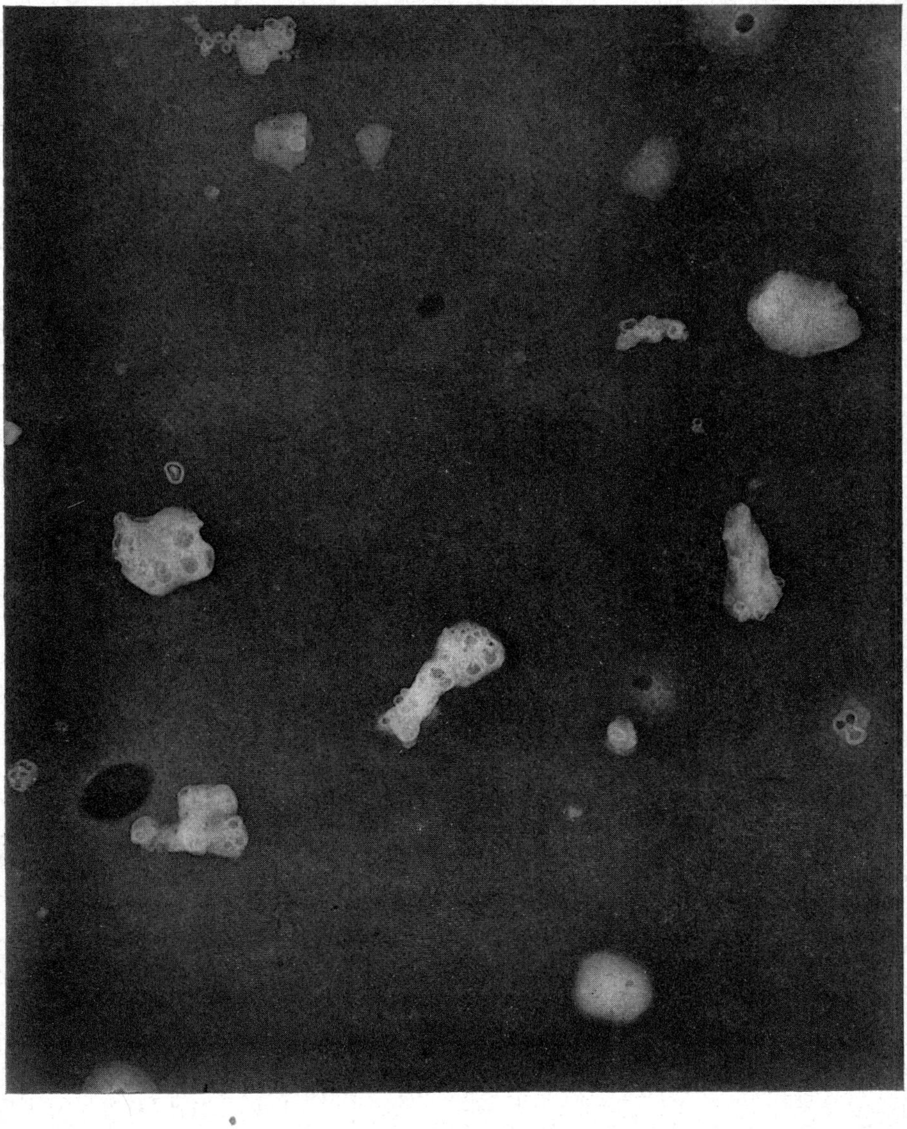

PLATE 3. Sample of smoke from a match impregnated with silver iodide. Note irregular, porous shapes of particles (64,600 ×). (Photo by E. Frank and J. P. Lodge, Jr.)

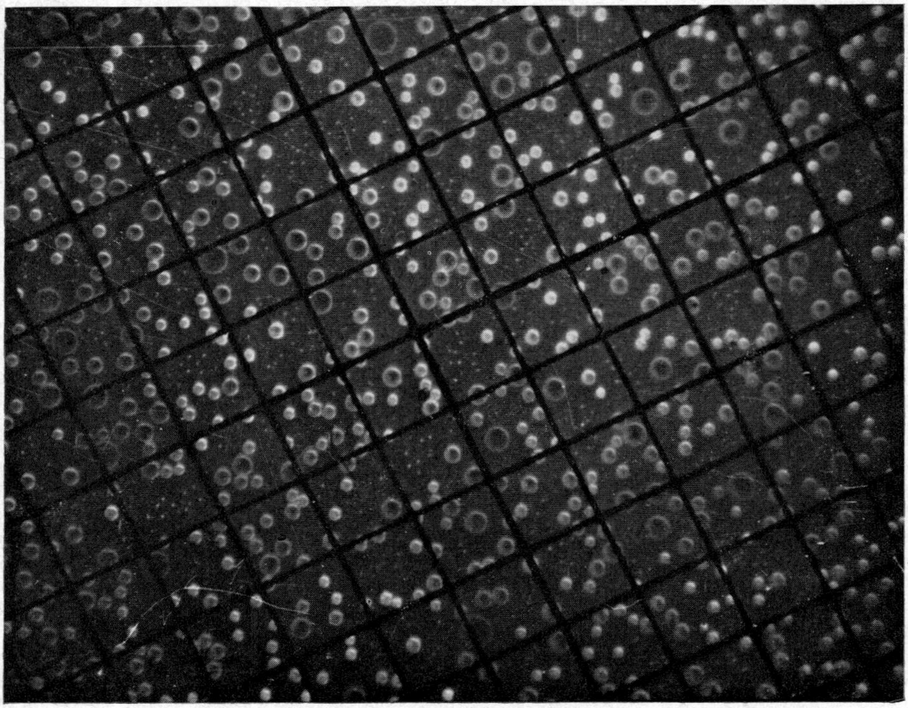

PLATE 4. A micrograph taken at 165× of 1–2 μ radius dioctyl phthalate droplets generated in the laboratory by a La Mer generator. (Photo by P. Brown.)

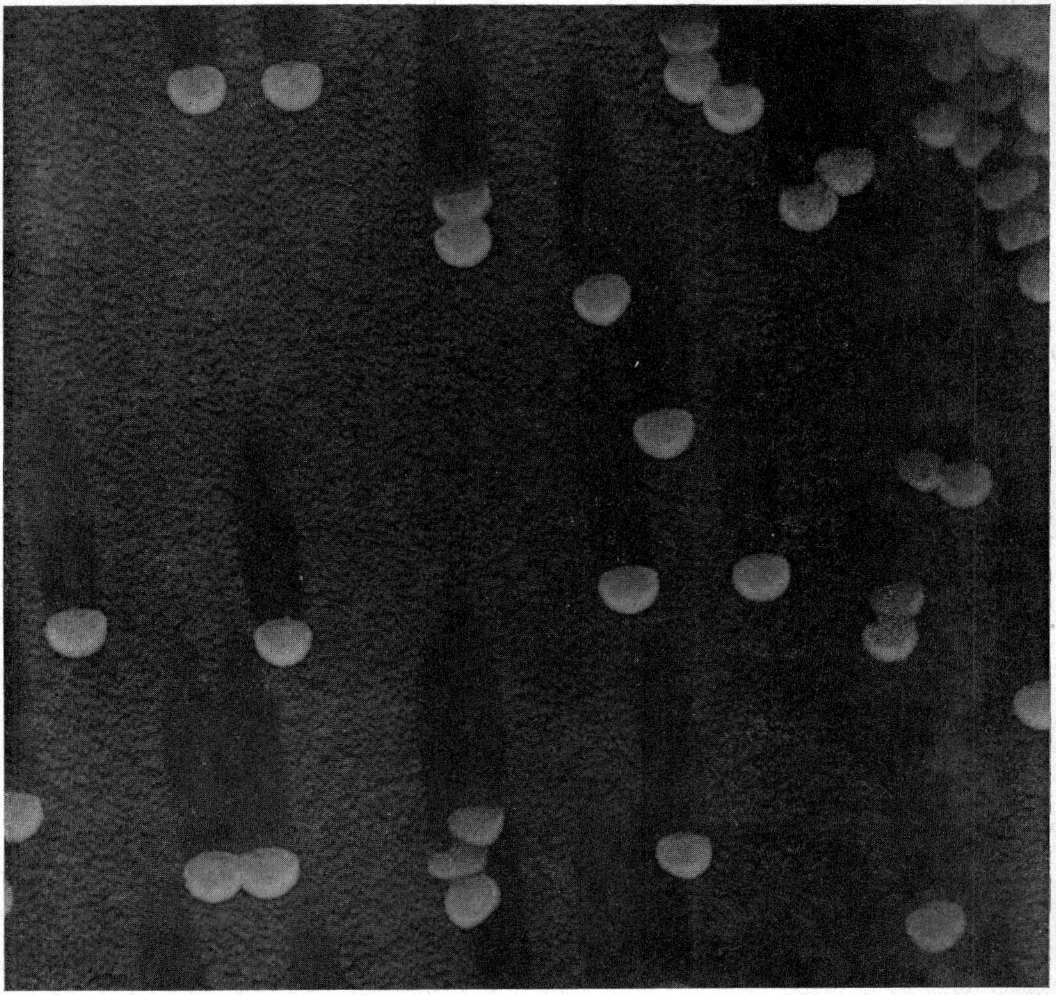

PLATE 5. Photomicrograph taken at 130,680× of an aerosol of atomized polystyrene latex suspension. Latex particles are 0.088 μ radius. (Photo by E. Frank and J. P. Lodge, Jr.)

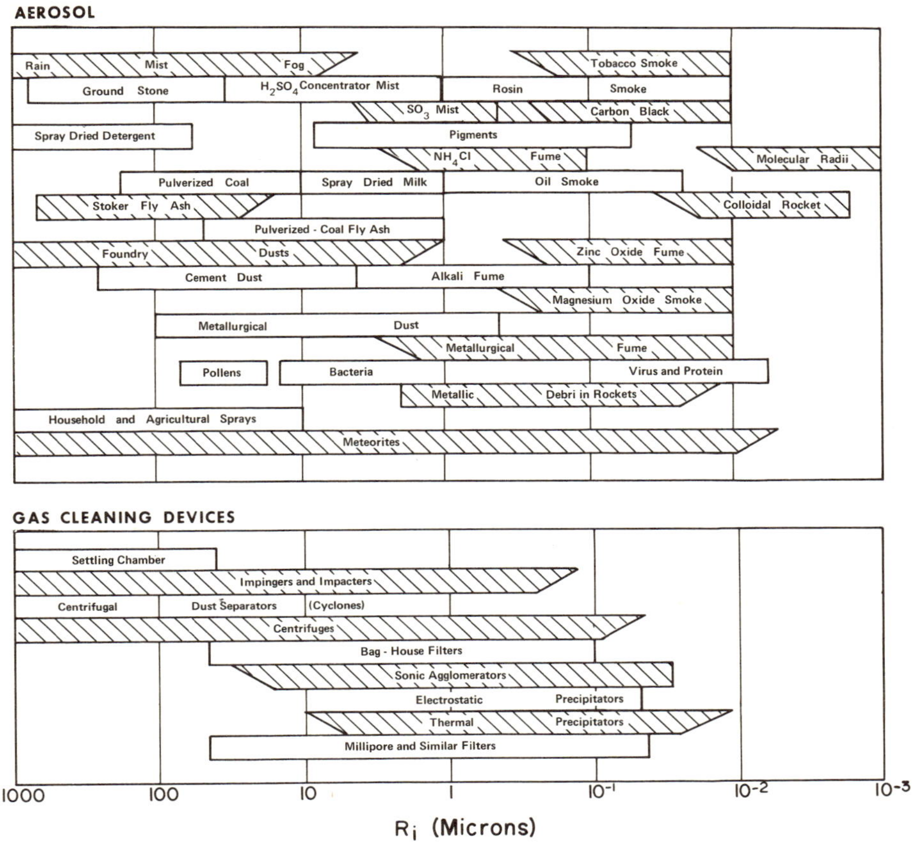

FIG. 1.2

colloidal suspension, a *hydrosol.* The term aerosol then implies qualitatively a degree of stability with respect to settling in a gravitational field. It has been found, of course, that hydrosols settle considerably more slowly than aerosols because of the larger viscosity of the suspending medium, and smaller density difference between particles and the medium. Thus the early analogy of Donnan and others is a rather crude one. Nevertheless, historically, the first criterion for defining an aerosol or an aerocolloidal system involves the "small" settling rate of suspended particles.

In describing aerosols, the settling rate alone cannot specify precisely such systems ranging from smokes to dusts and mists. The arbitrary use of colloidal stability implies that the suspension exists in an external force field, and that a large surface is available for catching particles moving under the influence of the force field. Such definitions are unduly loose in a modern scientific sense. For example, meteoritic material, including interplanetary dust, can certainly be called an aerocolloidal suspension. By earthly standards, these suspensions are stable to settling because they do not exist in a strong gravitational field. For example, using the settling velocity given by the classical formula $v_s = 2\pi\rho_i R_i^2 g/9\mu_G$, where ρ_i is the particle density, g the gravitational constant, and μ_G the viscosity of the medium, $v_s \approx 0.1$ cm sec^{-1} typically for particles of 1 μ radius in the Earth's environment. This would imply settling times of about an hour in an ordinary room. However, a cloud of meteors of diameters larger than one meter conceivably could exist for years before being trapped by any kind of a surface. Thus "stable" suspensions cannot depend only on settling rate as determined by particle size alone. Characteristic scales of length and time in a cloud's environment as well as the forces acting on particles have to be taken into account.

A second criterion sometimes used to describe an aerocolloidal system requires that the Brownian motion of particles be significant in the dynamical behavior of the particles. This restriction again implies a limit on the size of particles in earth bound clouds. For particles in air at normal pressures, Brownian motion becomes important for particles less than about 0.5 μ in radius. It is interesting to note, however, that the ideas of Brownian motion (thermal agitation) have also been applied to the dynamics of such gigantic entities as stars (see, for example, Chandrasekhar[(5)]).

To obtain certain results in the classical theory of particle settling and Brownian motion, the drag force of the viscous medium acting on particles has to be specified. The use of the Stokes' relation of the drag requires a third restriction on applicability of the classical theory of aerocolloidal systems. That is, the particle motion must be controlled by viscous forces of the fluid, or the Reynolds number for particles, $Re_i = \rho_G q_i R_i/\mu_G$, must be less than unity, where q_i is the particle velocity.

In a given gas, the Reynolds number restriction places an upper limit of the product velocity and the particle size. For example, at pressures of 760 mmHg, 10 μ particles cannot exceed velocities of 30 cm sec^{-1}, while 1 μ particles can have velocities up to 300 cm sec^{-1}.

The first three criteria frequently used for specifying aerocolloidal systems are essentially dynamic in nature. However, one can add a physicochemical limit on particle size. One of the interesting features of aerocolloidal systems lies in their unusual surface chemistry. Aerosols are unique from other multiphase mixtures in this respect because of their very high surface to volume ratio, S/V.†

It seems acceptable, in general, to workers in aerosol science that the four criteria listed above can specify the aerocolloidal "state", at least qualitatively. However, the first three limitations are not independent. Thus, for the purposes of this book, we shall adopt the following mutually independent criteria for defining an aerosol or an aerocolloidal system:

(1) The Reynolds number, $\rho_G q_i R_i/\mu_G = Re_i < 1$.
(2) The surface to volume ratio, $S/V \gg 10^3$ cm^{-1}.

Since we have avoided specifying limits on the particle concentration in describing aerocolloids, we can establish at this point that such systems can encompass, with (1) and (2), within our definitions, a broad range of particulate clouds. This range can include, in principle, systems of very high particle concentration like fluidized beds, an intermediate class like smokes and mists, and very dilute clouds like interplanetary or cosmic dusts.

B. Scope and Limitations

In this monograph, our aim is to explore the dynamical behavior of idealized aerosol particles in the light of modern developments in classical mechanics. The idealization that we refer to here assumes that the solid or liquid particles suspended in a gas can be modeled as macroscopically smooth, chemically inert, spherical bodies. Details about their surface and internal properties are unspecified microscopically, but not ignored. Such

† More useful as scaling parameter are dimensionless numbers. Such a dimensionless scale factor for expressing uniqueness of surface properties involving S/V is proposed in Chapter 2.

properties are accounted for arbitrarily in the theory only in terms of rather gross, empirical parameters.

The particles are suspended in a non-uniform, but un-ionized gas. The gas-particle system is considered to be near conditions of equilibrium so that macroscopic thermodynamic properties like temperature and pressure have meaning in specifying the state of the system. An attempt is made to regard the influences of fields of external force, and interactions between particles as well as the gas. In particular, the effects of electrostatic charging on the particles are taken into account at least in an elementary way, essentially by applying the Debye–Hückel theory of electrolytes (e.g. Hill,[6] also Balescu[7]).

In dealing with the dynamics of aerocolloidal systems, interactions between the particles and the suspending non-uniform medium will be treated, as well as interactions (collisions) between the particles themselves.

The molecular hypothesis and its logical extension, the kinetic theory, are used as the model of the suspending gas. Thus, the method employed in developing the theory for aerosol dynamics rests on the direct application of the principles of the kinetic theory of gases (e.g. Chapman and Cowling[8]).

The classical theory of aerosol behavior has evolved largely from considerations of continuum theory. However, it seems appropriate here to regard the present understanding of aerosol dynamics within the more general structure of statistical mechanics of gases, with certain very important limiting simplifications for the aerosol model in (a) the limit of behavior in the continuum regime, i.e. the ratio $\lambda_G/R_i \rightarrow 0$, and (b) the limiting regime of rarified gas dynamics where the aerosol particles may be assumed in many respects effectively as large molecules ($\lambda_G/R_i \geqslant 10$).

If the gas–particle system exists near equilibrium, then it is possible to simplify the theoretical treatment by making use of well known molecular transport coefficients of the fluid and the particles like the viscosity μ_G, the thermal conductivity k_G and the mass diffusivity D. Furthermore, we shall assume that the surface energy of particles can be given by the macroscopically defined variable, the surface tension σ.

In the next chapter, certain fundamental features of continuum dynamics and the theory of gases are discussed as they apply to aerosol clouds. The criteria for defining an aerocolloidal system are described in more detail. The formal structure and limitations of the theory are set up as preparation for examining a series of important dynamical problems in the succeeding chapters.

The sections that follow, Chapter 2 deals first with the behavior of single particles traveling through a gas, and reacting to various aerodynamic and external forces. The limiting regimes where continuum theory and free molecule theory apply are developed in Chapters 4 and 5. Then the known extensions of theoretical models to account for the transitional behavior between these regimes are investigated subsequently in Chapters 5 and 6.

Further discussion of the interaction between the suspending gas and single particles stems from consideration of heat and mass transfer to (and from) particles. Several aspects of this portion of aerosol dynamics are examined in Chapters 4, 5 and 6.

When one centers attention on the behavior of a few particles within a cloud of particles, further simplification of the theory is required. To deal with the diffusion of aerosol particles, for example, the particles are assumed to consist in a limiting way of large molecules in a gas mixture. Such an assumption allows the use of the well known equations of transport theory for estimating the diffusional rates of aerosols during their thermal agitation. Brownian and turbulent diffusion are examined in Chapter 7.

The last chapters of this monograph attempt to treat the formation and growth of aerosol particles in clouds. As one begins to consider clouds of particles, it becomes necessary to give up following individual bodies as they travel around in the gas. Instead, a tractable theory requires the specification of the statistical behavior of a large number of particles. Hence under these circumstances, the individual particles lose their identity, and the statistics of the entire cloud are calculated, to attain information about certain measurable macroscopic properties of the aerosol cloud.

Chapter 8 discusses the mechanics of disintegration of material to form particulate suspensions. Since very little theoretical work has been reported on this aspect of aerosol generation this section necessarily has to be limited in scope, and is rather qualitative.

In contrast to the limited effort expended on disintegration processes, there exists a great deal of work on the process of condensation of vapor to form clouds of aerosol. The discussion in Chapter 9 centers on the formation of aerosols by homogeneous and heterogeneous nucleation. Implicit in the theory of nucleation is the use of certain sweeping statistical mechanical assumptions about aerosols to simplify the mathematical treatment. These considerations will be outlined in Chapter 9 within the framework of the other sections of this book.

The final chapter, Chapter 10, examines the growth of aerosols by collision and coagulation of particles. Here again, several simplifying statistical assumptions are made to treat the collision dynamics. The limiting regimes of continuum behavior and free molecule flow may be treated relatively easily. However, the transitional regime of flow between these limiting realms remains inadequately specified with the present knowledge of the subject.

We should state at the beginning that we anticipate that the idealized picture of aerosol behavior set down here probably will have limited applicability to real aerocolloidal systems. The kinetic theory outlined in this monograph doubtlessly applies best to liquid droplets (Plate 4), or to latex spheres atomized into an aerocolloidal suspension in the laboratory (Plate 5). The theory's usefulness will be less for such complicated particulate suspensions as natural aerosols (see Plates 1 and 2), and smokes consisting of solid particles of very complex shapes (see Plate 3). It is recognized, however, that even the crudest kinetic theories were remarkably accurate in their predictions about the dynamical behavior of gases.[8] And more sophisticated, realistic models of gases using essentially the same computational procedures proved to be even more quantitatively applicable to the behavior of gases. An aerocolloidal system is considerably more complex than a gas mixture. However, we hope that the same kind of analysis used for gases applied to aerosols will prove its value for understanding at least qualitatively the behavior of these aerial dispersions. The experimental evidence currently available, though rather crude in many cases, suggests that the idealized theory of aerosol dynamics is quantitatively satisfactory for interpreting certain observational results, especially for single particles and for Brownian diffusion. Unfortunately, it appears that the ideal model becomes less satisfactory when attempting to follow the patterns of the formation and growth of large numbers of particles in clouds. However, the reported experimental results are too limited to provide a complete test of all the existing theory. There is certainly a need to undertake new experimentation to see where the theoretical development fails.

Although detailed discussion of experimental techniques is beyond the scope of this book, certain experimental results will be incorporated into the discussion where available to give the reader an indication of the success of the theory in predicting the behavior of aerosols.

CHAPTER 2

FUNDAMENTAL CONSIDERATIONS

As INDICATED in the introductory remarks, the diversity and complexity of aerocolloidal systems are notable. The study of these systems may involve particles of variable shape ranging in size from slightly larger than molecular dimensions to macroscopic dimensions. The number density of particles may vary in extremes from a few particles per cubic centimeter to values approaching those encountered in gases at normal pressures, $\sim 10^{19}$ particles per cubic centimeter. The particles can be solid, liquid, a metastable state involving (macroscopically) gases and condensed states, or the particles may exist as a combination of all three. Aerial dispersions also may exhibit unusual physicochemical characteristics as, for example, when the particles consist of biological fragments, bacteria, viruses, etc. Frequently the particles are charged, possess electrical and magnetic moments, and are possibly radioactive. In addition, the suspending gas may be in any arbitrary state of density, temperature and motion.

Owing to such complications it is necessary for a quantitative characterization of any aerocolloidal system to make use of a simplified model introduced qualitatively in Chapter 1. The model represents a considerable reduction in the complexities implicit in the diversity noted above. Nevertheless, it is believed that such a highly idealized model furnishes a fairly accurate description of the system's mechanical behavior in many instances. And in other cases the model provides a basis of comparison in much the same way that the model of ideal gases provides a basis of comparison with real gases. We shall see below, and in the following chapters, that even the simple model chosen here is somewhat complex in its quantitative character, and that our present understanding of the theory is incomplete.

According to our conceptual theory, an aerocolloidal system consists of solid or liquid particles which are smooth spherical bodies suspended in a gas, which is assumed to be described by the kinetic theory of dilute gases. In general, the physicochemical properties of the particles are described within this model by a number of empirical parameters.

The purpose of this chapter is to set forth in a quantitative manner the characteristics of aerial dispersions as provided for in the hard, smooth sphere picture. We shall begin with a more concise definition of an aerocolloidal system than considered earlier in Chapter 1. Then the basic mechanical parameters of such a system will be outlined in more detail. These parameters will be used to define certain different regimes of the idealized aerosol. Next, the viewpoint adopted here for the quantitative characterization of a particle–gas suspension will be presented together with a detailed discussion of the Boltzmann equation, which describes the suspending gas, and which determines the dynamics of the particles. This discussion includes a presentation of various solutions of the Boltzmann equation. These solutions are of principal importance when specification of the detailed dynamical behavior is required of one or two aerosol particles in the suspending gas. Finally, the chapter will conclude with a brief mention of the physicochemical characteristics of the particles themselves, with particular emphasis on the effect of the aerosol particle size on these characteristics. It will be seen that this particular aspect of aerosols, while potentially of great importance, is one about which little is known.

2.1. Characterization of Aerocolloidal Systems

A. Definition of an Aerocolloidal System

Formally one can regard a polydisperse system of aerosol particles in the general case as a multicomponent mixture with a very large number of components. With this view it is then possible to characterize the system by means of a very large set of macroscopic and microscopic parameters. By placing certain restrictions on the general case, there can be constructed from these basic parameters certain dimensionless parameters, which define regions of particular interest in aerosol physics.

In order to set out the basic parameters, it is necessary to begin construction in more quantitative detail of the model, considered as an assembly of many components making up the aerial suspension. First let us consider the implication of the notion of the "system". By the term "aerocolloidal system" one means an arbitrary element of volume filled with gas and suspended particles that may have a highly variable physicochemical nature. In characterizing such a "system" we introduce the following formal restrictions. As discussed previously, we assume that the particles are smooth, inert, rigid spheres. Furthermore, when two particles collide, there will be some probability of "reaction" in that the two particles will stick together and possibly coalesce to form a larger particle. This probability of sticking will be considered a known quantity.

Basic Dimensional Parameters

The basic dimensional parameters characterizing an aerosol described by the hard spherical model are listed in Table 2.1. The subscript i indicates a particular particle species i identified as having a radius R_i and a given charge $\pm Q_i$. The indexing i refers to a system based on the conservation of mass for unit density particles. That is, $R_i = i^{1/3}R_1$. The mean velocities $\mathbf{q}_i$, $\mathbf{q}_G$ are measured relative to a fixed reference frame. The accommodation coefficients of mass, momentum, and energy, denoted by α_{ci}, α_{mi}, and α_{ti}, respectively, are

TABLE 2.1. BASIC PARAMETERS FOR AEROSOL MECHANICS

	Particles	Gas
Number density	n_i	n_G
Mean temperature	T_i	T_G
Mean or mass average velocity	$\mathbf{q}_i$	$\mathbf{q}_G$
Mean free path	λ_{ij}, λ_{iG}	λ_G, λ_{Gi}
Individual mass	m_i	m_G
Individual radius	R_i	R_G
Individual charge	$\pm Q_i$	Very weakly ionized
Coagulation (or reaction) probability	ε_{ij}	0
Aerodynamic lengths	L_{vi}	$\bar{L}_{vG}$, L_G
Interaction parameter	$\mathcal{I}_{\mu ib}$	$\mathcal{I}_{\mu Gb}$
Accommodation coefficients of mass, momentum and energy	α_{ci}, α_{mi}, α_{ti}	

measures of the efficiency of exchange of these properties between the particle i and the suspending gas.

The various mean free paths represent the average distance in the system traveled by a particle of species i as, for example, between particular collisions with particles of species j, where j could also be a gas molecule G as indicated in Table 2.1. The aerodynamic lengths are characteristic distances associated with changes in the various number densities, temperatures, and velocities n, T, and $\mathbf{q}$. In the gas there may be more than one aerodynamic length according to the particular geometry, and the state of the gas volume under consideration.

We may consider a pure gas with mass m_G, or a multicomponent mixture in which m_G is the average molecular mass of the gas given by

$$m_G = \frac{1}{n_G} \sum_{\alpha} n_\alpha m_\alpha, \tag{2.1}$$

where n_α = the number density of molecular species α. The definition for an average molecular radius is introduced

$$R_G = \frac{1}{n_G} \sum n_\alpha R_\alpha. \tag{2.2}$$

The interaction parameters $\mathscr{I}_{\mu ib}$, and $\mathscr{I}_{\mu GB}$ can represent the mass, momentum, and energy accommodation coefficients for particles of type i and gas G respectively on boundary surface b of the system.

When the particle size is not negligible compared with the spatial variations of the system, there may be some difficulty in defining the mean parameters for the particles such as the number density, the temperature, the particle velocity, and the mean free path. This point will be discussed briefly later. For the present it will be assumed that n_i, T_i, $\mathbf{q}_i$, L_{ij} represent suitable averages for the particles of species i and j in the aerocolloidal system, and that they serve to characterize the particular class of particle in the same manner that similar properties are used to characterize the behavior of dilute gases.

To illustrate the scope of variability of parameters in Table 2.1, Table 2.2 shows typical ranges of the basic parameters as found for aerocolloidal systems of the type encountered in pollution of the earth's atmosphere. Some of the values are estimated from a recent compilation of experimental air pollution studies by Löbner[9].

TABLE 2.2. TYPICAL RANGES OF SOME OF THE BASIC PARAMETERS FOR AEROCOLLOIDAL SYSTEMS IN ATMOSPHERIC AEROSOLS
(Table compiled in part from information of A. Löbner[9])

	Particles	Air
Number density (number cm^{-3})	$\sim 10^2$–10^4	$\sim 10^{19}$
Mean temperature (°K)	$T_i \simeq T_G$	$\sim$240–310
Magnitude of mass average velocity q (cm sec^{-1})	$\sim 10^{-2}$–10^3	~ 0–10^3
Mean free path (cm)	$\lambda_{ij} \gtrsim 10^2$	$\lambda_G \sim 6 \times 10^{-6}$ (N_2)
Individual particle radius (cm)	$R_i \sim 10^{-7}$–10^{-3}	$R_G = 1.87 \times 10^{-8}$ (N_2)
Individual particle mass (gm)	$m_i \sim 10^{-13}$–10^{-9}	$m_G = 4.6 \times 10^{-23}$ (N_2)
Individual charge, number elementary charges	$Q_i \sim 0$–100[a]	Very weakly ionized single charge
Coagulation or reaction probability	$0 \leqslant c_{ij} \leqslant 1$	0
Aerodynamic length (cm)	$L_i \to \infty$	$L_{vG} > R_i$

[a] Based on estimates of Loeb[10]

From the numbers listed in Table 2.2, the basic parameters of aerosol particles may be distinguished easily from those of the gas, except for temperature and mean velocity. The equality of these two properties essentially implies that the entire particle–gas system is near equilibrium.

It is of interest to bear in mind, that in general, several orders of magnitude differences exist between particles and gas for such properties as number density, mean free path, and particle mass in aerosol systems normally encountered on the earth.

B. Mechanical Regimes of Aerocolloidal Systems

All of the characteristic dimensional parameters introduced in Table 2.1 would appear in a complete analytical description of an aerial suspension. Thus, these parameters permit a classification of such idealized systems in a quantitative manner, and now will be used to define in a more exact manner our conceptual model of an aerosol. We shall find that the parameters in Table 2.1 also will appear in sets of dimensionless groups that will be used to represent the various regimes of behavior of an aerosol.

Some Restrictions

Difficulties in giving precision to the term "aerosol" or its synonym "aerocolloidal systems" were suggested in Chapter 1. As noted in the previous discussion one usually thinks of an aerosol or an aerocolloidal system, practically speaking, as a collection of "small" particles suspended in a gas. However, the specification of the upper and lower bounds on the particle size really appears to be rather arbitrary.

We may consider as the truly basic unit of length in any aerosol the radius of a suspending gas molecule. Using customary definitions from the kinetic theory of dilute gases, this radius may be associated with an equivalent rigid sphere of the same mass as the gas molecule. Hence, the basic unit of mass in the aerosol is the mass of the suspending gas molecule m_G. Then the lower limit of particle size is specified arbitrarily for a gas with equivalent molecular radius R_G by the following conditions:

$$R_G/R_i < 1; \quad m_G/m_i \ll 1, \tag{2.3}$$

where subscript i indicates the corresponding properties of a species different from the suspending gas in the system (i.e. $i \neq G$). If these conditions hold, we shall refer to the system as an *aerocolloidal system*. These limits are entirely consistent with the criteria (1) and (2) on p. 6. It may be noted further that this definition, eqn. (2.3), does not distinguish between what is usually considered to be an aerosol particle and a large, heavy molecule. In fact, in the idealized concept of spherical aerosol particles, there can be no clear-cut distinction between the two bodies. A macro-molecule suspended in a gas qualifies in all practical respects as an aerosol particle. The distinguishing features of an aerosol particle appear to be the possible variability in its individual physicochemical properties, its relatively large size, and its mass change in time in the processes of evaporation or condensation. These features are in contrast to the nature of a gas molecule which has in this context an invariant structure or organization, except in the event of chemical reactions, and a relatively small, constant mass within our modelling assumptions.

As remarked further in Chapter 1, the upper limit of size of an aerosol particle is completely arbitrary. We simply choose the upper limit of particle size to be a radius of 30 μ. This upper size limit corresponds well with historical usage of the term "aerosol" particle, as implied by the restrictions (1) and (2) on p. 6.

The nature of the suspending gas must also be specified. We have restricted our discussion to un-ionized, or very slightly ionized suspending gases of moderate to low density. One could, of course, talk about aerocolloidal systems in which the suspending gas is highly ionized or dense, or both. Such systems deserve very careful discussion in their mechanical definitions. In this book we assume only very weak ionization in the suspending gas so that the molecule–molecule interactions are simple to analyze. Furthermore, no consideration is given here to suspensions involving dense gases.

For the purpose of developing the theory in this book, we require further that the suspending gas satisfy the following conditions:

$$R_G/\lambda_G \ll 1, \tag{2.4}$$

$$R_G/L_G \ll 1. \tag{2.5}$$

Here the aerodynamic length L_G is the distance over which there is a significant change in the state variables n_G, T_G, and $\mathbf{q}_G$ of the gas. These two conditions have been shown[11] to be the conditions for which the Boltzmann equation provides a satisfactory description of a real system made up of molecules having no important internal degrees of freedom. The Boltzmann equation is known[8] to provide an adequate description of dilute systems of monatomic molecules, and of polyatomic molecules for many physical processes. One notable exception is the case of heat transfer in assemblies of polyatomic molecules in which certain suitable approximations still can usually be made.[8]

Among other things, eqns. (2.4) and (2.5) imply that the suspending gas is not dense (the gas obeys the ideal gas law, $p = n_G kT$), and that the mean parameters of the gas are not changing rapidly with regard to the collision interval for the gas molecules. Many of the aerocolloidal systems of importance including those encountered in atmospheric processes, involve suspending gases which satisfy the requirements implicit in eqns. (2.4) and (2.5).

C. A Single Particle in an Infinite Gas

Inasmuch as many aerosols involve extremely small concentrations of particles as suggested in Table 2.1, it is frequently useful to consider the behavior of a single aerosol particle in a gas effectively infinite in extent. That is, other suspended particles, physical boundaries surrounding the gas, or other objects in the gas are taken to be located at very large distances from the single particle whose behavior is of interest. The magnitude of the relative distances between bodies for the single particle assumption to be valid is discussed further in Chapters 3 and 7, but in general one expects that such an assumption is valid if $R_i/L_{vi} \gg 10$.

To begin with, in specifying the behavior of a single particle, the gas is considered to be in an arbitrary un-ionized or weakly ionized state with dimensional parameters as given in Table 2.1. We remark, of course, that the dimensional parameters defined in terms of particle averages lose their significance in the case of a single aerosol particle. The single particle will have a particular velocity $\mathbf{q}_i$ and a particular temperature T_i whose definition must be approached with caution in any non-equilibrium condition. Furthermore, there will be no mean parameters such as number density n_i in the usual sense, since we have centred attention on a particular particle.

The terms λ_{ij}, c_{ij}, and L_{vi} also have no explicit meaning for the single isolated particle. However, it is possible to assign a mean free path λ_{iG} for the single particle.

For the present, the aerodynamic lengths L_{vG} in the gas phase,† will not be considered in detail. The influence of external fields of force also will be disregarded for the present. Later, brief mention will be made of the consequences of the inclusion of the aerodynamic lengths and of external fields of force. Within these restrictions, the remaining relevant parameters from Table 2.1 include only the mass m_i and radius R_i of the single particle traveling with a relative velocity $\mathbf{q}_i - \mathbf{q}_G$ to a gas with number density n_G and temperature T_G. The gas molecules have an individual mass denoted by m_G, a radius R_G, and a mean free path λ_G. Thus nine basic parameters, m_i, R_i, $|\mathbf{q}_i - \mathbf{q}_G|$, n_G, T_G, λ_G, m_G, R_G, and a single aerodynamic length L_G (eqn. 2.5), together with the accommodation coefficients of mass, momentum and energy, then serve to specify the dynamics of the idealized single particle–gas system.

TABLE 2.3. DIMENSIONLESS PARAMETERS FOR A SINGLE SPHERICAL PARTICLE SUSPENDED IN AN INFINITE GAS

Knudsen number	$Kn_i = \lambda_G/R_i$
Mach number	$Ma_i = \|\mathbf{q}_i - \mathbf{q}_G\|/\bar{v}_G$
Schmidt number	$Sc_i = R_i^2 n_G \lambda_G$
Brown number	$Br_i = \bar{v}_i/\bar{v}_G$
[a] Boltzmann Number I	$Bo_{\text{I}} = R_G/\lambda_G$
[a] Boltzmann Number II	$Bo_{\text{II}} = R_G/L_G$

[a] Because of the fundamental importance of these ratios in the kinetic theory of gases, we have named them for Boltzmann in recognition for his pioneering work in statistical mechanics, and for their relevance to the applicability of the Boltzmann equation.

Dimensionless Groups

From these nine basic parameters, six independent dimensionless groups can be formed in addition to the accommodation coefficients. The values of these six dimensionless parameters define various regimes of behavior of the single particle–gas system. The six dimensionless parameters are tabulated in Table 2.3.

The velocities $\bar{v}_G$ and $\bar{v}_i$ are the mean "thermal" speeds, defined as

$$\bar{v}_G = (8kT/\pi m_G)^{\frac{1}{2}}, \tag{2.6}$$

$$\bar{v}_i = (8kT/\pi m_i)^{\frac{1}{2}}. \tag{2.7}$$

The last two dimensionless numbers serve merely to specify the regime for the evolution of the gas. Implicitly, we have already chosen values of Bo_{I} and Bo_{II} in the requirements given by eqns. (2.4) and (2.5). These choices indicate that the gas can be described by the Boltzmann equation.

A brief discussion of the remaining dimensionless quantities Kn_i, Ma_i, Sc_i, and Br_i is presented below. The particular regimes defined by these numbers are also indicated. The accommodation coefficients will be mentioned in Chapter 5; they will not enter into the following discussion.

† Aerodynamic lengths L_{vG} other than that length L_G introduced in eqn (2.5) would be associated, for example with periodic or other unsteady motion of the gas.

The Knudsen Number

The first of the dimensionless parameters is the *Knudsen number*. This parameter provides a measure of the departure of the transport processes in the gas–particle system from the laws of continuum mechanics. In terms of the dimensional parameters defined in Section 2.1, the Knudsen number Kn_i for a single particle of radius R_i in an effectively infinite gas is

$$Kn_i = \lambda_G/R_i. \tag{2.8}$$

This Knudsen number is defined in the range

$$0 < Kn_i < \infty.$$

Only in the limit $Kn_i \to 0$ can the gas surrounding a particle be regarded as a *continuum*, and, consequently, can the transport processes to and from the particle be analyzed accordingly. The regime $Kn_i > 0$ is frequently termed *the non-continuum regime.*

The other extreme of the Knudsen number, in which $Kn_i \to \infty$ is usually referred to, after gas dynamics usage, as the *free-molecule regime.* In this regime, where the mean free path of gas molecules is very large relative to the particle radius, there are no intermolecular collisions between gas molecules in the vicinity of the particle, and in this same zone the gas molecules travel in the absence of external fields in straight line paths (see also Fig. 2.6). In this extreme case it can be assumed that gas molecules striking the particle have not been influenced by the particle. Also, in the free molecule regime, one may regard the particle, for many purposes, as a large molecule undergoing independent and binary collisions with the gas molecules.

The regime between the two extremes, $Kn \to 0$ and $Kn \to \infty$, is frequently referred to as *the transition regime.* Considerable difficulties, in comparison with the extremes, are found in analytically describing transport processes to a particle in this range of Knudsen number. For small values of Kn_i, continuum analysis has been applied with the *slip flow* boundary condition.[12] In this range, the mathematical procedures of applying slip conditions at the particle surface have been employed intuitively. The procedure of extending continuum

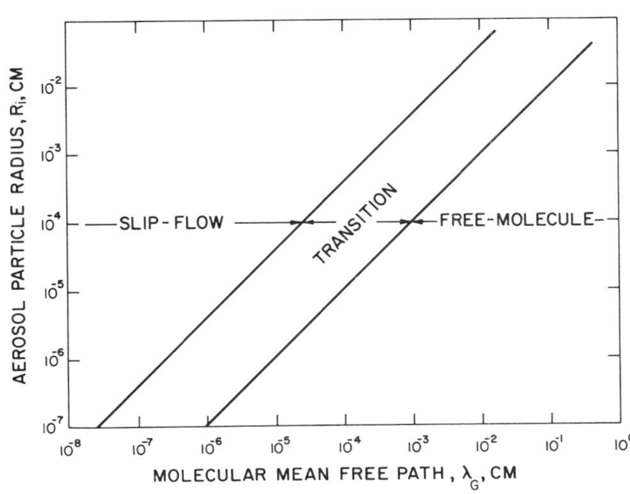

FIG. 2.1. Ranges of gas dynamic behavior for aerosol particles in terms of radius and mean free path.

theory by use of gas slippage at the particle surface is without rigorous basis, and hence no theoretical limits of Kn_i can be assigned to the regime called the *slip flow regime.*

In general, for the transition region where for practical purposes $0.25 < Kn_i < 10$, no simple analytical approach like the ones used in the slip flow regime appears to have been completely successful for describing particle dynamics. Figure 2.1 indicates schematically the various regimes defined by the Knudsen number.

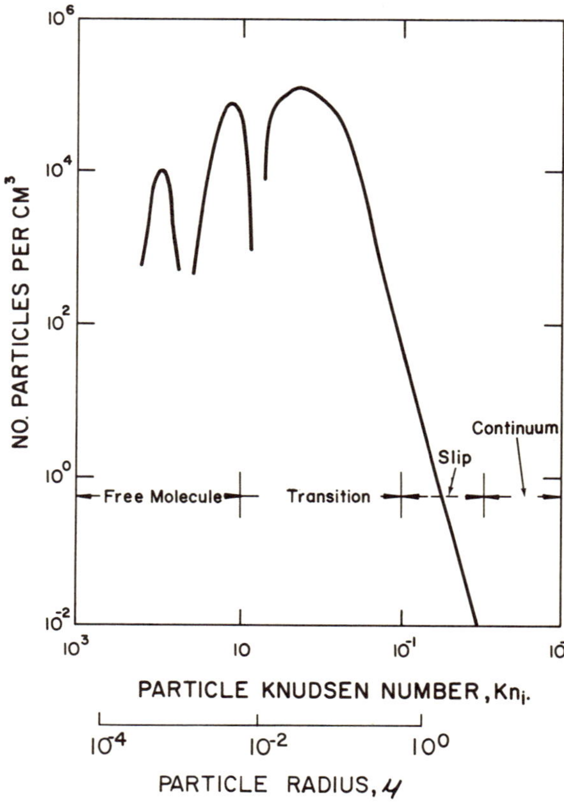

FIG. 2.2. Typical gas dynamic regimes of aerosols in the lower atmosphere (the troposphere).

Figure 2.2 shows values of Kn_i typically found in natural phenomena involving aerosols. It can be seen that in atmospheric aerosols one frequently deals with regimes which exist in non-continuum conditions. Many aerosols involved in natural phenomena exist in the transition regime of Kn_i where one encounters the greatest difficulty in determining the mathematical descriptions of transport processes to suspended particles.

In later chapters we shall try to indicate some of the procedures which have been worked out for the theoretical descriptions of the various Knudsen regimes shown in Fig. 2.1.

The Mach Number and its Relation to the Reynolds Number

When there is relative motion between the aerosol particle and the gas, a dimensionless parameter, which we have termed the *Mach number* (Ma_i) appears. In this development we use the following definition of the Mach number:

$$Ma_i = |\mathbf{q}_G - \mathbf{q}_i|/\bar{v}_G, \tag{2.9}$$

where $\bar{v}_G$ is the mean thermal speed of the gas molecules as given by eqn. (2.6).

The usual definition of the Mach number involves not $\bar{v}_G$ but, in its place, the velocity of sound in the gas. The velocity of sound depends on the gas pressure. However, the Mach number defined above is independent of the gas pressure. Therefore, this velocity ratio appears to be of more value in discussions of aerocolloidal systems than the Mach number as ordinarily defined.

The familiar dimensionless parameter, the Reynolds number, Re_i, which is commonly encountered in flow problems, can be defined in terms of the Mach number and Knudsen number introduced above. That is,

$$Re_i \approx 4Ma_i/Kn_i. \tag{2.10}$$

Thus, *either the Mach number or the Reynolds number* will be regarded as an independent parameter in a given problem involving relative motion between the single particles and the suspending gas, at least within the restrictions of our spherical particle model.

Actually, in the analysis of many aerosol problems including relative motion in the non-continuum regimes ($Kn_i > 0$), the Knudsen and Mach numbers appear as more natural dimensionless parameters than the Reynolds number. This is readily seen in the discussion of Section 2.2 and in Chapters 5 and 6.

In the continuum regime, by definition $Kn_i \to 0$ and, if $Sc_i \to \infty$ and $Br_i \to \infty$, then Re_i in this regime remains the sole relevant dimensionless parameter which is subject to arbitrary variation. Under these circumstances, the familiar results of classical aerodynamics (incompressible flow) apply within the continuum approximation. One then recognizes various regimes of (incompressible) gas flow which are characterized entirely by the Reynolds number.

If the incident gas flow to the sphere is steady, for $Re_i \ll 1$, the gas flow everywhere is steady. Then Stokes' law [eqn. (3.8)] gives the drag on the sphere in this case. In cases where Re_i is the order of unity, the approximations of Oseen[13] or Proudman and Pearson[14] apply as discussed in Section 3.1. For $Re_i > 1$, but less than some value, Re_i^{Cr}, one observes the development of a laminar boundary layer around the sphere with separation at some point producing a wake downstream from the sphere. The gas flow in this case is now unsteady but regular, and hence may still be described as laminar (see Section 3.1). At some value of the Reynolds number, Re_i^{Tur} the flow becomes unsteady and irregular. Re_i^{Tur} in principle can be obtained from stability analysis of the continuum equations describing the gas flow. However, at present an accurate measure of Re_i^{Tur} still must be obtained from experimental investigation. Measurements indicate that Re_i^{Tur} is of the order of 10^3–10^4 for flow around smooth spheres.[15]

In the near continuum regime in which $Kn_i < 0.01$, the continuum equations may still be taken to apply to the behavior of the system provided that the restrictions of Sc_i and Br_i, given above for $Kn_i \to 0$, are maintained.

One distinguishes the following five flow regimes for $Kn_i \leqslant 0.01$: (a) incompressible flow when $Ma_i \ll 1$, and the relative speed is small and the density of the gas may be assumed a constant; (b) sub-sonic flow in which $Ma_i \ll 1$ but the gas density varies; (c) the transitional flow where $Ma_i \approx 1$; (d) super-sonic flow with $Ma_i > 1$; and (e) hypersonic flow $Ma_i \gg 1$. These definitions ordinarily apply to Ma_i written with the speed of sound c_0. However, for gases at standard temperature and pressure $c_0 \approx \bar{v}_G$, so the regimes have essentially similar meanings here. For many aerial suspensions, one finds that $Ma_i \ll 1$. Hence, in these cases at very small Kn_i one has effectively incompressible flow, so that all the results from classical aerodynamics may be applied to determine the theory for the dynamics of the particle.

As Kn_i increases beyond values of ~ 0.01, it is not possible to identify uniquely particular flow regimes with values of Ma_i for the sphere–gas system owing to a lack of detailed experimental observations available at this time. The understanding of such possibly non-continuum flows is in a state of development, and is one of the frontier areas of gas dynamics classified as *rarified gas dynamics.* Fortunately for most aerosols of practical importance, the relative motion between the particle and the suspending gas is quite small. Therefore, one is *usually* concerned in the non-continuum regimes only with the variation of Kn_i with $Ma_i \ll 1$ and $Re_i < 1$.

The Significance of the Brown Number and the Schmidt Number

In the particle–gas system, the two dimensionless parameters, the Schmidt and Brown numbers, together with the Knudsen number, serve to characterize the Brownian motion of particles. In particle–gas systems, where it is necessary to consider the Brownian or diffusive motion of the particle, the Mach number generally is much less than one. In this incompressible regime, the flow is independent of Ma_i so that Ma_i does not enter as an important parameter.

The parameter we have called the Brown number, for Robert Brown's classic nineteenth-century work in Brownian motion is the velocity ratio

$$Br_i = \bar{v}_i/\bar{v}_G. \tag{2.11}$$

This dimensionless ratio accounts for the effect of the mass of the particle on its diffusive motion. The Brown number serves as an appropriate index for the transition from molecular behavior wherein $Br_i \approx 1$, to Brownian particle behavior in which $Br_i \ll 1$. For $Br_i \ll 1$, the well-developed theory of Brownian motion as outlined in Chapter 7 may be applied in describing the diffusive motion of a single particle.

The Schmidt number is defined in terms of the particle radius, the gas density and the mean free path of the gas

$$Sc_i = R_i^2 n_G \lambda_G. \tag{2.12}$$

Values of Sc_i signify the relative importance of molecular momentum transfer in the gas to diffusive transport of particles.† The importance of Brownian motion of the particles decreases with increasing Sc_i. For a particle of unit density in air at standard conditions, Brownian motion of the particle for most purposes can be neglected compared with convective transport of particles whenever $Sc \sim 10^6$–10^7. The Schmidt number defined in eqn. (2.12) does not have the form of the Schmidt number found in molecular diffusion studies. Normally Sc is defined in the following way:

$$Sc = \mu_G/\rho_G D_{AB}. \tag{2.13}$$

† In consideration of the diffusion of particles to boundary surfaces the Péclét number is useful and describes the relationship between convective and diffusive transfer of particles. The Péclét number is defined for particles as $q_i L/D_i$.

Here D_{AB} is the mutual diffusion coefficient of a particular molecular mixture, while ρ_G and μ_G are respectively mean density and viscosity coefficient of the gas mixture. Equation (2.13) is proportional to eqn. (2.12) if the expression for the diffusion coefficient of a free molecule aerosol is substituted in eqn. (2.13) and the viscosity coefficient expressed in terms of the mean free path† of the suspending gas is introduced. One easily could define Sc_i generally in terms of the Brownian diffusion coefficient D_i; however, the meaning of eqn. (2.12) seems more clear for gas–particle systems. If a Schmidt number after eqn. (2.13) would be defined for aerosols, such a definition would have to involve the Knudsen number plus a semi-empirical coefficient such as the momentum accommodation coefficient α_m. We have chosen to restrict the *basic* parameters to the simplest set possible. Hence, eqn. (2.12) is used for the definition of the Schmidt number Sc_i as it applies to aerosols.‡

D. Some Relevant Additional Dimensionless Parameters

If we relax the requirement of an infinite gas and steady motion of the gas at large distances from the particle, then various aerodynamic lengths L_{vG} or L_{vi} have to be taken into account. In particular, the lengths L_{vG} can be associated with the boundary dimensions of a finite particle–gas system, or they can represent classes of wavelengths associated with periodic gas motion as, for example, with the presence of acoustic waves in the gas.

In instances of shear imposed on the gas through the boundary surfaces of the system, one could define an additional Reynolds number differing from the one on p. 23 which would characterize the fluid motion alone. Thus the particle–gas system would be described by the Reynolds number involving a characteristic dimension of the boundary surfaces as well as by the particle parameters Kn_i, Ma_i, Sc_i, and Br_i. Similarly, there are other parameters which could appear as, for example, the Knudsen number associated with a characteristic dimension of the boundaries holding the fluid suspension.

It is clear, therefore, that each particle–gas system which is finite in extent has to be described not only by the five basic gas dynamical parameters plus the accommodation coefficients characterizing the particle–gas interaction, but also by as many additional parameters as required to characterize the various interactions between the gas, the particle, and the boundaries of the system.

E. Assemblies of Particles in a Gas

The dynamics of collection of particles in a gas should be interpretable partly by the set of basic parameters for single particles discussed above. In addition, the various averaged quantities for clouds as noted in Table 2.1 have to be incorporated in a treatment of particle assemblies. To speak meaningfully of the average particle parameters n_i, $\mathbf{q}_i$, T_i, λ_{ij}, we must assume that if these parameters change over a characteristic distance, then a volume L_G^3 of the system must contain a large number§ of particles of type i and j. Similarly, the parameters such as $\mathbf{q}_i$ and T_i no longer can be sufficient to provide complete information of the state of any particular species of particles if the properties of the suspending gas

† In the rigid sphere model for gases, $\mu_G = 0.499\rho_G\bar{v}_G\lambda_G$.

‡ The conventional definition of Sc used in (2.13) where $D_i = B_i kT$, B_i being the mobility, traditionally has been used in the literature describing convective diffusion of aerosols. Therefore this definition is assumed for the discussion with the Péclét number in Chapter 7.

§ In the statistical sense, a "large number" is one large enough so that the fluctuations about the average for temporal and spatial purposes may be neglected. Such conditions have been discussed in detail in defining similar properties for gases. See, for example, Jeans.[16]

change appreciably over the dimensions of the individual particles. For example, a particle would have an internal temperature gradient if the gas temperature changed markedly over a distance the order of the diameter of the particle. Hence one must use considerable caution in speaking of mean particle parameters in the usual gas kinetic sense.

With the perhaps critical assumption that the principal particle parameters n_i, $\mathbf{q}_i$, T_i, and λ_{ij} can be specified in a gas kinetic sense, and that there is variation in particle size and physicochemical nature (i.e. that the particles are both *polydisperse* and *polygenetic*), there will be a very large number of characteristic basic parameters necessary to describe completely even an idealized system of aerosol particles. If the particle parameters cannot be specified in the usual gas kinetic sense, then the number of particle parameters in general will increase considerably. A complete description involving such large sets of basic parameters is exceedingly difficult. In some cases such as fluidized beds of particulates or nozzle expansions at high speed, involving critical gases, this problem of a highly multivariate analysis must be dealt with at least in part for a satisfactory description of the system.

The Single Particle Regime

Many aerocolloidal systems permit considerable simplification if the following conditions are met:

$$\text{(a)}\ n_i/n_G \ll 1 \quad \text{for all } i, \tag{2.14}$$

$$\text{(b)}\ \lambda_G n_i^{\frac{1}{3}} \ll 1 \quad \text{for all } i, \tag{2.15}$$

$$\text{(c)}\ R_i n_j^{\frac{1}{3}} \ll 1 \quad \text{for all } i, j, \tag{2.16}$$

$$\text{(d)}\ n_i^{-\frac{1}{3}} L_{vi}^{-1} \ll 1 \quad \text{for all } i, \tag{2.17}$$

$$\text{(e)}\ Q_i Q_j \kappa / kT \ll 1 \quad \text{for all } i, j, \tag{2.18}$$

Here κ is the Debye reciprocal length $\kappa = [(4\pi\tilde{e}^2/kT) \sum_{i=1}^{p} n_i z_i^2]^{\frac{1}{2}}$.

Under these restrictions, all the particles can be considered to be essentially independent of each other.

The last ratio, which has been named the Debye number, just implies that the electrical interaction between the particles are weak and Coulombic in nature. With restriction (e), each individual particle in the collection can be characterized by the basic parameters Kn_i, Ma_i, Sc_i, and Br_i and the accommodation coefficients of mass, momentum, and energy, as described for the special case of the single particle in an infinite gas. The bulk of the theory of aerosol dynamics heretofore has confined itself to investigation of problems in which these restrictions apply. For convenience, we can then speak of aerocolloidal systems which satisfy eqns. (2.14)–(2.18) as in the *single particle regime*. When the aerocolloidal system is not in the single particle regime, many complications suggested above develop. If the system exhibits the phenomena of coagulation or deposition of particles on surfaces but satisfies conditions (2.14)–(2.17), we speak of such a system as being in the *quasi-single particle regime*.

Regimes outside the quasi-single particle regime can be identified for specific systems as for example the case of extremely large n_i, which may be termed *fluidized bed regime*. Other regimes might include highly charged particle-ionized gas regime (or a *thermonuclear regime*), a highly accelerated particle regime (or *rocket exhaust regime*), etc. None of these cases are well understood at this time, and we defer any detailed treatment of these regimes to possibly later volumes in this series. This book concerns itself almost entirely with the single particle regime. In the last three chapters, some approximate extensions of theories based on the single particle regime are considered for a few special examples of the behavior of assemblies of particles, with particular reference to particle generation, and particle coagulation.

TABLE 2.4. USUAL RANGES OF THE VALUES FOR RATIOS OF EQUATIONS (2.14)–(2.18) FOUND IN THE TROPOSPHERE FOR AEROCOLLOIDAL SYSTEMS

n_i/n_G	$\sim 10^{-13}$–10^{-19}
$\lambda_G n_i^{\frac{1}{3}}$	$\sim 10^{-3}$–10^{-5}
$R_i n_j^{\frac{1}{3}}$	$\sim 10^{-1}$–10^{-7}
$n_i^{-\frac{1}{3}} L_v^{-1}$	~ 0
$Q_i Q_j \kappa / kT$	$\sim 10^{-2}$–10^{-6}

To illustrate the extent of applicability of calculations based on the single particle regime, typical values of the ratios in eqns. (2.14)–(2.19) have been estimated for aerosols in the earth's atmosphere. These ranges are listed in Table 2.4 for such systems as clouds, fogs and urban aerosols. From the ranges shown in this table it is evident that a large class of aerosols of practical interest may be described by the basic mechanical parameters governing the single particle regime. This is an important conclusion, because it justifies in a way, on fundamental grounds, the applicability of the vast majority of work in aerosol mechanics up to the present time.

2.2. Quantitative Formulation of Aerosol Dynamics

The mechanical nature of an aerocolloidal system has been described qualitatively in terms of important characteristic parameters. We can now specify more quantitatively the dynamical theories applicable to the idealized system. These theories will permit, at least in principle, a detailed description of aerial suspensions as modeled in the hard sphere restrictions.

As in the previous section, one could treat an aerocolloidal system formally as a gas mixture which could be analyzed quantitatively by the methods of statistical mechanics. As a practical matter such a viewpoint and such a procedure introduces some rather difficult problems. First of all, the particles in an aerocolloidal system exhibit change in size by evaporation, by condensation, or by coagulation and coalescence. Such processes are difficult to describe in a practical way in a truly general statistical mechanical treatment. Even without the complications of processes of coagulation or coalescence, the occurrence of particle–particle collisions between spheres that are large relative to the mean free path of the gas requires that the description of particle motion account for the aerodynamic behavior of gas molecules lying between the moving bodies. Furthermore, in systems where the gas properties change over distances of the order of the particle diameter, a very large set of internal coordinates are necessary to determine the state of any given particle. Such examples serve to illustrate that the incorporation of the particles simply as components of a gas mixture into a statistical mechanical treatment may be fraught with great complexities. It is evident then that some approximations are necessary in describing an aerocolloidal system, and that one cannot expect to obtain as general a description of such systems as might be found from a rigorous statistical mechanical treatment.

In the discussion which follows we assume that the aerocolloidal system is in the single particle regime. The individual particles in the viewpoint we adopt here become, within restrictions discussed below, simply bodies to which transfer of mass, heat, and momentum occur primarily by the action of the suspending gas molecules. Along with this viewpoint it is necessary to remember, however, that the particles exhibit Brownian motion.

One of the distinguishing characteristics of an aerocolloidal system is the random motion, Brownian motion, displayed by the aerosol particles as a result of the fluctuations in the impacts of gas molecules with the particles. If the number of particles per unit volume is sufficiently great for a statistically meaningful sample but not great enough for there to be substantial particle–particle interaction, then it is possible to develop a stochastic model to describe the particle dynamics. The equation describing this stochastic model, the Fokker–Planck equation, will be discussed in detail in Chapter 7. The stochastic model, of course, cannot describe the exact dynamical behavior of any particular particle. Instead, the model gives the distribution of expected behavior of the particles. In this manner, for example, the diffusive motion of aerosol particles is related through a moment of the distribution to the mobility of a representative particle in the suspending gas.

Within the limits imposed by the assumption of non-interacting particles in the Fokker–Planck stochastic model, all the mean properties of a collection of aerosol particles are related almost solely† to the independent interaction of individual particles with the surrounding gas molecules. Therefore, there is an alternative viewpoint in the single particle regime to that which proceeds rigorously through a statistical mechanical formulation. In this more phenomenological approach, one may regard the particles as bodies in the gas to which transfer of mass, momentum, and energy occurs by means of the suspending gas. The gas–particle surface interactions are accounted for by the set of semi-empirical parameters we have termed the accommodation coefficients. As an extension of this concept, the transport of mass, momentum, and energy by collections of non-interacting particles to boundary surfaces is treated by a set of "gas-like" macroscopic conservation equations whose transport coefficients are determined approximately by those calculated for a single particle in the gas. Normally the effects of particle–particle collisions are ignored in such transport theories. In similar phenomenological theories of nucleation or of coagulation or coalescence, the particle–particle collisions are studied as a separate phenomenon, in much the same way as homogeneous, gas phase chemical reactions are sometimes investigated separately from the gaseous transport phenomena. This philosophy essentially represents what is meant by the term quasi-single particle regime.

While the viewpoint of regarding the particles as essentially closed internal boundary surfaces in the gas may present complications when particle–particle collisions must be accounted for, this concept generally proves to be simpler for many aerocolloidal systems

† There may be, for example, transfer of energy to the particles by electromagnetic radiation from the system boundaries so that in this instance the surrounding gas molecules are not the only agent determining the properties of the particles.

than the alternative of dealing with the particles and the gas molecules in any generalized statistical mechanical treatment.

A. The Physics of the Suspending Gas

Having chosen effectively a phenomenological approach to the dynamics of aerosols by means of an equation of the Fokker–Planck type, let us proceed to describe the quantitative relations which will be applied to calculate the gas–particle interactions. We discuss now the description of the gas in an aerocolloidal system in the presence of the particles and any other boundary surfaces which may exist. From the viewpoint adopted here, the gas-particle interaction and gas-boundary interactions appear as boundary conditions in determining the state of the gas.

The principal problem in constructing a quantitative description of this model of an aerocolloidal system is the determination of an evolution equation describing the state of the gas, and the solution of that equation for particular systems of interest.

In Section 2.1 we have placed restrictions on the nature of the gas in the model aerosol that we shall be concerned with in this monograph. These imposed restrictions make it possible to describe the state of the gas in terms of the Boltzmann integro-differential equation. In the subsequent sections, the Boltzmann equation, and various approximation schemes for its solution will be discussed.

Origins of the Boltzmann Equation

The Boltzmann equation is one of the early and more successful attempts at a resolution of a basic problem of statistical physics. Simply stated this problem is: Given the molecular constitution of matter, how can the macroscopic properties of the material be explained through the laws of motion and interaction of the constituent molecules?

The original equation of Boltzmann was derived on the basis of rather intuitive concepts which, however, have proven to be outstandingly successful when applied to the behavior of gases. Some of the recent effort in statistical physics has been concerned with the re-derivation of the Boltzmann equation on a less intuitive, more rigorous basis. Since the Boltzmann equation plays an essential role in the theory of aerosols, it is useful to examine its derivation briefly. Because there are many treatises on this subject, only the method of attack followed in derivations of the Boltzmann equation shall be outlined. Various references, e.g. (7), (17), and (18), may be studied for more complete details. In what follows we shall assume that the motion of the individual gas molecules is governed by the laws of classical mechanics. This assumption is sufficient for the purpose of introducing the basic procedures leading to the Boltzmann equation. We consider only a gas with a single chemical species. The formulation for a one-component gas easily can be generalized to include multi-component mixtures (e.g. Chapters 4 and 5).

With the conceptual picture of a large collection of gas molecules as classical mass points of number $\mathcal{N}$ having $3\mathcal{N}$ degrees of freedom. The motion in time of each molecule in this system traces out a trajectory in a six-dimensional phase space (μ space), or alternatively the entire system of $\mathcal{N}$ molecules traces out a trajectory in a $6\mathcal{N}$ dimensional phase space (Γ space) according to the Hamilton equations of motion

$$\dot{\mathbf{x}}_\alpha = \partial\mathscr{H}/\partial\mathbf{p}_\alpha, \qquad (2.19)$$

$$\dot{\mathbf{p}}_\alpha = \partial \mathscr{H}/\partial \mathbf{x}_\alpha, \; \alpha = 1, 2, \ldots, \mathscr{N}; \tag{2.20}$$

where $\mathbf{x}_\alpha$ and $\mathbf{p}_\alpha$ are the conjugated configuration and momentum vectors for a particular molecule α and the dot indicates the time derivative. $\mathscr{H}$ is the Hamiltonian;

$$\mathscr{H} = \sum_{\alpha=1}^{\mathscr{N}} [\mathbf{p}_\alpha^2/2m + U(\mathbf{x}_\alpha)] + \sum_{\alpha<\beta} \Phi(|\mathbf{x}_\alpha - \mathbf{x}_\beta|), \tag{2.21}$$

where U is a given external potential energy and Φ is the potential energy of interaction of all pairs $\alpha + \beta$ molecules in the gas. For a large number of identical systems, an *ensemble* of systems according to the terminology of Gibbs,[17] we can suppose in any region of Γ space a density of system points:

$$P^{(\mathscr{N})}(\mathbf{x}_1 \ldots \mathbf{x}_{\mathscr{N}}, \mathbf{p}_1 \ldots \mathbf{p}_{\mathscr{N}}, t). \tag{2.22}$$

If we assume there are no sources or sinks of molecules, then one can write an equation of continuity in phase space:

$$\frac{\partial P^{(\mathscr{N})}}{\partial t} + \nabla^{(\mathscr{N})} . (P^{(\mathscr{N})} \mathbf{V}^{(\mathscr{N})}) = 0, \tag{2.23}$$

where $\mathbf{V}^{(\mathscr{N})}$ is the velocity vector along the trajectory in Γ space with components $\dot{\mathbf{x}}_1 \ldots \dot{\mathbf{x}}_{\mathscr{N}}, \dot{\mathbf{p}}_1 \ldots \dot{\mathbf{p}}_{\mathscr{N}}$. From Hamilton's equation, it follows that

$$\nabla^{(\mathscr{N})} . \mathbf{V}^{(\mathscr{N})} = \sum_{i=1}^{\mathscr{N}} \left(\frac{\partial \dot{x}_i}{\partial x_i} + \frac{\partial \dot{p}_i}{\partial p_i} \right) = 0, \tag{2.24}$$

and therefore

$$\frac{\partial P^{(\mathscr{N})}}{\partial t} + \mathbf{V}^{(\mathscr{N})} . \nabla^{(\mathscr{N})} P^{(\mathscr{N})} = 0, \tag{2.25}$$

which indicates that the points move as an incompressible fluid in Γ space.

$P^{(\mathscr{N})}$ is the probability density function for finding one system of the ensemble in the volume element $d\mathbf{x}^{(\mathscr{N})}\, d\mathbf{p}^{(\mathscr{N})}$ around $\mathbf{x}^{(\mathscr{N})}\, \mathbf{p}^{(\mathscr{N})}$ in Γ space. However, it is usual to work not with $P^{(\mathscr{N})}$ but instead with a distribution, $f^{(\mathscr{N})}$, representing the probability of finding the points associated with the molecules of the system in any arbitrary order in the volume elements $dx_1, dp_1 \ldots dx_{\mathscr{N}}\, dp_{\mathscr{N}}$ around $\mathbf{x}_1\mathbf{p}_1 \ldots \mathbf{x}_{\mathscr{N}}\mathbf{p}_{\mathscr{N}}$ in Γ space. $f^{(\mathscr{N})}$ and $P^{(\mathscr{N})}$ are therefore simply related:

$$f^{(\mathscr{N})} = \mathscr{N}! P^{(\mathscr{N})} \tag{2.26}$$

Equation (2.25) may be rewritten, in terms of $f^{(\mathscr{N})}$

$$\frac{\partial f^{(\mathscr{N})}}{\partial t} + \sum_{\alpha=1}^{\mathscr{N}} \left\{ \left[\frac{\mathbf{p}_\alpha}{m_\alpha} . \nabla_{\mathbf{x}_\alpha} f^{(\mathscr{N})} \right] + (\mathscr{E}_\alpha + \mathscr{F}_\alpha) . \nabla_{\mathbf{p}_\alpha} f^{(\mathscr{N})} \right\} = 0, \tag{2.27}$$

where the external force on particle α, $\mathscr{E}_\alpha$ has been introduced along with $\mathscr{F}_\alpha$ the force on α owing to all the other molecules in the system. Equations (2.25) and (2.27) are both termed Liouville's equations.

One is not usually interested in $f^{(\mathcal{N})}$ but in the lower order distribution functions $f^{(s)}$ for s molecules in μ space:

$$f^{(s)}(\mathbf{x}^{(s)}, \mathbf{p}^{(s)}) = \frac{1}{(\mathcal{N} - s)!} \iint f^{(\mathcal{N})}(\mathbf{x}^{(\mathcal{N})}, \mathbf{p}^{(\mathcal{N})}) \mathrm{d}\mathbf{x}^{(\mathcal{N}-s)}\, \mathrm{d}\mathbf{p}^{(\mathcal{N}-s)}. \tag{2.28}$$

For the case of a dilute gas, the reduced distribution function $f^{(1)}(\mathbf{x}^{(1)}, \mathbf{p}^{(1)})$ is sufficient to define completely the state of the gas.

Integration of the Liouville equation gives an expression for the evolution of $f^{(1)}$:

$$\frac{\partial f^{(1)}}{\partial t} + \mathbf{x}_1 \cdot \nabla_{\mathbf{x}_1} f^{(1)} + \mathscr{E}_1 \cdot \nabla_{\mathbf{v}_1} f^{(1)} = \frac{1}{m_G} \iint \nabla_{\mathbf{v}_1} f^{(2)} \cdot \mathscr{F}_{12}\, \mathrm{d}\mathbf{x}_2\, \mathrm{d}\mathbf{v}_2\, \nabla_{\mathbf{v}_1} f^{(2)}. \tag{2.29†}$$

As before, m_G is the molecular mass. Also $\mathbf{v}_1 = \mathbf{p}_1/m_G$ and $\mathscr{F}_{12}$ is the force on molecule 1 owing to interaction with molecule 2 in the system. Equation (2.29) is not a closed equation inasmuch as higher order distributions like $f^{(2)}$ depend themselves on $f^{(3)}$, which depends on $f^{(4)}$, and so on. There is a hierarchy of equations from $f^{(1)}$ to $f^{(\mathcal{N})}$ which is usually termed the BBGKY hierarchy[(18)] after the work of Bogoliuboff,[(19)] Born,[(20)] Green,[(23)] Kirkwood,[(21)] and Yvon.[(22)]

It is at this point that perturbation procedures and various *ad hoc* assumptions have been introduced by various investigators[(11, 17, 18)] to derive the Boltzmann equation. Even very rigorous derivations[(11, 17)] still remain valid only in the limit of an infinite system.

B. Derivation of the Boltzmann Equation

A derivation of the Boltzmann equation is now briefly sketched which relies on physical intuition with consequent sacrifice of rigor.

Equation (2.29) for the distribution function is exact. For a dilute gas this equation contains much more detail than is necessary. In this case of a dilute gas one assumes properly that virtually all interactions are two-body collisions. The two-body collisions, in turn, are assumed to occur over an interaction distance r_0 between a colliding pair of molecules. Therefore for distance $r > r_0$ between the pair, there is no interaction. This means that for $r > r_0$, the motion of each member of the collision pair is independent and uncorrelated with the motion of the other member. Owing to this lack of correlation, the pair distribution $f^{(2)}$ is fractorizable, so that before collision, for $r > r_0$:

$$f^{(2)} = f_1^{(1)} f_2^{(1)}. \tag{2.30}$$

After collision, for $r > r_0$, correlations cease and therefore

$$f^{(2)} = f_1^{(1)\prime} f_2^{(1)\prime}. \tag{2.31}$$

During collision, $f^{(2)}$ satisfies the differential equation

$$(\mathbf{v}_2 - \mathbf{v}_1) \cdot \nabla_{\mathbf{r}} f^{(2)} + \frac{1}{m_G} \mathscr{F}_{12} \cdot (\nabla_{\mathbf{v}_1} f^{(2)} - \nabla_{\mathbf{v}_2} f^{(2)}) = 0, \tag{2.32}$$

where

$$\mathbf{r} = \mathbf{x}_2 - \mathbf{x}_1. \tag{2.33}$$

† Here the vector operators are used as $\nabla_{\mathbf{x}_1} f = \partial f/\partial \mathbf{x}_1$, $\nabla_{\mathbf{v}_1} f = \partial f/\partial \mathbf{v}_1$, etc.

Equation (2.32) follows from equation for $f^{(2)}$ in the BBGKY hierarchy where, with the dilute gas assumption, the terms involving $f^{(3)}$ are set equal to zero.

From eqn. (2.32)

$$\frac{1}{m_G}\mathscr{F}_{12}\,.\,\nabla_{\mathbf{v}_1}f^{(2)} = \frac{1}{m_G}\mathscr{F}_{12}\,.\,\nabla_{\mathbf{v}_2}f^{(2)} - (\mathbf{v}_2 - \mathbf{v}_1)\,.\,\nabla_{\mathbf{r}}f^{(2)}. \qquad (2.34)$$

Substituting this in eqn. (2.29),

$$\frac{\partial f^{(1)}}{\partial t} + \mathbf{v}_1\,.\,\nabla_{\mathbf{x}_1}f^{(1)} + \mathscr{E}_1\,.\,\nabla_{\mathbf{v}_1}f^{(1)} = \frac{1}{m_G}\iint\left\{\frac{1}{m_G}\mathscr{F}_{12}\,.\,\nabla_{\mathbf{v}_2}f^{(2)} - (\mathbf{v}_2 - \mathbf{v}_1)\,.\,\nabla_{\mathbf{r}}f^{(2)}\right\}d\mathbf{x}_2\,d\mathbf{v}_2, \qquad (2.35)$$

it is seen that the term

$$\frac{1}{m_G}\mathscr{F}_{12}\,.\,\nabla_{\mathbf{v}_2}f^{(2)} \qquad (2.36)$$

vanishes as a surface integral in velocity space. For the remaining term we convert from spatial volume integration in eqn. (2.35) to a surface integration over a sphere of radius r_0. Let $d\underline{\sigma}$ be the projection of an element of the surface of this sphere, ds, on to a plane normal to $\mathbf{g} \equiv \mathbf{v}_2 - \mathbf{v}_1$. Then by the theorem of Gauss, the right-hand side of eqn. (2.31) becomes

$$\int \{f^{(2)}(\mathbf{ds}\,.\,\mathbf{g} > 0) - f^{(2)}(\mathbf{ds}\,.\,\mathbf{g} < 0)\}\, g\, d\underline{\sigma}\, d\mathbf{v}_2. \qquad (2.37)$$

The Boltzmann equation then has the final form

$$\frac{\partial f}{\partial t} + \mathbf{v}\,.\,\nabla_{\mathbf{x}}f + \mathscr{E}\,.\,\nabla_{\mathbf{v}}f = \int \{f'f'_1 - ff_1\}\, g\, d\underline{\sigma}\, d\mathbf{v}_1, \qquad (2.38)$$

where we have used the factorization property, eqns. (2.30) and (2.31), and have dropped subscripts in the usual manner.[8] The term $d\underline{\sigma}$ is a collision cross-section which will be evaluated next from the mechanics of classical two-body collisions.

C. Mechanics of Two-body Collisions

In the preceding derivation of the Boltzmann equation, the two colliding molecules in this approximation were not considered to be casually connected until the molecules were within each other's sphere of influence where $r < r_0$. Inside this sphere of influence, the two molecules are assumed to interact according to the laws of classical mechanics. Analysis of the two-body collision according to these laws will permit an estimation of the collision cross-section in eqn. (2.38) and of the relation between the primed and unprimed molecular velocities in eqn. (2.38).

Let us consider only elastic collisions between classical non-relativistic particles as point masses which mutually interact by a spherically symmetric potential function. As remarked previously for systems of interest in kinetic theory, the classical rather than the quantum mechanical treatment of molecules is a satisfactory approximation for our purposes.

The motion of the two particles (molecules) 1 and 2 in the absence of a third body is described by Newton's laws:

$$\mathbf{F}_1 = m_1\ddot{\mathbf{r}}_1, \quad \mathbf{F}_2 = m_2\ddot{\mathbf{r}}_2 = -\mathbf{F}_1, \tag{2.39}$$

since the forces $\mathbf{F}_1$, $\mathbf{F}_2$ arise from the mutual interaction of the two particles 1 and 2. In eqn. (2.39), $\mathbf{r}_1$ and $\mathbf{r}_2$ denote the position vectors of molecules 1 and 2.

The equations of motion may be rearranged in the following manner:

$$m_1\mathbf{F}_1 = m_1m_2\ddot{\mathbf{r}}_1, \quad -m_1\mathbf{F}_1 = m_1m_2\ddot{\mathbf{r}}_2. \tag{2.40}$$

Therefore

$$\mathbf{F}_1 = \frac{m_1m_2}{m_1+m_2}\frac{\mathrm{d}^2}{\mathrm{d}t^2}(\mathbf{r}_1 - \mathbf{r}_2) \equiv \underline{\mu}\frac{\mathrm{d}^2}{\mathrm{d}t^2}(\mathbf{r}_1 - \mathbf{r}_2). \tag{2.41}$$

Taking the cross product with the vector $(\mathbf{r}_1 - \mathbf{r}_2)$,

$$(\mathbf{r}_1 - \mathbf{r}_2) \times \mathbf{F}_1 = \underline{\mu}(\mathbf{r}_1 - \mathbf{r}_2) \times \frac{\mathrm{d}^2}{\mathrm{d}t^2}(\mathbf{r}_1 - \mathbf{r}_2) = 0 \tag{2.42}$$

because the forces are central.

By a simple rearrangement,

$$(\mathrm{d}/\mathrm{d}t)\,[(\mathbf{r}_1 - \mathbf{r}_2) \times \mathrm{d}(\mathbf{r}_1 - \mathbf{r}_2)/\mathrm{d}t] - \mathrm{d}(\mathbf{r}_1 - \mathbf{r}_2)/\mathrm{d}t \times \mathrm{d}(\mathbf{r}_1 - \mathbf{r}_2)/\mathrm{d}t = 0. \tag{2.43}$$

From eqn. (2.43) we obtain on integration,

$$(\mathbf{r}_1 - \mathbf{r}_2) \times \mathrm{d}(\mathbf{r}_1 - \mathbf{r}_2)/\mathrm{d}t = \mathbf{K}, \tag{2.44}$$

where $\mathbf{K}$ is a vector of fixed direction and magnitude, independent of time. Therefore, molecules 1 and 2 are at all times moving in a plane perpendicular to $\mathbf{K}$. Furthermore, their center of mass moving with velocity $\mathbf{G}$ is constant owing to conservation of linear momentum

$$(m_1 + m_2)\,\mathbf{G} = m_1\mathbf{v}_1 + m_2\mathbf{v}_2 = m_1\mathbf{v}_1' + m_2\mathbf{v}_2', \tag{2.45}$$

where the primes indicate values of the molecular velocities after collision. Thus the plane to which $\mathbf{K}$ is normal moves with a constant velocity $\mathbf{G}$, as sketched in Fig. 2.3.

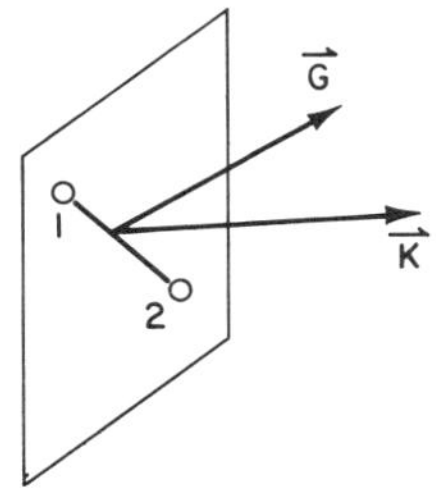

FIG. 2.3. Schematic diagram of the relation between the velocity of the center of mass G and two particles.

Consider the motion of the particles (molecules) as seen by an observer at the center of mass. In the center of mass reference system, the particle velocities are:

$$\mathbf{v}_{1\text{cm}} = \mathbf{v}_1 - \mathbf{G} = \frac{m_2}{m_1 + m_2}(\mathbf{v}_1 - \mathbf{v}_2) \equiv \frac{m_2}{m_1 + m_2}\mathbf{g}_{12}, \tag{2.46}$$

$$\mathbf{v}_{2\text{cm}} = \mathbf{v}_2 - \mathbf{G} = \frac{m_1}{m_1 + m_2}(\mathbf{v}_2 - \mathbf{v}_1) \equiv \frac{m_1}{m_1 + m_2}\mathbf{g}_{21}. \tag{2.47}$$

Therefore, in the center of mass coordinate system the molecular velocities are parallel or collinear and oppositely directed. At distances $\mathbf{r} \equiv |\mathbf{r}_1 - \mathbf{r}_2|$ much greater than the range of the interaction potential, the vectors $\mathbf{g}_{12}$ and $\mathbf{g}_{21}$ are in the direction of the trajectory of each respective molecule (molecules move in straight lines in the absence of an external force). Hence the molecular trajectories before and after interaction are parallel and separated by a distance b, the impact parameter, as drawn in Fig. 2.4.

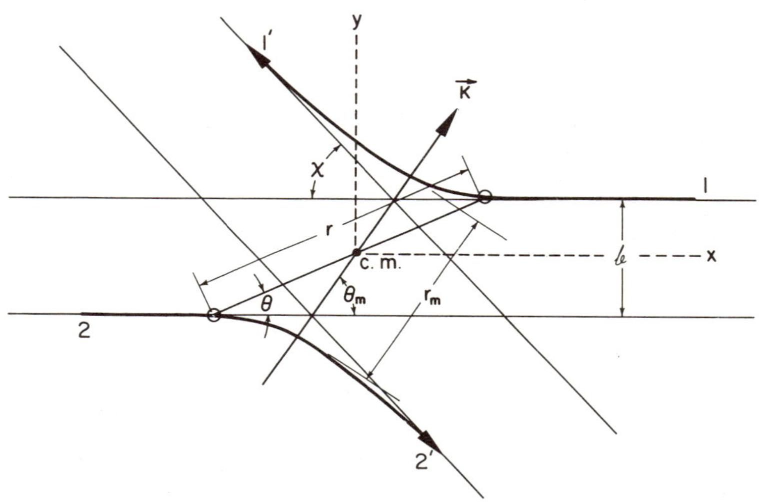

FIG. 2.4. Geometry of collision between two identical hard sphere molecules.

In the center of mass reference system the coordinates of the molecules for any separation distance r are:

$$(1) \qquad \begin{cases} x_1 = [m_2/(m_1 + m_2)]\, r\cos\theta, \\ y_1 = [m_2/(m_1 + m_2)]\, r\sin\theta, \end{cases} \tag{2.48}$$

$$(2) \qquad \begin{cases} x_2 = -[m_1/(m_1 + m_2)]\, r\cos\theta, \\ y_2 = -[m_1/(m_1 + m_2)]\, r\sin\theta. \end{cases} \tag{2.49}$$

The total kinetic energy W of the two molecules in the center of mass system then becomes

$$W = \tfrac{1}{2}m_1(\dot{x}_1^2 + \dot{y}_1^2) + \tfrac{1}{2}m_2(\dot{x}_2^2 + \dot{y}_2^2) = \tfrac{1}{2}\mu(\dot{r}^2 + r^2\dot{\theta}^2). \tag{2.50}$$

Similarly, the total angular momentum Γ is

$$\Gamma = m_1(x_1\dot{y}_1 - y_1\dot{x}_1) + m_2(x_2\dot{y}_2 - y_2\dot{x}_2) = \underline{\mu} r^2\dot{\theta}. \tag{2.51}$$

For $r \gg r_0$, the distance of interaction,

$$W = \tfrac{1}{2}\underline{\mu} g^2 \tag{2.52}$$

and

$$\Gamma = \underline{\mu} b g. \tag{2.53}$$

The equations for W and Γ are identical to those obtained by studying the motion of a fictitious mass $\underline{\mu}$ in the presence of a center of force at the center of mass.

The trajectories of this fictitious molecule are specified completely by the requirements of conservation of energy and angular momentum:

$$\underline{\mu} b g = \underline{\mu} r^2\dot{\theta}, \tag{2.54}$$

$$\tfrac{1}{2}\underline{\mu} g^2 = \tfrac{1}{2}\underline{\mu}(\dot{r}^2 + r^2\dot{\theta}^2) + \Phi(r), \tag{2.55}$$

where $\Phi(r)$ is the potential energy of interaction of the pair of molecules.

The deflection angle χ is of considerable importance, and can be calculated in the following way. We have from eqns. (2.54) and (2.55)

$$\dot{\theta} = bg/r^2, \tag{2.56}$$

$$\dot{r} = -g\left[1 - \left(\frac{b}{r}\right)^2 - \frac{\Phi(r)}{\frac{1}{2}\underline{\mu} g^2}\right]^{\frac{1}{2}}. \tag{2.57}$$

Eliminating time from eqns. (2.65) and (2.66),

$$\frac{\mathrm{d}r}{\mathrm{d}\theta} = -\frac{r^2}{b}\left[1 - \left(\frac{b}{r}\right)^2 - \frac{\Phi(r)}{\frac{1}{2}\underline{\mu} g^2}\right]^{\frac{1}{2}}, \tag{2.58}$$

which is the differential equation of the trajectory of the fictitious particle of mass $\underline{\mu}$ about the center of force. Equation (2.58) shows that the trajectory of each molecule is symmetrical about the line of closest approach r_m between the two molecules.

The distance of closest approach r_m is given by the largest solution of

$$1 - \left(\frac{b}{r}\right)^2 - \frac{\Phi(r)}{\frac{1}{2}\underline{\mu} g^2} = 0. \tag{2.59}$$

If the solution exists,

$$\theta_m = \int_{r_m}^{\infty} \frac{b}{r^2}\left[1 - \left(\frac{b}{r}\right)^2 - \frac{\Phi(r)}{\frac{1}{2}\underline{\mu} g^2}\right]^{\frac{1}{2}} \mathrm{d}r. \tag{2.60}$$

Since $\chi = \pi - 2\theta_m$,

$$\chi(b, g) = \pi - 2b\int_{r_m}^{\infty} r^{-2}\left[1 - \left(\frac{b}{r}\right)^2 - \frac{\Phi(r)}{\frac{1}{2}\underline{\mu} g^2}\right]^{\frac{1}{2}} \mathrm{d}r. \tag{2.61}$$

The molecular velocities before and after encounter are specified by the following relations:

$$\mathbf{v}_1' - \mathbf{v}_1 = 2[m_2/(m_1 + m_2)]\,(\mathbf{g}_{21}\,.\,\mathbf{k})\,\mathbf{k}, \tag{2.62}$$

$$\mathbf{v}_2' - \mathbf{v}_2 = 2[m_1/(m_1 + m_2)]\,(\mathbf{g}_{21}\,.\,\mathbf{k})\,\mathbf{k}, \tag{2.63}$$

where **k** is the unit vector in the direction $2 \rightarrow 1$ along the line r_m (see Fig. 2.4). From the geometry of the collision,

$$\mathbf{k}\,.\,\mathbf{g}_{21} = |\mathbf{g}_{21}|\cos\left[\tfrac{1}{2}(\pi - \chi)\right]. \tag{2.64}$$

The velocities after collision $\mathbf{v}_1'$ and $\mathbf{v}_2'$ are then determined completely by the velocities before collision $\mathbf{v}_1$ and $\mathbf{v}_2$ and the unit vector **k**.

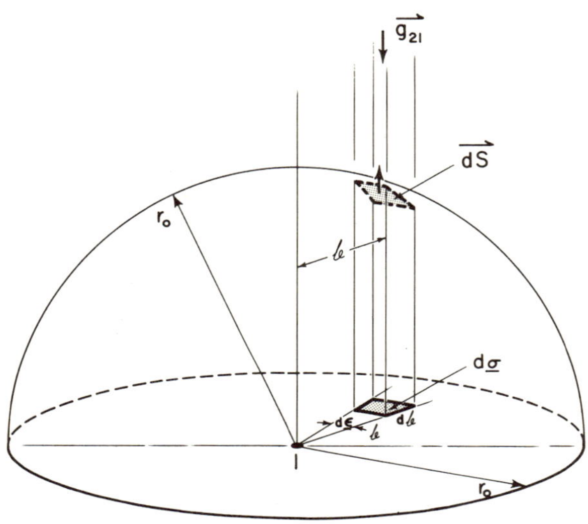

FIG. 2.5. Schematic diagram of the relations between collisional interaction parameters.

Figure 2.5 indicates schematically the relation between the collision cross-section $d\underline{\sigma}$, the impact parameter b, and the polar angle $\underline{\epsilon}$. From the geometry sketched in Figure 2.5,

$$d\underline{\sigma} = b\,db\,d\underline{\epsilon}. \tag{2.65}$$

The impact parameter b, usually will range from zero to some cut-off parameter corresponding to $b = r_0$, which in many cases can be infinity. The polar angle ϵ ranges from 0 to 2π.

One ordinarily defines a differential collision cross-section $I(g, \chi)$ for collision in terms of a solid angle $d\Omega = \sin\chi\,d\chi\,d\underline{\epsilon}$

$$d\underline{\sigma} = b\,db\,d\underline{\epsilon} = I(g, \chi)\,d\Omega. \tag{2.66}$$

The expression for χ in terms of the potential energy of interaction between a pair of molecules Φ is given by eqn. (2.61). The particular form of Φ affects only the relation between χ and the variables g and b, as indicated in eqn. (2.61).

D. Conservation Laws

For the collision of idealized spherical molecules, it is now possible, in principle, with the results of Section 2.2, A, B, and C, to solve the Boltzmann equation for f. It is more convenient to work with a renormalized distribution, $f(\mathbf{x}, \mathbf{v}, t)$ such that the zeroth velocity moment of the distribution is

$$\int f \, d\mathbf{v} = n_G(\mathbf{x}, t) \tag{2.67}$$

which gives the number density $n_G(\mathbf{x}, t)$ of molecules at position $\mathbf{x}$ at time t. The integral symbol in eqn. (2.67) signifies integration over all velocity space.

Other macroscopic observables follow as higher velocity moments of $f(\mathbf{x}, \mathbf{v}, t)$.

The mean gas velocity

$$\mathbf{q}(\mathbf{x}, t) = \frac{1}{n_G(\mathbf{x}, t)} \int f\mathbf{v} \, d\mathbf{v}. \tag{2.68}$$

The mean energy density

$$E(\mathbf{x}, t) = \tfrac{3}{2} n_G(\mathbf{x}, t) \, kT(\mathbf{x}, t) = \int f \tfrac{1}{2} m_G (\mathbf{v} - \mathbf{q})^2 \, d\mathbf{v} \tag{2.69}$$

for molecules with only translational communicable energy.

The observables n_G, $\mathbf{q}$, T specify the state of a gas at a particular point. Higher moments will enter into equations for the transfer of momentum and energy in a pure gas.

If we take directly the moments m, $m\mathbf{v}$, $\frac{1}{2}mv^2$ of the Boltzmann equation, eqn. (2.47), with the renormalized distribution f:

$$\frac{\partial f}{\partial t} + \mathbf{v} \, . \, \nabla_{\mathbf{x}} f + \mathscr{E} \, . \, \nabla_{\mathbf{v}} f = \iint \{f' f'_1 - f f_1\} \, g b \, db \, d\underline{\epsilon} \, d\mathbf{v}_1 \tag{2.70}$$

we obtain the equations of conservation of mass, momentum, and energy respectively. However, the various flux expressions which arise must either be related to the state variable of the gas n_G, $\mathbf{q}$, T or determined directly from knowledge of the distribution function f.

We now shall simply indicate the procedures for obtaining the various conservation equations.

When eqn. (2.70) is operated on, say by integration and multiplication by the operator, $\int d\mathbf{v} f \psi$, where ψ is some function of the molecular velocity one gets:

$$\frac{\partial}{\partial t} \int d\mathbf{v} \psi f + \int d\mathbf{v} \psi \mathbf{v} \, . \, \nabla_{\mathbf{x}} f + \int d\mathbf{v} \psi \mathscr{E} \, . \, \nabla_{\mathbf{v}} f = \iiint d\mathbf{v} \psi \, \{f' f'_1 - f f_1\} \, g b \, db \, d\underline{\epsilon} \, d\mathbf{v}_1. \tag{2.71}$$

It can be shown[8] that the right-hand side of eqn. (2.71) can be rearranged such that

$$\iiint d\mathbf{v} \psi \, \{f' f'_1 - f f_1\} \, g b \, db \, d\underline{\epsilon} \, d\mathbf{v}_1 = \tfrac{1}{4} \iiint d\mathbf{v} (\psi + \psi_1 - \psi' - \psi'_1)(f' f'_1 - f f_1) \, g b \, db \, d\underline{\epsilon} \, d\mathbf{v}_1. \tag{2.72}$$

Inasmuch as mass, momentum, and energy are conserved on collision,

$$\left.\begin{aligned} m + m_1 &= m' + m'_1, \\ m\mathbf{v} + m_1 \mathbf{v}_1 &= m'\mathbf{v}' + m'_1 \mathbf{v}'_1, \\ \tfrac{1}{2} m\mathbf{v}^2 + \tfrac{1}{2} m_1 \mathbf{v}_1^2 &= \tfrac{1}{2} m' \mathbf{v}'^2 + \tfrac{1}{2} m'_1 \mathbf{v}_1'^2. \end{aligned}\right\} \tag{2.73}$$

Then if ψ is any one of m, $m\mathbf{v}$, $(1/2)\, m\mathbf{v}^2$, the right-hand side of eqn. (2.71) is zero.

From the definitions of n_G, $\mathbf{q}$, T above, eqn. (2.71) has the following form for $\psi = m$, $= m\mathbf{v}$, $= \frac{1}{2}m\mathbf{v}^2$:

(a) $\psi = m$:

$$\frac{\partial \rho_G}{\partial t} + \nabla_{\mathbf{x}} \cdot \rho_G \mathbf{q} = 0, \tag{2.74}$$

where $\rho_G = m_G n_G$; m is set equal to molecular mass m_G introduced previously.

(b) $\psi = m\mathbf{v}$:

$$\rho_G \frac{D\mathbf{q}}{Dt} = -\nabla_{\mathbf{x}} . \bar{\bar{\Pi}} - \rho_G \mathscr{E}, \tag{2.75}$$

where

$$\bar{\bar{\Pi}} = m_G \int \mathrm{d}\mathbf{v} f(\mathbf{v} - \mathbf{q})(\mathbf{v} - \mathbf{q})$$

is the second order tensor termed the *pressure tensor*. Also the *substantial derivative* is defined as†

$$\frac{D}{Dt} \equiv \frac{\partial}{\partial t} + \mathbf{q} . \nabla_{\mathbf{x}} = \frac{\partial}{\partial t} + \mathbf{q} . \nabla.$$

(c) $\psi = \frac{1}{2}m(\mathbf{v} - \mathbf{q})^2$:

$$\frac{DT}{Dt} = -\frac{2}{3n_G k}\{\bar{\bar{\Pi}} . \nabla\mathbf{q} + \nabla . \mathbf{h}\}. \tag{2.77}$$

For molecules with only translatory communicable energy, $\mathbf{h}$ is the heat flux vector:

$$\mathbf{h} = \int f \tfrac{1}{2}m(\mathbf{v} - \mathbf{q})^2 (\mathbf{v} - \mathbf{q})\, \mathrm{d}\mathbf{v} \tag{2.78}$$

relative to the mean velocity.

We note again that these macroscopic conservation equations are essentially useless for practical calculations until the higher moments $\bar{\bar{\Pi}}$, and $\mathbf{h}$ are specified either in terms of n_G, $\mathbf{q}$, T, or calculated directly from f.

E. The Approach to Equilibrium

We now turn to the problem of finding solutions of the Boltzmann equation which will permit either use of the conservation equations by relating $\bar{\bar{\Pi}}$ and $\mathbf{h}$ to n_G, $\mathbf{q}$, T, or by direct determination of the distribution function f.

As a first step, let us consider the question of the form of f when the gas is at equilibrium. That is, f in this case is independent of time, and varies spatially, if at all, only according to some externally imposed force. The form of the distribution function f for equilibrium conditions was determined many years ago by Boltzmann by making use of the well known H theorem.[8] The derivation of Boltzmann's famous H theorem is summarized as follows:

Consider the Boltzmann equation in the form:

$$\frac{\partial f}{\partial t} = -\mathbf{v} . \nabla f - \mathscr{E} . \nabla_{\mathbf{v}} f + \iint (f'f'_1 - ff_1)\, g\, \mathrm{d}\ell\, \mathrm{d}\underline{\epsilon}\, \mathrm{d}\mathbf{v}_1.$$

Let

$$H(t) = \iint f \ln f\, \mathrm{d}\mathbf{x}\, \mathrm{d}\mathbf{v}. \tag{2.79}$$

† Here, and in subsequent expressions, $\nabla_{\mathbf{x}}$ is abbreviated to ∇.

Then

$$\frac{\mathrm{d}H}{\mathrm{d}t} = \iint \frac{\partial f}{\partial t} \ln f \,\mathrm{d}\mathbf{x}\,\mathrm{d}\mathbf{v} = \left(\frac{\mathrm{d}H}{\partial t}\right)_{\text{streaming}} + \left(\frac{\mathrm{d}H}{\mathrm{d}t}\right)_{\text{collisions}} \tag{2.80}$$

Applying Liouville's theorem for the case of an ideal gas,

$$\left(\frac{\mathrm{d}H}{\mathrm{d}t}\right)_{\text{streaming}} = 0. \tag{2.80}$$

Under these circumstances

$$\left(\frac{\mathrm{d}H}{\mathrm{d}t}\right)_{\text{collisions}} = \iiiint (1 + \ln f)(f'f'_1 - ff_1)\, g\ell \,\mathrm{d}\ell \,\mathrm{d}\underline{\epsilon}\,\mathrm{d}\mathbf{v}_1\,\mathrm{d}\mathbf{v}\,\mathrm{d}\mathbf{x}. \tag{2.82}$$

From eqn. (2.72) we have the rearrangement

$$\left(\frac{\mathrm{d}H}{\mathrm{d}t}\right)_{\text{collisons}} = \frac{1}{4}\iiiint \ln \frac{ff_1}{f'f'_1}(f'f'_1 - ff_1)\, g\ell \,\mathrm{d}\ell \,\mathrm{d}\underline{\epsilon}\,\mathrm{d}\mathbf{v}_1\,\mathrm{d}\mathbf{v}\,\mathrm{d}\mathbf{x}. \tag{2.83}$$

and this is zero *only* if

$$ff_1 = f'f'_1, \quad \text{or} \quad \ln f' + \ln f'_1 = \ln f + \ln f_1 \tag{2.84}$$

This condition is satisfied by the following form for f inasmuch as $\ln f$ from (2.84) is a summational invariant for encounters:

$$f = n_G \left(\frac{m_G}{2\pi kT}\right)^{\frac{3}{2}} \exp\left[-\frac{m_G}{2kT}(\mathbf{v} - \mathbf{q})^2\right], \tag{2.85}$$

and n, $\mathbf{q}$, T may be functions of $\mathbf{x}$ and t.

Equation (2.85) satisfies the equation

$$\frac{\partial f}{\partial t} + \mathbf{v} \,.\, \nabla f + \mathscr{E} \,.\, \nabla_{\mathbf{v}} f = 0 \tag{2.86}$$

only when $\mathbf{q} = 0$, T is constant, and

$$n_G = n_{G0} \exp\left[-\frac{m_G}{kT}\Phi(\mathbf{x})\right], \tag{2.87}$$

where $\Phi(\mathbf{x})$ is the potential function of the external force $\mathscr{E}$.

For special forms of $\mathscr{E}$ or Φ, eqn. (2.85) is not correct. Special solutions for these forms have been studied in some detail.[24]

We see then that for these forms eqn. (2.85), with eqn. (2.87) for an external field, is the form of the distribution function which gives the number of molecules in an equilibrium gas at time t with positions in the range $\mathrm{d}\mathbf{x}$ and velocities in the range $\mathrm{d}\mathbf{v}$. Inasmuch as eqn. (2.85) shows that f for an equilibrium gas is an even function of the molecular velocity, it is clear that transport processes, such as heat and momentum transfer, which are odd functions of the molecular velocity, do not exist in this situation. We must, therefore, look to solutions of the Boltzmann equation out of equilibrium in order to obtain results for such processes.

F. Some Solutions of the Boltzmann Equation for Extremes of Knudsen Number

No general solution of the complete, non-linear Boltzmann equation, (2.70), has been obtained as yet. For specifying at least approximate solutions, one usually turns either to perturbation methods suggested by parameters of smallness for certain physical approximations, or to approximations of the collision term. Several forms of solutions which have been attempted are outlined below. In later chapters, details also will be given for certain special problems relating to aerosol physics.

Suppose we rewrite the Boltzmann equation in dimensionless form by choosing suitable scaling factors:

$$\frac{\partial f^{\oplus}}{\partial t^{\oplus}} + \mathbf{v}^{\oplus} \,.\, \nabla^{\oplus} f^{\oplus} + \mathscr{E}^{\oplus} \,.\, \nabla_{\mathbf{v}^{\oplus}} f^{\oplus} = \frac{0{\cdot}707}{\pi} \frac{R_v}{\lambda_G} J^{\oplus}(f^{\oplus} f_1^{\oplus}). \tag{2.88}$$

The symbols appearing here are defined as follows:

$$\mathbf{v}^{\oplus} = (m_G/2kT)^{\frac{1}{2}}\mathbf{v}, \quad f^{\oplus} = f/n_G(m_G/2\pi kT)^{\frac{3}{2}}, \quad \mathbf{x}^{\oplus} = \mathbf{x}/R_v,$$
$$t^{\oplus} = t/(R/\bar{v}_G), \quad \mathscr{E}^{\oplus} = (R_v/\bar{v}_G^2)\,\mathscr{E}, \quad \text{and} \quad J^{\oplus}(f^{\oplus} f_1^{\oplus}) = \iint (f^{\oplus\prime} f_1^{\oplus\prime} - f^{\oplus} f_1^{\oplus})\, g^{\oplus}\, \mathrm{d}\underline{\sigma}^{\oplus}\, \mathrm{d}\mathbf{v}^{\oplus}{}_1, \tag{2.89}$$

where

$$\mathrm{d}\underline{\sigma}^{\oplus} = \frac{1}{d_0^2}\mathrm{d}\underline{\sigma}.$$

The term d_0 is the effective collision diameter of the gas molecules while $\bar{v}_G$ is the mean molecular speed. λ_G is taken as the mean free path for molecules modeled as rigid spheres:

$$\lambda_G = 0{\cdot}707/\pi n_G\, d_0^2, \tag{2.90}$$

with some number density n_G.

The scale factor R_v could be, for example, the radius of an aerosol particle. Again, it could be the radius of a tube through which gas is flowing. In any event, R_v/λ_G is recognizable as an inverse Knudsen number, Kn^{-1}. Therefore, for an aerosol particle in an effectively infinite gas, the qualitative discussion in Section 2.1 applies.

At this point it is worthwhile reconsidering our previous remarks about the significance of the Knudsen number, especially in the light of the solutions to the Boltzmann equation.

Solutions for the Free Molecule Regime

Where the mean free path λ_G is much larger than the scale factor for the system R_v, then

$$Kn^{-1} = R_v/\lambda_G \to 0 \tag{2.91}$$

and Eqn. (2.88) becomes

$$\frac{\partial f^{\oplus}}{\partial t^{\oplus}} + \mathbf{v}^{\oplus} \,.\, \nabla^{\oplus} f^{\oplus} + \mathscr{E}^{\oplus} \,.\, \nabla_{\mathbf{v}^{\oplus}} f^{\oplus} = 0. \tag{2.92}$$

That is, there are no collisions of gas molecules in this limit of eqn. (2.91), and the gas molecules follow trajectories as prescribed by eqn. (2.92).

In the instance of a single aerosol particle in an effectively infinite un-ionized gas, and in the absence of external fields, the surrounding gas molecules travel in straight line paths according to

$$\frac{\partial f^{\oplus}}{\partial t^{\oplus}} + \mathbf{v}^{\oplus} . \nabla^{\oplus} f^{\oplus} = 0. \tag{2.93}$$

See Fig. 2.6 for example. In the sketch, gas molecules miss each other and only strike the particle surface. Under this extreme condition, the particle does not alter the distribution function for the incoming gas molecules.

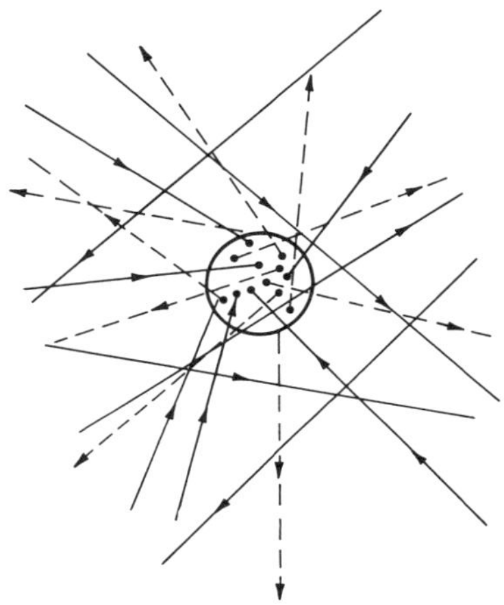

FIG. 2.6. Molecular trajectories before and after collision with a sphere in the free molecule regime.

The application of the Boltzmann equation and its solution for the transport of mass, momentum, and heat to a single particle in the free molecule regime is discussed more fully in Chapter 5.

Equation (2.92) is a simple first order, linear partial differential equation. Such an equation presents no fundamental difficulty to solution except for the introduction of complexities in geometrical boundary conditions.

Solutions for the Continuum Regime

In the other extreme, where the inverse Knudsen number

$$Kn^{-1} \to \infty, \tag{2.94}$$

the geometrical scale far exceeds the mean free path of the gas. In this limit, the collision integral

$$J^{\oplus}(ff_1) = 0, \tag{2.95}$$

and the gas is said to be in *isentropic flow*. When this condition exists, there is momentum flux, no heat flux, and eqns. (2.75) and (2.77) reduce to

$$\rho_G \frac{\mathrm{D}\mathbf{q}}{\mathrm{D}t} = -\rho_G \mathscr{E} \tag{2.96}$$

and

$$\frac{\mathrm{D}}{\mathrm{D}t}(\rho_G T^{-\frac{3}{2}}) = 0 \tag{2.97}$$

Equation (2.96) is the well known *Euler equation* for inviscid flow of fluids

The solution to the Boltzmann equation for the limiting case in which $J^{\oplus} = 0$ was found within the solution of Chapman and Enskog.[8] This is the solution of the Boltzmann equation for which the distribution function effectively is expanded in λ_G/R_v:

$$f = f^{(0)} + (\lambda_G/R_v)f^{(1)} + \dots \tag{2.98}$$

However, the Chapman–Enskog method of solution is more general than the perturbation solution expansion indicated by eqn. (2.98) because the Chapman–Enskog solution applies to an infinite gas without a physical boundary scale factor (i.e. $R_i \to \infty$).

In Enskog's solution,[8] f essentially is expanded in terms of a parameter θ:

$$f = \frac{1}{\theta}f^{(0)} + f^{(1)} + \theta f^{(2)} + \theta^2(f^{(3)}) + \dots. \tag{2.99}$$

On equating powers of θ, one obtains the following perturbation equations:

$$J(f^{(0)}f_1^{(0)}) = 0, \tag{2.100}$$

$$\frac{\partial f^{(0)}}{\partial t} + \mathbf{v}\,.\,\nabla f^{(0)} + \mathscr{E}\,.\,\nabla_{\mathbf{v}} f^{(0)} = J(f^0 f_1^{(1)}) + J(f^{(1)}f_1^{(0)}), \tag{2.101}$$

and so forth. Equation (2.100) is equivalent to $J^{\oplus} = 0$, from which the Euler relations stem. The Euler equations derive from the approximation that the solution to Boltzmann's equation only comes from $f^{(0)}$:

$$f^{(0)} = n(\mathbf{x})\left(\frac{m}{2\pi kT(\mathbf{x})}\right)^{\frac{3}{2}} \exp\left\{-\frac{m}{2kT(\mathbf{x})}(\mathbf{v} - \mathbf{q}(\mathbf{x}))^2\right\}.$$

It is found that the Enskog expansion is an expansion in the number density where

$$n_G = 1/\theta. \tag{2.102}$$

In the case of an infinite gas then the scale factor can be said to be the characteristic distance L_G over which n_G, $\mathbf{q}$, and T change appreciably; in other words, there exists usually, but not always, some distance much greater than the mean free path of the gas.

It is clear that the Chapman–Enskog method of solution without physical boundaries will lead to the same perturbation method with physical boundaries. The effect of physical boundaries must appear as surface or boundary conditions on the solution. However, in this approximation these boundary conditions cannot determine the form of the distribution function, and the boundary conditions cannot depend on the details of the gas molecule–boundary surface interaction. As we shall see, the Chapman–Enskog method of solution leads to just these results.

To proceed in finding the Chapman–Enskog solution either expansion, eqns. (2.98) or (2.99), may be employed. For generality and consistency with Chapman and Cowling[8], we shall use eqn. (2.99) in the subsequent discussion of this section. Following Chapman and Enskog, we write

$$f^{(1)} = f^{(0)}\Theta \tag{2.103}$$

so that eqn. (2.101) becomes

$$\frac{\partial f^{(0)}}{\partial t} + \mathbf{v} . \nabla f^{(0)} + \mathscr{E} . \nabla_{\mathbf{v}} f^{(0)} = f^{(0)} \int \mathrm{d}\mathbf{v}_1 f_1^{(0)} . \int \mathrm{d}\ell \, \mathrm{d}\underline{\epsilon} g \ell [\Theta' + \Theta_1' - \Theta - \Theta_1]. \quad (2.104)$$

The homogeneous equation

$$f^{(0)} \int \mathrm{d}\mathbf{v}_1 f_1^{(0)} \int \mathrm{d}\ell \, \mathrm{d}\underline{\epsilon} g \ell [\Theta' + \Theta_1' - \Theta - \Theta_1] = 0 \quad (2.105)$$

has the five solutions m_G, $m_G\mathbf{v}$, $\frac{1}{2}m_G\mathbf{v}^2$, the summational invariants. Equation (2.104) has a solution only if the left-hand side of the equation is orthogonal to the summational invariants ψ_i [cf. eqn. (2.73)]. From the Euler equations, including (2.74), (2.96), and (2.97), the macroscopic time derivatives can be expressed in terms of the macroscopic spatial derivatives, so that eqn. (2.101) becomes

$$f^{(0)}\{(\xi^2 - \tfrac{5}{2})(\mathbf{v} - \mathbf{q}) . \nabla \ln T + 2[\boldsymbol{\xi}\boldsymbol{\xi} - \tfrac{1}{3}\bar{\bar{I}}\xi^2] : \nabla \mathbf{q}\}$$
$$= f^{(0)} \int \mathrm{d}\mathbf{v}_1 f_1^{(0)} \int \mathrm{d}\underline{\sigma} g [\Theta' + \Theta_1' - \Theta - \Theta_1]. \quad (2.106)$$

Here $\bar{\bar{I}}$ is the unit tensor and

$$\boldsymbol{\xi} = (m_G/2kT)^{\frac{1}{2}} (\mathbf{v} - \mathbf{q}). \quad (2.107)$$

With the condition

$$\int f^{(0)} \Theta \psi_i \, \mathrm{d}\mathbf{v} = 0, \quad (2.108)$$

eqn. (2.106) has a unique solution.

It follows that this particular solution of eqn. (2.106) has the form

$$\Theta = -\frac{1}{n_G}\left(\frac{2kT}{m_G}\right)^{\frac{1}{2}} \mathbf{A} . \nabla \ln T - \frac{1}{n_G} \bar{\bar{B}} : \nabla \mathbf{q}, \quad (2.109)$$

where $\mathbf{A}$ and $\bar{\bar{B}}$ are functions only of the molecular velocity.

Substituting eqn. (2.109) into eqn. (2.106), $\mathbf{A}$ and $\bar{\bar{B}}$ are seen to be special solutions of the equations

$$f^{(0)}(\xi^2 - \tfrac{5}{2})\, \boldsymbol{\xi} = \frac{1}{n_G} \iint f^{(0)} f_1^{(0)} [\mathbf{A}_1 + \mathbf{A} - \mathbf{A}_1' - \mathbf{A}'] \, g \, \mathrm{d}\underline{\sigma} \, \mathrm{d}\mathbf{v}_1, \quad (2.110)$$

and

$$2f^{(0)}[\boldsymbol{\xi}\boldsymbol{\xi} - \tfrac{1}{3}\xi^2 \bar{\bar{I}}] = \frac{1}{n_G} \iint f^{(0)} f_1^{(0)} [\bar{\bar{B}}_1 + \bar{\bar{B}} - \bar{\bar{B}}_1' - \bar{\bar{B}}'] \, g \, \mathrm{d}\underline{\sigma} \, \mathrm{d}\mathbf{v}_1. \quad (2.111)$$

In order to determine $\mathbf{A}$ and $\bar{\bar{B}}$ an expansion procedure is used. We will illustrate the method by calculating the vector $\mathbf{A}$.

The conditions on $\mathbf{A}$ are given from eqns. (2.108) and (2.110). We assume that $\mathbf{A}$ can be expanded in a series of Sonine polynomials:

$$\mathbf{A} = \sum_{r=0}^{\infty} a_r \, S_{\frac{3}{2}}^{(r)}(\xi^2)\, \boldsymbol{\xi}. \quad (2.112)$$

With this expansion, eqn. (2.108) leads to the result

$$\int \mathrm{e}^{-\xi^2} \boldsymbol{\xi} . \mathbf{A} \, \mathrm{d}\xi = 2\pi a_0 \Gamma(\tfrac{5}{2}) = 0 \quad (2.113)$$

which obviously is satisfied by $a_0 = 0$. Here Γ is the gamma function. Substitution of the remaining terms of the expansion (2.112) into eqn. (2.110) leads to

$$\frac{1}{n_G}\int S_{\frac{3}{2}}^{(r)}(\xi^2)\,\boldsymbol{\xi}\,.\iint f^{(0)}f_1^{(0)}[\mathbf{A}_1 + \mathbf{A} - \mathbf{A}_1' - \mathbf{A}']\,\mathrm{d}\underline{\sigma}\,\mathrm{d}\mathbf{v}_1\,\mathrm{d}\mathbf{v} = \int f^{(0)}(\xi^2 - \tfrac{5}{2})\,\boldsymbol{\xi}\,.\,S_{\frac{3}{2}}^{(r)}(\xi^2)\,\boldsymbol{\xi}\mathrm{d}\mathbf{v}. \tag{2.114}$$

Referring to eqn. (2.112) one can now determine the set of coefficients a_r from the set of relations given by eqn. (2.114).

For the special case of Maxwellian molecules, which means that the molecules interact according to a force law of $F = \bar{K}r^{-5}$, the Chapman–Enskog solution is especially simple. For example, the coefficient of $\nabla \ln T$ in eqn. (2.109) is an eigenfunction of the operator J. In this case, $a_r = 0$ for $r > 1$, and

$$\pi a_1 \left(\frac{2\bar{K}}{m_G}\right)^{\frac{1}{2}} 0{\cdot}436 = 1. \tag{2.115}$$

In the Chapman–Enskog solution the heat flux in the second approximation, using eqn. (2.86), becomes

$$\mathbf{h}^{(1)} = -\frac{m_G}{2n_G}\left(\frac{2kT}{m_G}\right)^{\frac{1}{2}}\int f^{(0)}(\mathbf{v}-\mathbf{q})(\mathbf{v}-\mathbf{q})^2\,\mathbf{A}\,.\,\nabla \ln T\,\mathrm{d}\mathbf{v} \tag{2.116}$$

and the pressure tensor to this same approximation is

$$\Pi^{(1)} = -\frac{m_G}{n_G}\int f^{(0)}\,\bar{\bar{B}} : \nabla\mathbf{q}(\mathbf{v}-\mathbf{q})(\mathbf{v}-\mathbf{q})\,\mathrm{d}\mathbf{v}. \tag{2.117}$$

One, therefore, sees that the heat flux and pressure tensor to the second approximation are linear in the gradients $\nabla \ln T$ and $\nabla\mathbf{q}$ respectively. Hence the coefficients of $\nabla \ln T$ and $\nabla\mathbf{q}$ are identified with the macroscopic thermal conductivity k_G and the viscosity coefficient μ_G.

Substituting the expressions for $\mathbf{h}^{(1)}$ and $\bar{\bar{\Pi}}^{(1)}$ into the general conservation eqns. (2.75) and (2.85), we obtain the macroscopic equations

(Mass)

$$\frac{\partial \rho_G}{\partial t} + \nabla\,.\,\rho_G\mathbf{q} = 0. \tag{2.118}$$

(Momentum)

$$\rho_G\frac{\mathrm{D}\mathbf{q}}{\mathrm{D}t} = -\nabla\,.\,[p\bar{\bar{I}} - 2\mu_G\bar{\bar{e}}] + \rho_G\mathscr{E}, \tag{2.119}$$

where elements of the tensor of $\bar{\bar{e}}$ in cartesian coordinates are

$$e_{xx} = \tfrac{1}{3}\left(2\frac{\partial q_x}{\partial x} - \frac{\partial q_y}{\partial y} - \frac{\partial q_z}{\partial z}\right), \tag{2.120}$$

$$e_{xy} = \tfrac{1}{2}\left(\frac{\partial q_y}{\partial x} + \frac{\partial q_x}{\partial y}\right), \tag{2.121}$$

etc.

(Thermal energy)

$$\frac{\mathrm{D}T}{\mathrm{D}t} = -\frac{2}{3n_Gk}\{[p\bar{\bar{I}} - 2\mu_G\bar{\bar{e}}] : \nabla\mathbf{q} + \nabla\,.\,k_G\nabla T\}. \tag{2.122}$$

Equation (2.119) is the well known *Navier–Stokes equation*. Like the Euler terminology for eqns. (2.74), (2.75) and (2.77), the name is often linked to eqn. (2.122) too. Equations (2.118), (2.119), and (2.122), of course, are the usual continuum equations of aerodynamics and hydrodynamics.

The feature of the Chapman–Enskog procedure is that the transport coefficients k_G, μ_G, and $D_{\alpha\beta}$ for a gas mixture (see Chapter 4), are calculated from assumed molecular force laws. The special nature of this solution is seen from eqn. (2.106), for example, which ensures to this approximation, that the distribution function can depend only on n, $\mathbf{q}$, and T, and can depend on time and spatial variables only implicitly through n, $\mathbf{q}$, and T. Of course, this is exactly the implication of the intuitive or experimental hydrodynamic procedures using a phenomenological approach to fluid dynamics.

Obviously the expansion procedure indicated by Chapman and Enskog can be carried out to higher orders as has been done by several investigators.(8, 25) However, the applicability of these higher approximations to fluid dynamic problems remains an open question. In order to proceed into the slip flow regime, for example, it seems necessary to follow quite a different procedure.(12)

G. Solutions which Apply to the Knudsen Transition Regime

In describing the evolution of the fluid state in the free molecule and continuum regimes the great detail on the evolution of the state of the gas furnished by the Boltzmann equation simply is discarded.

However, the description of systems where the Knudsen number is finite represents a much more subtle and non-trivial use of the Boltzmann equation. In the range where the Knudsen number is finite, there is a transition from purely free-molecular or collisionless flow to aerodynamic flow. The theoretical analysis of this transition regime is a difficult one.

Unfortunately, the complex nature of the non-linear Boltzmann integro-differential equation has prevented up until now, even numerically, a general solution of the equation in complete form. It has been necessary instead to resort to a variety of approximation schemes in an effort to describe phenomena in the Knudsen transition region.

There have been applied, in fact, such an assortment of approximation schemes in the literature that it would be unwieldy to explore each one in detail here. Various references such as Street(12) may be consulted in which approximate solutions of the Boltzmann equation are obtained. Here we shall give a brief summary of several approximations which have been widely applied, particularly in the calculation of the transfer processes to spheres in a rarefied gas. These techniques we term the BGK(26) and the moment approximations.(25, 27) Before discussing these techniques, we first note the complications introduced by the presence of physical boundaries and the boundary conditions usually applied in rarefied gas dynamics.

Boundary Conditions for $Kn > 0$

One of the complications in describing the transport processes in the vicinity of a boundary surface in a rarefied gas is that the physical boundary introduces an essential discontinuity in the velocity distribution function f of the gas molecules. At the boundary surface one must distinguish two classes of molecules and, hence, two distinct distribution functions. The two classes are those molecules hitting the surface and those molecules reflecting from the surface. This two-sided nature of the distribution function f causes much of the difficulty in approximation schemes.

Boundary conditions for solution of the Boltzmann equation near a surface are usually introduced in terms of an empirical accommodation coefficient α. A typical boundary condition for the distribution function f would be of the following form:

$$f^{+} = \alpha f^{(0)+} + (1 - \alpha) f^{-}(\mathbf{v}'), \tag{2.123}$$

where

$$\mathbf{v}' = \mathbf{v} - 2\mathbf{n}(\mathbf{v} \cdot \mathbf{n}). \tag{2.124}$$

f^{+} denotes the distribution function for those molecules leaving the surface and f^{-} the distribution function for molecules hitting the surface at a particular point. Equation (2.123) states that the molecules leaving the surface are composed of two groups: a fraction α which leave the surface in equilibrium with the surface, and a fraction $(1 - \alpha)$ which are specularly reflected and have their component of velocity normal to the surface $\mathbf{v} \cdot \mathbf{n}$ reversed. In many instances investigators have distinguished various accommodation coefficients: tangential momentum, normal momentum, and energy accommodation coefficients. This topic will be discussed further in Chapter 5.

The BGK Approximation

Perhaps the most widely used method for obtaining results for transport processes in rarefied gases has been the so-called BGK approximation or model, after the investigators Bhatnagar, Gross, and Krook, who developed the model in its present form.(26)

The BGK approximation involves use of a relaxation model for the collision operator of the Boltzmann equation. The relaxation model of the Boltzmann equation is, in terms of dimensionless quantities, defined previously [Eqn. (2.89)].

$$\frac{\partial f^{\oplus}}{\partial t^{\oplus}} + \mathbf{v}^{\oplus} \cdot \nabla^{\oplus} f^{\oplus} + \mathscr{E}^{\oplus} \cdot \nabla_{\mathbf{v}\oplus} f^{\oplus} = \frac{\hat{\delta} R_{\mathrm{v}}}{\lambda_G} (f^{(0)\oplus} - f^{\oplus}). \tag{2.125}$$

Here $f^{(0)\oplus}$ is the dimensionless local Maxwellian distribution function:

$$f^{(0)\oplus} = \frac{1}{\bar{n}_G (m_G/2\pi k \bar{T})^{\frac{3}{2}}} n_G(\mathbf{x}) \left(\frac{m_G}{2\pi k T(\mathbf{x})} \right)^{\frac{3}{2}} \exp \left\{ \frac{-m_G}{2kT(\mathbf{x})} [\mathbf{v} - \mathbf{q}(\mathbf{x})] \right\}, \tag{2.126}$$

where $\bar{n}_G$ and $\bar{T}$ are some arbitrary reference number density and temperature and $n_G(\mathbf{x})$, $T(\mathbf{x})$, and $\mathbf{q}(\mathbf{x})$ are the local number density, temperature, and velocity respectively:

$$n_G(\mathbf{x}, t) = \int \mathrm{d}\mathbf{v} f(\mathbf{x}, \mathbf{v}, t), \tag{2.127}$$

$$\mathbf{q}(\mathbf{x}, t) = \frac{1}{n_G(\mathbf{x}, t)} \int \mathrm{d}\mathbf{v}\, \mathbf{v} f(\mathbf{x}, \mathbf{v}, t), \tag{2.128}$$

and

$$T(\mathbf{x}, t) = \frac{m_G}{3k n_G(\mathbf{x}, t)} \int \mathrm{d}\mathbf{v} [\mathbf{v} - \mathbf{q}(\mathbf{x})]^2 f(\mathbf{x}, \mathbf{v}, t). \tag{2.129}$$

In eqn. (2.125), $\hat{\delta}$ is a constant which can be chosen to make the Chapman–Enskog viscosity or heat conductivity agree with that given by the collisional model of Maxwell molecules. The ratio of the viscosity to the heat conductivity is not given correctly by this model.

Choosing the viscosity, $\hat{\delta}$ has the value

$$\hat{\delta} = 8/5\pi^{\frac{1}{2}}. \tag{2.130}$$

For the heat conductivity for monatomic molecules

$$\hat{\delta} = 16/15\pi^{\frac{1}{2}}. \tag{2.131}$$

Despite the shortcoming of using an arbitrary constant $\hat{\delta}$, the "collision term"

$$\frac{\hat{\delta}R_v}{\lambda_G}(f^{(0)\oplus} - f^{\oplus}) \tag{2.132}$$

satisfies for a single component gas the conservation laws given previously and provides decay by collisions to a Maxwellian distribution function.

Generalized models with several relaxation times, as opposed to the BGK model with a single relaxation time, have been proposed which enable these models to give correctly both the viscosity and the heat conductivity. These models are somewhat more complicated, and so have not been as generally studied as the BGK model. Some analogous models have also been proposed for gas mixtures.

Although the BGK model has a simpler form than the full collision term of the Boltzmann equation, the solution of eqn. (2.125) still involves generally a solution of coupled non-linear integral equations for the functions n_G, $\mathbf{q}$, T. So in spite of the simplification introduced by the BGK model, solution of eqn. (2.125) is difficult, and it has been necessary to employ numerical techniques or iterative procedures to obtain analytical results.

The BGK model appears to be a useful approximation which has been shown to give reasonable agreement with experiment in a variety of problems in gas dynamics.[28, 29] The application of the BGK model to the description of transfer processes to aerosol particles will be discussed in further detail in Chapter 5.

Moment Methods

Various moment methods have been employed as approximations to the solution of the Boltzmann equation. We will discuss two moment methods which are fairly representative of what may be expected from such analytical techniques.

The approaches we discuss make use of Maxwell's equations of transfer. These equations can be developed as follows. Consider a function ψ as any function of the molecular velocity $\mathbf{v}$. We apply the following operator to the Boltzmann equation:

$$\int d\mathbf{v}\,\psi.$$

The resultant equation is termed the Maxwell integral equation of transfer, given essentially by eqn. (2.71):

$$\frac{\partial}{\partial t}\left(\int d\mathbf{v} f\psi\right) + \nabla.\left(\int d\mathbf{v} f\mathbf{v}\psi\right) + \int d\mathbf{v} f\mathscr{E}\,.\,\nabla_{\mathbf{v}}\psi = \iiint \psi(f'f'_1 - ff_1)\, g\, d\underline{\sigma}\, d\mathbf{v}_1\, d\mathbf{v}. \tag{2.71a}$$

For $\psi = 1$, $m\mathbf{v}$, $\frac{1}{2}m(\mathbf{v} - \mathbf{q})^2$, we obtain, of course, the conservation equations discussed earlier.

Maxwell's integral equations of transfer were first employed in rarefied gas dynamic studies by Grad[25] in a procedure which has been termed Grad's thirteen-moment method.

In this method Grad assumes the distribution function has the following form in a rectangular cartesian coordinate system:

$$f(\mathbf{x}, \mathbf{v}, t) = f^{(0)}(\mathbf{x}, \mathbf{v}, t)\left[1 + \frac{\Pi_{ij}v_iv_j}{2pRT} - \frac{h_iv_i}{pRT}(1 - v^2/5RT)\right]. \qquad (2.133)\dagger$$

$f^{(0)}(\mathbf{x}, \mathbf{v}, t)$ is the local Maxwellian distribution function, R is the gas constant, p is the scalar pressure, and Π_{ij} and h_i are the components of stress and heat fluxes respectively, which in Grad's formulation, become dependent variables along with n, $\mathbf{q}$, and T. Equation (2.133) is now substituted in eqn. (2.71) and ψ is taken to be successively m_G, $m_G\mathbf{v}$, $m_Gv^2/2$, $m_G\mathbf{vv}$, and $m_G\mathbf{v}\,(v^2/2)$. This results in thirteen partial differential equations which include as the first five, the conservation equations, and determine the thirteen dependent variables.

Grad also developed boundary conditions which are similar to the slip flow boundary conditions discussed in Chapter 6.

One of the principal objections to the application of the thirteen-moment equations to problems of rarefied gas dynamics is that the distribution function chosen, eqn. (2.133) is not "two-sided". Hence it is to be expected that this procedure will probably not be very good for near-free molecule flows where the distribution function f^+ for molecules directed away from a surface is quite different from f^-, the function for molecules impinging on the surface. Another objection is that the assumed form of the distribution function, eqn. (2.133) is restricted by the use of polynomials in $\mathbf{v}$. Forms of the distribution function found from analytic solutions of the BGK model reveal in many geometries dependence on $\mathbf{v}$ which is highly nonlinear and not amenable to representation by the simple polynomial relations assumed by Grad.

With these objections to the thirteen-moment procedure in view, Lees[27] proposed a moment method which ensured the distribution function would be "two-sided" or bimodal. We illustrate Lees' method for Couette flow.‡ Lees proposed the following distribution functions:

$$f_1(\mathbf{x}, \mathbf{v}, t) = n_1(\mathbf{x}, t)\left[\frac{m_G}{2\pi kT_1(\mathbf{x}, t)}\right]^{\frac{3}{2}} \exp\left\{\frac{-m_G}{2kT_1(\mathbf{x}, t)}(\mathbf{v} - \mathbf{q}_1(\mathbf{x}, t))^2\right\}, \quad \text{for } v_x > 0, \qquad (2.134)$$

and

$$f_2(\mathbf{x}, \mathbf{v}, t) = n_2(\mathbf{x}, t)\left[\frac{m_G}{2\pi kT_2(\mathbf{x}, t)}\right]^{\frac{3}{2}} \exp\left\{\frac{-m_G}{2kT_2(\mathbf{x}, t)}(\mathbf{v} - \mathbf{q}_2(\mathbf{x}, t))^2\right\}, \quad \text{for } v_x < 0, \qquad (2.135)$$

where x is taken as the direction perpendicular to the two parallel planes in the Couette problem. Lees, therefore, uses a bimodal distribution made up of two Maxwellian distribution functions f_1 and f_2. The ten parameters n_1, T_1, $\mathbf{q}_1$, n_2, T_2, $\mathbf{q}_2$ are to be determined by the following moment procedure. The functions f_1 and f_2 are substituted in eqn. (2.71) and ψ is taken successively to be m_G, $m_G\mathbf{v}$, m_Gv^2, and selected terms of $m_G\mathbf{vv}$ and $[\mathbf{v}(m_Gv^2/2)]$. In this way ten equations for the ten parameters are obtained. By identifying the parameters n_1, T_1, $\mathbf{q}$, n_2, $\mathbf{q}_2$ with the number density, temperature, and velocity of the molecules reflected from the appropriate surface, boundary conditions for the differential moment equations

† In this relation, the subscripts are used as dummy indices to specify cartesian coordinates x, y, and z. This notation is used here to conform with cartesian tensor notation, (e.g. Jeffreys,[29]) often used in fluid dynamics. This special usage should not be confused with such indexing for particle species normally applied elsewhere in the book.

‡ Couette flow corresponds to the steady shearing motion in a gas sandwiched between two parallel plates.

are obtained. Therefore, the parameters can all be determined, and any moment of the distribution function

$$f = f_1(v_x > 0) + f_2(v_x < 0) \tag{2.136}$$

can be found. For example, in the Couette problem, the velocity profile is readily calculated:

$$\mathbf{q}(\mathbf{x}, t) = \frac{1}{n_G(\mathbf{x}, t)}\left\{ \int_{v_x > 0} d\mathbf{v}\, \mathbf{v} f_1 + \int_{v_x < 0} d\mathbf{v}\, \mathbf{v} f_2 \right\}. \tag{2.137}$$

Objections to Lees' moment method have been raised. As in all moment methods, the initial choice of the form of the distribution function is completely arbitrary. One can never be certain, therefore, what the results mean physically without experimental confirmations.

It would appear[30] that the moment methods, including the ones we have just discussed are particularly inaccurate when the Knudsen number is large; that is, for gas flows near the free molecule regime.

In summary, one can say that the BGK model, though highly approximate, nevertheless has proven to be very useful for calculations of rarefied gas flows over the whole Knudsen regime, $0 \leqslant Kn \leqslant \infty$. Its advantage is its simplicity and its ability to reproduce the important features of rarefied gas flows such as bimodal distribution functions in a non-arbitrary fashion. The moment methods, on the other hand, appear to be of principal interest only for cases of small Knudsen number. In this regime, however, an entirely different procedure can be applied, which has been termed the *slip-flow procedure*. This method uses a phenomenological extension of boundary conditions associated with the Navier–Stokes equations so that it does not entail solution of the Boltzmann equation. Instead, eqns. (2.118)–(2.122) are applied to a particular problem, and the effect of the rarefied nature of the flow appears through use of slip-flow boundary conditions. These conditions, for example, allow for slip of the gas over the surface and a temperature jump at the surface. The slip-flow procedure will be discussed in more detail in Chapter 6. Despite its widespread use, the slip-flow procedure is difficult to justify in a rigorous manner, and it remains a rather intuitive process which often may be misleading.

Progress in the field of rarefied gas dynamics is very rapid at present so that one may expect considerable improvements in a few years over the relatively crude methods such as the ones discussed above for obtaining information on rarefied gas flows.

2.3. Physicochemical Properties of Aerosol Particles

The description of the dynamics of aerosol particles requires knowledge of various physiochemical properties of the particles themselves. Although these properties are specified empirically in existing mechanical models, their nature must still be known. For example, the description of the dynamics may entail describing *a priori* for the particle the mass density, surface energy, freezing point (if liquid), heat of vaporization or sublimation, solubility, heat of adsorption for gases, vapor pressure, viscosity or elastic properties, thermal conductivity, diffusivity of components, magnetic and electric properties, chemical reactivity, radioactivity, and momentum and energy accommodation coefficients.

If the measurement of the particle composition is possible, and the additional assumption is introduced that the substance making up the aerosol particle possesses the same characteristics as the substance in macroscopic amounts, which we term here the *bulk state*, then obviously the various physical properties of the particle, such as those mentioned above, are either known or easily determinable by standard techniques. These techniques generally will not involve direct measurements on aerosol particles. This additional assumption of bulk state behavior, however, may not be justified generally for small aerosol particles, as is well known. The physical properties of small aerosol particles of a given substance can be quite different from the corresponding properties of the same substance in the bulk state.†

Deviations in behavior of small aerosol particles from the bulk state are widely recognized for many physical properties such as vapor pressure, freezing point, and crystal structure. Yet there has been a lack of a systematic classification of these deviations. Also, although deviations are expected to occur for small particles, there have been few experimental measurements of such deviations owing partly to the difficulties of such measurements.

We may classify deviation from the bulk properties of the properties of small aerosols as either *extrinsic* or *intrinsic*. *Extrinsic deviations* are associated with characteristics of aerosols which are not inherent, but are caused by external agencies such as the mode of formation of the particle or the absence of phase transition nuclei in the particles. Thus, extrinsic deviations are associated more with a lack of control and knowledge in the particle generation process, than with any fundamental cause. *Intrinsic deviations* may occur in several ways. One type of intrinsic deviation is associated with the radius of curvature of small particles. For example, it is well known that a liquid droplet with given radius has a higher vapor pressure than that found with another droplet of larger radius. For sufficiently small particles, another type of intrinsic deviation occurs. Here the intermolecular energy of interaction of molecules making up a very small particle is altered by the fact that a given molecule does not interact with an extremely large number of other molecules, but instead can interact only with a limited number of other molecules in the particle. Further, if the aerosol particle is still smaller, almost of molecular size, molecular fluctuations will be so large that it is probably no longer meaningful to speak of the physical properties of a particle of such small size in the usual macroscopic sense.

One type of extrinsic deviation is found in the lowering of the freezing point(32) or the raising of the boiling point from that for the bulk state for small liquid droplets. Such effects are usually attributed to the absence of phase transition nuclei. The absence of such nuclei stems from the fact that the bulk material from which the aerosol particles are formed probably contains, for example, only a small number of nuclei per unit volume, so that there is only a very small probability that any small aerosol particle will contain even one nucleus. This circumstance results in the situation that nearly all aerosol particles formed by vapor condensation and subsequent cooling well below the melting point of the parent material, are likely to be in a metastable liquid state. As an example, sodium chloride has been observed to exist as a relatively stable sub-cooled liquid at room temperature, hundreds of degrees below the melting point of crystalline sodium chloride.‡

Another example of an extrinsic deviation recently has been reported by Jäger and Kieschke.(38) In examining absorption spectra from freshly formed aerosols composed of $Fe(CO)_5$, 30–200 Å in size, they found spectra corresponding to excited states of carbon monoxide, and bands possibly associated with the triplet transition (O—O) $d^3 \rightarrow b^3\Sigma^+$. The excitation had energy levels of 7–9 eV, suggesting that the excitation was not due to chemical reactions or incident photons only. It is possible that the spectral absorption also was related to gas molecules adsorbed on the iron surface, or to large surface energy of the

† We may define bulk state here in terms of the state of a substance in a particle with $R \gtrsim 10\ \mu$.

‡ Of course, small particles also exhibit supercooling caused by the curvature of the particle.

small particles. When the small particles coagulate, or surface crystallites are relocated, large amounts of energy may be released and transferred to gas molecules adsorbed on the particle surface. In this way, certain high energy levels may be populated in a manner differing from that predicted by thermal equilibrium.

Basic extrinsic properties of natural and man-made aerosols are beginning to be examined in more detail. For example, Corn and Reitz(39) have reported some interesting measurements of specific surfaces and densities of samples of urban aerosols taken in the Pittsburgh area. They found that the specific surface would increase by a factor of about two after degassing these samples at 200°C. Thus natural aerosols may easily contain a variety of adsorbed molecular species which can contribute potentially to anomalies in expected behavior of the particles, especially on deposition in the lungs of animals.

Other types of extrinsic deviation are found in the special properties imparted small particles by the manner of preparation of the particles. For example, production of small particles by grinding in a colloid mill has been observed to produce alteration in the heat of adsorption of gases by the particles.(33) In general, the method of particle production may introduce or abstract defects or microscopic impurities which differ from what is found in the parent material. Often extrinsic and intrinsic deviations for the same physical property, as in the example of super-cooling of sodium chloride cited above, may both occur in a particle. This fact makes the study of intrinsic deviations most difficult.

Intrinsic deviations are perhaps most widely known through the effect of the radius of curvature of small particles on many physical properties such as vapor pressure, freezing point, surface tension, heat of evaporation, etc. Intrinsic deviations not directly associated with the radius of curvature, have been observed by X-ray crystallographic studies of very small crystallites ($R_i < 10^{-6}$ cm). In these studies, the observed lattice spacings in the small crystallites differed significantly from the lattice spacings observed for the bulk state of the parent material.(36) The effect of such alterations on various physical properties have never been studied. In general, one expects that for particle of radius less than the order of 0·01 μ, intrinsic deviations of this sort must occur; however, it is obviously very difficult to observe

TABLE 2.5. DEVIATIONS OF PHYSICAL PROPERTIES OF SMALL PARTICLES FROM BULK STATE EITHER OBSERVED EXPERIMENTALLY OR PREDICTED IN THEORY

	Reference
Extrinsic deviations	
Freezing temperature	(32)
Condensation temperature	(32)
Physical adsorption	(33) (38)
Specific surface area and density	(39)
Intrinsic deviations	
Vapor pressure	(31)
Heat of evaporation	(31)
Freezing temperature	(31)
Surface tension, or surface energy	(31) (38)
Solubility	(35)
Equilibrium between a floating particle and a surface film	(31)
Azeotropic composition	(31)
Chemical equilibrium in a heterogeneous reaction involving the particle	(31)
Optical properties	(34)
Crystal structure	(36)
Work function	(37)

such deviations experimentally. The result is that there is virtually no information available other than that suggested above.

A recent illustration of an attempt to look for deviations in surface energy of small spheres of about 0·05 μ in radius was reported by Blackman *et al.*[40] They found for spherical particles of silver, lead, and bismuth evaporated under a microscope, that the surface energy σ in these three cases was essentially that of the bulk material. This seems consistent with the calculations for water listed in Table 2.6.

In Table 2.5 are given various observed or conjectured deviations classified according to the scheme introduced here.

TABLE 2.6. VARIATION OF SOME PROPERTIES OF AEROSOL PARTICLES WITH SIZE. TABULATION OF THERMODYNAMIC VARIABLES AS A FUNCTION OF PARTICLE RADIUS GIVEN IN MICRONS

Ratio of property to same property for bulk state (particle radius > 10 μ)	>10	10	1	0.1	0.01	0.001	Reference
Vapor pressure of water p/p_0 at 5°C	1.0000	1.0001	1.0012	1.012	1.124	3.32	(31)
Heat of vaporization of water at 4°C, λ/λ_0	1.0000	0.999994	0.99994	0.9994	0.994	0.94	(31)
Surface tension of water at 4°C, σ/σ_0	1.0000	—	0.9996	0.996	0.968	0.755	(31)
Solubility of crystalline substances	1.0000	—	—	—	~1	~2	(35)

In Table 2.6 are some examples of the magnitude of some of these deviations. It can be noted that from Table 2.6 there is very little information at all on the intrinsic deviations, which might reasonably be anticipated to occur for the transport properties of small particles. The entire subject of intrinsic deviations perhaps represents the most poorly understood area in the physics and chemistry of aerosols.

CHAPTER 3

DYNAMICS OF SINGLE PARTICLES IN THE CONTINUUM APPROXIMATION

THE general treatment of the dynamics of spherical bodies suspended in a fluid characterized as a continuum restricts analysis to the condition where the Knudsen number of the body Kn_i approaches zero. It has been found in practice that the results based on this restriction may be applied to particle behavior under conditions where $Kn_i \ll 1$.

In general, the exchange of momentum between the gas and the particle involves strong interaction with heat and mass transfer to the particle. Thus the forces acting on particles in non-uniform gases may be tied closely to both temperature and to concentration gradients as well as to velocity gradients in the suspending medium. The assumption that the Knudsen number is zero greatly simplifies the calculation of particle motion, especially in a gas that is non-uniform, i.e. in a non-equilibrium gas containing gradients of velocity, temperature, or concentration of (gaseous) components. Under such circumstances, temperature gradients will not affect momentum transfer directly, and the particles will only experience momentum transfer in non-uniform gas mixtures resulting from a drift of diffusing gas components associated with *Stefan flow*. Since the latter effect is generally rather weak, the aerodynamic forces in the regime $Kn_i \to 0$ may be estimated to a good approximation by applying the classical continuum theory of a pure viscous fluid. The nature of this simplification will become more apparent with the subsequent discussion in Chapters 5 and 6.

In this chapter we shall deal only with aspects of momentum transport to spherical particles in the continuum approximation. To emphasize the conceptual decoupling of momentum transfer and heat and mass transfer, the latter phenomena for $Kn_i \to 0$ will be examined separately in Chapter 4. Since the transport mechanisms generally cannot be isolated completely in a non-uniform gas under conditions where non-continuum theory applies, the three different rate processes will be treated together in discussing the free molecule regime and transitional regimes where $Kn_i > 0$.

3.1. Steady Rectilinear Motion in a Homogeneous Gas

A. Stokes' Solution

Perhaps no other individual piece of information has had such wide application to the theory of aerosols than Stokes' solution for a solid particle moving steadily through a stagnant medium of infinite extent. Stokes' classical result is so vital to the development of the continuum theory for aerosol mechanics that it is worthwhile to review these findings in some detail.†

When bodies move slowly enough through a viscous fluid, the Reynolds number observed for the particle, Re_i, will be smaller than unity. Under these conditions one finds the influence

† For a more complete discussion of low Reynolds number hydrodynamics, see Happel and Brenner.[41] For a discussion of mathematic perturbation methods as $Re_i \to 0$, see also Van Dyke.[42]

of fluid inertia small compared with the influence of fluid friction, and consequently the inertial terms are customarily neglected in this range. Such a restriction greatly simplifies the equations of motion of the fluid. Proceeding from the Navier–Stokes equation, eqn. (2.119), we assume that the gas is incompressible, or $Ma \to 0$, and the flow is steady. Equation (2.118) becomes

$$\nabla . \mathbf{q} = 0, \tag{3.1a}$$

and in the absence of external forces ($\mathscr{E} = 0$), eqn. (2.119) reduces to

$$\mathbf{q} . \nabla \mathbf{q} = -\frac{1}{\rho_G} \nabla p + \frac{\mu_G}{\rho_G} \nabla^2 \mathbf{q}, \tag{3.1b}$$

where p and $\mathbf{q}$ are defined respectively previously as the gas pressure and the mean average velocity. In eqn. (3.1b), the inertial terms on the left side contain velocity to the second order, while the frictional terms, the second term on the right side, are first order in velocity. Hence for very slow motion where the velocity approaches zero, the inertial terms diminish more rapidly than the frictional terms. It also is clear that the frictional terms must dominate the fluid motion if the viscosity becomes sufficiently large.

Then for steady flow under conditions where $Re_i \to 0$, the equation governing incompressible gas motion further reduces to

$$\nabla p = \mu_G \nabla^2 \mathbf{q}, \tag{3.2}$$

where p is the pressure in the fluid and $\mathbf{q}$ is the velocity vector of the fluid. If we take the divergence of eqn. (3.1b), then we find that

$$\nabla . \nabla p = \nabla^2 p = 0. \tag{3.3}$$

Thus, as a first approximation, the pressure field for a moving fluid in which the viscous terms predominate may be described by solutions to Laplace's equation. Solutions to this expression [eqn. (3.3)] for a variety of different boundary conditions are well known from the theory of inviscid fluids and from electrostatics.

G. C. Stokes first solved in 1851 the problem of the steady passage of a solid particle of radius R_i through a viscous medium of infinite extent. To give a solution for this problem we shall adopt Prandtl's[43] and Lamb's[44] approach

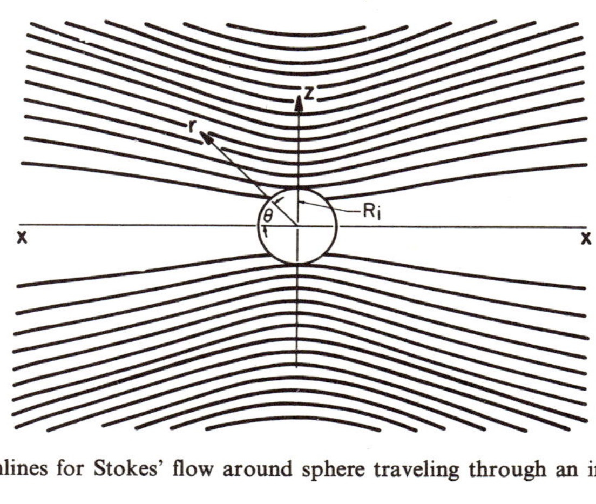

FIG. 3.1. Streamlines for Stokes' flow around sphere traveling through an infinite medium.

Suppose we consider the particle fixed in space with fluid streaming past it. Then the velocity of the fluid relative to the particle at a value of the radial coordinate **r** of infinity will equal U_∞, the particle velocity seen by an observer.

One anticipates that the pressure should be higher on the surface of the particle where fluid approaches it than on the side where fluid leaves the particle. The simplest solution for the pressure field is

$$p - p_\infty = -\frac{cx}{r^3}, \tag{3.4}$$

where c is a constant, and

$$r = \sqrt{(x^2 + y^2 + z^2)}.$$

Substituting the pressure distribution given by eqn. (3.4) into eqn. (3.1) and making use of the equation of continuity [Eq. (3.1a)], one can determine that the velocity field is given by

$$\begin{aligned} q_r &= \frac{1}{2} U_\infty \cos\theta \left(\frac{R_i}{r}\right)^2 \left[\left(\frac{R_i}{r}\right) - 3\left(\frac{r}{R_i}\right) + 2\left(\frac{r}{R_i}\right)^2\right] \\ q_\theta &= \frac{1}{4} U_\infty \sin\theta \left(\frac{R_i}{r}\right)\left[3 + \left(\frac{R_i}{r}\right)^2 - 4\left(\frac{r}{R_i}\right)\right] \end{aligned} \tag{3.5a}$$

and

$$c = \tfrac{3}{2}\mu_G U_\infty R_i, \tag{3.6}$$

where u, v, and w are the components of fluid velocity in the x, y, and z directions. In finding eqns. (3.5a) and (3.6), use is made of the free stream condition that $\mathbf{q} = \mathbf{U}_\infty$ as $\mathbf{r} \to \infty$, and the condition that $\mathbf{q} = 0$ at the surface of the particle where $r = R_i$.

Having the solution for slow fluid flow past a sphere fixed in space, we can derive the case of interest for the steady motion of the sphere through a stagnant fluid. This is done simply by superimposing a uniform flow of $-U_\infty$ on the solution, eqn. (3.5a). The velocity components in spherical coordinates for this latter case are given by [41]:

$$\begin{aligned} q_r &= -\frac{1}{2} U_\infty \cos\theta \left(\frac{R_i}{r}\right)^2 \left[\frac{R_i}{r} - \frac{3r}{R_i}\right], \\ q_\theta &= -\frac{1}{4} U_\infty \sin\theta \left(\frac{R_i}{r}\right)\left[3 + \left(\frac{R_i}{r}\right)^2\right]. \end{aligned} \tag{3.5b}$$

The streamlines corresponding to the slow steady motion of a rigid sphere through a stagnant fluid are shown in Fig. 3.1.

With eqn. (3.5a) the components of stress near the sphere are:

normal $$p_{rr} = -\frac{3}{2}\frac{\mu_G U_\infty}{R_i}\left(\frac{R_i}{r}\right)^2 \cos\theta.$$

tangential $$p_{r\theta} = \frac{3}{2}\frac{\mu_G U_\infty}{R_i}\left(\frac{R_i}{r}\right)^4 \sin\theta.$$

The total drag force $\mathscr{D}^c$, or the resistance to the motion of the sphere can be calculated from the components of the stress. Integrating the stresses over the sphere surface S,

$$-\mathscr{D}^c = \int [(-p_{rr})_{r=R_i} + (p_{r\theta})_{r=R_i}]\, dS = \underset{\text{(form drag)}}{2\pi\mu_G R_i U_\infty} + \underset{\text{(frictional drag)}}{4\pi\mu_G R_i U_\infty} \tag{3.7}$$

Thus the resultant force acting on the sphere as given by Stokes is

$$\left.\begin{aligned} \mathscr{D}^c &= 6\pi\mu_G R_i U_\infty, \\ \text{for} \qquad Re_i &= \frac{\rho_G|\mathbf{U}_\infty|R_i}{\mu_G} \to 0. \end{aligned}\right\} \tag{3.8}$$

In many problems involving the steady motion of aerosols, the velocity of a particle is assumed to be proportional to the force acting on the body. The constant of proportionality is called the mobility B_i^c. From eqn. (3.8), for the case where $Kn \to 0$ and $Re_i \ll 1$,

$$B_i^c = (6\pi\mu_G R_i)^{-1}. \tag{3.9}$$

Stokes' Approximation and Inertial Effects

Unless the Reynolds number characteristic of slow flow is identically zero, the inertial effects in the fluid will exist even though they may be exceedingly small. The assumption, attributed to Stokes, of completely disregarding the inertial term $\mathbf{q}\,.\nabla\mathbf{q}$ in the Navier–Stokes equations essentially buries forever the possibility of accounting for the influence of fluid inertia. As the Reynolds number of the particles approaches unity, the neglect of the inertial term becomes less and less valid. Oseen, in his book,[13] extended this criticism farther by suggesting that, in fact, one could never neglect the inertial terms in the theory no matter how small the Reynolds number (see also Happel and Brenner[41]). Oseen noted that Stokes' solution has the form $\mathbf{q} = \mathbf{U}_\infty + U_\infty R_i 0(r^{-1})$ at large distances from the particle. Thus, as $r \to \infty$, $\nabla^2\mathbf{q} \sim U_\infty R_i 0(r^{-3})$, and $\mathbf{q}\,.\nabla\mathbf{q} \sim U_\infty^2\, R_i 0(r^{-2})$, where $0(r^{-2})$ means the order of . . . , and the ratio of the inertial term to the viscous term far from the sphere then becomes

$$\frac{\rho_G \mathbf{q}\,.\,\nabla\mathbf{q}}{\mu_G \nabla\mathbf{q}} \sim 0\left(\frac{rU_\infty\rho_G}{\mu_G}\right), \quad r \to \infty\,.$$

This implies that the local Reynolds number based on the distance r can become very large at large r despite the fact that the particle Reynolds number Re_i is much less than unity. In other words, Stokes' implicit assumption that the inertial terms are negligible everywhere cannot be consistent with his form of the velocity field. Oseen then noted that in fact Stokes' solution in the limit $\rho_G u R_i/\mu_G \to 0$ becomes incorrect only at $r/R_i \to \infty$. To rectify this difficulty, Oseen suggested that Stokes' result should serve principally as a first approximation to the actual pattern of flow. The second approximation could be deduced by using the linearized equations obtained by introducing the streaming velocity $\mathbf{U}_\infty$ into the equations of motion,

$$\mu_G\nabla^2\mathbf{q} - \nabla p = \rho_G\mathbf{U}_\infty\,.\,\nabla\mathbf{q},$$
$$\nabla\,.\,\mathbf{q} = 0.$$

Oseen's results gave the following form for the drag:

$$\mathscr{D}^c = 6\pi\mu_G R_i \mathbf{U}_\infty[1 + \tfrac{3}{8}Re_i + 0(Re_i^2)]. \tag{3.10}$$

According to Oseen's calculations, the streamlines around a sphere which travels through the fluid differ markedly from Stokes' result, as indicated in Fig. 3.2. Even the weak effect of the inertial terms show up strongly in modifying the flow field around the sphere. Stokes' solution has symmetry fore and aft of the body (Fig. 3.1), while Oseen's results indicate that asymmetry should exist in the streamline patterns far from the sphere (Fig. 3.2). If $\rho_G U_\infty r/\mu_G < 1$, the flow in the neighborhood of the sphere given by Oseen is virtually the same as found by Stokes, and essentially the same rule for the drag force is obtained, as indicated in eqn. (3.10). However, at larger distances from the body the velocity distribution becomes quite different, even for values of $\rho_G U_\infty r/\mu_G \approx 1$. In Oseen's flow field there exists a paraboloidal domain of disturbance extending behind the sphere where the velocity of motion decays as the reciprocal of the distance r as shown in Fig. 3.2. In this zone, the fluid forms a wake behind the body.

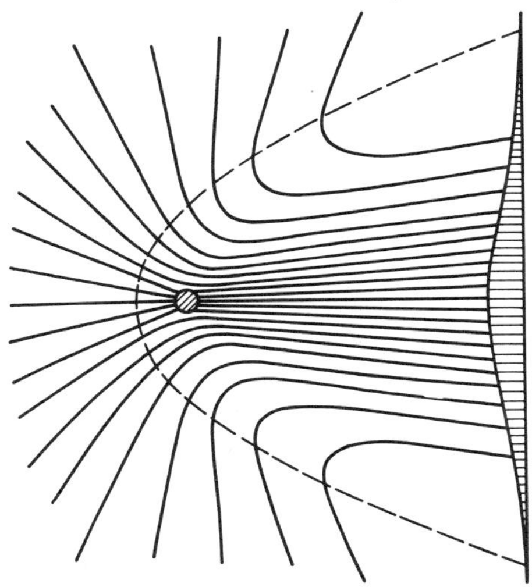

FIG. 3.2. Streamlines around a sphere moving from right to left through an infinite medium based on Oseen's approximation. (From H. Schlichting[67], p. 98.)

If the Reynolds number Re_i approaches zero, the paraboloid of asymmetry in Fig. 3.2 would move away from the body towards infinity, and the Oseen flow in effect would transform to the Stokes' regime in Fig. 3.1.

Since Oseen's work there has been considerable argument about the interpretation of Oseen's approximation in the light of predictions expected from solutions of the Navier–Stokes equations. Using the term $\mathbf{U}_\infty . \nabla\mathbf{q}$, a reasonably good approximation of the flow field should result at larger distances from the sphere. However, the approximation will be a poor one near the body as $\mathbf{q} \to 0$. Hence in the Reynolds number expansion for the drag force, eqn. (3.10), the second term may not be accurate. Furthermore, Oseen's analysis does not provide any suggestion of a systematic method for seeking higher order approximations for the flow around the sphere.

The incomplete picture of the role of inertial effects in the slow motion of bodies through suspending fluids led to further efforts in the field and to extensions summarized by Proudman and Pearson.[14] These workers have indicated that Oseen's results provide a "zeroth" approximation to exact solutions of the Navier–Stokes equation at low Reynolds number.[42] Thus Oseen's approximation justifies theoretically Stokes' result, but it does not appear to furnish higher order corrections. Curiously enough, the alternate perturbation scheme of Proudman and Pearson fortuitously arrives at a relation for the drag force in the form of Reynolds number expansion that is the same as eqn. (3.10) up to the second term. Higher order terms, however, differ from Oseen's results. It is interesting that Proudman and Pearson's higher order effects have not been verified in experimental observations for the drag coefficient of spheres (see, for example, Maxworthy[46]).

Based on the extensive work reported for the steady viscous flow around a spherical body, it appears that Stokes' result, at least for the drag force, is well justified in the range of particle Reynolds numbers much less than unity. Experimental results confirm this conclusion. Typical of the older results for the drag coefficient reported by Goldstein,[45] defined as

$$C_D = \frac{\mathscr{D}^c}{\frac{1}{2}\rho_G U_\infty^2 \pi R_i^2}, \tag{3.11}$$

are shown in Fig. 3.3 up to Re_i of 5. The agreement between the theory and the experiments is remarkably good except at the higher values of Reynolds number. A more recent review by Maxworthy[46] indicates a similar correspondence between theory and experiment. However, Maxworthy finds that in the range $Re_i > 0.1$, Oseen's solution is as good or possibly better than Proudman and Pearson's results. Maxworthy's data, plotted in the form of the ratio of measured drag to Stokes' drag minus one vs. Reynolds number, revealed that the observed drag on a sphere approached Oseen's theoretical result below $Re_i \simeq 0.4$.

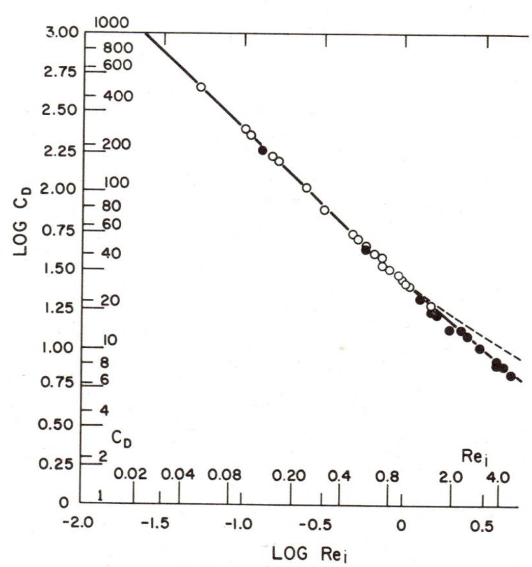

FIG. 3.3. Drag coefficients as a function of Reynolds number for spheres moving through an infinite medium. ---- Theoretical (Oseen). ● Liebster and Schiller[408]. ○ Schmiedel[409].

However, Stokes' theory appeared to be exact only in the limit $Re_i \to 0$.† Oseen's result generally involves a small correction to the Stokes' drag for $Re_i < 0.1$ so that, for practical purposes, the latter relation generally continues to be used in aerosol work.

Liquid Droplets

When small liquid droplets pass through a viscous fluid, the drag problem becomes more complicated than Stokes' problem because of the possibility for circulation of liquid inside the droplet. The steady translational motion of droplets in a fluid was first investigated theoretically by Rybczynski[(47)] and by Hadamard.[(48)]

In general, a droplet of liquid traveling in an immiscible fluid will tend to maintain its spherical shape as a result of the interfacial tension acting at the boundary between the two fluids. If the motion of the droplet is sufficiently small, the stresses tending to deform the droplet will be overpowered by the action of the interfacial tension, and, as a first approximation, the droplet will remain spherical.[(43)] The flow field in both fluids and the resistance on the droplet then may be estimated within the Stokes' approximation for the following boundary conditions:

$$\mathbf{q} \to \mathbf{U}_\infty, \qquad r \to \infty,$$

and the radial component of velocity in both fluids is bounded by

$$q_r = 0 \quad \text{at} \quad r = R_i.$$

The tangential velocity component q_θ must be continuous at the interface, i.e. taking subscript i to mean inside the droplet and o to mean outside the droplet,

$$(q_\theta)_o = (q_\theta)_i \quad \text{at} \quad r = R_i.$$

In addition, it is assumed that the pressure distribution on the droplet is given by conditions of thermodynamic equilibrium

$$p_i = p_o + \frac{2\sigma}{R_i},$$

where σ is the surface tension of the droplet surface. Using these conditions, Rybczynski and Hadamard found that the Stokes' drag was modified by a factor involving the ratio of viscosities of the two fluids, or

$$\mathscr{D}^c = 6\pi\mu_G R_i \mathbf{U}_\infty \left[\frac{1 + \frac{2}{3}(\mu_o/\mu_i)}{1 + (\mu_o/\mu_i)}\right]. \tag{3.12}$$

We can readily see that as μ_i becomes very large (for virtually solid particles), eqn. (3.12) reduces to eqn. (3.8). For the case of water droplets in air, the correction to the Stokes' relation amounts to about 0.6 per cent at 20°C. Therefore, for most practical considerations of aerocolloidal systems, one may disregard this correction.

† The very precise measurements reported recently by Pruppacher and Steinberger[(419)] over the range $0.0005 < Re_i < 5$, indicate that Oseen's solution is approached only below $Re_i \simeq 0.0005$. These data show that Oseen's theory for the drag on a sphere is in error by 2 per cent or less below $Re_i \simeq 0.1$, but by as much as 10 per cent in the range $Re_i \simeq 1.0$.

B. Effects of External Forces and of Non-uniformities in the Gas

So far we have considered only the steady motion of a solid or liquid particle that does not interact with an equilibrium gas or is not influenced by body forces. In addition to the resistance of the viscous fluid that acts on a particle, other forces may develop, including those resulting from external fields of forces like electrical fields or gravitational fields.

Settling Velocity in a Gravitational Field

Perhaps the most common example of a calculation involving an external force is the estimate of the terminal settling velocity for small particles in a gravitational field.

For the Stokes' drag given in eqn. (3.8) we can now calculate the settling velocity of small particles in a gas. Balancing the drag $\mathscr{D}^c$ with the buoyancy force acting on the particle,

$$(\rho_i - \rho_G)\mathscr{g}\tfrac{4}{3}\pi R_i^3 = 6\pi\mu_G R_i U_\infty$$

and then the terminal settling velocity becomes

$$U_\infty = v_s^c = \frac{2}{9}\frac{R_i^2(\rho_i - \rho_G)\mathscr{g}}{\mu_G}, \quad \begin{matrix} Re_i \to 0. \\ Kn_i \to 0. \end{matrix} \tag{3.13}$$

Since the particle density will normally exceed the gas density by 10 or more, eqn. (3.13) reduces to

$$v_s^c \approx \frac{2}{9}\frac{R_i^2\rho_i\mathscr{g}}{\mu_G}, \quad \begin{matrix} Re_i < 1, \\ Kn_i \to 0. \end{matrix} \tag{3.14}$$

Very careful and thorough measurements of the settling velocity of particles in liquids agree with predictions of eqn. (3.13) within the error of the experiment.(49) Results for the fall rate of particles of radii of 10 μ–30 μ also have been found to agree within less than 0.5 per cent of the theory.(2) Thus Stokes' formula for the resistance acting on solid particles traveling through a fluid in the range where the Knudsen number for the particle is much less than unity, represents one of the more accurate physical laws in fluid dynamics. The feeling of confidence in this theory has led to widespread application of Stokes' solution to the dynamical behavior of colloidal systems.

Effect of Introducing Walls at a Finite Distance from the Particle

Whenever particles are suspended in fluids in a confined space, the drag on the particles is modified by the confining walls. The increase in resistance for spheres traveling in a bound volume results from a combination of effects stemming from the fact that the boundary conditions at $\mathbf{r} \to \infty$ in Stokes' problem are modified, a fluid flow is induced in a confined space by the passage of particles, and a shearing motion near walls may be present in the fluid. A correction for the Stokes' drag has been worked out for various cases of spheres moving in a continuum near boundaries of different geometries. Happel and Brenner,(41) for example, give analytical solutions based mostly on the method of reflections in the form

$$\frac{\mathscr{D}^c}{6\pi\mu_G R_i U_\infty} = \frac{1}{1 - K(R_i/l) + 0(R_i/l)^3}, \tag{3.15}$$

where K is a constant, and l is the distance from the surface to the center of the particle. Some typical K values are listed in Table 3.1.

It is of interest to note, in particular, that the drag coefficient increases as a particle approaches a plane wall. A solution [41] for the continuum motion of a particle as it moves in a perpendicular direction towards a plane wall indicates that its resistance actually approaches infinity for very small distances away from the surface. Happel and Brenner's

TABLE 3.1. TYPICAL VALUES OF THE COEFFICIENT K IN EQN. (3.15) FOR A SPHERE TRANSLATING PAST A RIGID BODY (FROM HAPPEL AND BENNER[41])

Geometry of wall	Location of sphere center	Direction of motion	K
Circular Cylinder	Cylinder axis	Axial	2.104
Plane parallel walls	Midway between walls	Parallel to walls	1.004
	Half-way to center line between walls	Parallel to walls	0.653
Single plane wall		Parallel to wall	0.563
		Perpendicular to wall	1.125

TABLE 3.2. CORRECTION OF STOKES' DRAG FOR A PARTICLE MOVING PERPENDICULARLY TOWARD A PLANE SURFACE IN A CONTINUUM. CALCULATION BASED ON HAPPEL AND BRENNER'S [41] EQN. (7–4.41).

A. *Solid Plane Surface*

l/R_i	$\mathscr{D}^c/6\pi\mu_G R_i U_\infty$
1	∞
1.54	3.04
3.76	1.41
10.1	1.12
∞	1

B. *Plane Free Surface*

l/R_i	$\mathscr{D}^c/6\pi\mu_G R_i U_\infty$
1	∞
1.54	1.97
3.76	1.25
10.1	1.08
∞	1

results for different values of the ratio l/R_i are shown in Table 3.2. In Table 3.2A are listed the ratio of $\mathscr{D}^c/6\pi\mu_G R_i U_\infty$ for a sphere approaching normally a rigid solid surface. Table 3.2B gives similar results for a sphere falling towards a planar free surface. These results indicate that a sphere can never come into contact with a surface (or another particle) within the continuum theory (see also Section 10.3). Since it is known that particles can actually collide as they come together, the continuum calculations have to be corrected. Actual contact between bodies can be obtained if one takes into account (a) the influences of the non-continuum aspects of the squeezing out of gas as the particles approach each other (for $l/\lambda_G \rightarrow 0$), and (b) possible dominance of electrical or van der Waals' forces as particles approach each other.

Experimentally, the drag on individual spheres in a cloud of particles has been found to increase as particles are packed more closely together. Corrections for the change in resistance of spheres in a dilute cloud have been estimated in the empirical form similar to eqn. (3.15):

$$\mathscr{D}^c = 6\pi\mu_G R_i \mathbf{U}_\infty(1 + K'\phi^n).$$

Here K' and n are coefficients, the order of unity, and ϕ denotes the volume fraction of dispersed phase. Various experimental and theoretical studies of this effect are treated in detail by Happel and Brenner[41] and Fuchs.[2]

Within the Stokes' approximation, the influence of walls on viscous motion of particles is relatively small, as expected, if the ratio of particle size to the characteristic size of the enclosing volume is less than one-tenth.

Aerodynamic forces between neighboring particles must also be taken into account in calculating the relative motion between such bodies. This question is crucial to the calculation of collision rates for falling spheres in a continuum, and is discussed further in Section 10.3.

Motion in a Uniform Electrical Field

If charged particles pass through an electrical field a body force analogous to gravity exerts an influence in the particles. Using a balance of the electrical force and the drag force similar to the one in the previous section, the terminal velocity of a particle in an electric field is given as

$$v_E = Q_i\mathscr{E}_0 B = Q_i\mathscr{E}_0/6\pi\mu_G R_i, \tag{3.16}$$

where Q_i denotes the net charge on the particle, and $\mathscr{E}_0$ is the strength of the electrical field.

Control of particle motion using an electric field has been used extensively some years ago in experiments to determine the amount of charge on aerosols. Particles can be suspended in a vertically oriented electric field indefinitely by balancing the settling velocity due to gravity with an upward directed electric force (induced by properly adjusting the sign of the gradient in electrical potential).

Using eqns. (3.16) and (3.14), we see that the velocity of particles in an electric field can balance the gravitational settling velocity when

$$v_s = v_E$$

or, for the continuum case,

$$\frac{2}{9}\frac{R_i^2 g\rho_i}{\mu_G} = Q_i\mathscr{E}_0/6\pi\mu_G R_i,$$

giving the field required to balance exactly gravity:

$$\mathscr{E}_0 = \frac{4}{3}\frac{\pi R_i^3 g\rho_i}{Q_i}. \tag{3.17}$$

By observing a particle moving in a vertical electric field over range of particle size and pressure, one can determine the value of the mobility over a wide range of particle radii. The settling velocity v_s can be found by switching the electric field off. The fall rate can then be observed for different directions of the field $\mathscr{E}_0$, giving velocities equivalent to

$v_s + v_E$ and $v_s - v_E$. Then v_E can be determined. By charging the particles with various magnitudes of charge, one can vary the charge $Q_i(=z\tilde{e}$, z = number of unit charges $\tilde{e}$). Then B_i can be estimated from eqn. (3.16).

One of the best-known experiments in classical physics involving the behavior of particles in an electrical field is that of Millikan,[50] who observed the fall of oil droplets to determine the value of the unit of charge $\tilde{e}$.

Slippage of Gas Around Particles

The experiments of Millikan and those of others revealed that the mobility of particles followed closely eqn. (3.9) for ratios of the mean free path of the suspending medium λ_G to particle radius R_i approaching zero. However, if either the gas pressure or the particle radius were reduced, significant departure from the form of eqn. (3.9) was observed. The experiments suggested that the apparent viscous drag decreased with increasing values of $Kn_i(= \lambda_G/R_i)$. The increase in mobility under these circumstances was linked to the slip of gas at the surface of the body. In other words, the tangential component of velocity $\mathbf{q}_{r=R_i}$ relative to the surface, may not be zero. Such fluid slip was mentioned in Chapter 2, and can arise in different ways as we shall see later in Chapter 6. The "slippage" of gas along a surface is inherently a non-continuum effect; thus, its theoretical analysis involves conditions where the parameter Kn_i is finite.

The experimental work at the turn of this century indicated that for small values of Kn_i the mobility relation should be corrected empirically by the relation

$$B_i = (1 + \tilde{A}Kn_i)/6\pi\mu_G R_i, \tag{3.9a}$$

where $\tilde{A}$ is assumed a constant. The coefficient on the right-hand side, $1 + \tilde{A}Kn_i$, is called the *Stokes–Cunningham correction.*

Using eqn. (3.9a) we may use Millikan's scheme above to determine mobilities of particles as a function of Knudsen number. Following Fuchs,[2] if we take the square root of the gravitational settling velocity with the Stokes–Cunningham correction, multiply by v_E and solve for Q_i, we find that

$$\begin{aligned}\tilde{e} = Q_i &= \left[\frac{18\pi v_E}{\mathscr{E}_0}\left(\frac{\mu_G^3 v_s}{2\rho_i g}\right)^{\frac{1}{2}}\right](1 + \tilde{A}Kn_i)^{-\frac{3}{2}} \\ &= Q_i'(1 + \tilde{A}Kn_i)^{-\frac{3}{2}}.\end{aligned} \tag{3.18}$$

Here Q_i' is the charge on the particle assuming Stokes' law applies. In the case of oil droplets, all of the terms in brackets may be evaluated independently by the experiment. If the charge on the droplets is varied a small amount, corresponding to a unit or two or charge by radiation from X-rays or γ-rays, then

$$\Delta Q_i = (z_1 - z_2)\,\tilde{e},$$

where $(z_1 - z_2)$ is most frequently ± 1, less often ± 2, etc. Similarly, the change in the term in brackets containing v_E of eqn. (3.18), Q_i', is given as $(z_1 - z_2)\,\tilde{e}'$. By taking the highest common factor for several ΔQ_i's the value of the unit of charge assuming Stokes' law $\tilde{e}'$ can be evaluated.

Then, from eqn. (3.18),

$$\tilde{e}'^{\frac{2}{3}} = \tilde{e}^{\frac{2}{3}}(1 + \tilde{A}Kn_i). \tag{3.19}$$

If values of $\tilde{e}'^{\frac{2}{3}}$ are plotted for various observations as a function of Kn_i for small $\tilde{\mathbf{A}} \cdot Kn_i$, a straight line should appear, according to eqn. (3.19). The slope of the line should be $\bar{A}\tilde{e}^{\frac{2}{3}}$ and the intersection with the ordinate axis at $Kn_i = 0$ should be $\tilde{e}$. Essentially by this method the value of the electronic charge, and the validity of the Stokes–Cunningham correction for small Kn_i were first determined accurately.

The methods described above were applied over a range of Kn_i much wider than the small values involved in the Stokes–Cunningham correction. Using droplets of 5 μ radius or larger to avoid fluctuations by Brownian motion, the resistance of particles was measured in gases up to $Kn_i = 134$. The experimental results indicated that the Stokes–Cunningham correction applied with $\tilde{A}$ = constant only up to $Kn_i \approx 1$. Beyond this range, the constant $\tilde{A}$ was found to depend on Kn_i[50, 51]

$$\tilde{A} = \hat{A} + \hat{B}\exp(-\hat{C}Kn_i^{-1}), \tag{3.9b}$$

where $\hat{A}$, $\hat{B}$ and $\hat{C}$ are constants, given, for example, by Davies[52] as

$$\hat{A} = 1.257, \quad \hat{B} = 0.400, \quad \hat{C} = 1.10.$$

Theoretical estimations of the correction for Knudsen number involve application of the non-continuum theory of gases spanning a transitional regime where $0.5 < Kn_i \leqslant 10$, and the free molecule regime in which $Kn_i > 10$. Calculation of the constants in eqn. (3.9b) based on theoretical arguments are described in Sections 5.3 and 6.2.

Non-uniformities in the Suspending Gas

When temperature gradients or gradients in concentration of chemical components of the gas exist, aerosol particles have been observed to experience forces directly associated with these gradients. The aerosol motion resulting from gaseous non-uniformities became known to physicists toward the end of the nineteenth century. Particles were observed, for example, to be repelled by heated surfaces. And so-called "dust free layers" were found in air located very near surfaces wetted with a volatile material even though no marked temperature difference existed between the surface and the surrounding gas.

The particle drift in gases induced by thermal effects involves *radiometric forces*. Thermally induced forces acting on aerosol particles arise from the non-uniform heating of a body, either caused by temperature gradients in the suspending gas, or by absorption in the particle of incident radiation. The former case results in motion of particles called *thermophoresis*, while the latter gives rise to a movement called *photophoresis*. Both of these phenomena depend in one way or another on departures from the viscous boundary condition given by Stokes that the fluid velocity relative to the particle surface, at the surface, is zero. Thus in the limit $Kn_i = 0$, no thermal force can develop. Unless there is a condition of thermal slip, the thermophoretic effect or photophoretic effect will not appear, as illustrated by the remarks in Chapter 6. Analysis of the radiometric forces for the regime where $Kn_i \gtrsim 0.25$ is described in Chapter 5.

In contrast to the non-continuum requirements for the radiometric forces, particles can experience a motion in gas mixtures with concentration gradients both by continuum and non-continuum effects. When gradients in concentration of gaseous components exist, a mass flow of one or more components may result from different mean velocities of molecules in the mixture relative to the mass average flow of the gas. This mean flow, called *Stefan flow*, produces an impulse on particles suspended in the gas which is essentially a continuum effect (see also Section 4.4). In analogy to slippage of gas along a particle resulting from temperature gradients in non-continuum regimes, a similar slippage can develop from gradients in concentration of gaseous components. Like the radiometric forces, this latter effect can develop only when $Kn_i > 0$. Drift of particles in a non-uniform, isothermal gas mixture by Stefan flow, or by slippage of gas molecules at the particle surface has been called *diffusiophoresis* in the literature. Such a drift has been involved in reducing the number of aerosol particles in the neighborhood of an evaporating droplet or liquid, causing a "dust free space" to be observed. Because diffusiophoresis by Stefan flow is closely related to heat and mass transfer to spheres, this effect will be described in more detail in the next chapter. The non-continuum aspects of diffusiophoresis are treated in Chapters 5 and 6.

3.2. Accelerated Motion of Particles in a Moving Medium

A. Generalized Equations of Motion

When an aerosol particle travels in a moving medium the particle generally will tend to lag behind the flow of the fluid. It is customary† to assume that the Stokes relation for the particle drag may be used for the unsteady movement of particles in such a system, so that the frictional drag is given as

$$\mathscr{D}^c = 6\pi\mu_G R_i(\mathbf{q} - \mathbf{q}_i). \tag{3.20}‡$$

Then within the continuum approximation the particle inertial force may be written as[55]

$$\frac{4}{3}\pi R_i^3 \rho_i \frac{d\mathbf{q}_i}{dt} = 6\pi\mu_G R_i(\mathbf{q} - \mathbf{q}_i) + \frac{4}{3}\pi R_i^3 \rho_G \frac{d\mathbf{q}}{dt} + \frac{2}{3}\pi R_i^3 \rho_G \times \left(\frac{d\mathbf{q}}{dt} - \frac{d\mathbf{q}_i}{dt}\right) + 6R_i^2\sqrt{(\pi\rho_G\mu_G)} \int_{t_0}^{t} \frac{dt'\,[(d\mathbf{q}/dt') - (d\mathbf{q}_i/dt')]}{(t - t')^{\frac{1}{2}}} + \mathbf{F}_e + \mathbf{F}_{NU}, \quad Kn_i \to 0,\; Re_i \ll 1. \tag{3.21}$$

In eqn. (3.21), t_0 denotes the starting time for particle acceleration, and $\mathbf{q}$, as before, represents the gas velocity near the particle, but far enough away not to be disturbed by the relative motion of the body ($\mathbf{q} = \mathbf{U}_\infty$).

The terms in eqn. (3.21) have the following meanings: the term on the left side is the resultant force acting on the particle. The first term on the right is the viscous resistance given by Stokes' law. The second term is due to the pressure gradient in the gas surrounding the particle, caused by acceleration of the gas by the particle. The third term denotes the force required to accelerate the apparent mass of the particle relative to the ambient gas. And the fourth term, *the Basset term*, accounts for the effect of the deviation in gas flow pattern from steady state. The last two terms are the forces resulting from external potential fields $\mathbf{F}_e$ and from non-uniformities in the suspending medium $\mathbf{F}_{NU}$. In the case where $Kn_i \to 0$, the only force involved in this term will be that associated with the Stephan current in a diffusing mixture of gases (see also Section 4.2).

In general, most environmental conditions applicable to aerosols will be such that the

† See, for example, Oseen[13] or Basset.[53] Levich[54] has given a further justification for the use of the Stokes' drag given in eqn. (3.8).

‡ To simplify the formulas, the superscript c is dropped in subsequent discussion in this chapter, with the understanding that the formulas developed apply to the continuum regime.

second, third, fourth and sixth terms on the right of eqn. (3.12) will be negligible. However, the Basset term, which increases the effective resistance in the particle, may become important if the particle is accelerated suddenly to a high velocity by strong external forces. For aerosol systems, Fuchs[(2)] and Hinze[(55)] have indicated that this term normally remains small compared with the Stokes' resistance.

Particle Motion in a Turbulent Medium

The motion of aerosols in a turbulent medium will generally depend on the particle concentration as well as the particle density, shape, and size relative to the characteristic (length) scale of turbulence in the gas. If the particle concentration is large, there will be direct interaction between the particles through collisions and through the influences of particles on flow of gas near the particles (see also Sections 7.4 and 10.2). For example, Hinze[(55)] has suggested that the motion of particles relative to the fluid may create additional turbulence, and cause added dissipation of kinetic energy in the fluid, thus damping some of the turbulent agitation initially present. If, however, the particle concentration is very low, the interference of neighboring particles may be neglected, and the body may be treated as if it were alone in the medium.

When the particle is large compared with the scale of turbulence, defined, for example, by eqn. (7.161), the principal influence of turbulence on the particle will be the increase of its apparent flow resistance. In this case, the particle will tend to follow the slower, large scale motions (eddies) in the fluid. Alternatively, if the particles are very small compared with the smallest scale of turbulence, their motion will be influenced by all components of the turbulence in the gas. The latter situation frequently characterizes the conditions of aerosol motion in turbulent gas streams.

Since the flow resistance of a suspended particle is largely viscous in nature, the motion of a single body will be coupled to the fluid behavior, especially for the smaller sized eddies of turbulence. The presence of a particle will increase viscous dissipation in the turbulent medium. The additional dissipation will be caused by the ever present lag in turbulent movement between the particle and the fluid. Since this lag increases with higher frequency agitation, i.e. higher wave number components of the spectrum of kinetic energy in the turbulent fluid, one expects that the suspension of particles may produce changes in the high wave number range of the spectrum[(55)] (see also Kuchanov and Levich[(56)]).

Very little is known about the interaction between particle motion and the turbulent agitation in the surrounding fluid. Most work has been devoted to finding conditions where the particle motion in a turbulent medium can be specified by a linear equation of the form of (3.21). Generally, the balance of forces on an aerosol moving in a "turbulent" manner in a turbulent fluid will be much more complicated than (3.21) because of the random nature of particle and fluid velocity fields, as well as the relative velocity $\mathbf{q}_i - \mathbf{q}$ which will vary irregularly both in time and in space. It is possible, however, to derive a form equivalent to eqn. (3.21) for the acceleration of spherical particles (where $Re_i \to 0$ and $Kn_i \to 0$) within the following set of assumptions:

(1) The turbulence in the fluid is homogeneous and steady (see also Section 7.4).
(2) The domain of turbulence is infinite in extent.
(3) The particle is smaller than the smallest (wave) length characterizing turbulent eddies, but large enough to disregard any displacement by Brownian motion.
(4) External forces acting on the particle originate from a potential field.

(5) No forces associated with gaseous non-uniformities other than velocity gradients are considered.

(6) During the motion of the particle, the medium surrounding the particle will consist of the same fluid elements.

Hinze[55] has noted at the outset that the first four assumptions are physically reasonable, and may actually be satisfied. However, the sixth assumption can never be realized in a real field of turbulence. If fluid elements containing a small particle could be isolated as an undeformable entity, then it might be feasible to satisfy assumption (6) provided the fluid element was larger than the amplitude of motion of the suspended body relative to the fluid (no overshooting). It is known that fluid elements undergoing turbulent motion are twisted and stretched into long thin threads. Thus it is not reasonable to expect the particle to maintain a static environment in this stretching process. Moreover, negligible overshooting requires that the density ratio ρ_G/ρ_i approaches unity. Therefore, the application of the following idealized model of particle acceleration in a turbulent medium should be considered with caution.

We may begin by assuming that Eqn. (3.21) applies at any instant to the motion of an aerosol in the turbulent gas. Corrsin and Lumley[57] first suggested that eqn. (3.21) could be written for motion in a turbulent medium, especially if the second term on the right takes care of spatial inhomogeneities in fluid velocity $\mathbf{q}$. This term comes from the pressure drop in the ambient gas, given by the Navier–Stokes equation as

$$-\nabla p = \rho_G\left[\frac{\partial \mathbf{q}}{\partial t} + \mathbf{q}\cdot\nabla\mathbf{q}\right] - \mu_G\nabla^2\mathbf{q}. \tag{3.22}$$

Following the discrete particle, we further note that the operator

$$\frac{\mathrm{d}}{\mathrm{d}t} = \frac{\partial}{\partial t} + \mathbf{q}_i\,.\,\nabla \tag{3.23}$$

holds since we are following a discrete particle.

Making use of the relation

$$\frac{\partial \mathbf{q}}{\partial t} + \mathbf{q}\,.\,\nabla\mathbf{q} = \frac{\mathrm{d}\mathbf{q}}{\mathrm{d}t} + [\mathbf{q} - \mathbf{q}_i]\,.\,\nabla\mathbf{q}, \tag{3.24}$$

Eqn. (3.21) may be rearranged with eqns. (3.23)–(3.24) to read

$$\frac{\mathrm{d}\mathbf{q}_i}{\mathrm{d}t} = \frac{3\rho_G}{2\rho_i + \rho_G}\left[\frac{\mathrm{d}\mathbf{q}}{\mathrm{d}t} - \frac{2}{3}\nu_G\nabla^2\mathbf{q}\right] + \frac{2}{2\rho_i + \rho_G}\left\{\frac{9\mu_G}{2R_i^2}[\mathbf{q} - \mathbf{q}_i]\right.$$
$$\left. + \rho_G[\mathbf{q} - \mathbf{q}_i]\,.\,\nabla\mathbf{q}\right\} + \frac{9}{(2\rho_i + \rho_G)R_i}\sqrt{\left(\frac{\rho_G\mu_G}{\pi}\right)}\int_{t_0}^{t}\mathrm{d}t'\,\frac{(\mathrm{d}/\mathrm{d}t')\,[\mathbf{q} - \mathbf{q}_i]}{(t - t')^{\frac{1}{2}}} + \mathbf{F}_e. \tag{3.25}$$

Here ν_G is the kinematic viscosity of the gas μ_G/ρ_G. Hinze[55] has noted that this equation loses its non-linear character only if

$$\left|\rho_G\frac{\partial \mathbf{q}}{\partial \mathbf{r}}\right| \ll \frac{\mu_G}{4R_i^2}, \text{ or } \left|\frac{4R_i^2}{\nu_G}\frac{\partial \mathbf{q}}{\partial \mathbf{r}}\right| \ll 1. \tag{3.26a}$$

This condition is met if the particle is small, or if $|\partial\mathbf{q}/\partial\mathbf{r}^{\oplus}| \ll 1$, where $\mathbf{r}^{\oplus} = \mathbf{r}/R_i$. In addition, the equation remains first order and spatially independent only if the viscous part of the first term on the right side can be disregarded. This requires that

$$\left|\mathbf{q}_i \frac{\partial \mathbf{q}}{\partial \mathbf{r}}\right| \gg \left|v_G \frac{\partial^2 \mathbf{q}}{\partial \mathbf{r}^2}\right| \quad \text{or} \quad \left|\frac{\mathbf{q}_i(\partial \mathbf{q}/\partial \mathbf{r})}{v_G(\partial^2 \mathbf{q}/\partial \mathbf{r}^2)}\right| \gg 1. \tag{3.26b}$$

From the restrictions of (3.26 a and b), the ratio

$$\left|\frac{\mathbf{q}_i}{4R_i^2(\partial \mathbf{q}/\partial \mathbf{r})^2}\right| \ggg 1, \tag{3.27}$$

or in terms of $\mathbf{r}/R_i$, $|\partial^2\mathbf{q}/\partial\mathbf{r}^{\oplus 2}|$ must be very small.

Within the limitations of relations (3.26)–(3.27) and the assumptions (1)–(6), eqn. (3.25) resumes a linear character in time dependent terms only, which is equivalent to eqn. (3.21).

Summary of Equations

Noting that $\rho_i \gg \rho_G$ in aerosol systems, eqn. (3.21) can then be rewritten as

$$\frac{d\mathbf{q}_i}{dt} + \mathscr{A}(\mathbf{q}_i - \mathbf{q}) = \mathscr{B}\frac{d\mathbf{q}}{dt} + \mathscr{C}\int_{t_0}^{t} dt' \frac{(d/dt')[\mathbf{q} - \mathbf{q}_i]}{(t - t')^{\frac{1}{2}}} + \mathbf{F}'_e + \mathbf{F}'_{\mathrm{NU}}, \tag{3.28}$$

where $\mathbf{F}' = \mathbf{F}/m_i$, and the coefficients are:

$$\mathscr{A} = \frac{9\mu_G}{2\rho_i R_i^2}, \quad \mathscr{B} = \frac{3\rho_G}{2\rho_i}, \quad \mathscr{C} = \frac{9}{2\rho_i R_i}\sqrt{\left(\frac{\rho_G \mu_G}{\pi}\right)}$$

In analogy to the well known Langevin equation in the theory of Brownian motion (see also eqn. 7.29), we can write eqn. (3.28) as (neglecting $\mathbf{F}_{\mathrm{NU}}$):

$$\frac{d\mathbf{q}_i}{dt} + \mathscr{A}(\mathbf{q}_i - \mathbf{q}) = \hat{\mathbf{F}}'(t) + \mathbf{F}'_e \qquad \begin{matrix} Re_i \to 0, \\ Kn_i \to 0, \end{matrix} \tag{3.29}$$

where $\hat{F}'(t)$ denotes an "external" perturbation function (force per unit mass) acting on the accelerating particle. In eqn. (3.29) the terms including coefficients $\mathscr{B}$ and $\mathscr{C}$ have been incorporated into the perturbation function $\hat{F}'(t)$. This expression may be applied either to particles carried along in a laminar flow or in a turbulent flow of suspending gas, provided one bears in mind the theoretical limitations discussed in relation to turbulence.

We note that the factor $\mathscr{A}$ is the reciprocal of the product $B_i m_i$. This parameter has dimensions of (time)$^{-1}$. Thus the product $B_i m_i$ is a characteristic time scale, as noted in the next section.

For a system of cartesian coordinates eqn. (3.29) has the following components:

$$x \text{ component: } \frac{d\,q_{ix}}{dt} + \mathscr{A}(q_{ix} - q_x) = \hat{F}'_x + F'_{ex}. \tag{3.29a}$$

$$y \text{ component: } \frac{d\,q_{iy}}{dt} + \mathscr{A}(q_{iy} - q_y) = \hat{F}'_y + F'_{ey}. \tag{3.29b}$$

$$z \text{ component: } \frac{d\,q_{iz}}{dt} + \mathscr{A}(q_{iz} - q_z)\,\hat{F}'_z + F'_{ez}. \tag{3.29c}$$

These equations provide the essential framework for development of the mechanical theory for aerosols. Various classes of solutions to these relations are of primary importance to design and development of devices for analysis and removal of aerosols from gases.

B. Relaxation Time to Reach Steady Motion

One of the important questions in the theory of aerosols bears on the characteristic time required for particles to reach steady motion. To find such a time, we can apply the results of two simple mechanical problems analyzed in terms of the solutions to eqn. (3.29). Perhaps the simplest problem is the deceleration of a particle in a stationary gas without the influence of external forces. Assuming that the particle travels in the x direction, and $F_{NUx} = 0$ or $F_{ex} = 0$, eqn. (3.29a) becomes

$$\frac{\mathrm{d}\, q_{ix}}{\mathrm{d}t} + \mathscr{A}\mathbf{q}_{ix} = 0 \tag{3.30}$$

when there is no perturbation function $\hat{F}'(t)$. For the initial condition, $q_{ix} = q_0$ at $t = 0$, the solution of eqn. (3.30) is

$$q_{ix} = q_0 \exp(-\mathscr{A}t), \tag{3.31}$$

and the distance traveled by the particle is

$$x = \int_0^t q_{ix}\, \mathrm{d}t' = \mathscr{A}^{-1} q_0[1 - \exp(-\mathscr{A}t)]. \tag{3.32}$$

A similar problem involves calculating the time required to reach terminal speed in an external force field. Assuming that the force F_e only has one component F_{ez}, a constant, the terminal speed of a particle moving in the z direction will be given as a solution of (3.29c) with the initial condition $q_{iz} = 0$ at $t = 0$:

$$q_{iz} = v_s[1 - \exp(-\mathscr{A}t)], \tag{3.33}$$

where $v_s = 3\mathscr{A}^{-1}F_{ez}/4\pi\rho_i R_i^3$. If the external force is the Earth's gravitational field, $v_s = g\mathscr{A}^{-1}$. The distance traveled by the particle before reaching its terminal velocity is

$$z = \int_0^t q_{iz}\, \mathrm{d}t' = v_s t + v_s\mathscr{A}^{-1}(e^{-\mathscr{A}t} - 1). \tag{3.34}$$

The significance of the coefficient $\mathscr{A}^{-1}$ now becomes clear. This constant may be identified as the characteristic relaxation time for the particle which we have been seeking. In the first of the two simple problems cited above, the velocity has slowed to $1/e$ its initial value at $t = 1/\mathscr{A}$. In the second case, the particle velocity differs from v_s by $1/e$ by $t = 1/\mathscr{A}$. The introduction of the coefficient $1/\mathscr{A}$ then represents a "natural" parameter for the general behavior of aerosol mechanics. This also may be seen easily in complex solutions to the components of acceleration taken from the vector equation [eqn. (3.29)], and its components eqns. (3.29a)–(3.29c). As an example, for the z component, assuming that $\mathbf{F}_{NU} = 0$ and $\mathbf{F}'_{ez} = g$, Tchen[58] has given the general solution for $q_{iz}(t)$;

$$q_{iz}(t) = \frac{1}{\theta}\int_0^\infty \exp(-\mathscr{K}\xi)\sin(\theta\xi)\underline{\Phi}(t - \xi)\, \mathrm{d}\xi, \tag{3.35}$$

where

$$\mathscr{K} = \mathscr{A}(1 - \tfrac{3}{2}\mathscr{B}) \approx \mathscr{A} \tag{3.35a}$$

$$\theta^2 = \mathscr{A}^2 - \mathscr{K}^2 = \pi\mathscr{A}\mathscr{C}^2(1 - \tfrac{3}{4}\mathscr{B}) \approx \frac{9\rho_G\mathscr{A}^2}{2\rho_i} \tag{3.35b}$$

$$\underline{\Phi}(t) = \mathscr{A}^2[g\mathscr{A}^{-1} + q_z(t)] + \mathscr{A}(1 - 2\mathscr{B})\frac{dq_z}{dt} + \mathscr{B}\frac{d^2q_z}{dt^2} - \mathscr{C}(\mathscr{B} - 1)\int_0^\infty \frac{[d^2q_z/dt^2(t - \tau)]\,d\tau}{\sqrt{\tau}} \tag{3.35c}$$

$$\approx \mathscr{A}g + \mathscr{A}^2 q_z + \mathscr{A}\frac{dq_z}{dt} + \frac{3\rho_G}{\rho_i}\frac{d^2q_z}{dt^2} + \sqrt{\frac{9\rho_G\mathscr{A}}{2\rho_i\pi}}\int_0^\infty \frac{(d^2q_z/dt^2)(t - \tau)\,d\tau}{\sqrt{\tau}}. \tag{3.35d}$$

We note from Tchen's analysis that the general dynamical behavior of a particle then should depend on the purely physical and geometrical parameters, including the scales of the relaxation time, $\mathscr{A}^{-1}$, the ratio of the particle density, and the gas density ρ_i/ρ_G. In addition, the particle will respond to time varying perturbations stemming from the dynamics of the gas surrounding the particle. These perturbations will depend on the fluid velocity, the fluid acceleration, and the rate of change in fluid acceleration in the neighborhood of the particle.

The Significance of the Basset Term

For 10 μ radius particles of unit density in air at 20°, $\mathscr{A}^{-1} = 1.2 \times 10^{-3}$ sec, and $\rho_i/\rho_G \cong 10^3$. Hence only under conditions of very rapid changes in the velocity of the suspending medium do the particles feel a significant increase in apparent resistance from the Basset term, and the term containing the change in fluid accelerations.

The significance of the Basset term also can be estimated for the limiting case where the suspending medium is stationary far from the particle (i.e. $\mathbf{q} = 0$, $r \to \infty$). For the case where the particle is initially at rest, Boggio[59] has found a general solution to eqn. (3.29c), which reads

$$q_{iz}(t) = \frac{\mathscr{A}}{\Lambda_1 - \Lambda_2}\int_0^{t^\oplus} \Psi(\tau^\oplus)\{\exp[\Lambda_1(t^\oplus - \tau^\oplus)] - \exp[\Lambda_2(t^\oplus - \tau^\oplus)]\}\,d\tau^\oplus + F'_{ez}(0)/\mathscr{A}(\Lambda_1 - \Lambda_2)\exp(\Lambda_1 t^\oplus) - F'_{ez}(0)/\mathscr{A}(\Lambda_1 - \Lambda_2)\exp(\Lambda_2 t^\oplus), \tag{3.36}$$

where Λ_1 and Λ_2 are the roots of the characteristic equation $\Lambda^2 + (2 - \sqrt{(9\rho_G/2\rho_i)})\Lambda + 1 = 0$. For small values of $\sqrt{(9\rho_G/2\rho_i)} = \beta_0$, the roots may be approximated by

$$\Lambda_1 = -1 + i\sqrt{(9\rho_G/2\rho_i)}; \quad \Lambda_2 = 1 - i\sqrt{(9\rho_G/2\rho_i)}.$$

Then eqn. (3.36) becomes, for constant $F'_{ez} = g$:

$$q_{iz}(t^\oplus) = v_s\left\{1 - \exp(-t^\oplus)\left[\cos(\beta_0 t^\oplus) - \frac{\beta_0}{2}\sin(\beta_0 t^\oplus) + \frac{1}{\sqrt{\pi}}\sin(\beta_0 t^\oplus)\int_0^{t^\oplus}\frac{\exp(\tau^\oplus)\cos(\beta_0\tau^\oplus)\,d\tau^\oplus}{\sqrt{\tau^\oplus}} - \frac{1}{\sqrt{\pi}}\cos(\beta_0 t^\oplus)\int_0^{t^\oplus}\frac{\exp(\tau^\oplus)\sin(\beta_0\tau^\oplus)\,d\tau^\oplus}{\sqrt{\tau^\oplus}},\right]\right\} \tag{3.37}$$

where

$$t^{\oplus} = \mathscr{A}t \quad \text{and} \quad \tau^{\oplus} = \mathscr{A}\tau.$$

Fuchs[(2)] has calculated numerically $q_{iz}(t^{\oplus})$ $(= \mathscr{A}^{-1}q_{ix})$ from (3.37) and compared these values with the result without including the Basset term. Fuch's result for particles in air 0°C and 1 atm, is shown in Fig. 3.4. Without the Basset term, the solution for the velocity $q_{iz}(t^{\oplus})$ given in eqn. (3.37) reduces to

$$q_{iz}(t^{\oplus}) = v_s[1 - \exp(-t^{\oplus})].$$

This expression is shown by the dotted line in Fig. 3.4. The solid line in this figure corresponds to the exact solution as given by eqn. (3.37). From comparison of these curves, we find that the Basset term causes only a few per cent maximum deviation between the approximate result and that of eqn. (3.37). The reduction in velocity, $q_{iz}(t^{\oplus})$, depends only on β_0,

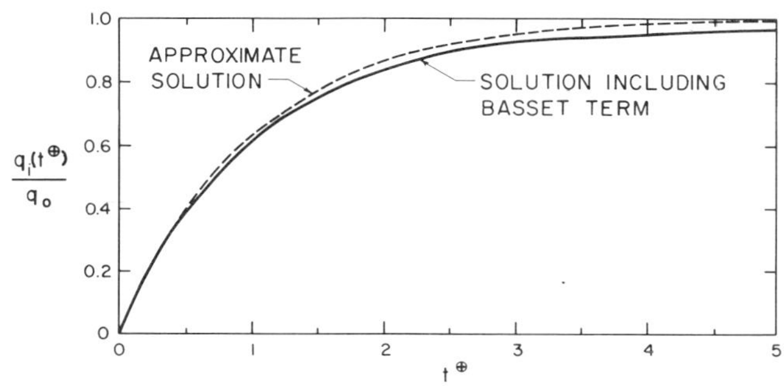

FIG. 3.4. Effect of the Basset term on two simple problems of a particle motion. (a) Approach to terminal velocity of a falling sphere q_i/q_0, (b) Deceleration of a particle in a viscous fluid, $1 - q_i/q_0$. Here, $q_0 = v_s$, or the initial particle velocity. (From Fuchs,[(2)] p. 74.)

which is proportional to the ratio of densities of the particle and the fluid. Neither the particle size nor the gas viscosity have an influence here. Thus, for particles in gases, the deviation in the velocity $q_{iz}(t^{\oplus})$ calculated with or without the Basset term is negligible for most practical purposes. A similar conclusion may be drawn for oscillatory motion of a particle in a gas. The resistance of the gas on the particle can be considered very nearly non-inertial, and equivalent to the Stokes' values determined for conditions of steady flow.

The case of particle deceleration during translation also can be solved with the Basset term included. If the velocity difference $(q_0 - q_{iz}(t))$ is introduced in place of $q_{iz}(t)$, and $g(= F'_{ez})$ is replaced by q_0, the initial particle velocity, a relation equivalent to eqn. (3.29c) may be written for the particle motion:

$$\frac{d(q_0 - q_{iz})}{dt^{\oplus}} + (q_0 - q_{iz}) - q_0 + \left(\frac{9\rho_G}{2\pi\rho_i}\right)^{\frac{1}{2}} \int_0^t \frac{d(q_0 - q_{iz})\, d\tau^{\oplus}}{d\tau^{\oplus}(t^{\oplus} - \tau^{\oplus})} = 0. \tag{3.38}$$

With the initial condition $[q_0 - q_{iz}(t^{\oplus})] = 0$ at $t^{\oplus} = 0$, eqn. (3.37) becomes a solution to eqn. (3.38) and the deceleration problem. By replacing the ordinate in Fig. 3.4 with the variable $[(1 - q_{iz}(t^{\oplus}))/q_0]$, we obtain from the figure the exact time required to slow a particle down to an arbitrary velocity along with the time based on the approximation of neglecting the Basset term. In the case of deceleration; however, the error in the particle velocity introduced by setting the Basset term zero is larger in the sense that the time required to halt the moving particle may be much larger than the approximate estimate. This is readily seen in Fig. 3.4 wherein the tail of the exact solution as $t^{\oplus}$ gets large and never reaches unity, while the approximate solution goes asymptotically to one by $t^{\oplus} \simeq 5$.

Fuchs[2] has noted that the stopping distance,

$$L_s^c = \mathscr{A}^{-1} q_0 = \frac{2 q_0 R_i^2 \rho_i}{9 \mu_G}, \tag{3.39}$$

given as an approximation as the area under the dotted line curve, would be much less than the more accurate area under the solid curve. For practical purposes, however, the distance given by eqn (3.39) is normally taken for estimating the distance traveled by a particle before it reaches exceedingly slow creep in a stagnant fluid.

It may be seen from Fig. 3.4 that after the time exceeds $\mathscr{A}^{-1}$ by a substantial amount, the particle velocity is approximately that of the terminal settling velocity. In addition, we note that the second term in eqn. (3.34) essentially becomes zero. Hence, the particle travels a distance equivalent only to $v_s t$ after $(\mathscr{A}^{-1} t) \gg 1$. This idea is often used for calculating the behavior of particles in a force field after times substantially surpassing the relaxation time, $\mathscr{A}^{-1}$. This also implies that in the absence of an external field, the particle will travel with the suspending gas after $\mathscr{A}^{-1} t \gg 1$. In other words, this restriction indicates no relative movement between a gas in laminar flow and the cloud of aerosol particles.

There are some cases where the particle velocity has been found proportional to the external force $\mathbf{F}_e$ acting on the particle. A relation analogous to eqn. (3.8) holds where

$$\mathbf{q}_i(t) = B_i \mathbf{F}_e(t) \tag{3.40}$$

in which the coefficient of proportionality is the mobility B_i. This class of particle motion has been named *quasi-stationary*[2] (see also Chapter 4).

C. Curvilinear Motion of Particles

The estimation of trajectories of particles traveling in curvilinear motion can be carried out, in principle, by integration of eqn. (3.29). Steady curvilinear motion of aerosols develops in many circumstances. Some cases include (a) particle flow in horizontally oriented streams

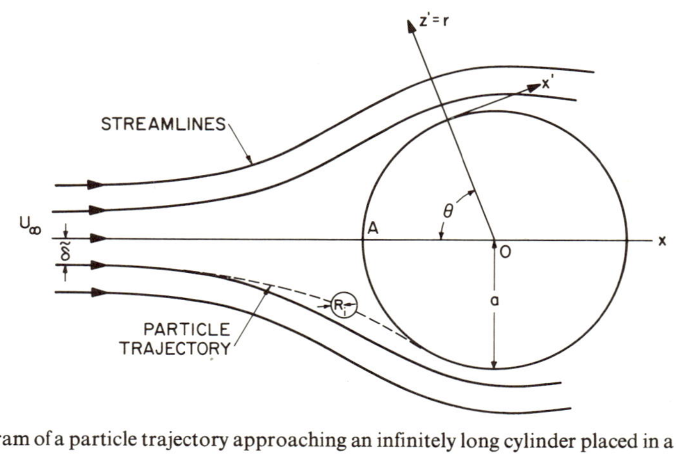

FIG. 3.5. Diagram of a particle trajectory approaching an infinitely long cylinder placed in a uniform flow.

of gas in a gravity field, (b) flow around obstacles, and (c) spinning motion in cyclones or centrifuges. These three cases are of great practical importance because of their relevance to the design of dust removal devices.

Because of inertia, a trajectory of a particle having a mass greatly exceeding the molecules of surrounding gas will depart from the path of the gas molecules in curvilinear flow. A typical example of such differences in flow patterns is drawn for the case of flow around a cylinder in Fig. 3.5. Typical paths of aerosol particles are shown in dotted lines. These are compared with the paths of fluid particles drawn in solid lines.† The departure of the particles from streamline flow may be sufficient to allow the particle to deposit on the curved surface.

In general the trajectory of any particle in the field of gas flow may be estimated by integrating the equation of motion for the particle over a time period defined in terms of the ratio of distance of travel to the particle velocity $|\mathbf{r}/\mathbf{q}_i|$. If the particles are large enough to disregard Brownian motion, eqn. (3.29) may be applied.‡ Such integrations of eqn. (3.29) require, of course, the knowledge of the flow field of the suspending gas, and one must assume that the particle velocity equals the fluid velocity at some distance $\mathbf{r}$ far from the collecting body.

Where there is concern for deposition of particles on surfaces, one seeks the limiting or critical trajectory of particles that will just touch the collecting body. When particles are removed by the influence of gravity during horizontal flow in a channel, the limiting trajectory would be the one that hits the bottom at the end of the channel. For a curved surface like the cylinder the critical trajectory is one just tangent to the surface as indicated in Fig. 3.5. A number of calculations to specify such limiting trajectories have been performed for surfaces of different geometries. We shall illustrate these results by considering only the case of flow past flat surfaces, and flow around cylinders or spheres.

The application of the theory of curvilinear motion to particle sampling by centrifugation will not be discussed. The use of centrifugal devices and cyclones has been treated in some detail by Fuchs[(2)] and Green and Lane.[(4)]

Oscillation of Particles in a Varying External Field of Force

To illustrate the use of eqn. (3.29) for calculating particle paths, let us begin by considering the case of forced oscillations of a particle in a stagnant medium. Suppose we want to estimate the velocity and trajectory of a particle falling through a horizontally oriented, periodically varying field of force. Disregarding the effect of the inertial terms, the vertical component of velocity is

$$q_{iz} = g\mathscr{A}^{-1}, \tag{3.41}$$

and the horizontal component is given in terms of the horizontal acceleration

$$m_i \frac{\mathrm{d}\, q_{ix}}{\mathrm{d}t} = F_{ex}(t) - \frac{q_{ix}}{B_i} \tag{3.42}$$

For a sinusoidally varying force,

$$F = F_0 \sin(2\pi \tilde{f} - \phi),$$

† Fluid particle paths generally trace the movement of the same identifiable parcel of fluid, while streamlines do not. However, fluid particle paths and streamlines remain the same in cases of steady fluid flow (see, for example, Prandtl and Tietjens[(60)]).

‡ In all of the subsequent calculations in this section, Brownian motion of particles is neglected.

where F_0 is a constant, $\tilde{f}$ is the frequency of oscillation, and ϕ is the phase angle. The solution for the horizontal component of velocity reads

$$q_{ix} = \frac{F_0 B_i}{[1 + (\mathscr{A}^{-1} 2\pi \tilde{f})^2]^{\frac{1}{2}}} \{\sin (2\pi \tilde{f} t - \phi) + \exp(-\mathscr{A} t) \sin \phi\}, \tag{3.43}$$

where the phase angle ϕ, is given by $\tan \phi = \mathscr{A}^{-1} f$.

The transient term, the second term within the brackets, decays quickly to zero, and a steady oscillation develops as

$$q_{ix} = q_{F_0} \sin (2\pi \tilde{f} t - \phi), \tag{3.44}$$

where q_{F_0} is given by the coefficient in eqn. (3.43). In the first approximation then, the amplitude and phase of oscillation are determined by the mobility B_i and the "viscous" relaxation time $\mathscr{A}^{-1}$ of the particle.

Equation (3.44) may be integrated again to find the horizontal displacement of the particle for the condition of steady oscillation:[2]

$$x = -\frac{F_0 B_i \cos (2\pi \tilde{f} t - \phi)}{2\pi \tilde{f} [1 + (2\pi \tilde{f} \mathscr{A}^{-1})^2]^{\frac{1}{2}}}. \tag{3.45}$$

The vertical displacement at time t is found from integration of (3.41), yielding

$$z = \mathscr{A}^{-1} g t. \tag{3.46}$$

Hence, the trajectory of the particle is sinusoidal, but 90° out of phase with the imposed force field.[2] The x component of the path of the particle reads

$$x = \frac{-F_0 B_i \cos ([(2\pi \tilde{f} z \mathscr{A})/g] - \phi)}{2\pi \tilde{f} [1 + (2\pi \tilde{f} \mathscr{A}^{-1})^2]^{\frac{1}{2}}}. \tag{3.47}$$

Some years ago, Abbott[61] has verified experimentally these theoretical results by photographing the paths of 30 μ water droplets in a horizontally oriented electrical field. In a 60 Hz field, the particles were observed to be sinusoidal, and phase shift, ϕ was determined to be within 1–2° of the predicted value. The difference has been attributed to differences in Stokes' law for horizontal and vertical motion[2] but there may be small inertial effects that cause deviations in the apparent resistance as compared to the value based on the Stokes' mobility $(6\pi\mu_G R_i)^{-1}$.

Fuchs[2] has pointed out that the oscillating particle technique can be used to determine simultaneously both the size and the charge of aerosol particles. By determining the distance between peaks in photographs of oscillating particle streaks, one can calculate from eqn. (3.47) a value of $\mathscr{A}^{-1}$. From this the particle radius may be found. The horizontal velocity q_{ix} equals $B_i \mathscr{E}_0 Q_i$, where $\mathscr{E}_0$ is the field strength and Q_i is the charge on the particle. The tangent of the local angle of inclination of the particle trajectories equals q_{iz}/q_{ix}, which in turn equals $g m_i \mathscr{E}_0 Q_i$. Hence, with independent knowledge of ρ_i and R_i, one can calculate the net charge on the particles.

D. Curvilinear Motion and Particle Deposition

Possibly the most extensive studies of accelerated motion of particles suspended in a continuum have been centered on the steady curvilinear motion of the particles near surfaces. Curvilinear motion in a steady state implies, of course, a particle acceleration as a result only of the convective term in the equations of motion. That is, the acceleration of the particle in a system of coordinates following the particle should be given by $d\mathbf{q}_i/dt = \mathbf{q}_i \cdot \nabla \mathbf{q}_i$ because the time rate of change of $\mathbf{q}_i$ with respect to a fixed spatial location $\partial \mathbf{q}_i/\partial t = 0$ by definition.

Precipitation of Aerosols in Parallel Flow Past a Plane Surface

To illustrate the application of simplified forms of eqn. (3.29) to the mechanics of deposition, consider first the problem of an aerosol cloud traveling horizontally through a straight channel in a gravity field. The gas flow is considered to be two-dimensional, laminar, and horizontally uniform. The flow is low enough in speed that the vertical convection currents are negligibly small compared with the gravitational settling velocity. The vertical motion of the particles in this stream will be given by eqn. (3.33), and the vertical displacement by eqn. (3.46). If the residence time of the channel $L/\overline{U}(L$ = channel length, $\overline{U}$ = mean velocity taken over the cross-section of the channel) is very long compared to the relaxation time $\mathscr{A}^{-1}$, then the transient term in eqn. (3.29c) may be dropped, or $q_{iz} = v_s$.

Considering only the region of the channel where the fluid velocity is fully developed, we can write the horizontal component of fluid velocity q_x as

$$q_x = \left(\frac{6z}{d} - \frac{3z^2}{d^2}\right)\overline{U}, \tag{3.48}$$

where d is the channel width.

If the inertia of particles is disregarded in the horizontal component, the particle trajectories can be determined by the equations

$$\frac{dx}{dt} = q_x, \quad \frac{dz}{dt} = q_z - q_{iz} = q_z - v_s. \tag{3.49}$$

The differential equation for the particle trajectories then may be obtained by eliminating dt from the expressions in (3.49):

$$\frac{dx}{-q_x} = \frac{dz}{q_z - v_s}. \tag{3.50}$$

Introducing the stream function for the fluid velocity distribution, ψ, then[2]

$$q_x = \frac{\partial \psi}{\partial z}, \quad q_z = -\frac{\partial \psi}{\partial x},$$

or

$$-\frac{dx}{\partial \psi/\partial z} = \frac{dz}{(\partial \psi/\partial x) + v_s}, \tag{3.51}$$

or

$$-v_s\, dx = d\psi. \tag{3.52}$$

Over the entire length of tube L,

$$v_s L = \psi_0 - \psi_L, \tag{3.53}$$

where ψ_0 and ψ_L are values of the stream function at the location of the particle at the beginning and end of the channel. From the definition of the stream function, this function has to equal the volumetric flow rate of fluid per unit width between the bottom surface and an arbitrary streamline. When ψ_L is equal to zero, eqn. (3.53) gives the trajectory corresponding to the boundary between the region of flow where particles settle out and the region where particles do not settle out. If the total flow rate per unit width is V'_G then the fraction of total number particles that are removed on the bottom surface η will be given by the ratio[2]

$$\eta = \psi_0 / V'_G = v_s L \Big/ \int_0^d q_z(z)\, dz = v_s L / d\bar{U}. \tag{3.54}$$

The parameter η is called the *precipitation or collection efficiency*. In this simple model it does not depend on the distribution of fluid velocity, but only on the residence time $L/\bar{U}$.

A similar analysis for the deposition of particles moving through a round tube yields essentially the same relation as eqn. (3.53), except for a coefficient of proportionality. For tubes of radius R_T,

$$\eta = \frac{3 v_s L}{8 R_T \bar{U}}. \tag{3.55}$$

For the case of viscous flow in tubes, Natanson[2] and independently, Thomas[62] have derived a more accurate formula,

$$\eta = \frac{2}{\pi}\left\{2\zeta(1 - \zeta^{\frac{2}{3}})^{\frac{1}{2}} + \arcsin \zeta^{\frac{1}{3}} - \zeta^{\frac{1}{3}}(1 - \zeta^{\frac{2}{3}})^{\frac{1}{2}}\right\}, \tag{3.56}$$

where

$$\zeta = \frac{3 v_s L}{8 R_T \bar{U}}.$$

Thomas' formula differs from Natanson's, but they may be shown to be equal with rearrangement. Under circumstances where $\zeta = 1$, eqn. (3.56) reduces to eqn. (3.55).

The application of an external field of force (an electrical field) in addition to gravity may also be incorporated in these models of tube or channel flow. The precipitation of aerosols in an electrically charged horizontal or vertical channel has been discussed, for example, by Fuchs.[2] Such devices have been used for determining particle mobilities.

Inertial Deposition of Particles Streaming at an Angle Past Surface

The fact that particles suspended in flowing gases may trace paths differing from the surrounding gas has been used extensively in the design of mechanical samplers. Regardless of their geometry, the removal of particles in these devices depends on inducing curvilinear flow in the aerosol stream. Perhaps most commonly used are instruments called *impingers* or *impactors*. These rely on efficient aerosol deposition from a jet of gas made to turn sharply past a plate placed normal to the gas stream (Fig. 3.6). Impinging devices of this kind largely rely on obtaining laminar flow across the collecting surface because flow separation, or eddying might greatly reduce the efficiency of collection. Simple obstacles like cylinders and spheres also have been of interest as collectors. The body of revolution traced by limiting particle trajectories hitting a sphere is drawn in Fig. 3.7. Impacting trajectories for the cylinder have been referred to above, and are shown in Fig. 3.5.

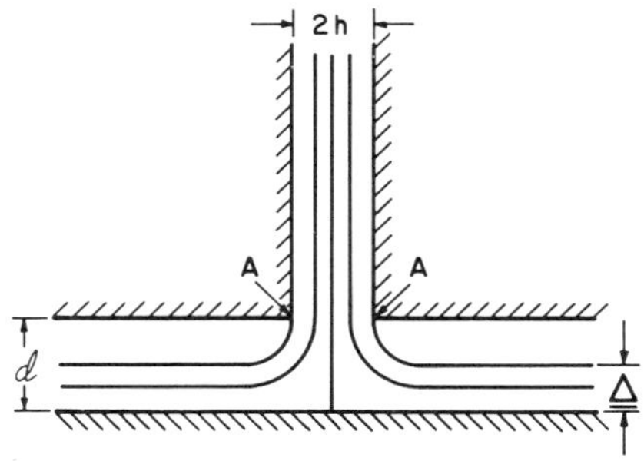

FIG. 3.6. Trajectories of particles from a jet impinging on a flat surface. (After Fuchs[2].)

The analysis of inertial deposition on curved surfaces usually involves the assumption that the particles are ideally points in the gas stream but have masses larger than the gas. Trapping of particles as estimated by this model is called *impaction*. A second effect may increase deposition on curved surfaces. If the finite size of the particles is taken into account, the limiting trajectories drawn in Fig. 3.5 or 3.7 will be displaced outwards slightly, giving rise to larger collection efficiencies. This effect is called *interception*.

The efficiency of collection of particles on obstacles η generally depends on the speed of the gas suspension, the size of the obstacle, and the mass and size of the particles. For impinging devices η also depends on the distance from the nozzle to the surface d and the nozzle width, $2h$ (Fig. 3.6). Roughly, $\eta \approx L_s/h$ for impingers.[2] For conditions of 100% efficiency of collection the impinging device would trap all particles in an area equivalent to the cross-sectional area of the jet. Similarly, $\eta = 100$ per cent for cylinders and spheres would require that the body remove all particles from an area of the gas stream equal to the

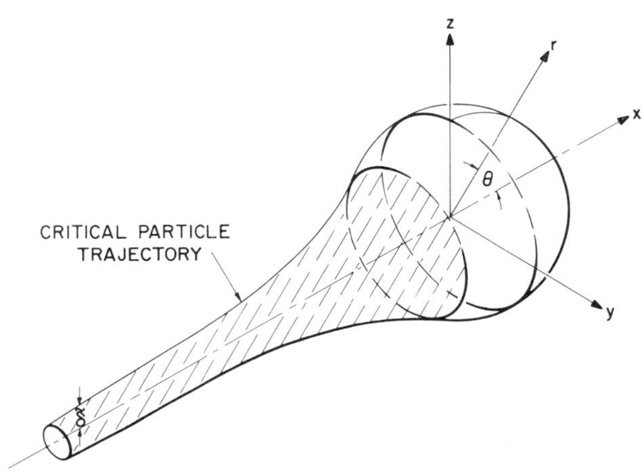

FIG. 3.7. Body of revolution made by critical trajectories of particles impacting on a sphere.

cross-sectional area of the obstacle. From the drawing of streamlines and particle trajectories in Figs. 3.5 and 3.7, this is not true in general because of the required flow around the collector.

We can define the collection efficiency in terms of the distance δ, indicated in Figs. 3.5 and 3.7 from the axis of the obstacle to the outermost trajectory of particles that strike the surface. For example, in the case of the sphere in Fig. 3.7,

$$\eta = \frac{\tilde{\delta}}{a},$$

where a is the radius of the cylinder.

Trajectories of particles in a moving fluid may be estimated from a step-by-step integration of the equations of motion, neglecting $F(t)$. For steady flow of gas around a body like the one in Fig. 3.5, for example, the integration would be carried out on the equation

$$m_i \frac{\mathrm{d}^2\mathbf{r}}{\mathrm{d}t^2} = 6\pi\mu_G R_i \left(\mathbf{q} - \frac{\mathrm{d}\mathbf{r}}{\mathrm{d}t}\right), \tag{3.57a}$$

where $\mathbf{r}$ is the radius vector taken from the center of the body, 0. Equation (3.57a) written in dimensionless form becomes

$$Stk \frac{\mathrm{d}^2\mathbf{r}^{\oplus}}{\mathrm{d}t^{\oplus 2}} + \frac{\mathrm{d}\mathbf{r}^{\oplus}}{\mathrm{d}t^{\oplus}} = \mathbf{q}^{\oplus}(\mathbf{r}^{\oplus}), \tag{3.57b}$$

where $\mathbf{r}^{\oplus} = \mathbf{r}/R_i$, $\mathbf{q}^{\oplus} = \mathbf{q}/U_\infty$, and $t^{\oplus} = U_\infty t/R_i$. The dimensionless number, called the Stokes number, is

$$Stk = \frac{U_\infty}{\mathscr{A}a} = \frac{L_s}{a} = \frac{2Re_f \rho_i R_i^2}{9\rho_G a^2} = \frac{2U_\infty \rho_i R_i^2}{9\mu_G a}. \tag{3.57c}$$

Here, $Re_f = U_\infty a/\nu_G$.

The Stokes number is equivalent to the ratio of the stop distance [eqn. (3.39)] to the characteristic radius of the obstacle. The Stokes number in this case is the only criterion, other than geometry, that determines similarity† for this problem.

To calculate the trajectories $\mathbf{r}^{\oplus}(t)$, the fluid velocity field $\mathbf{q}^{\oplus}(t)$ is assumed known. The outermost trajectories that just touch the body then may be determined assuming $R_i/a \ll 1$, and neglecting the influence of the particles on flow very near the obstacle. That is, the particles are assumed to be points suspended in the gas.

When $Stk \to 0$, or as the particle mass becomes small, the particles tend to follow streamlines, or in steady flow, $\mathbf{q}^{\oplus} \approx \mathrm{d}\mathbf{r}^{\oplus}/\mathrm{d}t$. In this case the particles will never reach the body within the above approximation. Thus, one expects that a certain minimum value of the Stokes number should exist above which the particle inertia dominates over the particle entrainment in the gas, and particles will begin to strike the obstacle. Below the critical value of the Stokes number, particles will not reach the body but will follow the gas around the body. The existence of the critical value of the Stokes number was apparently first established by Langmuir.(64)

For the limiting case of flow at very high Reynolds number Re_f, based on characteristic dimensions of the obstacle, Levin(65) has found analytically several general expressions for the critical value of the Stokes number. Assuming that the viscous boundary layer

† Principles of similarity are widely used in fluid dynamics problems. Some other examples are discussed in Chapters 7 and 10 (see also Bridgman(63)).

surrounding the body is very thin ($Re_f \gg 1$), Levin estimated values of Stk_{cr} from solutions of eqn. (3.57b) in the limit of inviscid, or potential flow of the gas. From knowledge of boundary layer theory,(67) we may expect that this would be a good first approximation for conditions of $Re_f \gg 1$. The flow field for this regime should be given approximately by inviscid flow around the obstacle except for the viscous region very near the surface of the obstacle.

In the region near the forward stagnation point A, for example, on the cylinder in Fig. 3.5, the fluid velocity given by inviscid theory is approximately

$$\left.\begin{aligned} q_{z'}^{\oplus} &= -\hat{\alpha} z'^{\oplus}, \\ q_{x'}^{\oplus} &= \hat{\beta} x'^{\oplus}, \end{aligned}\right\} \tag{3.58}$$

where $\hat{\alpha}$ and $\hat{\beta}$ are constants. As shown in Fig. 3.5, the intrinsic coordinate $x'^{\oplus}$ refers to the distance measured along the surface from the stagnation point and $z'^{\oplus}$ is the coordinate normal to it. Using these relations for the dimensionless fluid velocity, the differential equation for the component of motion along the stagnation streamline in the direction of the stagnation point A has the form

$$Stk \frac{d^2 r_{z'\oplus}^{\oplus}}{dt^{\oplus 2}} + \frac{dr_{z'\oplus}^{\oplus}}{dt^{\oplus}} + q_{z'}^{\oplus} = Stk \frac{d^2 r_{z'\oplus}^{\oplus}}{dt^{\oplus 2}} + \frac{dr_{z'\oplus}^{\oplus}}{dt^{\oplus}} + \hat{\alpha} z'^{\oplus}, \tag{3.59}$$

The solution of the expression is(65)

$$r_{z'\oplus} = c_s \exp(\Lambda_{1,2}\, t^{\oplus}), \tag{3.60}$$

where $\Lambda_{1,2}$ are roots of the characteristic equation, and are equal to(65)

$$\Lambda_{1,2} = -\frac{1}{2Stk}[1 \pm (1 - 4\hat{\alpha} Stk)^{\frac{1}{2}}]. \tag{3.61}$$

If $Stk > Stk_{cr} = (4\hat{\alpha})^{-1}$, then the roots are complex. The particles will reach the stagnation point where $z'^{\oplus} = 0$ at the time t_s defined by the condition

$$A_1 z_1'^{\oplus}(t_s) + A_2 z_2'^{\oplus}(t_s) = 0. \tag{3.62}$$

This equation has a solution corresponding to a finite value of time t_s at any real finite value of the arbitrary constants A_1 and A_2. Therefore, the particle for which $Stk > 1/4\hat{\alpha}$ moving along the stagnation streamline will reach the obstacle in a finite time. However, if $Stk < 1/4\hat{\alpha}$ the roots $\Lambda_{1,2}$ will be real, and eqn. (3.62) has no solution for finite time t_s. Then particles can never reach the surface of the body even along the stagnation streamline according to theory. The particle having $Stk < 1/4\,\hat{\alpha}$ cannot strike the surface from streamlines other than the central one because all other locations of flow have a tangential component of velocity, except for the stagnation streamline.

Levin's theory then gives the critical value of Stk in terms of the parameter $\hat{\alpha}$, which differs with the surface geometry of the obstacle. Based on potential theory, Stk_{cr} can be estimated for several bodies of simple shape, as tabulated in Table 3.3.

Though attempts to generalize Levin's results for inertial deposition have been made recently by Yurev,(66) the method of Levin does not provide results for cases where viscosity has to be considered (finite Re_f) because $q_x^{\oplus}$ $(=q_{z'}^{\oplus})$ cannot always be approximated by eqn. (3.58) near the stagnation point. However, in conditions of flow where boundary layer theory may be applied, $Re_f \ggg 1$, the form of (3.58) is frequently used as a first approximation to the viscous flow field. A few calculations have been carried out to determine the collection efficiencies of obstacles of simple geometry including those in Table 3.3. However, the general solution for particle motion at all Reynolds numbers, based on a characteristic length of the obstacle, is not available. Approximations have been restricted to the limiting cases of potential flow, and viscous flow.

TABLE 3.3. LIMITING VALUES OF THE STOKES NUMBER FOR COLLECTION OF PARTICLES ON BODIES. ESTIMATES BASED ON POTENTIAL FLOW NEAR A STAGNATION POINT (AFTER LEVIN[65]).

Obstacle in flow	Stk_{cr}
1. Infinitely long ribbon of width $2a$	
(a) Without separation	1/4
(b) With separation	$4/(\pi + 4)$
2. Elliptical cylinder, major to minor axis ratio X	$1/4(1 + X)$
3. Circular cylinder	1/8
4. Sphere	1/12
5. Circular disc	$\pi/16$
6. Flat jet striking plane at right angles with $(d/2h) \to \infty$	$2/\pi$

Impingement on a Flat Surface

In the case where a jet of air impinges on a smooth flat surface of infinite extent the calculations for a simplified model assuming uniform inviscid flow parallel to the walls of a nozzle, and parallel to the impinged surface have been reported by Davies and Aylward.[68] With the geometry shown in Fig. 3.6, Davies found a relation between the ratio of the nozzle and the surface d and the ratio of h to the distance Δ where a streamline originating from the nozzle edge would be located. Davies and Aylward's result is

$$\frac{d}{h} = \frac{[(h/\Delta)^2 + 1] \ln [(h/\Delta) + 1]}{\pi(h/\Delta)^2 (h/\Delta - 1)} + \frac{1}{(h/\Delta)}. \tag{3.63}$$

Using this relation, Davies and Aylward calculated the collection efficiency η for such a model assuming the geometry in Fig. 3.6, and assuming several values of (h/Δ) by estimating step by step over limiting particle trajectories. Two examples of these estimates are shown graphically in Fig. 3.8. If $h/\Delta = 1$, the ratio d/h is infinity, indicating an exceedingly narrow jet, or a very large distance from the impinged surface.

The theoretical results of Davies and Aylward are compared with experimental studies of Ranz and Wong[69] and May[70] in Fig. 3.8. Ranz and Wong used glycerine aerosols of 0.2–0·7 μ radius and studied their deposition of flat surfaces from slits of 0.2–0.7 mm widths or from round orifices 1–9 mm in diameter. Their data shown in Fig. 3.8 as dots and a curve correspond to results where the gas velocity was 10–180 m sec^{-1} in the nozzles. May's data were reported for a four-staged impactor using slits of different widths. Curves of the second and third stages of the staged impactor are shown in Fig. 3.8.

One can readily see from the sample of experimental results and theory that the agreement between the two is not very satisfactory. The critical value of Stk from experiments, or extrapolations of May's curves is about 0.2. However, Levin's theory (Table 3.3) gives a value for Stk_{cr} or $2/\pi$ or 0.637. Davies' calculations indicate that $Stk_{cr} \approx 0.3$ for $h/\Delta \approx 1$, and $Stk_{cr} = 0.2$ for $h/\Delta \approx 5$. Although Davies' and Aylward's theory brackets the experimental results in the low range of η, the experimental points in the upper range suggest that a departure, corresponding to values of η smaller than the theoretical, would develop for $Stk > 0.5$.

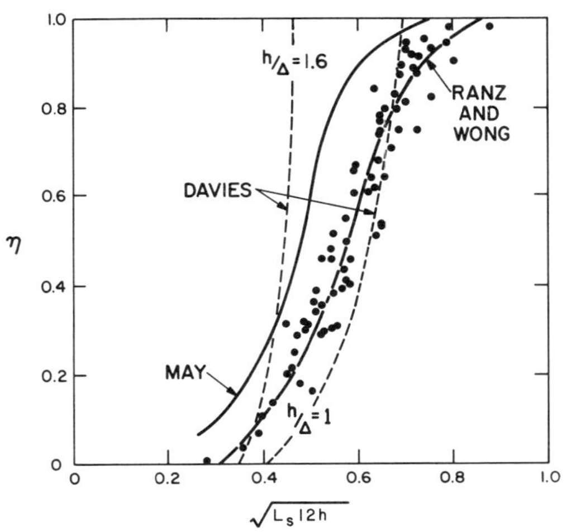

FIG. 3.8. Collection efficiency curves for simple impingement devices.

Fuchs[2] has suggested that the difference between experiment and theory for high η may be the result of eddy formation at the surface in the diverging jet stream. The eddies would tend to break up the streamline flow over the impinging surface, causing less deposition than expected from the theory. To confirm this explanation, one would also like to possess a better estimate of the influence of the viscous flow near the surface. The diverging pattern of air flow makes exact calculation of the aerodynamics for this case difficult, but approximate solutions based on boundary layer theory are available, for example, in Schlichting's book.[67]

If we considered the aerodynamic results given by Schlichting for the case of a divergent jet, we find that at high Reynolds number, the viscous layer forming at the surface is constant with distance from the origin 0 corresponding to the intersection of the centerline of the jet with the plane surface. A measure of the thickness of this viscous layer crudely corresponds to the displacement of streamline of inviscid flow upwards away from the surface. Qualitatively, this would imply that at relatively large values of Stokes number (particle trajectories for particles originating near the wall of the nozzle), the paths of particles would be displaced far enough away from the surface so that they might not strike the surface, whereas the simple theory might predict removal. This effect would occur without formation of eddies as considered by Fuchs. This would lower η in the range of larger Stk, as noted in the experiments. A quantitative calculation using the flow field given by boundary layer theory would be welcome to try to understand the failure of the simple mechanical arguments to explain the experimental observations.

In the case of impaction on curved objects in a uniform, parallel flow, the agreement between theory and experiment is little better than the impinging jet case. The two simple examples of deposition on a sphere and on an infinitely long cylinder have been examined in some detail in the literature.

Deposition on Spheres and Cylinders

Theoretical results for impaction on a sphere have been carried out for example by Langmuir and Blodgett[64] for viscous flow and for potential flow. They assumed for these computations that the field of gas flow around the sphere was given by equations for viscous flow by Stokes' solution for $Re_f < 1$ (see eqns. 3.5 and 3.5a). Calculations for the potential flow regime incorporated the solution given by Lamb[44] as

$$\frac{q_x}{U_\infty} = 1 + \frac{1 - 3\cos^2\theta}{2(r/a)^3}, \quad \frac{q_z}{U_\infty} = -\frac{3}{4}\frac{\sin 2\theta}{(r/a)^3}, \tag{3.64}$$

where θ is the angular dependence shown in Fig. 3.7.

The efficiency of collection for these limiting cases reads:

(1) (viscous flow)
$$\eta = 1 + \frac{0.75 \ln (2\,Stk)^{-2}}{Stk - 1.214}. \tag{3.65}$$

(2) (inviscid flow) $Stk \geqslant 0.2$
$$\eta = \frac{(Stk)^2}{(Stk + 0.25)^2}. \tag{3.66}$$

The curves corresponding to these experiments are shown in Fig. 3.9. In this case, Langmuir found that Stk_{cr} for viscous flow was 1.214, but for potential flow $Stk_{cr} \approx 0.0834$. The latter value substantially agrees with Levin's estimate (Table 3.3), and agrees satisfactorily with newer calculations of Dorsch *et al.*[71] in the limit $Re_i < 1$. As expected, the efficiency curve for viscous flow around a sphere lies lower than the curve for potential flow for all

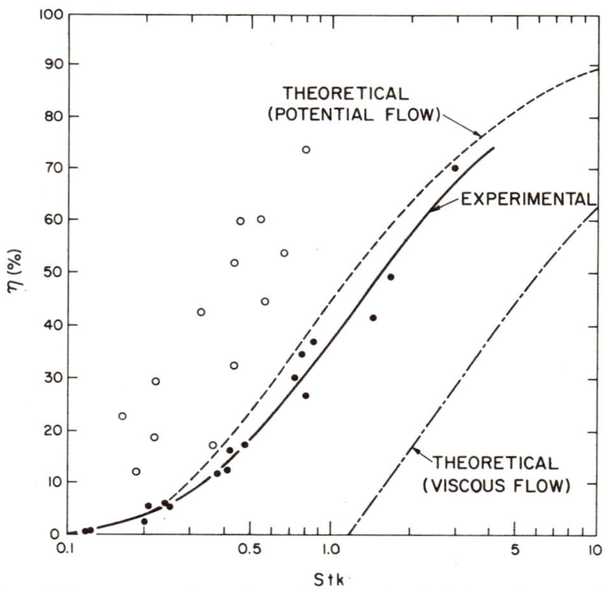

FIG. 3.9. Collection efficiency curves for spheres based on inertial impaction mechanism. ○ Data of Ranz and Wong.[69] ● Data of Walton and Woolcock.[72]

values of the Stokes number. One anticipates that solutions for η based on viscous flow at $Re_f > 1$ would lie in between these two limits. However, no quantitative estimates have been made to account for the effect of viscosity other than the limit of $Re_f < 1$.

Relatively few experimental results dealing with inertial impaction on spheres have been reported. The most frequently cited, those of Ranz and Wong,[69] are shown in Fig. 3.9 with more recent data of Walton and Woolcock.[72] Ranz and Wong's data were obtained for 0.18–0.65 μ spheres impinging on a 900 μ sphere fixed in a wind tunnel ($Re_f \approx 715$–5700). Their results show somewhat higher values of η for a given Stk than even the inviscid theory predicts. Walton and Woolcock's observations, however, for 2 μ aerosols of methylene blue impinging on water drops at $a = 0.25$–1.0 mm ($Re_f = 70$–870), show much better agreement with the theory. Their data lie just under the curve for potential flow as expected qualitatively.

The departure from the theory of Ranz and Wong's earlier data may be the result of an inadequate experimental procedure for determining η. It is curious, however, that their results for impaction on cylinders using the same procedures as in the sphere agree much better with theory (see below).

At about the same time Langmuir reported his findings for impaction in spheres, he also dealt with the problem of impaction on infinitely long cylinders. Using the solution for potential flow around cylinders (Lamb[44]), Langmuir and Blodgett[73] found by numerical integration that the efficiency curve had essentially similar form for cylinders and spheres. Making use of the solution of Lamb[44] for inviscid flow around a cylinder,

$$\frac{q_x}{U_\infty} = \frac{1 - \cos 2\theta}{(r/a)^2}, \quad \frac{q_z}{U_\infty} = -\frac{\sin 2\theta}{(r/a)^2}. \tag{3.67}$$

Langmuir and Blodgett integrated the equations of motion to determine limiting trajectories of particles. Their results for cylinders showed that

$$\left.\begin{aligned} \eta &= \frac{(Stk)^2}{(Stk + 0.06)^2} \qquad Stk > 0.08 \\ \eta &= 0, \qquad\qquad\qquad Stk < 0.08. \end{aligned}\right\} \tag{3.68}$$

Langmuir and Blodgett's calculation yielded a value of Stk_{cr} of about 0.08. This is somewhat lower than Levin's approximate estimate of 0.125 from Table 3.3.

The influence of viscosity on the determination of the impaction efficiency of cylinders has been examined by several workers. Davies,[74] for example, has calculated values of η for $Re_f = 0.1$ using Oseen's flow field:[13]

$$\begin{aligned} q_x &= U_\infty\,[\ln(r/a) - 0.5\,(1 - (a/r)^2)\cos 2\theta]/[2.002 - \ln 2Re_f], \\ q_z &= U_\infty\,[-0.5\,(1 - (a/r)^2)\sin 2\theta]/[2.002 - \ln 2Re_f]; \end{aligned} \tag{3.69a}$$

or in polar coordinates, near the cylinder:

$$\begin{aligned} q_r &= U_\infty \left\{ \frac{\ln\left(\dfrac{r}{a}\right) - 0.5\left[1 - \left(\dfrac{a}{r}\right)^2\right]}{2.002 - \ln 2\,Re_f} \right\} \cos\theta \\ q_\theta &= -\,U_\infty \left\{ \frac{\ln\left(\dfrac{r}{a}\right) + 0.5\left[1 - \left(\dfrac{a}{r}\right)^2\right]}{2.002 - \ln 2\,Re_f} \right\} \sin\theta \end{aligned} \tag{3.69b}$$

His results suggest that the impaction efficiency curve in the limit of viscous flow around the cylinder lies much lower than the difference found between viscous flow and potential flow around a sphere. Davies gives the empirical formula

$$\eta = 0.16\left[\frac{R_i}{a} + \left(0.25 + 0.4\frac{R_i}{a}\right) Stk - 0.0263\left(\frac{R_i}{a}\right) Stk^2\right], \quad Re_f = 0.1, \qquad (3.70)$$

which is shown in Fig. 3.10 as a comparison with Langmuir and Blodgett's curve for potential flow around the cylinder.

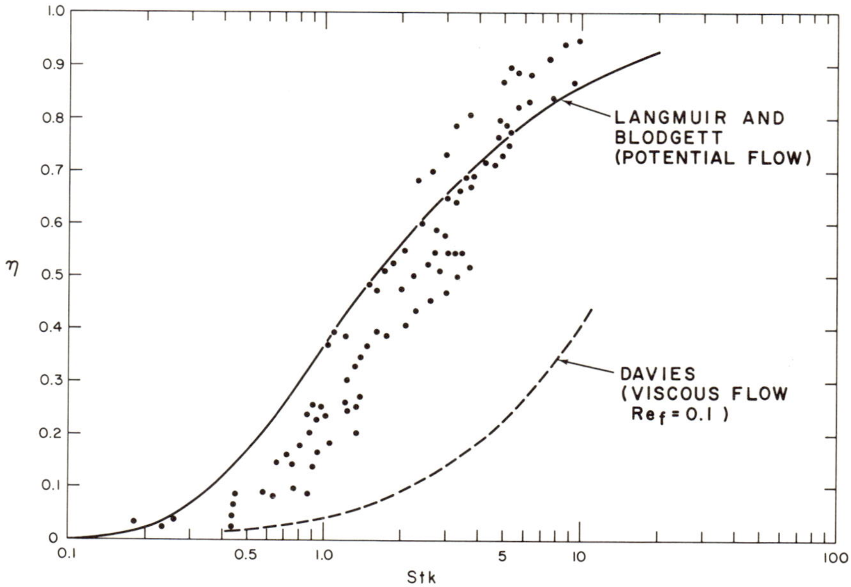

FIG. 3.10. Collection efficiency curves for cylinders. ● Experimental data of Ranz and Wong.(69)

More detailed computer calculations for the case of potential flow have been reported by Brun *et al.*(75) They have published efficiency curves as calculated from extensive estimates of η by computer. Like the data of Dorsch *et al.*,(71) Brun *et al.* sought to find the influence of departure from Stokes' resistance in the trajectories of particles (i.e. the effect of $Re_f > 1$). In the limit where $Re_f \gg 1$, the estimates of Brun *et al.* approach those of Langmuir and Blodgett. In the low range of *Stk*, up to 40, values of η predicted by Langmuir and Blodgett are about 10 per cent lower than those of Brun *et al.* However, above $Stk \approx 1$, Langmuir and Blodgett's values are 10 per cent larger than the newer calculations.

There exist relatively few experiments with which to test the impaction theory for deposition on cylindrical surfaces. Perhaps most frequently cited are the data of Ranz and Wong,(69) who investigated the impaction of nearly monodisperse sulfuric acid mists of mean radius 0.18–0.65 μ on wires of $Re_f = 55$–5000. A sampling of their data is shown in Fig. 3.10. The results are quite scattered, but below $Stk \approx 2$, they lie systematically below the curve predicted for potential flow. In view of the experimental difficulties in obtaining these data, the agreement between the experiments and the theory is considered at least qualitatively satisfactory.

Errors in the Impaction Efficiency Calculation

The computations above have modeled particle deposition in idealized terms of points of finite mass traveling roughly with the fluid without taking account of the finite size of particles, the influence of the viscous boundary layer (at relatively large Re_f), or Brownian motion. The last condition is discussed in detail with the results of Brownian diffusion in Chapter 7.

If one considers the effect of finite particle size, the collection efficiency generally should increase over the value calculated above. This readily can be seen from the diagram in Fig. 3.11. In this case, a point particle traveling at the streamline a distance $\tilde{\delta}$ away from the stagnation streamline will pass the body, but a finite sized particle would not. The interception effect then depends on the ratio R_i/a or $\hat{K}$.

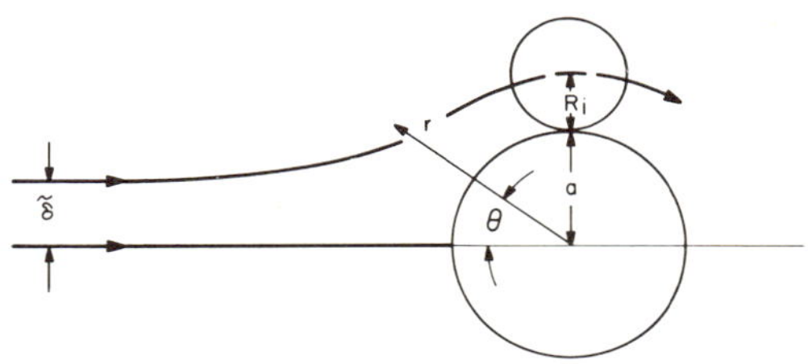

FIG. 3.11. Interception of particles by a cylinder.

In the extreme case where the particle has sufficient mass to move in a straight line despite curvilinear flow of gas ($Stk \to \infty$), the increase in η is[2]

$$\Delta\eta = \frac{R_i + a}{a} - 1 = \hat{K} \tag{3.71}$$

for a cylinder, and

$$\Delta\eta = \left(\frac{R_i + a}{a}\right)^2 - 1 = \hat{K}^2 + 2\hat{K} \approx 2\hat{K} \tag{3.72}$$

for a spherical obstacle.

In cases where Stk is relatively small, particles will at least approximately follow the fluid streamlines. The interception increment $\Delta\eta$ can then be estimated for the limiting configurations of potential flow and Stokes' flow around the obstacle.

For potential flow around a cylinder, the equation for an arbitrary streamline is $\psi = (r/a - a/r)\sin\theta = \text{const}, \tilde{\delta}$. At $r/a \to \infty$, the constant is $(r/a)\sin\theta$. Thus the constant $\tilde{\delta}$ can be identified with the distance from the stagnation streamline to the corresponding streamline at an infinite distance from the cylinder. When the angle θ, defined in Fig. 3.6, is $90°$ $\tilde{\delta} = (r/a - a/r)$. Then geometrically, a particle of finite size will give

$$\tilde{\delta} = (1 + \hat{K}) - 1/(1 + \hat{K}).$$

This corresponds to the incremental increase in η by interception since the particle running along this streamline would just touch the cylinder, while a particle of smaller radius would not. A tabulation of values of η for limiting flows around cylinders and spheres is shown in

Table 3.4. In a similar way, other examples of $\Delta\eta$ by interception may be estimated from the equations of streamlines for different geometries.

As in the case of diverging flow of a jet on a flat surface, the development of a viscous layer around curved surfaces can cause an error in estimating η. Since the boundary layer grows from the forward stagnant point at the equator poleward on a curved surface, the streamlines corresponding ideally to potential flow are displaced outward slightly as they pass around the body. At sufficiently high Re_f, the flow around the body may separate and an eddying wake may form to the lee of the body. Separation of flow is accompanied by a reversal of flow, or a return to the body on the leeward side. Thus it is possible to impact particles on the near side of the body as well as on the forward side. This is not possible, of course, from the limiting cases of potential flow or Stokes' flow. These fields of fluid flow are symmetrical about the obstacle as noted, for example, by eqns. (3.64) and (3.67).

TABLE 3.4. INCREMENTAL CHANGES IN COLLECTION EFFICIENCIES FOR SPHERES AND CYLINDERS RESULTING FROM THE INTERCEPTION EFFECT, $\hat{K} \ll 1$ (FROM FUCHS[2]).

Body	$\Delta\eta$
A. Potential flow	
(1) Sphere	$(1+\hat{K})^2 - \dfrac{1}{(1+\hat{K})} \approx 3\hat{K}$
(2) Cylinder	$(1+\hat{K}) - \dfrac{1}{(1+\hat{K})} \approx 2\hat{K}$
B. Viscous flow, $Re_f < 1$	
(1) Sphere	$(1+\hat{K})^2 - \frac{3}{2}(1+\hat{K}) + \dfrac{1}{2(1+\hat{K})} \approx \frac{3}{2}\hat{K}^2$
(2) Cylinder	$\left[(1+\hat{K})\ln(1+\hat{K}) - \dfrac{\hat{K}(2+\hat{K})}{2(1+\hat{K})}\right](2.002 - \ln 2Re_f)^{-1}$

Despite the possibility of deposition on the lee side in the separation region, the experimental observations have indicated a fairly good agreement with the theory based on limiting flow fields. This implies that most of the particles hit obstacles like cylinders and spheres on the forward side near the stagnation point. This largely has been verified by theoretical estimates of local values of η by Dorsch *et al.*[71] and Brun *et al.*[75]

Because the streamlines near a curved impacting surface are displaced slightly by the viscous region, the collection efficiency will be reduced a little below that of the value calculated for potential flow. This effect qualitatively should be the same magnitude as the interception effect but opposite in sign. Thus Fuchs[2] has suggested that the two opposing influences may nearly cancel each other, at least for small values of the ratio R_i/a or δ°/a, where δ° is the displacement thickness of the boundary layer (defined by Schlichting[67]).

Fuchs has estimated very crudely the reduction in η expected by the influence of the boundary layer. Particle trajectories which determine η tend to hit the surface of a cylinder or a sphere at azimuthal angles of $\theta = 60$–$80°$. In this range, the ratio $\delta^\circ/a \approx Re_f^{-\frac{1}{2}}$. Arguing that the reduction in η then roughly equals δ°/a, η may depend weakly on the Reynolds number of the body. No experimental evidence apparently is available to verify this conclusion.

E. Deposition of Particles in Electrical Fields

For many years it has been known that particulate matter can be removed quite readily by electrically charged screens or bars placed in the flowing suspension. Electrical precipitators using such configurations have proved to be very effective devices for removing particles from stack gases, for example. The efficiency of removal of such units depends on several factors including the magnitude of the imposed electric field $\mathscr{E}_0$, the degree of particle charging, and the flow rate in the stream. Some theoretical studies have been reported which elucidate the role of electrical forces in aiding deposition, particularly in surface of simple geometry. These investigations have examined particle impaction for limiting cases of potential and viscous flow without allowing for particle inertia. That is, only the regime below Stk_{cr} has been examined, where the ratio R_i/a is small, and the velocity of the suspension is relatively small.

Electrostatic Scattering

Before taking up the questions of the deposition of electrically charged aerosols on surfaces, let us examine briefly the scattering by electrical forces of an aerosol cloud in a stagnant medium. If the particles in the cloud are the same size, and Brownian agitation is neglected, the intensity of the electrical field $\mathscr{E}_0$ induced by the charged particles will be governed by the expression[(442)]

$$\nabla \cdot \mathscr{E}_0 = 4\pi NQ, \tag{3.73}$$

where N is the number of particles having a charge Q of the same sign.

The field intensity vector $\mathscr{E}_0$ is related to the velocity vector of the particles by eqn. (3.16) as

or

$$\left.\begin{aligned} \mathbf{q}_i &= B_i \mathscr{E}_0 NQ \\ \nabla \cdot \mathbf{q}_i &= 4\pi NQ^2 B_i \end{aligned}\right\} \tag{3.74}$$

Since

$$\nabla \cdot \mathbf{q}_i = -\frac{1}{N}\frac{dN}{dt}, \tag{3.75}$$

then

$$-\frac{1}{N}\frac{dN}{dt} = 4\pi NQ^2 B_i. \tag{3.76}$$

After a time t, the particle concentration will decrease by the amount determined by the integral of eqn. (3.76) or

$$\frac{1}{N} - \frac{1}{N_0} = 4\pi Q^2 B_i t. \tag{3.77†}$$

From these equations it follows that a cloud of aerosol particles having the same sign of charge is uniformly expanded by the action of its own induced electrical field. The rate of scattering or dispersal, given by the rate of decrease in concentration N, depends neither on the size nor on the shape of the space occupied by the cloud. The aerosol cloud increases

† Note the striking analogy between the decay in particle concentration by electrostatic scattering and coagulation as given by eqn. (10.109).

uniformly in volume while retaining its original shape at a rate equal to $-(1/N)(\mathrm{d}N/\mathrm{d}t)$. If the aerosol is confined in a vessel, the fractional rate of deposition on the walls will be given by eqn. (3.77) as

$$N = \frac{N_0}{1 + 4\pi Q^2 B_i N_0 L/\bar{U}} \tag{3.78}$$

if deposition results from electrostatic dispersion only.

Suppose the charging of the aerosol cloud is partly of opposite sign. The collection contains N_+ of positively charged particles and N_- of negatively charged particles, each bearing the same (absolute) magnitude of charge Q. If $N_+ > N_-$, the net charge per unit volume in the cloud is $(N_+ - N_-)\,Q$. Then from the reasoning above, we find that[2][442]

$$-\frac{1}{N_+}\frac{\mathrm{d}N_+}{\mathrm{d}t} = 4\pi Q^2 B_i(N_+ - N_-) \tag{3.79a}$$

and

$$-\frac{1}{N_-}\frac{\mathrm{d}N_-}{\mathrm{d}t} = 4\pi Q^2 B_i(N_+ - N_-). \tag{3.79b}$$

Thus under these conditions Fuchs,[2] and Pich[442] have indicated that the negatively charged particles (or the smaller number of particles of one sign of charge) are pressed towards the middle of the cloud where their concentration actually increases. The charge density $(N_+ - N_-)\ Q$ in this central zone decays asymptotically until the nucleus is nearly neutral, and consequently is stable. Simultaneously the scattering in the external portion of the cloud will continue.

When a cloud of particles of different size, and carrying charge of both signs are considered, eqn. (3.79) has to be generalized.[2] For the ith species of particle, where

$$N = \sum_{i=1}^{\infty} n_i, \quad R_i = i^{\frac{1}{3}} R_i,$$

and Q_i is the charge on the ith particle, then

$$-\frac{\mathrm{d}n_i}{\mathrm{d}t} = 4\pi B_i n_i Q_i \sum_{i=1}^{\infty} n_i Q_i = 4\pi B_i Q_i n_i N \bar{Q}_i, \tag{3.80}$$

where $\bar{Q}_i$ is the average charge per particle. The decay in total particle concentration may be estimated from (3.80). That is,

$$-\frac{\mathrm{d}\sum_{i=1}^{\infty} n_i}{\mathrm{d}t} = -\frac{\mathrm{d}N}{\mathrm{d}t} = 4\pi N \bar{Q}_i \sum_{i=1}^{\infty} B_i n_i Q_i = 4\pi N^2 \bar{Q}_i \overline{B_i Q_i}, \tag{3.81}$$

where $\overline{B_i Q_i}$ is the product of the mobility and the net charge averaged over all particles in the cloud. It is not possible, in general, to integrate Eqs. (3.80) or (3.81) since $\bar{Q}_i$ and $\overline{B_i Q_i}$ are time dependent parameters which involve solution of the set of equations in (3.80). If, however, the mobilities and the charges B_i and Q_i are nearly the same, and $\bar{Q}_i$ or $\overline{B_i Q_i}$ vary little in time, eqn. (3.76) may be applied as a crude approximation to the solution to eqn. (3.81).

Deposition on Curved Obstacles in a Uniform Flow

Under conditions where deposition is largely controlled by the electrical forces induced between the impacting obstacle and particles, the deposition rate or the flux to the obstacle can be determined in a general way. Since the electrical field is non-divergent or solenoidal, the particle concentration N in the suspension will remain a constant. For conditions of steady state, the particle concentration near the surface of the obstacle must remain the same as the value far away. Thus the total flow of particles towards the surface of the body will equal the integral of the flux of particles to the surface over the entire surface.[76]

In the absence of Brownian diffusion, the total number of particles hitting the surface is

$$\tilde{\Phi}_i = \mathrm{N} \int_S (q_r + B_i F_r)\, \mathrm{d}S, \tag{3.82}$$

where q_r is the component of flow velocity normal to the surface of the body (the radial component for spheres or cylinders), F_r is the component of external force directed normal to the surface, and $\mathrm{d}S$ is an element of surface area. In the case of an electrical field, $F_r = \mathscr{E}_{0r} Q$, where $\mathscr{E}_{0r}$ is the radial component of $\mathscr{E}_0$, the field strength, and Q_i is the charge on the particles.

If the particles are very small, the component flow velocity q_r will be negligible, and the integral in eqn. (3.82) becomes

$$B_i Q_i \int_S \mathscr{E}_{0r}\, \mathrm{d}S = 4\pi B_i Q_i Q',$$

where Q' is the total charge on the collector surface. Hence, the efficiency of collection can be written as the net flux j_i divided by the total number of particles flowing past the body, or

$$\eta = \frac{j_i}{N U_\infty S_0} = -\frac{4\pi Q_i Q' B_i}{U_\infty S_0}, \tag{3.83}$$

where S_0 is the cross-sectional area of the obstacle normal to the direction of the free stream flow U_∞.

Electrical Deposition on Spheres and Cylinders

For non-polarized spheres of radius a, charged oppositely to the particles,[2]

$$\eta = \frac{4 Q_i Q' B_i}{U_\infty a^2}, \tag{3.84}$$

and for cylinders,

$$\eta = \frac{2\pi B_i Q_i Q'}{U_\infty a}. \tag{3.85}$$

The case of deposition from aerosol clouds on spheres by electrical forces has been considered in more detail by Kraemer and Johnstone.[77] Based on particle trajectories, they calculated the curves for η as a function of parameters characterizing different electrostatic forces for potential flow around a conducting sphere. Interception was taken into account, but inertial effects (impaction) were neglected. Assuming that the parameters $(4/3)\pi R_i^3 N \ll 1$, and $[(4/3)\,\pi R_i^3]\, A_0^2/a^2 = 1$, where A_0 is the radius of the aerosol cloud, computations were made for situations including (a) Coulombic force between a charged sphere and charged particles, (b) force between unipolar charged aerosols and the charge induced by them on a grounded sphere, and (c) the induction force between a charged sphere and uncharged particles. The forces of electrical interaction in (*a–c*) can be written in dimensionless form, normalized to Stokes' resistance:

(a) Coulomb interaction $$K_E^{\oplus} = \frac{Q_i Q'}{6\pi a^2 R_i \mu_G U_\infty} \tag{3.86a}$$

(b) Induction forces to grounded sphere

$$K_G^{\oplus} = \frac{Q_i^2 N A_0^2}{3\mu_G R_i U_\infty a}. \tag{3.86b}$$

(c) Induction to charged sphere

$$K_T^{\oplus} = \frac{(\epsilon_s - 1) R_i^2 Q_i^2}{(\epsilon_s + 2) 3\pi a^5 \mu_G U_\infty}, \tag{3.86c}$$

where ϵ_s is the dielectric constant of the spherical collector.

A typical theoretical curve given by Kraemer and Johnstone is drawn in Fig. 3.12. These workers performed several experiments with dioctylphthalate aerosols of mean radius 0.27–0.59 μ. The particles were either charged with units of the same sign (unipolar) or

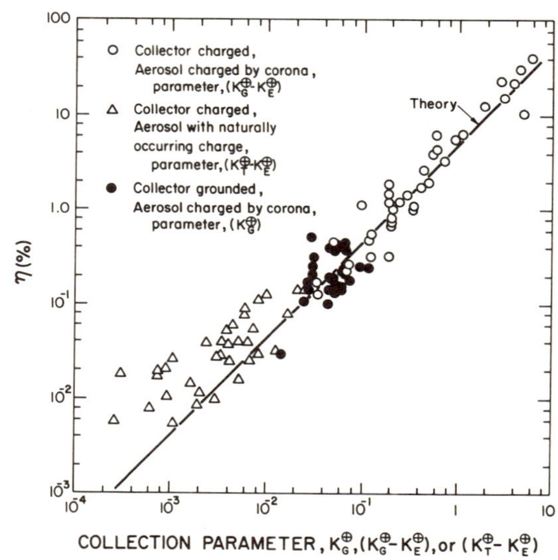

FIG. 3.12. Collection efficiencies for particles deposited on spheres by electrostatic forces. (After Kraemer and Johnstone.[77])

with different sign (bipolar) at average values $\bar{Q}_i$ of 0.15–137 units, and were collected on spheres of radius 0.33–0.55 cm. The free stream flow U_∞ was 1.5–6.9 cm sec^{-1}. Three different sequences of experiments were run in which one of the dimensionless forces $K^{\oplus}$, listed above, predominated over the others. These data are shown with theoretical curve in Fig. 3.12.

Kraemer and Johnstone determined that the interception effect tended to increase the value of η for a given value of $K^{\oplus}$ in the low ranges of $K^{\oplus}$. In particular, for ratios of $R_i/a \approx 0.1$ corresponding to their experimental data, the efficiency curves with interception included should bend upward from the theoretical curve in Fig. 3.12 at $K^{\oplus} \lesssim 0.1$. It appears that the significant departure from theory without including interception comes below $K^{\oplus} \approx 10^{-2}$.

Deposition resulting from electrical fields generated around a cylinder has been analyzed theoretically by Natanson[78] and later by Zebel.[79] These workers have considered examples similar to those discussed for spheres by Kraemer and Johnstone.[77] Using the theory of electrostatics to determine the electrical field around a dielectric cylinder, they estimated limiting particle trajectories for potential and viscous flow around the cylinder. These calculations were limited both to neglecting inertial ($Stk_{cr} \leqslant 0.125$), and interception effects. However, polarization and image forces were taken into account.

Natanson found that the collection efficiency for deposition of charged particles on an oppositely charged cylinder could be expressed in terms of eqn. (3.84) regardless of the limiting field of gas flow.

Natanson's calculations also indicated that two regimes of deposition could exist. Under conditions where the charge on the cylinder is large and the flow velocity is small, the radial component of particle velocity from the vector sum of the flow velocity and the velocity due to the electrical forces will be directed inward over the entire surface. Under these circumstances, eqn. (3.84) is found, and for $U_\infty \neq 0$, where $\eta \gg 1$,

$$\eta = \left[6 \left(\frac{\epsilon_c - 1}{\epsilon_c + 2} \right) \frac{B_i Q'^2 R_i^3}{U_\infty a^3} \right] \cdot \tag{3.87}$$

When the flow velocity is higher and the cylinder charge Q' is weaker, deposition should occur only on the upstream side of the obstacle. That is, $\eta \gg 1$,

$$\eta = 2\pi \frac{(\epsilon_i - 1)}{(\epsilon_i + 2)} \frac{B_i Q'^2 R_i^3}{U_\infty a^3}. \tag{3.88}$$

Under these conditions, the collection efficiency does not depend on the flow field.

If deposition of charged particles occurs on an uncharged cylinder,

$$\eta = \frac{3\pi}{2} \left[\frac{(\epsilon_i - 1)}{(\epsilon_c + 2)} \frac{B_i Q'^2}{U_\infty a^2} \right]^{\frac{1}{3}} \tag{3.89}$$

for potential flow, and

$$\eta = \left[\frac{(\epsilon_i - 1) Q_i^2 B_i}{(\epsilon_i + 2) U_\infty a^2 (2.002 - \ln 2Re_f)} \right]^{\frac{1}{3}} \tag{3.90}$$

for viscous flow assuming Oseen's flow field, given by eqn. (3.69).

Zebel[79] considered the problem of deposition of dielectric spheres on a dielectric cylinder on which an external electrical field $\mathscr{E}_0$ is imposed. Solutions for the cases of potential flow and viscous flow are given for the "point" particle trajectories. The flow fields given by eqns. (3.67) and (3.69) were used. In the limiting case of a metallic cylinder the capacitative parameter $\epsilon_c' \equiv (\epsilon_c - 1)/(\epsilon_c + 1)$ approaches zero, and an analytical solution for particle trajectories may be obtained as a function of the ratio $\bar{F}$ of the particle velocity induced by the electric field at the surface of the cylinder, to the flow velocity, U_∞

$$\tilde{F} = 2\left(\frac{\epsilon_i - 1}{\epsilon_i + 2}\right)\left(\frac{\epsilon_c - 1}{\epsilon_c + 1}\right)\frac{R_i^3 \mathscr{E}_0^2 B_i}{a U_\infty}. \tag{3.91}$$

A typical example of a trajectory for potential flow is shown for one orientation of the external field in Fig. 3.13a.

The trajectory patterns for potential flow display singularities in the mathematical solutions. The physical meaning of the singular points is that the force resulting from the electrical field is just balanced by the fluid resistance acting on the particle, so that a particle ideally will remain at this point indefinitely. In reality, the equilibrium is unstable, however, so that the smallest perturbation will cause the particle to move away from the singularity.

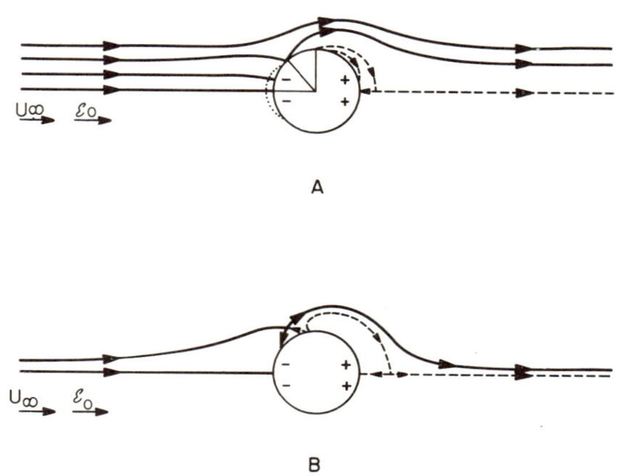

FIG. 3.13. Trajectories of uncharged particles influenced by the electrical forces while passing a dielectric cylinder. (a) Potential flow where $\tilde{F} = \sqrt{2}$, and $(\epsilon_c - 1)/(\epsilon_c + 1) = 0$; (b) viscous flow $\epsilon_c = 4$, $(\epsilon_c - 1)/(\epsilon_c + 1) = 0.6$, $(2.002 - \ln 2Re_f)\tilde{F} = 1$. (From Zebel.[79])

As the parameter $\epsilon_c' \to 0$, the singular points will be found on the cylinder surface at an angle $\theta = 45°$, and $\tilde{F} \leqslant \sqrt{2}$ for inviscid flow. This is shown in Fig. 3.13a for the case $\tilde{F} = \sqrt{2}$. With increase in $\tilde{F}$, $\tilde{F} > \sqrt{2}$, the singular points move farther and farther away from the obstacle surface. When the case of viscous flow around the cylinder is examined, the singular points are always found away from the cylinder, as shown in Fig. 3.13b. The explanation of this lies in the fact that the local flow velocity decreases rapidly to zero near the obstacle in contrast to potential flow where the tangential flow increases near the surface.

The existence of the singular points in the particle trajectories is crucial to the evaluation of the collection coefficient η. In the limit $\epsilon_c' \to 0$, the critical trajectories are just the trajectories approaching the singular point. For ideal (potential) flow,

$$\eta = \frac{\tilde{F}}{2}, \quad \text{for } \tilde{F} \leqslant \sqrt{2}; \quad \text{or} \quad \eta = (\text{const})_{\text{sing}}, \quad \tilde{F} > \sqrt{2}. \tag{3.92}$$

The $(\text{const})_{\text{sing}}$ is determined from the spatial coordinate of the singular point. This relation is shown in Fig. 3.14. Because the singular point moves away from the cylinder at $\tilde{F} > \sqrt{2}$,

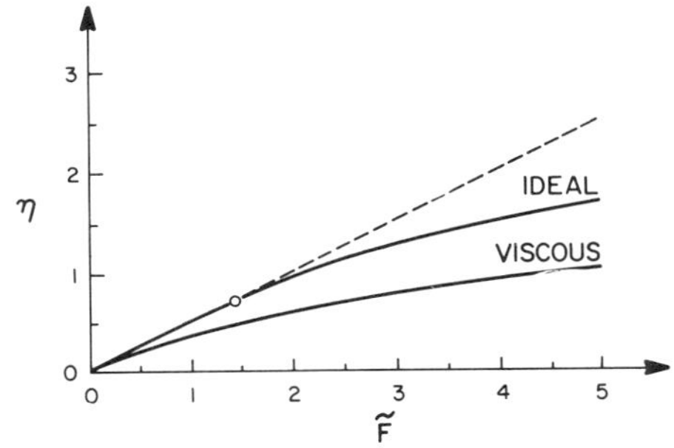

FIG. 3.14. Collection efficiency η for uncharged particles as a function of $\tilde{F}$. Ideal and viscous flow with $Re_f = 0.068$. (From Zebel.[79])

the particle may take two alternate branches away from the singularity. At least one of the branches may not intersect the cylinder. Hence η will always be lower than $\tilde{F}/2$ for $\tilde{F} > \sqrt{2}$.

For viscous flow,

$$\eta = \frac{(\text{const})_{\text{sing}}}{(2.002 - \ln 2\, Re_f)} \tag{3.93}$$

For conditions where $[(2.002 - \ln 2Re_f)\, \tilde{F}] \ll 1$, the singular point is approximately located on the cylinder surface at $\theta = 45°$. Then

$$\eta = \frac{\tilde{F}}{2}, \quad [(2.002 - \ln 2Re_f)\, \tilde{F}] \ll 1, \quad \epsilon_c' \to 0. \tag{3.94}$$

This relation appears to be analogous to the limiting result found by Natanson[78] for the case $\epsilon_c = 0$.

When $\epsilon_c' \neq 0$, the collection efficiency varies linearly with this parameter, the slope depending on the velocity ratio $\tilde{F}$.

Zebel also has calculated the trajectories for charged particles interacting with a charged cylinder. The trajectories are determined as function of the parameter

$$\hat{n} = \frac{[(\epsilon_c - 1)/(\epsilon_c + 1)]\, \tilde{G} - 1}{\tilde{G} + 1}, \tag{3.95}$$

where

$$\tilde{G} = \frac{\mathscr{E}_0 Q_i B_i}{U_\infty}. \tag{3.96}$$

This ratio, the particle velocity in the electric field to the free stream velocity, can be negative, unlike the ratio $\tilde{F}$. Some interesting trajectories for potential flow are shown in Fig. 3.15. For all values of $\hat{n} < -1$, a "dust free space" surrounds the cylinder and no deposition occurs (Fig. 3.15a). In this example $\tilde{G} = -0.5$ if $\epsilon_c' = 1$. If $\mathscr{E}_0$ is taken as positive here, the particles have to carry negative charge for $\tilde{G}$ to be negative.

Under circumstances where $-1 < \hat{n} < 1$, no singularities in the trajectories occur outside the cylinder. The example in Fig. 3.15b shows the paths of particles when $\hat{n} = 0.5$. In this case, if $\epsilon_c' > 0.5$, $\tilde{G} > 0$, then the particles are positively charged. Similarly, if $\epsilon_c' < 0.5$, $\tilde{G} < 0$, and the particles are negatively charged.

If $\hat{n} = 0$, the remarkable pattern shown in Fig. 3.15c develops. Here $\tilde{G} = \epsilon_c'$, and the aerodynamic flow field and the particle flow induced by the electric field exactly compensate for each other, and the trajectories become straight lines parallel to $\mathscr{E}_0$ and pass through the obstacle. The electrical effects just balance the aerodynamic effects and η is identically one in this case.

Over the range of $-1 \leqslant \hat{n} \leqslant 1$,

$$\eta = 1 + \hat{n} = \frac{\mathscr{E}_0 Q_i B_i}{U_\infty}\left[\frac{(\epsilon_c - 1)/(\epsilon_c + 1) + 1}{(\mathscr{E}_0 Q_i B_i/U_\infty) + 1}\right] \tag{3.97}$$

and

$$\eta = \tilde{G}\left[1 + \frac{(\epsilon_c - 1)}{(\epsilon_c + 1)}\right], \quad \text{positive } \tilde{G} \ll 1. \tag{3.98}$$

The computations made for viscous flow indicate that the "dust free space" can appear near the cylinder for cases where $\tilde{G}(2.002 - \ln 2Re_f) < 0$. When this parameter exceeds

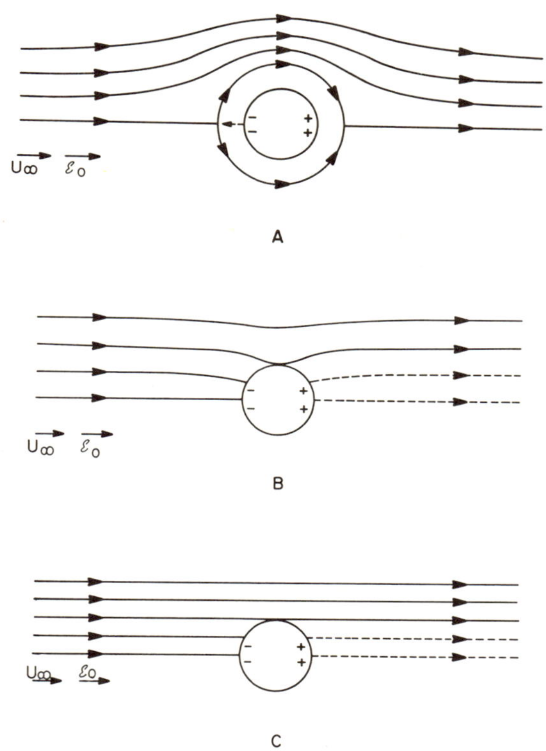

FIG. 3.15. Ideal trajectories for charged particles in potential flow passing a charged dielectric cylinder. (a) Dust free space for negatively charged particles moving around a cylinder when $\hat{n} = -3$; (b) particle paths when the particles are positively charged $\hat{n} = 0.5$ and $\tilde{G} > 0$, $(\epsilon_c - 1)/(\epsilon_c + 1) > 0.5$, and (c) particle paths when the aerodynamic effects just balance the electrical forces, $\hat{n} = 0$. (From Zebel.[79])

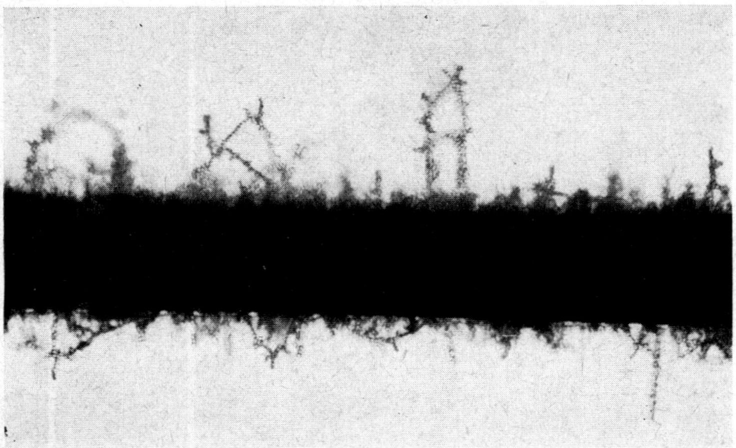

PLATE 6. Tree-like branches of charged 1.3 μ radius polystyrene spheres deposited from an air-stream on a circular cylinder. (Photo by D. Hochrainer.)

zero, the critical trajectory ends on the cylinder at $\theta = 90°$, and the collection coefficient estimated for viscous flow becomes

$$\eta = \tilde{G}\left[\frac{(\epsilon_c - 1)/(\epsilon_c + 1) + 1}{\tilde{G} + 1}\right], \quad \tilde{G} > 0. \tag{3.99}$$

Again, for situations where the velocity ratio is small ($\tilde{G} \ll 1$), η becomes independent of the aerodynamic flow field.

Zebel(80) has also examined the theory for deposition of charged and uncharged particles on an electrically polarized sphere. The results for this case are very similar to those described here for the cylinder. Since he has included Brownian diffusion in this treatment, discussion of these results will be deferred until Chapter 7 where diffusional deposition problems are outlined.

Recently, Hochrainer *et al*(177) have substantially verified experimentally most of Zebel's(79) theoretical conclusions regarding the influence of electrical forces on particles traveling past a cylinder. Their experiments showed that the particles velocities calculated from Zebel's analysis agreed with observations for the tangential velocity component at $\theta = 90°$. However, there were certain discrepancies at the forward stagnation point where the radial velocity component of the particles differed markedly from the theoretical estimate, particularly in the presence of attractive electrical forces. The authors were unable to explain this anomalous result in the light of the other measurements agreeing with the theory.

In conclusion, we remark in passing that the deposition efficiency of charged, solid aerosol particles on a charged body can only be predicted qualitatively by aerodynamic–electrostatic theory at present. Because the theoretical calculations have to assume that particles are deposited uniformly in such a way that the surface remains geometrically the same, the theory cannot account for observed deposits such as that in Plate 6. Here charged polystyrene spheres have collected on a cylindrical fiber to form tree-like protrusion from the cylinder surface. These branches grow preferentially, extending far out into the air flow.

CHAPTER 4

HEAT AND MASS TRANSFER TO SINGLE PARTICLES IN A CONTINUUM

THE rate of transfer of scalar quantities like thermal energy and material to a suspended body is of wide academic interest. From a practical standpoint, however, such transport processes have their principal importance to aerosol behavior in estimating the rate of evaporation or growth of the small particles in the presence of a condensable vapor. In this chapter we shall consider first the simpler cases of scalar transport when the boundary of the sphere remains fixed radially. Then the more complicated cases of diffusional processes will be treated in which the particle size either increases with vapor absorption (condensation), or decreases by evaporation.

In the following sections the particle motion is assumed to be either zero relative to the fluid, or the movement is assumed to be governed by Stokes' law.

4.1. The Macroscopic Equations for Heat and Mass Transfer in a Continuum Regime

Like the case of momentum transport discussed in Chapter 3, the calculation of energy and mass transport to a spherical body in a continuum may be accomplished through macroscopic or phenomenological equations combined with suitable boundary conditions. These partial differential equations are derived from the laws of conservation of matter, momentum, and energy. As we saw in Chapter 2, no fundamental difficulties are encountered in their derivation for dilute gases from the kinetic theory of gases. Similarly, these equations may be obtained from hydrodynamic theory, combined with thermodynamics. In so doing, the so-called thermodynamic fluxes (i.e. chemical reaction rates, mass fluxes of components, heat flux, and viscous stresses) have to be related to thermodynamic driving forces acting on elements of the fluid medium. Although the means of finding such relations are known through the theory of irreversible thermodynamics (e.g. DeGroot and Mazur[81]), there exists some diversity of formulation of relation between fluxes and driving forces, especially in the older literature.

A. Laws of Conservation of Matter and Energy

In this chapter we shall again make use of the "continuum" transport equations derived in Chapter 2 from approximate solutions to the Boltzmann equation. The equations of interest here are:

Continuity

$$\frac{\partial \rho_G}{\partial t} + \nabla \cdot \rho_G \mathbf{q} = 0, \tag{2.118}$$

and *conservation of energy*

$$\frac{\partial T}{\partial t} + \mathbf{q}\,.\,\nabla T = \frac{-2}{3n_G k}\{[p\bar{I} - 2\mu_G\bar{\bar{e}}] : \nabla\mathbf{q} + \nabla\,.\,k_G\nabla T\}. \tag{2.122}$$

Since these equations are written for pure, monatomic gases, let us first expand them to include multicomponent systems. After preliminary remarks about the certain assumptions involved in simplified forms of the macroscopic transport equations, solutions to problems of heat and mass transfer in spherical geometry will be outlined.

There have been a number of articles published recently that summarize the derivations of the macroscopic equations for multicomponent systems. Of these, perhaps the reviews of Hirschfelder, Curtiss, and Bird,(17) Bird,(82) and Merk(83) are most useful. To illustrate how the macroscopic equations are derived from continuum theory, we shall adopt an outline based on Merk's comments as an alternative to further development of the derivation in Chapter 2, based on kinetic theory. In the following formulation, eqns. (2.118) and (2.122) are expanded to include the terms for non-uniformities in a gas mixture as well as chemical reactions in the gas.

Conservation of Matter

Suppose a gaseous system composed of n_α components is considered. The number density of molecules of component α is n_α, and the mass of this component is m_α. Then the mass density is given by

$$\rho_\alpha = n_\alpha m_\alpha. \tag{4.1}$$

In a fixed reference system, the velocity of the α species is defined as $\mathbf{q}_\alpha$. Then the law of conservation of component α is

$$\frac{\partial \rho_\alpha}{\partial t} = -\nabla\,.\,(\rho_\alpha \mathbf{q}_\alpha) + G_\alpha, \tag{4.2}$$

where G_α is the mass of component α produced per unit time and unit volume by chemical reaction. By assuming that the system is closed, no material enters or leaves the system so that

$$\sum_{\alpha=A}^{n} G_\alpha = 0. \tag{4.3}$$

If the index α is summed over n in eqn. (4.2), with eqn. (4.3) the equation of continuity, eqn. (2.118) results. We note, of course, that the mass density of the gaseous system is

$$\rho_G = \sum_{\alpha=1}^{n} \rho_\alpha, \tag{4.4}$$

and the mass average velocity of the gas then is

$$\mathbf{q} = \sum_{\alpha=A}^{n} c_\alpha \mathbf{q}_\alpha, \tag{4.5}$$

where $c_\alpha = \rho_\alpha/\rho_G$.

In most problems in fluid dynamics, the equations of motion are written in terms of the barycentric system, i.e. the mass average or mean velocity is the reference velocity of the fluid. Thus the operator for the substantial derivative

$$\frac{\mathrm{D}}{\mathrm{D}t} \equiv \frac{\partial}{\partial t} + \mathbf{q}\,.\,\nabla$$

allows the equation of continuity (2.118) to be rewritten as

$$\frac{\mathrm{D}\rho_G}{\mathrm{D}t} + \rho_G \nabla\,.\,\mathbf{q} = 0. \tag{4.6}$$

Diffusional processes relative to the average motion of a system may be specified by a mass flux. For any component, the mass flux of a gaseous component α is written

$$\mathbf{j}_\alpha = \rho_\alpha(\mathbf{q}_\alpha - \mathbf{q}). \tag{4.7}$$

Using eqn. (4.5) and (4.7) we find that

$$\sum_{\alpha=A}^{n} \mathbf{j}_\alpha = 0. \tag{4.8}$$

Using eqns. (4.6), (4.7), and (4.2), the equation for conservation of species then becomes

$$\rho_G \frac{Dc_\alpha}{Dt} = -\nabla\,.\,\mathbf{j}_\alpha + G_\alpha. \tag{4.9}$$

This relation may also be written in terms of molar densities, and molar velocities, as illustrated later in Chapter 5.

Conservation of Energy

Bearing in mind that a monatomic gas' internal energy per unit mass $\hat{U}$ is $3n_G kT/2$, eqn. (2.122) may be rewritten, with use of tensor identities, as

$$\rho_G \frac{\mathrm{D}\hat{U}}{\mathrm{D}t} = -\nabla\,.\,\mathbf{h} - p(\nabla\,.\,\mathbf{q}) - 2\mu_G \bar{\bar{e}} : \nabla\mathbf{q}. \tag{4.10}$$

For a gas mixture one accounts for the work done on the system by external forces acting on diffusing components by adding a term on the right-hand side of eqn. (4.10).

$$\rho_G \frac{\mathrm{D}\hat{U}}{\mathrm{D}t} = -\nabla\,.\,\mathbf{h} - p(\nabla\,.\,\mathbf{q}) - 2\mu_G \bar{\bar{e}} : \nabla\mathbf{q} + \sum_{\alpha=A}^{n} (\mathbf{j}_\alpha\,.\,\mathbf{F}_\alpha). \tag{4.11}$$

Equation (4.11) states that the rate of change in the internal energy of the elemental volume equals the sum of (a) the net heat lost or gained by the flux term (term 1 on right), (b) the rate of work done on the element (term 2 on the right), (c) the heat developed by the viscous stresses (the third term on the right), and lastly (d) the heat gained by work of external forces $\mathbf{F}_\alpha$ acting on components in the system.

In aerodynamics, eqn. (4.11) usually has the name of the *thermal energy equation*. The heat flux **h** here generally includes the conductive term, along with additional terms accounting for radiative heating $\mathbf{h}_r$ plus the flux carried by mass fluxes of components

$$\mathbf{h} = -k_G \nabla T + \mathbf{h}_r + \sum_{\alpha=A}^{n} \hat{H}_\alpha \mathbf{j}_\alpha, \tag{4.12}$$

where $\hat{H}_\alpha$ is the partial specific enthalpy of species α.

Equation (4.11) derived on the basis of continuum theory has somewhat less restrictions in validity as contrasted to eqn. (2.122) derived from kinetic theory.[83]

B. Relations between Fluxes and Driving Forces

To relate the diffusional fluxes to the thermodynamic forces, macroscopic theory relies on the thermodynamics of irreversible processes.† Accordingly we have

$$\mathbf{j}_\alpha = \beta_{\alpha 0}\mathbf{X}_0 + \sum_{\alpha=A}^{n} \beta_{\alpha\gamma}\mathbf{X}_\gamma, \tag{4.13}$$

$$\mathbf{h} = \beta_{00}\mathbf{X}_0 + \sum_{\alpha=A}^{n} \beta_{0\gamma}\mathbf{X}_\gamma, \tag{4.14}$$

where $\beta_{0\gamma}$ are the phenomenological coefficients, and the thermodynamic force vectors are given by

$$\mathbf{X}_0 = -\nabla \ln T, \tag{4.15a}$$

$$\mathbf{X}_\gamma = -[T\nabla(\hat{\mu}_\alpha/T) + \hat{H}_\alpha \nabla \ln T - \mathbf{F}_\alpha], \tag{4.15b}$$

$\hat{\mu}_\alpha$ denoting the (partial specific) chemical potential of species.

Two important relations between the coefficients $\beta_{\alpha\gamma}$ have been determined that considerably reduce the number of unknowns in eqns. (4.13 and 4.14). Onsager[81] has determined that

$$\beta_{\alpha\gamma} = \beta_{\gamma\alpha}(\alpha, \gamma = 0, 1, 2 \ldots n). \tag{4.16}$$

And making use of the fact that the production of entropy by heat conduction and by diffusion must be positive,

$$\sum_{\alpha=A}^{n} \beta_{\alpha\gamma} = 0 \tag{4.17}$$

It follows from this pair of expressions that there should be only $\frac{1}{2}n(n+1)$ independent phenomenological coefficients.

The next step in the discussion requires that we identify the phenomenological coefficients $\beta_{\alpha\gamma}$ with the ordinary diffusional coefficients. Restricting the arguments to a mixture of ideal gases, and applying the definition of the chemical potential, we find that[83]

$$\mathbf{j}_\alpha = -p\sum_{\alpha=A}^{n} \frac{\beta_{\alpha\gamma}}{\rho_\alpha}\mathbf{d}_\alpha^a - D_\alpha^T \nabla \ln T. \tag{4.18}$$

Here we have separated the diffusion by thermal gradients from the mechanical gradients. The term $\mathbf{d}_\alpha^a$ is

$$\mathbf{d}_\alpha^a = \nabla(n_\alpha/n_G) + \rho_\alpha\left(V_\alpha - \frac{1}{\rho_G}\right)\nabla \ln p - \frac{\rho_\alpha}{p}\left(\mathbf{F}_\alpha - \frac{1}{\rho_G}\sum_{\beta=A}^{n} \rho_\beta \mathbf{F}_\beta\right), \tag{4.19}$$

where V_α = partial specific volume of species α. Hence the mechanical part is divided into

† These types of relations have been used recently by Derjaguin *et al.*[85,86] in attempting to elucidate the relations for the thermal force and the diffusiophoretic force.

three parts; that of ordinary diffusion $\nabla\rho_\alpha$, pressure diffusion ($\nabla \ln p$), and diffusion by external forces. The diffusion coefficients in the multicomponent system are then *defined* by the relation

$$\mathbf{j}_\alpha = \frac{n_G^2}{\rho_G} \sum_{\substack{\alpha=A \\ \alpha\neq\gamma}}^{n} m_\alpha m_\gamma D_{\alpha\gamma}^a \mathbf{d}_\gamma^a - D_\alpha^T \nabla \ln T, \tag{4.20}$$

where $D_{\alpha\gamma}^a$ are the ordinary diffusion coefficients and D_α^T denotes the coefficient of thermal diffusion.

In the absence of radiation,† the heat flux also may be written in a form analogous to (4.20):

$$\mathbf{h} = -\beta_{00}\nabla \ln T - p \sum_{\alpha=A}^{n} \frac{\beta_{0k}}{\rho_k} \mathbf{d}_\gamma^a. \tag{4.21}$$

Transport in a Binary System

If only an ideal two component system is of interest, the relation (4.20) is greatly simplified. It reduces to

$$\mathbf{j}_A = -\rho_G D_{AB} \mathbf{d}_A - D_A^T \nabla \ln T, \tag{4.22}$$

where

$$\mathbf{d}_A = \nabla c_A + \frac{(m_B - m_A)}{\frac{1}{2}(m_A + m_B)} c_A c_B \left[\nabla \ln p - \frac{\mathbf{F}_A - \mathbf{F}_B}{p(V_A - V_B)}\right]. \tag{4.23}$$

Similarly, the expression for the heat flux h in a binary system may be found:[83]

$$\mathbf{h} = -k_G \nabla T + p(V_A - V_B) \frac{D_A^T}{2\rho_A \rho_B D_{AB}} \frac{\mathbf{j}_A(m_A + m_B)}{(m_A - m_B)}. \tag{4.24}$$

C. Maxwell's Diffusion Laws for Multicomponent Systems

The analysis in some studies have found it more convenient to use forms of the mass fluxes outlined many years ago by Maxwell.[89] Merk[83] has derived Maxwell's relations in a straightforward way from eqn. (4.20) by introducing the diffusion velocity

$$\mathbf{v}_\alpha = \mathbf{q}_\alpha - \mathbf{q}. \tag{4.25}$$

It is simpler if the net velocity associated with thermal diffusion is split in such a way that

$$\mathbf{v}'_\alpha = \mathbf{v}_\alpha + \frac{D_\alpha^T}{\rho_\alpha} \nabla \ln T. \tag{4.26}$$

Maxwell's coefficients of diffusional resistance $C_{\alpha\beta}$ are defined such that

$$D_{\alpha\beta}^m = 1/C_{\alpha\beta}. \tag{4.27}$$

† Unless otherwise specified $\mathbf{h}_r$, the heat flux by radiation is disregarded in subsequent discussion.

Substitution of (4.27) and (4.26) into (4.19) gives

$$\mathbf{d}_\alpha^a = \sum_{\alpha=A}^{n} C_{\alpha k} c_\alpha^\circ c_\gamma^\circ (\mathbf{v}_\gamma' - \mathbf{v}'), \tag{4.28}$$

where the circled quantities are molar concentrations, $c_\alpha^\circ = m_G c_\alpha / m_\alpha$. This equation is essentially Maxwell's diffusion law.

Merk has noted that for ideal gases, Curtiss and Hirschfelder showed that the first approximation to Maxwell's diffusional resistances are equal to the binary diffusivities, or

$$1/C_{\alpha\beta} \approx D_{\alpha\beta} \quad \text{(ideal gas)}. \tag{4.29}$$

The binary diffusion coefficients are generally known whereas the multicomponent diffusivities are not. Therefore, eqn. (4.29) is a very important relation which cannot be deduced from the thermodynamics of irreversible processes.

With eqn. (4.28) the heat flux may be rewritten in terms of Maxwell's law as[83]

$$\mathbf{h} = -k_G \nabla T + \frac{kT}{n_G} \sum_{\alpha,\gamma} \frac{n_\gamma}{D_{\alpha\gamma}^m} \frac{D_\alpha^T}{m_\alpha} (\mathbf{q}_\alpha - \mathbf{q}_\gamma), \tag{4.30}$$

where the thermal conduction of the gas is given by

$$k_G = \frac{\beta_{00}}{T} - \frac{k}{n_G} \sum_{\alpha,\gamma} \frac{n_\alpha n_\gamma}{D_{\alpha\gamma}^m} \frac{D_\alpha^T}{\rho_\alpha} \left(\frac{D_\alpha^T}{\rho_\alpha} - \frac{D_\gamma^T}{\rho_\gamma} \right).$$

D. Simplification of the Macroscopic Equations of Conservation

The set of partial differential equations governing the conservation of matter and thermal energy are much too complex to be tractable by present analytical techniques. Therefore, it is customary to simplify them drastically for application to gas mixtures near equilibrium. Suppose we assume:

(1) all the Maxwell diffusivities are equal;
(2) the gas mixture is ideal, and consists only of two components A and B;
(3) the external forces exerted on all components are equal;
(4) pressure diffusion and thermal diffusion are negligible;
(5) $Re_i \to 0$, $Ma_i \to 0$ such that viscous dissipation is small and the gas is treated as incompressible.
(6) No chemical reactions are present.

The conservation equations then reduce to a much simpler form:
continuity of species:

$$\frac{\partial \rho_\alpha}{\partial t} + \mathbf{q} \cdot \nabla \rho_\alpha = D_{\alpha\beta} \nabla^2 \rho_\alpha, \tag{4.31}$$

and *conservation of energy*:

$$\frac{\partial T}{\partial t} + \mathbf{q} \cdot \nabla T = \frac{k_G}{\rho_G C_p} \nabla^2 T. \tag{4.32}$$

These relations and eqn. (3.1) with appropriate boundary conditions are those customarily used to describe gaseous transport processes to spheres.

If we denote ζ as either ρ_α or T, these equations may be written in the spherical coordinates (r, θ, ϕ) follows†:

$$\frac{\partial \zeta}{\partial t} + q_r \frac{\partial \zeta}{\partial r} + \frac{q_\theta}{r} \frac{\partial \zeta}{\partial \theta} + \frac{q_\phi}{r \sin \theta} \frac{\partial \zeta}{\partial \phi} = \hat{\mathscr{D}} \left[\frac{1}{r^2} \frac{\partial}{\partial r} \left(\frac{r^2 \partial \zeta}{\partial r} \right) + \frac{1}{r^2 \sin \theta} \frac{\partial}{\partial \theta} \left(\sin \theta \frac{\partial \zeta}{\partial \theta} \right) + \frac{1}{r^2 \sin \theta} \left(\frac{\partial^2 \zeta}{\partial \phi^2} \right) \right], \tag{4.33}$$

where $\hat{\mathscr{D}} = D_{AB}$ or $k_G/\rho_G C_p$.

The boundary conditions usually specified for the problems involve either the flux at the spherical surface, or the concentration ζ at r equal to R_i, or infinity. That is,

$$\left. \begin{aligned} &\text{(a)}\ \zeta = f_i(\theta, \phi) \quad \text{at } r = R_i, \\ &\text{or}\quad \zeta = f_2(\theta, \phi) \quad \text{at } r \to \infty. \end{aligned} \right\} \tag{4.34}$$

$$\text{(b)}\ \left[4\pi r^2 \hat{\mathscr{D}} \left(\frac{\partial \zeta}{\partial r} \right) \right]_{r=R_i} = f_3(\theta, \phi). \tag{4.35}$$

If there exists symmetry with respect to the angles θ and ϕ, these conditions reduce to forms like:

$$\left. \begin{aligned} &\text{(c)}\ \zeta = \zeta_s \quad r = R_i, \\ &\text{or} \\ &\text{(d)}\ \zeta = \zeta_\infty \quad r \to \infty. \end{aligned} \right\} \tag{4.36‡}$$

$$\text{(e)}\ \left[4\pi r^2 \left(\frac{\partial \zeta}{\partial r} \right) \right]_{r=Ri} = \text{const.} \tag{4.37‡}$$

General solutions to eqn. (4.33) with such boundary conditions may be found in treatises like Carslaw and Jaeger.[87]

4.2. Diffusional Processes to a Particle in a Stagnant Gas

The simplest class of solutions to eqn. (4.33) is those restricted to steady or stationary state in a stagnant medium§ where $\partial \zeta / \partial t = 0$. Equation (4.33) then reduces to

$$\nabla^2 \zeta = 0 \tag{4.38}$$

under steady state conditions. If the flux condition, (4.37), at the spherical surface is set equal to zero and we fix the concentration ζ at $r \to \infty$, we obtain by integration of (4.38) the trivial solution

$$\zeta = \text{const.} = \zeta_r \tag{4.39}$$

everywhere.

† Here ϕ is the angular coordinate taken about the axis normal to the axis around which θ is measured.

‡ These boundary conditions may not be strictly correct when the exchange processes take place at an appreciable rate. A more rigorous expression uses the kinetics of molecular processes near the spherical surface, with the accommodation coefficients. For mass transfer, see, for example, Schrage.[88]

§ These solutions also apply to cases where the particle is in uniform translation with the same velocity as a uniformly flowing gas.

On the other hand, if the concentration at the particle surface is fixed as ξ_s, and at $r = \infty$, as ζ_∞, integration of (4.38) yields

$$\frac{\zeta - \zeta_\infty}{\zeta_s - \zeta_\infty} = \frac{R_i}{r}. \tag{4.40}$$

If we make use of eqn. (4.40) to evaluate the net transfer of heat or material to the surface of a droplet, we find that

$$\left[4\pi r^2 \hat{\mathscr{D}} \left(\frac{\partial \zeta}{\partial r}\right)\right]_{r=R_i} = 4\pi R_i \hat{\mathscr{D}}(\zeta_\infty - \zeta_s). \tag{4.41}$$

Thus the boundary condition (4.37) also has to be satisfied in this case.

A. Growth by Condensation and Evaporation

In considering the growth of droplets in a supersaturated vapor, eqn. (4.41) conforms to a relation derived by Maxwell in 1877.[(89)]

$$\tilde{\Phi}^c_{Am} = 4\pi R_i D_{AB}(\rho_{A\infty} - \rho_{As}), \tag{4.42}$$

where component A is the condensable vapor.

Noting that $\rho_A = p_A M_A/RT$, Maxwell's equation may be rewritten in the form

$$\tilde{\Phi}^c_{Am} = \frac{4\pi R_i D_{AB} M_A}{RT}(p_{A\infty} - p_{As}), \tag{4.42a}$$

where M_A is the molecular weight of the species A. For evaporation into a drier medium, the expression for the evaporation rate at steady state would be the same as eqn. (4.42a) except that the sign of the pressure difference would be reversed.

Quasi-stationary Processes

The applicability of Maxwell's equation is somewhat limited in considering droplet growth or evaporation. Strictly speaking, mass transfer to a small droplet cannot be a steady process because the radius changes, causing a change in the evaporation rate. However, when the difference between the vapor density far from the droplet and at the droplet surface is very small, one can assume (see below) that the rate given by eqn. (4.42a) holds at any instant. That is, the evaporation of growth of the particle by diffusion proceeds as a *quasi-stationary* process, in analogy to the definition on p. 66 (Chapter 3).

Since the rate of change in mass of the droplet equals $\tilde{\Phi}^c_m$, the equation for mass transfer may be rewritten with

$$\frac{dm_i}{dt} = \tilde{\Phi}^c_m$$

as

$$R_i \frac{dR_i}{dt} = \frac{1}{2}\frac{dR_i^2}{dt} = \frac{D_{AB}}{\rho_L}(\rho_{A\infty} - \rho_{As}), \tag{4.43}$$

where ρ_L is the liquid density. Thus the surface area $S = 4\pi R_i^2$ changes as

$$\frac{dS}{dt} = \frac{8\pi D_{AB}}{\rho_L}(\rho_{A\infty} - \rho_{As}). \tag{4.44}$$

Upon integration of these equations, we find that

$$R_0^2 - R_i^2 = \frac{2D_{AB}}{\rho_L}(\rho_{A\infty} - {}_{As})\,t \tag{4.45}$$

and

$$S_0 - S = \frac{8\pi D_{AB}}{\rho_L}(\rho_{A\infty} - \rho_{As})\,t, \tag{4.46}$$

where subscript 0 refers to the initial conditions of the particle. Thus in the quasi-stationary state, a single droplet growing or evaporating by diffusion of vapor should display a linear variation with time.

Influence of Droplet Curvature

In the case of vapor transport to the surface of very small droplets, the correction of vapor pressure for surface curvature has to be taken into account. As an example, consider small droplets evaporating in an atmosphere saturated with the vapor of the liquid. Application of the Kelvin–Gibbs equation (9.16) indicates that, for small values of the ratio of the vapor pressure at the surface p_{As} to that far from the droplet, $p_{A\infty}$,

$$\frac{p_{As} - p_{A\infty}}{p_{A\infty}} = \frac{\rho_{As} - \rho_{A\infty}}{\rho_{A\infty}} \approx \frac{2\sigma M_A V_{AL}}{R_i kT} \tag{4.47}$$

where V_{AL} is the volume of an A molecule in the liquid phase, and $p_{A\infty}$ refers to the vapor pressure of A in equilibrium with a flat surface of liquid at the same temperature as the droplet. If the concentration jump at the droplet surface is disregarded (see Section 6.2A), the vapor concentration near the droplet becomes, with eqn. (4.40)

$$\rho_A \approx \rho_{A\infty} + (\rho_{As} - \rho_{A\infty})\frac{R_i}{r} = \rho_{A\infty}\left(1 + \frac{2\sigma M_A V_{AL}}{rkT}\right). \tag{4.48}$$

In other words, the concentration distribution maintains a constant value during diffusional transfer, and, as Fuchs[90] has noted, steady state conditions only develop. The rate of evaporation or growth also remains a constant:

$$\tilde{\Phi}^c_{Am} = 8\pi\rho_{A\infty}\frac{\sigma M_A D_{AB} V_{AL}}{kT}. \tag{4.49}$$

This relation evidently was derived first by Nestle (see also Fuchs[90]).

B. Temperature Changes during Evaporation or Condensation

Strictly speaking, either evaporation or condensation of droplets has to be accompanied by thermal changes resulting from release or absorption of latent heat. The simultaneous transfer of thermal energy and material to or from a droplet can, of course, be calculated within the continuum approximation from eqns. (4.31 and 4.32). If steady heat conduction in the gas only is considered, the temperature distribution around the droplet is given by eqn. (4.40), and the corresponding heat flux has the form given by eqn. (4.41).

For the case of steady condensation or evaporation in a medium whose temperature is

T_∞ as $r \to \infty$, the heat transferred to or from the particle equals that evolved during condensation or consumed during evaporation, or

$$H = \tilde{\Phi}^c_{Am}\lambda_v, \tag{4.50}$$

where λ_v is the latent heat of vaporization of the liquid. Using eqns. (4.42), (4.40) and (4.41), we link the temperature field and the vapor concentration by the relation

$$T_s - T_\infty = \frac{\lambda_v D_{AB}}{k_G}(\rho_{As} - \rho_{A\infty}) \tag{4.51}$$

or

$$T_s - T_\infty = \frac{\lambda_v D_{AB}}{k_G R\overline{T}}(p_{As} - p_{A\infty}), \tag{4.51a}$$

where $\overline{T}$ is a mean value of temperature between the extremes of T_s and T_∞. These relations also were derived by Maxwell in his 1877 work.[89] And they indicate that evaporation or growth by condensation of vapor under quasi-stationary conditions is accompanied by temperature changes that are independent of size.

The ratio $\chi = (T_\infty - T_s)/(\rho_{As} - \rho_{A\infty})$ is often called the *psychrometric ratio*. We note that if the Lewis number, $Le = (k_G/\rho_G C_p D_{AB})$, is unity, the dimensionless ratios

$$\frac{\rho_A - \rho_{A\infty}}{\rho_{As} - \rho_{A\infty}} = \frac{T - T_\infty}{T_s - T_\infty} \tag{4.52}$$

provided the boundary conditions for T and ρ_A at the droplet surface and far from the droplet are the same. The droplet temperature during condensation or evaporation is nearly uniform because the heat influx and heat utilization both occur at the droplet surface. Then the appropriate boundary conditions ρ_{As} and T_s are constant since ρ_{As} is fixed by T_s. Then from (4.52) the distributions of vapor density ρ_A and temperature remain in a fixed ratio equal to χ.

Variation in Vapor Diffusivity

When temperature changes occur in the vapor, variation in the vapor diffusivity also develops. Therefore Maxwell's relation for the growth rate should be modified to account for this influence. The diffusivity of vapor in a gas empirically varies with temperature as

$$\frac{D_{AB}}{D_{AB\infty}} = \left(\frac{T}{T_\infty}\right)^n. \tag{4.53}$$

Generally the thermal conductivity of a gas does not vary as strongly with temperature as the mass diffusivity. Hence, if it is assumed that the conductivity of the surrounding gas remains constant, then differentiating eqn. (4.40),

$$\frac{dr}{r^2} = \frac{dT}{R_i(T_\infty - T)}.$$

With substitution of this expression into a generalized form for the steady transfer rate $\tilde{\Phi}^c_{Am}$,

$$\tilde{\Phi}^c_{Am} = 4\pi(\rho_{As} - \rho_{A\infty})\Bigg/\int_r^\infty \frac{dr}{D_{AB}(r)\,r^2},$$

we find that [90]

$$\tilde{\Phi}^c_{Am} \approx 4\pi D_{AB} R_i (\rho_{As} - \rho_{A\infty}) \left[\frac{T^n_\infty}{T_\infty - T_s} \left(\frac{1}{T_s^{n-1}} - \frac{1}{T_\infty^{n-1}} \right) \right]^{-1}. \tag{4.54}$$

From the kinetic theory of gases, the exponent $n \approx 2$, and [90]

$$\tilde{\Phi}^c_{Am} \approx 4\pi D_{AB} R_i (\rho_{As} - \rho_{A\infty}) \frac{T_s}{T_\infty} = 4\pi R_i (\rho_{As} - \rho_{A\infty})(D_{ABs} D_{AB\infty})^{\frac{1}{2}}. \tag{4.55}$$

Correction for Radiation

The theory for quasi-stationary heat and mass transfer to a droplet neglect the effect of thermal radiation in the heat flux **h**. The net contribution per unit area of droplet due to surface radiation is given by

$$h_r = \epsilon_r \sigma_{st} (T^4_\infty - T^4_s), \tag{4.56}$$

where ϵ_r is the correction for emissivity not equal to a black body [ϵ_r (black body) = 1]. The ratio of the radiative flux to the net conductive flux then crudely equals

$$\frac{h_r}{h_c} = \frac{R_i \epsilon_r \sigma_{st} (T^4_\infty - T^4_s)}{k_G (T_\infty - T_s)} \approx 4 R_i \epsilon_r \sigma_{st} T^3_\infty. \tag{4.57}$$

For air at 290°K, this ratio is

$$\frac{h_r}{h_c} \approx 2.2 \epsilon_r R_i. \tag{4.57a}$$

Thus, as Fuchs[90] has pointed out, radiation may be disregarded for droplets of radii of 0.05 cm or less.

C. Non-steady State Transport Processes

In general it may not be possible to assume that diffusional transport to particles takes place in the steady state. Fuchs,[90] for example, has outlined certain rough considerations for the limitations of the quasi-stationary process in evaporation or condensation. It is clear from his remarks and those of Philip,[91] for example, that one must be careful not to apply the steady state approximations indiscriminately to aerosol behavior (see also the numerical solutions of Cable and Evans[92]).

It is possible to set very crude limits on the characteristic time after which steady state conditions will apply. In the diffusional processes, the relaxation time for changes should be the order of R_i^2/D_{AB}. For variations on scales of this magnitude or less, one would not expect the approximation of steady state to be valid.

In the case of evaporation or condensation of droplets, the characteristic time associated with change of radius is the order of $R_i^2/D_{AB}\zeta^{\oplus}$, where

$$\zeta^{\oplus} = \frac{\rho_A - \rho_{A\infty}}{\rho_{As} - \rho_{A\infty}}. \tag{4.58}$$

If the dimensionless differences in vapor density are small, the time for the diffusional process to approach a steady state is small compared with the time for an appreciable change in radius of the droplet. Thus in most cases relevant to evaporation a growth of aerosol particles, we expect qualitatively the diffusion process will closely approximate the quasi-stationary case except during a short initial stage.

To get a more quantitative idea of the limitations of the steady state concept, we have to consider solutions to the time dependent equations like eqn. (4.33). Let us again restrict our remarks to conditions where there are no angular asymmetries. However, we note in passing that growth by condensation frequently takes place on a nucleus of dust (see also Chapter 8) wherein asymmetries must be present at least initially in the growth process. Under these circumstances there may be at least an initial period of time in which vapor diffusion will occur with essentially one-dimensional character. Bearing this in mind, it is of interest to consider a generalized diffusion problem governed by

$$\frac{\partial \zeta}{\partial t} = \hat{\mathscr{D}} r^{1-m} \frac{\partial}{\partial r}\left(r^{m-1}\frac{\partial \zeta}{\partial r},\right) \quad r \geqslant R_i. \tag{4.59}$$

This expression retains the dependence on a radial coordinate r, but incorporates more general dimensions. For example, spherical symmetry involves $m = 3$, a cylindrical system requires $m = 2$, and in a flat surface, $m = 1$.

For growth or evaporation, the boundary conditions of interest are

$$\zeta = \rho_A = \rho_{A\infty}, \quad r > R_i, \quad t = 0, \tag{4.60a}$$

$$\zeta = \rho_A = \rho_{As}, \quad r = R_i, \quad t \geqslant 0, \tag{4.60b}$$

$$(\rho_A - \rho_{As})\frac{\mathrm{d}R_i}{\mathrm{d}t} = D_{AB}\left(\frac{\partial \rho_A}{\partial r}\right)_{r=R_i}, \tag{4.60c}$$

$$r = R_0, \quad t = 0. \tag{4.60d}$$

In the case where $r = 0$ at $t = 0$, there exists an exact solution to eqn. (4.59), which is given by Philip[91] as

$$\zeta^{\oplus} = 1 - \frac{\Gamma(1 - n/2, \eta_0^2)}{\Gamma(1 - n/2, \eta_0'^2)}, \quad \eta_0 \geqslant \eta_0'. \tag{4.61}$$

Here $\eta_0 = r/(4D_{AB}t)^{\frac{1}{2}}$, $\eta_0' = R_i/(4D_{AB}t)^{\frac{1}{2}}$, and Γ is the incomplete gamma function,

$$\Gamma(\beta, x) = \int_x^\infty \exp(-\gamma)\,\gamma^{\beta-1}\,\mathrm{d}\gamma.$$

The dimensionless radius is given by the relation

$$\exp(\eta_0'^2)(\eta_0')^m \Gamma\left(1 - \frac{m}{2}, \eta_0'^2\right) = \frac{\rho_{A\infty} - \rho_{As}}{\rho_L - \rho_{As}} = \theta.$$

The corresponding solution for the steady state is, in analogy to eqn. (4.40),

$$\zeta = (r/R_i)^{2-m}, \quad r \geqslant R_i, \tag{4.62}$$

$$R_i = \left[\frac{(2m-4)\,D_{AB}(\rho_{A\infty} - \rho_{As})\,t}{(\rho_L - \rho_{As})}\right]^{\frac{1}{2}}. \tag{4.63}$$

For most applications to the behavior aerosols, the ratio θ is the order of 10^{-6} or less. Hence the incomplete gamma function may be approximated by one or two terms of a power series. Philip[91] gives the following relations for the dimensionless radius based on the leading term in an expansion for the gamma function:

$$m = 1, \quad \eta_0' = \frac{2\theta^{\frac{1}{2}}}{\pi^{\frac{1}{2}}}\left[1 + \frac{2\theta}{\pi}\right], \tag{4.64a}$$

$$m = 3, \quad \eta_0' = 2^{\frac{1}{2}}\left[1 + \left(\frac{\pi}{8}\right)^{\frac{1}{2}}\theta^{\frac{1}{2}}\right] \tag{4.64b}$$

In contrast, the quasi-stationary state theory predicts

$$m = 1, \quad \eta_0' = 0. \tag{4.65a}$$

$$m = 3, \quad \eta_0' = (2)^{\frac{1}{2}}. \tag{4.65b}$$

Thus the quasi-stationary case deviates from the exact solution even in the approximation using only one term of the power series expansion for η_0'. In practice, however, it is clear that for conditions encountered in aerocolloidal system where $\theta \lesssim 10^{-6}$, this deviation is unimportant.

Growth of Particles of Finite Radius

The continuum model for describing particle growth by diffusion of vapor, strictly speaking, is applied to cases where a particle of finite initial radius exists. Thus the foregoing problem is largely of academic interest. We now must be concerned with the properties of solutions to the diffusion equations with initially finite radius. Unfortunately, the only available exact solution for this case is the constant radius problem. This situation represents an intermediate circumstance between increase in radius for condensation, and shrinking of radius for evaporation.

The solution to eqn. (4.33) with symmetry in θ and ϕ angles, and with boundary conditions given as:

$$\begin{aligned} &\text{(a)}\ \zeta = \zeta_\infty, \quad r > R_i, \quad t = 0, \\ &\text{(b)}\ \zeta = \zeta_s, \quad r = R_i, \quad t > 0; \end{aligned} \tag{4.66}$$

has been reported, for example, by Carslaw and Jaeger,[87] and the net transfer rate to the surface is

$$\tilde{\Phi}_{Am}^c = 4\pi R_i D_{AB}(\rho_{As} - \rho_{A\infty})\left[1 + \frac{R_i}{(\pi D_{AB} t)^{\frac{1}{2}}}\right] \tag{4.67}$$

Here we can readily see the significance of the characteristic time $t_d = R_i^2/D_{AB}$. Where the (real) time exceeds t_d by a factor of 10 or more, the steady transfer rate

$$\tilde{\Phi}_{Am}^c = 4\pi R_i D_{AB}(\rho_{As} - \rho_{A\infty}) \tag{4.68}$$

approximates the exact theoretical rate very well for the case where $R_i = \text{const}$.

Using the two exact solutions to the diffusion equation, [eqn. (4.61)], Philip [91] has estimated the approximate solution for the moving radius case for a particle of finite radius. In the zero initial radius problem, where diffusion of vapor is considered,

$$\tilde{\Phi}_{Am}^c = 4\pi R_i D_{AB}(\rho_{As} - \rho_{A\infty})\left[1 + \left(\frac{\pi\theta}{2}\right)^{\frac{1}{2}}\right]. \tag{4.69}$$

Thus from eqns. (4.67) and (4.69), we can write the rate of transfer as

$$\tilde{\Phi}^c_{Am} = \tilde{\Phi}^c_{Qs} + \tilde{\Phi}^c_{\text{int}} + \tilde{\Phi}^c_{\text{trans}}, \qquad (4.70)$$

where $\tilde{\Phi}^c_{Qs}$ denotes value for quasi-stationary conditions, $\tilde{\Phi}^c_{\text{int}}$ denotes the part associated with movement of the droplet boundary, and $\tilde{\Phi}^c_{\text{trans}}$ is the portion representing the initial one-dimensional nature of the process of interfacial transfer. In the constant radius case, $\tilde{\Phi}^c_{\text{int}}$ is zero. For the moving interface, $\tilde{\Phi}^c_{\text{int}}$ is equal to $4\pi D_{AB}(\rho_{As} - \rho_{A\infty})\, R_i/(\pi\theta/2)^{\frac{1}{2}}$. On the other hand, the effect of initial one-dimensionality of the process is zero in eqn. (4.69), but equal to $4\pi D_{AB}(\rho_{As} - \rho_{A\infty})\, R_i^2/(\pi D_{AB} t)^{\frac{1}{2}}$ for eqn. (4.67).

For the case where the particle has an initially finite radius R_0, Philip's perturbation solution for the condensation contains flux additives of the exact solutions above:

$$\tilde{\Phi}^c_{Am} = 4\pi D_{AB}(\rho_{As} - \rho_{A\infty})\, R_i \left[1 + \left(\frac{\pi\theta}{2}\right)^{\frac{1}{2}}\right] + \frac{4\pi D_{AB}(\rho_{As} - \rho_{A\infty})\, R_0^2}{(\pi D_{AB} t)^{\frac{1}{2}}}. \qquad (4.71)$$

At small times, eqn. (4.71) appears as a perturbation of eqn. (4.67) with an accounting of the interfacial movement. At large times eqn. (4.71) may be treated as a perturbation to eqn. (4.69), taking into consideration the initial one-dimensionality.

Introducing the dimensional radius $R^{\oplus} = R_i/R_0$, and the dimensionless time $t^{\oplus} = D_{AB}\theta t/R_0^2$, integration of (4.71) yields[91]

$$R^{\oplus} = (1 + 2t^{\oplus}) + \theta^{\frac{1}{2}} \frac{(2/\pi)^{\frac{1}{2}} \sinh^{-1}(2t^{\oplus}) + (\pi/2)^{\frac{1}{2}} t^{\oplus}}{(1 + 2t^{\oplus})^{\frac{1}{2}}}. \qquad (4.72)$$

The quasi-stationary solution for growth by condensation encompasses only the first term on the right-hand side of this expression.

In the case of evaporation, only the sign of terms in the dimensionless concentration θ changes because $\rho_A > \rho_{A\infty}$ in this case. Then the arguments using a perturbation analysis may proceed as in the above case of condensation. When there is a retreat of the interface by evaporation, there is no exact solution available to fix the value of $\tilde{\Phi}^c_{\text{int}}$. However, Philip assumes that one can take $\tilde{\Phi}^c_{\text{int}}$ for evaporation as minus that for condensation by assuming conditions of local continuity. An analysis given by Philip[91] by an alternate approximation to the perturbation confirms the validity of this approximation. With this approximation applied to the perturbation method, the leading terms solution for the dimensionless radius of an evaporating droplet gives

$$R^{\oplus} = (1 - 2t^{\oplus})^{\frac{1}{2}} - \theta^{\frac{1}{2}} \frac{(2/\pi)^{\frac{1}{2}} \sin^{-1}(2t^{\oplus}) - (\pi/2)^{\frac{1}{2}} t^{\oplus}}{(1 - 2t^{\oplus})^{\frac{1}{2}}}. \qquad (4.73)$$

Again the quasi-stationary solution yields only the first term on the right-hand side of (4.73).

To demonstrate how the error in the quasi-stationary approaches enters in, it is convenient to introduce the parameter Ê such that $\mathrm{E}\theta^{\frac{1}{2}}$ is the leading term in the fractional error of the quasi-stationary value of $t^{\oplus}$. Rearranging eqns. (4.72) and (4.73), we obtain

(a) *condensation*:

$$\hat{\mathrm{E}}(R^{\oplus}) = \frac{(8/\pi)^{\frac{1}{2}} \cosh^{-1} R^{\oplus}}{R^{\oplus 2} - 1} + \left(\frac{\pi}{2}\right)^{\frac{1}{2}}, \qquad (4.74)$$

(b) *evaporation*:

$$\hat{E}(R^{\oplus}) = \frac{(8/\pi)^{\frac{1}{2}} \cos^{-1} R^{\oplus}}{1 - R^{\oplus}} - \left(\frac{\pi}{2}\right)^{\frac{1}{2}}. \tag{4.75}$$

These expressions are plotted in Figs. 4.1 and 4.2 along with the results of an earlier analysis of Fuchs.[90] The result of Eqn. (4.74), drawn in Fig. 4.1, agrees with the exact relations in the limits of $R^{\oplus} \to 1$ and $R^{\oplus} \to \infty$. It seems likely that the error in the intermediate range of $R^{\oplus}$ also would be relatively small. Philip notes that Fuchs' older

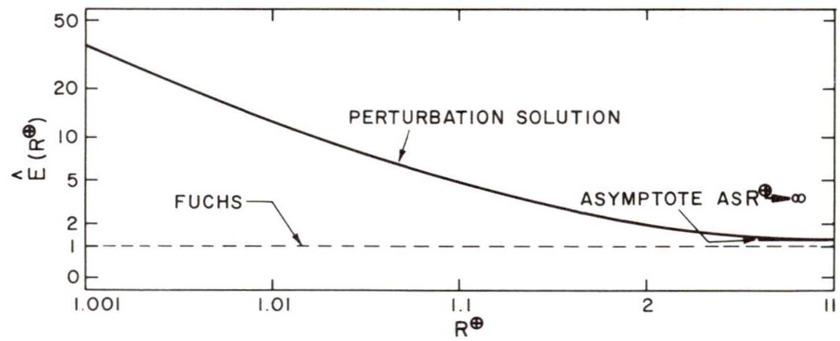

FIG. 4.1. Fractional error between quasi-stationary solution and Philip's perturbation solution for condensing droplet. (After Philip.[91])

analysis provided an estimate of the error $\hat{E} \approx 1$. By centering attention on the interfacial movement and neglecting the apparent initial one-dimensionality, Fuchs' estimate only applies to conditions of error in the limit $R^{\oplus} \to \infty$. However, for smaller values of $R^{\oplus}$, the fractional error $\hat{E}$ in the quasi-stationary solution as estimated by Philip seems to be much larger than the value predicted by Fuchs.

The estimated error in the quasi-stationary solution for evaporation as drawn in Fig. 4.2, shows how it should increase with increasing dimensionless size. The best estimate of

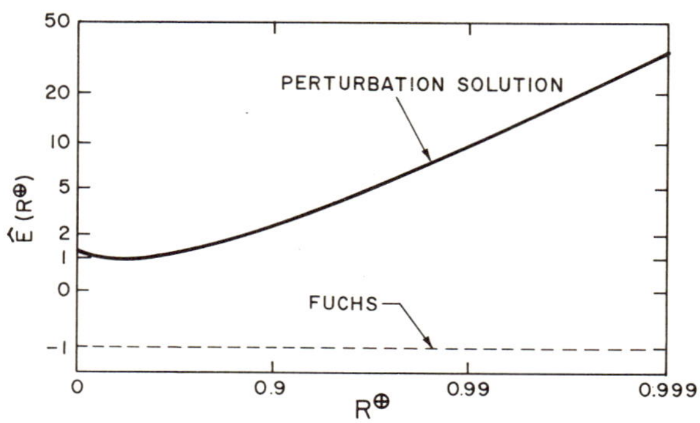

FIG. 4.2. Fractional error between quasi-stationary solution and Philip's perturbation solution for evaporating droplet. (After Philip.[91])

$\hat{E}(R^{\oplus})$ in this case should be the value at $R^{\oplus} \to 1$. The approximate analysis of Fuchs[90] is again substantially different for small times. Further, Philip suggests that the disregarding of the one-dimensionality effect by Fuchs leads to a fundamental "error" in the wrong sign of $\hat{E}$. The one-dimensionality contribution introduced by Philip far exceeds the negative effect of the interfacial movement.

Aside from the question of sign in the fractional error, there is a second important distinction between condensation and evaporation. For condensation, the contribution of the one-dimensionality effect becomes negligible as $R^{\oplus}$ increases, while this effect predominates for all values of $R^{\oplus}$ in evaporation. This contradicts Fuchs' conclusion that both the true rate of evaporation and condensation exceeds the estimate given by the quasi-stationary approximation.

Philip has checked the results from his perturbation analysis by examining approximate solutions to the diffusion equation with a continuity preserving boundary condition at large values of the radial coordinate. That is, instead of forcing the constant flux condition at large values of the radial coordinate, the exact solution, $\Phi^c_{Am} \propto \exp(-r^2/4D_{AB}t)$ is roughly approximated by setting the boundary condition $\rho_{A\infty}$ at a finite (moving) radius such that, for all time, the total flux of vapor to the droplet interface equals the rate of change in vapor content of the atmosphere surrounding the particle. The results of this alternate solution for $\hat{E}(R^{\oplus})$ are very similar to those obtained by the perturbation method. This independent check gives some confidence in Philip's conclusions.

Condensation or Evaporation with a Varying Density at the Particle Surface

Under certain circumstances diffusional growth of particles may involve a variation in time of the surface condition. We have discussed as an illustration the possibility of modification in vapor pressure as a result of the change in surface curvature. Another example is the influence of soluble salts on the growth of a droplet, on a soluble nuclei, say in the Earth's atmosphere (see also Chapter 9). It is possible, of course, that either one or both of these effects may eliminate growth or evaporation.

It is of interest to investigate the general properties of diffusional transfer with a radius dependence, and a time varying surface condition. Some years ago Fuchs[90] attempted to treat this question by considering the condition

$$\zeta = \zeta_s + f(t), \quad r = R_i, \quad t \geqslant 0 \tag{4.76}$$

in addition to the equations governing the constant radius problem, i.e. eqns. (4.33) and (4.36). The solution to this set of equations with eqn. (4.76) is known, and has been given by Carslaw and Jaeger[87] as

$$\zeta - \zeta_s = \frac{2}{\pi^{\frac{1}{2}}} \frac{R_i}{r} \int\limits_{(r-R_i)/2(D_{AB}t)}^{\infty} \left[t - \frac{(r-R_i)}{4D_{AB}\gamma^2} \right] \exp(-\gamma^2)\, d\gamma. \tag{4.77}$$

In deducing the rate of transfer at the surface $\bar{\Phi}^c_{Am}$, Fuchs evidently made an error leading to an incorrect form. Philip[91] has reported the correct relation to be

$$\bar{\Phi}^c_{Am} = 4\pi D_{AB} R_i \left[\hat{f}(t) + \frac{R_i \hat{f}(+0)}{(\pi D_{AB} t)^{\frac{1}{2}}} + \frac{R_i}{(\pi D_{AB})^{\frac{1}{2}}} \int_0^t \frac{\hat{f}'(t-\gamma)}{\gamma^{\frac{1}{2}}}\, d\gamma \right]. \tag{4.78}$$

where $\hat{f}'$ denotes the derivative of the function $\hat{f}(t)$.

After evaluating the integral in eqn. (4.78), Philip finds that[91]

$$\tilde{\Phi}^c_{Am} = 4\pi D_{AB} R_i \left[\hat{f}(t) + \frac{R_i}{(D_{AB}t)^{\frac{1}{2}}} \left\{ \frac{\hat{f}(+0)}{\pi^{\frac{1}{2}}} + \frac{\Gamma(\beta + 1)}{\Gamma(\beta + \frac{1}{2})} [\hat{f}(t) - \hat{f}(+0)] \right\} \right] \tag{4.79}$$

provided that the function $\hat{f}$ has a form

$$\hat{f} = \hat{f}(+0) + At^{\beta}, \quad \beta > 0.$$

If $A > 0$, Philip showed that

$$\pi^{\frac{1}{2}} \leqslant \frac{(D_{AB}t)^{\frac{1}{2}}}{R_i} \cdot \frac{\tilde{\Phi}^c_{\text{trans}}}{\tilde{\Phi}^c_{Qs}} \leqslant \frac{\Gamma(\beta + 1)}{\Gamma(\beta + \frac{1}{2})}$$

in contrast to the problem specified by the constant condition at the surface,

$$\pi^{\frac{1}{2}} = \frac{(D_{AB}t)^{\frac{1}{2}}}{R_i} \cdot \frac{\tilde{\Phi}^c_{\text{trans}}}{\tilde{\Phi}^c_{Qs}}$$

Here $\tilde{\Phi}^c_{\text{trans}}$ represents the last term in brackets on the right-hand side of eqn. (4.79). Because the ratio of gamma functions in eqn. (4.79) is approximately $(\beta + \frac{1}{4})^{\frac{1}{2}}$, it can be shown that for β as large as 20, the ratio $\tilde{\Phi}^c_{\text{trans}}/\tilde{\Phi}^c_{Qs}$ will not exceed about five times that for the constant surface condition problem. This result suggests that perturbations of the quasi-stationary solution for the diffusional rate resulting from short period fluctuations in ρ_{As} may be only a little larger than those in the constant surface condition case. Philip's conclusion disagrees with Fuchs' remark that the quasi-stationary result is an unsatisfactory approximation for situations where $\beta > \frac{1}{2}$. Although the short time fluctuations in surface condition appear to modify the quasi-stationary results, much longer time changes may have a stronger influence. This point seems to need further clarification.

Experimental Studies of Evaporation and Condensation

Experimental verification of transport processes to spheres in a stagnant medium is marred by difficulties in avoiding spurious effects like natural convection resulting from density gradients in the gas. In the case of diffusion of vapor to a body, accurate comparisons between the experimental observations and the theoretical predictions have been clouded, e.g. by inadequate knowledge of the diffusivity.

A wide variety of experiments have been conducted, especially for the growth and evaporation of droplets by vapor transport. These investigations and their limitations have been reviewed in some detail by Fuchs.[90] They generally fall into two classes, those working with droplets on a substrate or suspended from fibers or wires. and those using particles in free fall. The usefulness of these studies has suffered from the above mentioned difficulties as well as from additional problems involved in accounting for heat flow in the suspension device for example, and difficulties in estimating accurately rates of growth of freely falling droplets. Most of the experiments involving free fall have made use of the Millikan cell technique.

Upon examination of the reported experimental results as reported by Fuchs,[90] it is evident that these observations serve only to qualitatively confirm the theoretical predictions. Despite this reservation, the extensive measurements of transport processes in other systems where the experimental difficulties are not so severe appear to provide a reasonably good confirmation of the general applicability of the macroscopic theory for such processes occurring in the continuum regime. Perhaps the weakest link between theory and experiment in the case of growth and evaporation, at this time, is the lack of observational knowledge about the range of practical applicability of the quasi-stationary approximation. Unfortunately, such information will be exceedingly difficult to obtain in the laboratory.

4.3. Diffusional Processes to a Particle Moving Relative to the Gas Phase

Whenever a body moves relative to the surrounding gas, it becomes necessary to account for transport by convection as well as by processes of molecular transfer. If the coordinate system is attached to the moving sphere we again may use eqns. (4.33)–(4.37) as the relations governing the transport processes. In contrast to the case of the stagnant medium, however, the convective terms, the last three terms on the right-hand side of eqn. (4.33) have to be retained. Suppose we substitute the dimensionless concentration or temperature

$$\frac{\zeta - \zeta_\infty}{\zeta_s - \zeta_\infty} = \zeta^\oplus \tag{4.80}$$

into eqns. (4.31) and (4.32). We can make these equations dimensionless by introducing a characteristic length, say the radius of the sphere R_i, and a characteristic time R_i/U_∞:

$$\frac{\partial \zeta^\oplus}{\partial t^\oplus} + \mathbf{q}^\oplus . \nabla^\oplus \zeta^\oplus = \frac{1}{Pe} \nabla^{\oplus 2} \zeta^\oplus, \tag{4.81}$$

where $t^\oplus = U_\infty t/R_i$, $\nabla^\oplus = R_i \nabla$, $\mathbf{q}^\oplus = \mathbf{q}/U_\infty$, $\mathbf{r}^\oplus = \mathbf{r}/R_i$, and the Péclét number, $Pe = U_\infty R_i/\hat{\mathscr{D}}$. The Péclét number is the characteristic ratio for convective diffusion. It represents a ratio of the rate of transport by fluid motion to the rate of transport by molecular processes. This number is, of course, the product of two other fundamental ratios introduced previously, i.e.

$$Pe = Re_i . Pr, \quad \text{or} \quad Pe = Re_i . Sc. \tag{4.82}$$

Under conditions of steady state where $\partial \zeta^\oplus/\partial t = 0$. the transport of heat and matter should depend on the Peclet number alone. It then follows from dimensional reasoning that

$$\left.\begin{aligned} &\text{(heat transfer)} \quad T^\oplus = f_1(\mathbf{r}^\oplus, t^\oplus, Re_i, Pr) \\ &\text{(mass transfer)} \quad \rho_A^\oplus = f_2(\mathbf{r}^\oplus, t^\oplus, Re_i, Sc). \end{aligned}\right\} \tag{4.83}$$

Generally, in dealing with transport processes by convection, one customarily uses the heat or mass transfer coefficients $\hat{h}$ or $\hat{k}$. These are defined as:

(heat transfer)

$$\hat{h} \equiv \frac{1}{4\pi R_i^2(T_s - T_\infty)} \int_S \left[-k_G \left(\frac{\partial T}{\partial r} \right)_{r = R_i} \right] \mathrm{d}S, \tag{4.84}$$

(mass transfer)

$$\hat{k} \equiv \frac{1}{4\pi R_i^2(\rho_{As} - \rho_{A\infty})} \int_S \left[-D_{AB} \left(\frac{\partial \rho_A}{\partial r} \right)_{r = R_i} \right] \mathrm{d}S, \tag{4.85}$$

where S is the surface area of the sphere. If we utilize the dimensionless ratio introduced above in eqn. (4.81), we obtain two dimensionless numbers; the Nusselt number Nu and the Sherwood number Sh (the Nusselt number for mass transfer). For spheres, these parameters are then defined as:

(heat transfer)

$$Nu \equiv \frac{2\hat{h}R_i}{k_G} = \frac{1}{2\pi}\int_{S^\oplus} -\left(\frac{\partial T^\oplus}{\partial r^\oplus}\right)_{r^\oplus=1} dS^\oplus; \tag{4.86}$$

(mass transfer)

$$Sh \equiv \frac{2\hat{k}R_i}{D_{AB}} = \frac{1}{2\pi}\int_{S^\oplus} -\left(\frac{\partial \rho_A^\oplus}{\partial r^\oplus}\right)_{r^\oplus=1} dS^\oplus. \tag{4.87}$$

From these two relations it follows that for conditions in which there exists no relative motion between the sphere and the medium ($Pe \to 0$),

$$Nu = Sh = 2. \tag{4.88}$$

And with the dimensionless ratios defined above, one also recognizes by dimensional reasoning from eqns. (4.86) and (4.87) that

$$Nu = f_3(Re_i, Pr, t^\oplus)$$

and (4.89)

$$Sh = f_4(Re_i, Sc, t^\oplus).$$

The problem of determining the solutions for convective transport of heat and material to spheres in the continuum regime now reduces to the determination of the functional relations in eqn. (4.89)

A. Convective Transport at Conditions of Steady State

The problem of determining the relations between the Nusselt number and the Péclét numbers for a sphere suspended fully in a continuum in the Stokes' regime is a difficult one even for the case of steady state. The bulk of the theoretical investigations thus have confined their attention to this problem.

We can formulate the basic equation for transport of energy or matter by rewriting eqn. (4.33) in dimensionless form, and setting $\partial\zeta^\oplus/\partial t = 0$:

$$\left.\begin{aligned} Pe\left[q_r^\oplus \frac{\partial \zeta^\oplus}{\partial r^\oplus} + \frac{q_\theta^\oplus}{r^\oplus}\frac{\partial \zeta^\oplus}{\partial \theta} + \frac{q_\phi^\oplus}{r^\oplus \sin\theta}\left(\frac{\partial \zeta^\oplus}{\partial \phi}\right)\right] &= \nabla^{\oplus 2}\zeta^\oplus \\ = \frac{1}{r^{\oplus 2}}\frac{\partial}{\partial r^\oplus}\left(r^{\oplus 2}\frac{\partial \zeta^\oplus}{\partial r^\oplus}\right) + \frac{1}{r^{\oplus 2}}\frac{\partial}{\partial \sin\theta}\frac{\partial}{\partial \theta}\left[\sin\theta\frac{\partial \zeta^\oplus}{\partial \theta}\right] &+ \frac{1}{r^{\oplus 2}\sin^2\theta}\left(\frac{\partial^2 \zeta^\oplus}{\partial \phi^2}\right). \end{aligned}\right\} \tag{4.90}$$

with the boundary conditions

$$\zeta^\oplus = 1 \quad \text{at} \quad \mathbf{r}^\oplus = 1 \quad \text{and} \quad \zeta^\oplus = 0 \quad \mathbf{r}^\oplus \to \infty.$$

The relations for the velocity vector $\mathbf{q}^\oplus$ may be found from the Stokes' solution, eqn. (3.5). We note, of course, that there is symmetry with respect to the angle ϕ so that only components $q_r^{\oplus}$ and $q_\theta^{\oplus}$ are of interest (see also Fig. 3.1). From the discussion above, the limit for the Nusselt number as $Pe \to 0$ has the value of 2. In addition, boundary layer theory has shown that:[93]

$$Nu \to (2)^{\frac{1}{3}}[0.991\,Pe^{\frac{1}{3}}] \tag{4.91†}$$

as $Pe \to \infty$. Hence the complete solution to this problem for all values of Pe must satisfy these lower and upper restrictions.

Solutions to the Convective Diffusion Problem for Small Pe.

One of the first to investigate this question was Fuchs.[94] He assumed that the conditions of steady flow in the Stokes' regime would imply that the concentration or temperature distribution around the translating sphere would differ little from the no flow case. The difference could be expressed as a proportionately small factor $U_\infty\tilde{\epsilon}$, proportional to U_∞. Then, assuming for simplicity that $\zeta_\infty \to 0$,

$$\zeta = \zeta_s\frac{R_i}{r} + U_\infty\tilde{\epsilon}. \tag{4.92}$$

Setting the net flux of heat or matter equal to zero,

$$\hat{\mathscr{D}}\nabla^2\zeta - \mathbf{q}\,.\,\nabla\zeta = 0,$$

and substituting (4.92) into this relation, we find that

$$\hat{\mathscr{D}}U_\infty\nabla^2\tilde{\epsilon} + \mathbf{q}\,.\left(\frac{R_i\mathbf{r}}{r^3} + \mathbf{U}_\infty\,.\,\nabla\tilde{\epsilon}\right) = 0. \tag{4.93}$$

Because the second term in the brackets of (4.93) has been assumed smaller than the first term, we may write

$$\nabla^2\tilde{\epsilon} \approx \frac{R_i\mathbf{q}\,.\,\mathbf{r}}{\hat{\mathscr{D}}r^3U_\infty}.$$

The fluid flow in the Stokes' regime is symmetrical relative to the equatorial plane of the sphere. Thus $\nabla^2\tilde{\epsilon}$ has to have values equal but opposite in sign at all points symmetrically located around this plane. If one makes an analogy to potential theory, the sphere must be uniformly polarized in distribution of temperature or concentration in the direction of fluid flow. This led Fuchs to conclude that the gradient in $\tilde{\epsilon}$ and consequently the change in $\nabla\zeta$ caused by the flow would have the same magnitude and direction at corresponding locations on the forward and rear surfaces. In other words, this almost trivial result states that the diffusional transfer remains unchanged and equal to the motionless case when $Re_i \ll 1$.

A later analysis by Krönig and Bruijsten[95] suggested that, in analogy to the expansion for the drag coefficient, eqn. (3.10), the Nusselt number could be written in terms of the Péclét number as

$$Nu = 2 + Pe + \frac{581}{480}Pe^2 + \ldots \tag{4.94}$$

† In succeeding sections the parameter Nu will be taken to imply also the Sherwood number with suitable redefinition of the Péclét number.

for small values of Pe. This result was obtained by substitution of a series form

$$\zeta^{\oplus} = \sum_{n=0}^{\infty} (Pe)^n \zeta_n^{\oplus} \tag{4.95}$$

into eqn. (4.90) The functions $\zeta_n^{\oplus}$ were determined by the recurrence formula

$$\nabla^2 \zeta_n^{\oplus} = q_r^{\oplus} \frac{\partial \zeta_{n-1}^{\oplus}}{\partial r^{\oplus}} + \frac{q_\theta^{\oplus}}{r^{\oplus}} \frac{\partial \zeta_{n-1}^{\oplus}}{\partial \theta},$$

Breiman,[96] on the other hand, concluded at about the same time that for small Pe

$$Nu = 2 + Pe + Pe^2 \ln 2Pe - 0.1336 Pe^2 + \ldots \tag{4.96}$$

This relation disagrees with eqn. (4.94) beginning in the second order terms. The difference in these two solutions for the same mathematical problem apparently lies in the methods of approximation used to evaluate the distribution of $\zeta^{\oplus}$. For example, the method of Krönig and Bruijsten involved a classical perturbation expansion of the convection equation, while Breiman applied a technique using a Green's function to transform the problem into an integral equation.

The perturbation method using eqn. (4.95) was also tried by Frisch.[97] However, incorrect assumptions about the concentration and velocity distribution led him to a result

$$Nu = 2 + O(Pe)^2,$$

which must be rejected.

Until recently, the relation for the Nusselt number at intermediate Péclét numbers was not known except for approximate solutions like Friedlander's[98] based on boundary layer theory.

The discrepancies in solutions to eqn. (4.90) have been re-examined by Acrivos and Taylor.[99] They noted that the approach originally taken by Krönig and Bruijsten[95] leads to the first term of the expansion in eqn. (4.95), or

$$\zeta_0^{\oplus} = \frac{1}{r^{\oplus}},$$

but the higher order terms in the series do not vanish at infinity. Even worse, for $n \geqslant 2$, the functions diverged. The reason for the failure in the normal expansion procedure lies in the fact that no matter how small Pe is, there is a region far from the spherical surface wherein the conduction and convection terms become the same order of magnitude. This prevents eqn. (4.95) from being a completely adequate approximation to the function $\zeta^{\oplus}$. This kind of difficulty is essentially the same as that encountered in attempting to construct an accurate solution for the flow field around the sphere at low Reynolds number. We recall from Chapter 3 that the straightforward perturbation correction of Stokes' solution leads to the problem of a divergence at $r \to \infty$. This serious deficiency in Krönig and Bruijsten's approach led Acrivos and Taylor to seek a solution for Nu by the method of singular perturbation expressions, as discussed, for example, by Proudman and Pearson.[14]

This technique requires the construction of an inner and outer expansion for the distribution $\zeta^{\oplus}$ such that the inner solution satisfies the boundary condition at the sphere, $r = R_i$; the outer solution vanishes as $r \to \infty$; and the two solutions match identically at some arbitrary distance from the surface (both solutions remain bounded at $Pe \to 0$).

Construction of the solution by the perturbation method of Proudman and Pearson[14] has yielded the relation:[99]

$$Nu = 2 + Pe + Pe^2 \ln Pe + 0.0608\, Pe^2 + \frac{Pe^3}{4} \ln (2Pe) \ldots \tag{4.97}$$

for $Pe < 1$.

This expression is quite similar to the series given by eqn. (3.10) for the drag coefficient. It represents an expression which will allow estimation of the Nusselt number in terms of the Péclét number up to $Pe \approx 0.5$. Like the case of the drag force discussed in Chapter 3, the first two terms of the series remain the same as obtained previously by classical methods. However, the quadratic terms and higher order terms become different. As Acrivos and Taylor have remarked, moreover, their singular perturbation method offers the advantage over the other techniques in that the results may be checked stepwise during the complicated analysis through the requirement for matching the inner and outer expansions somewhere in the space around the sphere.

Numerical values of the Nusselt number are tabulated over the range $0.1 \leqslant Pe \leqslant 0.5$ for the solutions of Krönig and Bruijsten, Breiman, and Acrivos and Taylor in Table 4.1. There is significant deviation between these results at values of Pe as low as 0.1. However, the calculations of Acrivos and Taylor should represent the more accurate values over the entire range of Péclét number up to 0.5.

Influence of Increasing Reynolds Number

The results presented by Acrivos and Taylor constitute a solution to the convective diffusion problem for vanishingly small Reynolds number. It is of interest to obtain an estimate of the variation in Nusselt number when a more realistic velocity profile than the Stokes' is used. A complete analysis of this question using the perturbation method becomes exceedingly cumbersome. However, Acrivos and Taylor[99] have examined the changes in functional dependence of Nu on Pe for a general velocity profile in the limits $Pe \to 0$, and $Pe \to \infty$.

As long as the velocity distribution in the fluid meets the requirement

$$q_r/q_\infty \to \cos\theta \quad \text{as} \quad \frac{r}{R_i} \to \infty$$

and

$$q_\theta/q_\infty \to -(1-\cos^2\theta)^{\frac{1}{2}} \quad \text{as} \quad \frac{r}{R_i} \to \infty,$$

TABLE 4.1. COMPARISON BETWEEN VARIOUS APPROXIMATE SOLUTIONS OF THE CONVECTIVE TRANSPORT EQUATION FOR LOW PÉCLÉT NUMBERS

Pe	Nusselt numbers		
	Acrivos and Taylor Eqn (4.97)	Krönig and Bruijsten Eqn. (4.94)	Breiman Eqn. (4.95)
0.1	2.084	2.112	2.115
0.2	2.165	2.248	2.231
0.3	2.259	2.409	2.334
0.4	2.379	2.594	2.414
0.5	2.534	2.803	2.467

it was found that[99]

$$Nu = 2 + Pe + O(Pe^2 \ln 2\, Pe).$$

In other words, the functional dependence of *Nu* on *Pe* will remain the same for the first two terms in the expansion.

In the limiting case where $Pe \to \infty$, arguments about the asymptotic behavior of laminar boundary layers are applied (e.g. Morgan and Warner[100]). In the resulting expression,

$$Nu = 1.24\, Pe^{\frac{1}{3}} \left[1 + \frac{Re_i}{8} + \frac{3}{40} Re_i^2 \ln 2\, Re_i + 0(Re_i^2)\right]. \quad (4.98a)\dagger$$

Using the formula given by Proudman and Pearson for the drag coefficient, similar to eqn. (3.10), we find that

$$Nu = 1.25\, Pe^{\frac{1}{3}}(C_{DS}/C_{DS})^{\frac{1}{3}}, \quad Pe \to \infty \quad (4.99)$$

where the ratio (C_D/C_{DS}) denotes the correction to Stokes' relation for the drag coefficient. Equation (4.99) includes a Reynolds number correction to the limiting form of the Nusselt number formula up to but not necessarily incorporating terms of $0(Re_i^2)$.

Acrivos and Taylor have remarked that this represents a rather interesting result. Equation (4.99) along with eqn. (4.97) allow us to infer that the functional relation between *Nu* and *Pe* as estimated using the Stokes' distribution for fluid velocity is less sensitive to an increase in Reynolds number than the Stokes' drag coefficient discussed in Section 3.1.

B. Quasi-stationary Evaporation and Growth of a Moving Droplet

The steady state solutions to the convective diffusion equation represent as before a quasi-stationary case for the evaporation or growth of droplets moving relative to their suspending medium. We anticipate that in the extreme of $Pe \to 0$ that the limitations discussed previously for quasi-stationary diffusional growth should also apply here. There are no experimental data to verify this conclusion, and extensive theoretical calculations on this subject have not been made. However, it seems reasonable to assume that the quasi-stationary assumption will apply in practice to the growth and evaporation of moving droplets for small *Pe*, and small values of the dimensionless concentration $\zeta^{\oplus}$.

For quasi-stationary conditions the net transfer rate of vapor to or from a moving sphere is

$$\tilde{\Phi}_m^c = 4\pi R_i D_{AB}(\rho_{As} - \rho_{A\infty}) \left[1 + \frac{Pe}{2} + \frac{Pe^2}{2} \ln 2Pe + \ldots, \right] Pe < 1. \quad (4.100)$$

The term in square brackets is often called the *wind constant*, f_w, which marks the increase in evaporation (condensation) rate with relative movement between the particle and the medium.

The fall in temperature of an evaporating droplet also may be estimated from the Nusselt number:

$$\frac{T_\infty - T_0}{\rho_{As} - \rho_{A\infty}} = \frac{\lambda_v D_{AB}}{k_G} \left[\frac{1 + (Re_i . Sc) + (Re_i . Sc)^2 + \ldots}{1 + (Re_i . Pr) + (Re_i . Pr)^2 + \ldots}\right].$$

† This dependence on *Re* differs from the expression for *Nu* at $Re \gg 1$ given by Fuchs.[90] His eqn. (11.22) is

$$Nu = 2\tilde{\beta} Pe^{\frac{1}{3}} Re_i^{\frac{1}{6}} \text{ for } Re_i > 100 \quad (4.98b)$$

($\tilde{\beta}$ is a constant). Of course, this Range of Re_i is not of interest to us in this book.

Because the Lewis number, $Sc/Pt \approx 1$, for water droplets in a gas, the temperature change by release of latent heat during particle growth is affected very little by convection.

Since the Péclét number is proportional to R_i, it follows from eqn. (4.100) that the rate of change in radius squared with time will no longer be constant as estimated in eqn. (4.43). Instead, the slope of a plot of R_i^2 vs, time for an evaporating or growing droplet should tend to decrease with increasing time in the quasi-stationary case.

Experimental Observations of Droplet Evaporation

Experimental observations to confirm the theoretical estimates of condensation or evaporation on droplets moving relative to their medium are equally as difficult to obtain as in the case of no particle motion. Because of the limitations of small Reynolds number, one must operate experiments at flow rates $\sim 0.2\,\text{cm sec}^{-1}$ with 0.5 mm particles, for example. In this range of flow, the accuracy of results are often severely reduced by the effects of natural convection, etc.

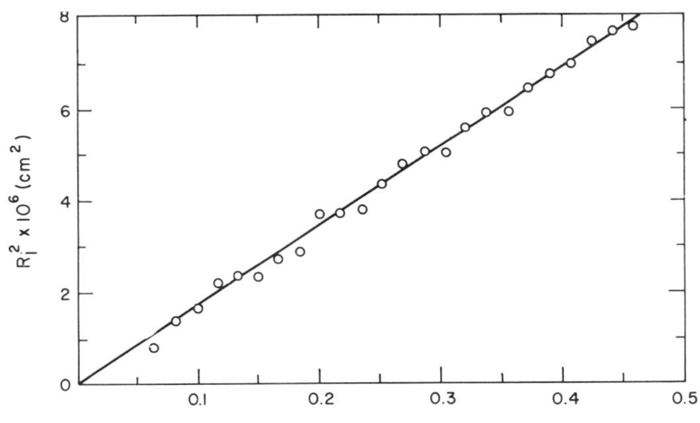

FIG. 4.3. Growth of a freely falling droplet of ethyl alcohol in hydrogen inside a Wilson cloud chamber. (From Hasen.[101])

Only a scant number of experiments have been reported for the diffusional transport to freely falling droplets in the Stokes' range. One of the first set of measurements was made by Hasen,[101] who observed the growth of drops of ethyl alcohol in hydrogen gas in a Wilson cloud chamber. The nominal supersaturation ratio in the chamber was 1.15. By photographing droplets at a rate of 30 photos per second the growth rate could be traced. A plot of the square of the radius of growing droplets as a function of time as estimated from Hasen's experiments is drawn in Fig. 4.3. Though reliable results can only be obtained after 0.2 sec, an extrapolation of the experimental curve to zero time passes through the origin. The experimental points yield a reasonably good straight line when plotted as R_i^2 vs. t. This is to be anticipated for the theory based on quasi-stationary growth. These data correspond to particles of radii 10–30 μ, with $Pe \approx 0.017$ to 0.50. The departure from a straight line in Fig. 4.3 from eqn. (4.100) is small for this range of Pe, and these results are in qualitative agreement with the theory. However, the growth as estimated from the slope of the line suggests that the condensation is somewhat slower than the theoretical rate.[90]

Hasen's method employed in the cloud chamber cannot be very accurate so that qualitative agreement between theory and experiment is probably all that can be expected.

Some recent studies of the growth of micron sized water droplets in air of Kassner's colleagues, as interpreted by Lund and Rivers,[102] and results of Schuster[103] (see also Section 9.3) suggest that the quasi-stationary theory can be applied quantitatively to experiments in cloud chambers.

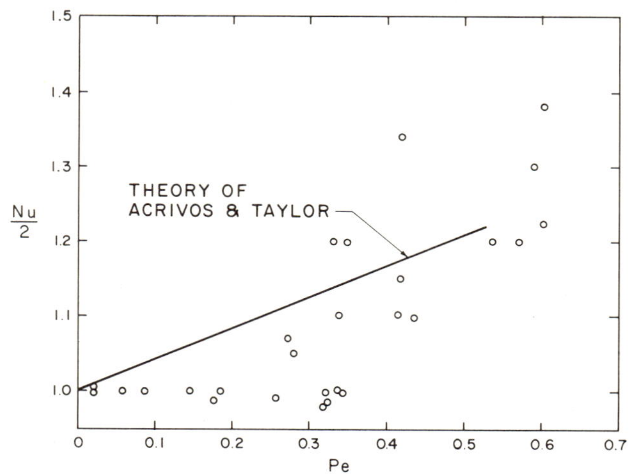

FIG. 4.4. Wind constant versus Péclét number for the evaporation of freely falling water droplets in air. (Data of Kinzer and Gunn.[104])

A set of experiments designed to evaluate the evaporation rate of small water droplets also has been reported by Kinzer and Gunn.[104] These measurements were made in the 5–70 μ radius droplets at temperatures of 0–40°C and 10–100 per cent relative humidity. The droplets were allowed to evaporate during free fall in long columns. By measuring variations of the rate of fall of the droplets and accounting for temperature changes, their reduction in size by evaporation could be calculated. These experiments had the advantage over the cloud chamber studies in that the vapor concentration of the surrounding medium was accurately measured and not calculated. Some results of Kinzer and Gunn's measurements are shown in Fig. 4.4. Here the wind constant is plotted as a function of Péclét number. There is considerable scatter in the experimental data above $Pe \approx 0.25$. However, the measurements exhibit a qualitative trend similar to the theoretical curve. For small values of Pe, i.e. $Pe \to 0$, the value of $f_w (= Nu/2)$ equals unity as predicted by the theory. Yet the experimental points do not coincide well with the theoretical curve based on eqn. (4.97) even for small Péclét numbers. Instead of the nearly linear increase in f_w as predicted by the theory for $Pe < 1$, a sharp break in the experimental curve is found at $Pe \approx 0.3$.

Nusselt numbers for mass transfer to spheres have been calculated empirically from Frössling's equation[90]

$$Nu \approx 2(1 + \tilde{\beta} Pr^{\frac{1}{3}} Re_i^{\frac{1}{2}}), \tag{4.101}$$

which reduces to the correct value of Nu when $Pe \to 0$, but does not correctly fit either the Krönig and Bruijsten equation or Acrivos and Taylor's more recent calculations for

$Pe < 1$. Most experiments for heat and mass transfer at Reynolds numbers greater than one-hundred give $\tilde{\beta} = 0.33$. However, Kinzer and Gunn's results for $Pe < 1$, suggest that $\tilde{\beta} \approx 0.65$. Because of this discrepancy and the inconsistencies between (4.100) and (4.97), for example, it would be desirable to repeat evaporation experiments on moving droplets in the range of small Péclét number.

4.4. Mass Transfer Coupled to Particle Motion

From Onsager's reciprocal relations, eqn. (4.14), one generally can expect a coupling between momentum and mass or heat transfer on the basis of irreversible thermodynamics, The only interaction of this kind presently known for the continuum regime, where $Kn_i \rightarrow 0$, is particle motion in a gas mixture with concentration gradients. This motion is related to *Stefan flow* of the gas, named for its discoverer, J. Stefan.[105] Because of this phenomenon, mentioned earlier in Chapter 3, it is possible for large particles to exhibit diffusiophoresis in a non-uniform gas mixture in the absence of external force fields. The magnitude of the diffusiophoretic force can be estimated within the continuum approximation by applying the transport equations summarized in Section 4.1. To see how this kind of particle motion comes about, let us consider first the case of a non-volatile non-absorbing particle suspended in a binary gas mixture. The diffusion field is imposed externally on the particle, say by a liquid surface at a different temperature from the gas. Since the particle is non-absorbing, the normal component of the gas velocity at the particle surface is zero, and the tangential component will also equal zero at the surface by the Stokes' condition. Under this set of conditions, the particle will tend to be at rest with respect to the center of mass of the surrounding gas mixture. This means that the particle will move along the velocity vector $\mathbf{q}_s$ of the Stefan current.[106, 107]

Consider the binary mixture, where the subscripts A and B will refer to the two components. In a binary system, the sum of the mass flux of component A and B is zero, or, from eqn. (4.8)

$$\mathbf{j}_A + \mathbf{j}_B = 0. \tag{4.102}$$

For a system essentially at constant pressure,†

$$\mathbf{j}_A = \rho_A(\mathbf{q}_A - \mathbf{q}) \approx - D_{AB}\nabla\rho_A \tag{4.103}$$

and

$$\mathbf{j}_B = \rho_B(\mathbf{q}_B - \mathbf{q}) \approx - D_{AB}\nabla\rho_B. \tag{4.104}$$

In addition, we note from eqn. (4.4) that

$$\rho_A + \rho_B = \rho_G. \tag{4.105}$$

From equations (4.102)–(4.104)

$$\mathbf{q}_s = \mathbf{q} = \frac{\rho_A\mathbf{q}_A + \rho_B\mathbf{q}_B}{\rho_G}. \tag{4.106}$$

The velocity of the Stefan current $\mathbf{q}_s$ is then identified as the mass average velocity in eqn. (4.106).[108] If the entire gas mixture is quiescent, $\mathbf{q} = 0$, with the Stefan current becoming zero. In the situation where component B diffuses through stagnant A, the velocity $\mathbf{q}_A$ is zero, another simple relation is derived between the concentration of the diffusing

† If the total pressure were exactly constant, there could be no mass flow, $\mathbf{q}$.

component B and the Stefan velocity $\mathbf{q}_s$. Using the relations between $\mathbf{j}_A$ and the concentration gradient of component A,

$$\mathbf{q}_s \approx \frac{-D_{AB}\nabla\rho_B}{\rho - \rho_B} = \frac{-D_{AB}\nabla c_B}{c_A}. \tag{4.107}$$

This theoretically predicted condition evidently would correspond to Facy's[109] observation of the drift of particles away from a free evaporating surface unbounded for large distances away from the liquid surface. The velocity of particle motion then equals the mass flow of gas (component B) towards the surface that just compensates for the diffusional flux of component A away from the surface.

For the condition where aerosols are either repelled from, or attracted to an evaporating or condensing spherical droplet, eqn. (4.107) becomes

$$\mathbf{q}_s \approx \frac{D_{AB}(c_{Bs} - c_{B\infty})\, RM_B}{c_A r^2 M_A}, \tag{4.108}$$

where the concentration gradient ∇c_B is approximated by the difference in vapor concentrations at the droplet surface, c_{Bs}, and the vapor concentration very far from the droplet, $c_{B\infty}$. The distance r^2 is the square of distance from the center of the droplet, M_B denotes the molecular weight of the diffusing vapor.

When the particle itself is undergoing phase transition, there is a mass flow of vapor normal to the droplet, but the boundary condition involving tangential diffusional flow (slip) has to be zero for the continuum approximation. In the first approximation, where $\rho_A/\rho_B \ll 1$, and $Re_i \ll 1$, the concentration distribution of vapor can be calculated independently of the solution to the aerodynamic problem equivalent to Stokes' analysis. For this case, Derjaguin and Dukhin[106] have solved the problem in which a volatile particle composed of A moves steadily in an infinite medium, whose gaseous component B is everywhere at rest with respect to a reference fixed on the droplet, while the imposed vapor stream of A passes through the gas B, and is maintained around the droplet by an external field of vapor.

The steady state solution for the concentration of vapor in such a system is found by fixing the concentration gradient at the surface, giving

$$\nabla\rho_A = \text{const.}, \quad r = R_i.$$

Having fixed the concentration gradient, the velocity field around the particle may be estimated, assuming that $\mathbf{q}_A = 0$. Boundary conditions on the surface of the sphere are:

$$\begin{aligned} &\text{tangential component:} \quad q_\theta(R_i, \theta) = 0; \\ &\text{normal component:} \quad q_r(R_i, \theta)\,\rho_B(R_i, \theta) - D_{AB}\left(\frac{\partial\rho_A}{\partial r}\right)_{r=Ri} = 0, \end{aligned} \tag{4.109}$$

where θ is the azimuthal angle over the sphere. The Stokes' approximation of the equation of motion now may be solved with eqn. (4.109), yielding the components of the tangential stress and normal stress acting on the particle surface. The force acting on the volatile particle in the externally imposed diffusion field is

$$\mathbf{F}_d^c = -\,6\pi\mu_G R_i D_{AB}(\nabla\rho_A)_{\text{ext}}/\rho_B. \tag{4.110}$$

The direction of motion of the particle then depends on the gradient in imposed concentration. The diffusiophoretic force in eqn. (4.110) is entirely equivalent to a "secondary" Stefan current set up at the surface of the volatile droplet. This force does not correspond to the diffusiophoretic force associated with a slip at the particle surface in the sense that slippage is not involved directly.

An analogy between the polarization of the electrical field around a sphere has been drawn to the distortion of concentration distribution around an evaporating or condensing droplet traveling through an external gradient in concentration.[107] The "diffusion polarization" results in a recoil force $\mathbf{F}_d^c$ opposing the externally generated Stefan flow, that is greater on the side of higher rate of evaporation, the side where the concentration of vapor is lower. Thus the volatile particle tends to align its motion with the direction of the imposed concentration gradient.

The direction of particle motion extends towards increasing concentration of vapor. If the imposed concentration gradient is set up by a phase change of the same material as the droplet, say from an evaporating surface, the motion of diffusional polarization is exactly compensated for by an opposing Stefan flow from that surface. Hence within the continuum approximation, the droplet will remain stationary in the zone of concentration gradient in an isothermal gas mixture. This is seen readily by converting eqn. (4.110) to a diffusion velocity by the mobility relationship and comparing this velocity with the Stefan flow $\mathbf{q}_s$ in eqn. (4.107).

When the temperature of the droplet is not the same as the surrounding medium, the balance of Stefan flow between droplet and an evaporating surface apparently does not take place. To account for the temperature difference, the problem of heat transfer between the droplet and the gas also must be incorporated into the model. Solving this problem for the conditions of no slip on the particles above, Dukhin and Derjaguin[110] found that the diffusiophoretic force predicted with heat and diffusion polarization is, in the case of interaction of two droplets 1 and 2,

$$\mathbf{F}_d^c = \frac{6\pi\mu_G D_{AB} R_2}{\rho_B}\left[A + \frac{\lambda_v D_{AB} A + k_p B}{(k_{p1} + 2k_{p2}\lambda_v D_{AB}\rho_{As}(T')/\Omega')}\frac{\rho_{As}(T)}{\Omega'}\right], \tag{4.111}$$

where A and B are constants, giving the concentration and temperature gradients at $r \to \infty$, and at the azimuthal angle $\theta = \pm 90°$,

$$\rho_B = \rho_{B\infty} + Ar\,.\cos\theta, \quad A = (\partial\rho_B/\partial r\cos\theta)_{\theta=\pm 90°}, \quad r \to \infty,$$

$$T = T_\infty + Br\,.\cos\theta, \quad B = (\partial T/\partial r\cos\theta)_{\theta=\pm 90°}, \quad r \to \infty.$$

T' is the psychrometric temperature, and $\Omega' = (b' + T')/2$, b' = constant given by the form of the equilibrium vapor pressure curve of component A, and λ_v is the latent heat of vaporization of the material.

The total diffusion force has to include the Stefan current from the externally imposed diffusion field, $F_{dE}^c = -6\pi\mu_G R_2 D_{AB} A/\rho_A$. The net force acting on the particle 2 then is the sum of the evaporational component from the droplet and externally imposed Stefan flow:[110]

$$\mathbf{F}_{dE}^c + \mathbf{F}_d^c = -\frac{12\pi\mu_G D_{AB}\rho_{As}(T')\left[\lambda_v D_{AB}(\nabla\rho_A)_{\text{ext}} + k_G(\nabla T)_{\text{int}}\right]}{(k_{p1} + 2k_{p2} + 2\lambda_v D_{AB}\rho_{As}(T')/\Omega')\,\Omega'\rho_B}. \tag{4.112}$$

The direction of the net diffusional force depends on the term in brackets, $[\lambda_v D_{AB}(\nabla\rho_A)_{ext} + k_{p1}(\nabla T)_{int}]$, where $(\nabla T)_{int}$ is the internal temperature gradient in the droplet 2. When the surface temperature is below psychrometric temperature, like a thawing ice particle, other particles are attracted. However, particles are repelled in cases of freezing particles where the surface temperature is above the psychrometric temperature from the release of latent heat during freezing.

Very little experimental information has become available that can verify the conclusions about the continuum motion of particles by Stefan currents in a diffusion field. The qualitative observations of Facy(109) tend to confirm the direction of migration as predicted by the theory. However, direct attempts to measure the diffusive force exerted between two evaporating droplets, or between non-volatile spheres and a surface in phase transition apparently have been obscured by the spurious influences of convection (see also Fuchs(2) and Derjaguin *et al.*(108)). Interestingly enough, experimental data, though limited, too, are more plentiful for verifying the diffusiophoresis theory in the non-continuum regime (see Chapters 5 and 6).

CHAPTER 5

TRANSFER PROCESSES TO SINGLE PARTICLES THE FREE MOLECULE AND TRANSITION REGIONS OF KNUDSEN NUMBER

THE existence of gas dynamic effects which cannot be explained by continuum fluid mechanics has been known for a very long time indeed (e.g. A Fresnel[111]). Only within the past few years, however, have non-continuum effects in gas dynamics been investigated systematically using the tools of modern kinetic theory. The current state of non-continuum gas dynamics relevant to aerosol physics already has been surveyed in Chapter 2.

In this chapter some current approaches to the description of the important transfer processes to single aerosol particles in the transition and free molecule regions of Knudsen number Kn_i will be summarized.

In accordance with the discussion of Chapter 2, the transition region of Knudsen number for a single aerosol particle in an effectively infinite medium is defined approximately by present theoretical and experimental evidence by the bounds: $0.25 \lesssim Kn_i \lesssim 10$. The exact lower limit in this bound will usually depend on the particular transfer process involving the aerosol particles.

The free molecule region comprises the region $Kn_i \to \infty$. However, for many transfer processes only a small error will be introduced by making use of the free molecule description of transfer processes when $Kn_i \geqslant 10$.

In natural and technological phenomena involving aerosol particles, one frequently deals with transfer processes in the region $0.25 \leqslant Kn_i \leqslant \infty$. For example, in pollution of the lower atmosphere by particulate matter, the particles in the sub-micron range of size are by far the most numerous and constitute perhaps the greatest potential health and economic hazard. The initial, and in many cases later stages of formation of clouds, fogs, and smokes involve transfer processes to aerosol particles in these transition and free molecule regimes. Virtually all atmospheric processes involving particles in the upper troposphere and higher are in these regimes of Knudsen number. Other more specific instances of particles in these regimes for atmospheric pressure are illustrated in Fig. 1.2.

We shall discuss here some general methods for calculation of transfer processes to single aerosol particles from moments of the molecular velocity distribution function. Typical results for the free molecule region, for which the theory of transfer is exact, will be given. For the transition region there is no definitive theory, owing to the complications in this region. The present knowledge is incomplete. Consequently, we shall principally summarize some results which have been obtained using iterative techniques and the BGK model of the Boltzmann equation discussed in Chapter 2.

5.1. Transfer Processes to Single Particles from Moments of the Molecular Velocity Distribution Function

In this exposition it is presumed that the reader is familiar with the discussion of the

Boltzmann equation and its application to transfer processes in rarefied gas dynamics. We restrict ourselves to the consideration of a single aerosol particle suspended in an effectively infinite gas. The state of the gas at large distances from the particle can be specified in accordance with the particular transport process to be described. For emphasis one of the primary restrictions of this book is restated, namely that the state of the suspending gas is adequately described by the Boltzmann equation for the molecular velocity distribution function, $f(\mathbf{v}, \mathbf{x}, t)$ for a single gas molecule. Then for a single-component gas

$$\frac{\partial f}{\partial t} + \mathbf{v} \cdot \nabla f + \mathscr{E} \cdot \nabla_{\mathbf{v}} f = J(f f'), \tag{5.1}$$

where $f(\mathbf{v}, \mathbf{x}, t)$ is the number of molecules with velocity $\mathbf{v}$ at a point $\mathbf{x}$ in physical space at time t. The restriction, re-stated above, implies that transfer processes to the particle can only occur by molecular impacts on the particle surface. The intermolecular forces exerted by gas molecules at a distance from the surface of the particle cannot contribute to the transfer processes. If, for example, the suspending fluid were a dense gas or a liquid, it would be necessary to account for this contribution to some of the transfer processes.

A solution of the Boltzmann equation, eqn. (5.1), for proper boundary conditions, gives the velocity distribution function $f(\mathbf{v}, \mathbf{x}, t)$. If $f(\mathbf{v}, \mathbf{x}, t)$ can be found in this way, then the average of the rate of transfer of any function of the molecular velocities can be obtained by taking the appropriate moment in velocity space of this distribution function (cf. Chapter 2).

In macroscopic observations one is interested in the average of the fluxes (rates of transfer per unit area) of molecules or mass, momentum, and energy, i.e. in the molecular or mass flux, the heat flux to a particle, and the net force acting on a particle. In aerosol physics, the total rates of transfer to an aerosol particle are of primary importance. These total rates are obtained simply by integrating the appropriate fluxes over the surface S of the particle. This procedure gives

$$\tilde{\Phi} = \int \mathrm{d}S\mathbf{n} \, . \left\{ \int_{+} \mathrm{d}\mathbf{v} \, \mathbf{v} f^{+} + \int_{-} \mathrm{d}\mathbf{v} \, \mathbf{v} f^{-} \right\}. \tag{5.2}$$

for the total rate of molecular (mass) transfer $\tilde{\Phi}$.

$$\mathbf{F} = -\int \mathrm{d}S\mathbf{n} \, . \left\{ \int_{+} \mathrm{d}\mathbf{v} m_G \mathbf{v}\mathbf{v} f^{+} + \int_{-} \mathrm{d}\mathbf{v} \, m_G \mathbf{v}\mathbf{v} f^{-} \right\} \tag{5.3}$$

for the net force $\mathbf{F}$ acting on the particle, and

$$H = \int \mathrm{d}S\mathbf{n} \, . \left\{ \int_{+} \mathrm{d}\mathbf{v} \tfrac{1}{2} m_G v^2 \mathbf{v} f^{+} + \int_{-} \mathrm{d}\mathbf{v} \tfrac{1}{2} m_G v^2 \mathbf{v} f^{-} \right\} \tag{5.4}$$

for the total rate of transfer of translational energy H to the particle by the gas molecules. For polyatomic molecules, one must add in the rate of transfer of the other molecular energies, including internal energy modes. It is not yet possible to do this in a completely rigorous manner.

The parameters f^{+} and f^{-} indicate the velocity distribution functions of the gas at the particle surface. These functions, f^{+} and f^{-}, are defined for molecules for which $\mathbf{v} \, . \, \mathbf{n} > 0$ and $\mathbf{v} \, . \, \mathbf{n} < 0$ respectively, and the integrations are so restricted. $\mathbf{n}$ is a unit vector normal to a surface element $\mathrm{d}S$.

In principle, it is possible to calculate these rates of transfer for arbitrary Knudsen number from knowledge of the distribution functions f^{+} and f^{-}. However, the Boltzmann equation must be solved to obtain f^+ and f^-, and a general solution of the Boltzmann equation has not yet been obtained. Hence for very small Kn_i one seeks "normal" solutions of the Boltzmann equation and applies the Navier–Stokes equations with slip boundary conditions if necessary (cf. Chapters 2 and 3). For very large Kn, the free molecule region, considerable simplification arises and all the transfer processes can be described exactly as will be shown below.

The function f^+, as has been noted, represents the velocity distribution function for molecules leaving the particle surface. Hence f^+ at the surface represents a boundary condition on the general solution of the Boltzmann equation for a particle in an effectively infinite gas. The Boltzmann equation, obviously, can give us no information on the form of this boundary condition. Some law must be obtained which will give the distribution of reflected gas molecules. Most generally, one could describe the reflection mechanism by a probability function $\tilde{P}(\mathbf{v}, \mathbf{v}_1)$, where $\tilde{P}(\mathbf{v}, \mathbf{v}_1)\,d\mathbf{v}$ gives the probability that a gas molecule hitting the surface with velocity $\mathbf{v}_1$ would be reflected with a velocity lying in the range $(\mathbf{v}, d\mathbf{v})$. $f^+(\mathbf{v})$ at the surface would be given by the formal expression

$$f^+(\mathbf{v}) = -\underline{\int} \frac{\mathbf{n}\,.\,\mathbf{v}_1}{\mathbf{n}\,.\,\mathbf{v}} \tilde{P}(\mathbf{v}, \mathbf{v}_1) f^-(\mathbf{v}_1)\,d\mathbf{v}_1. \tag{5.5}$$

Unfortunately, at present, there is little information of a sort which could lead us to the detail necessary for obtaining an analytical expression for $\tilde{P}(\mathbf{v}, \mathbf{v}_1)$. At the present time, no one can offer any real improvement over the original prescription of the boundary condition proposed by Maxwell[(89)] and Smoluchowski[(112)], discussed already in Chapter 2.

These boundary conditions assume

$$f^+(\mathbf{v}) = \alpha_v f^{(0)+}(\mathbf{v}) + (1 - \alpha_v) f^-(\mathbf{v}'), \tag{5.6}$$

where $f^{(0)+}$ is the Maxwellian distribution of molecules leaving a particle surface element.

$$f^{(0)+} = n_G^+(m_G/2\pi k T^+)^{\frac{3}{2}} \exp\left[-m_G v^2/2kT^+\right]. \tag{5.7}$$

The velocity

$$\mathbf{v}' = \mathbf{v} - 2\mathbf{n}(\mathbf{n}\,.\,\mathbf{v}) \tag{5.8}$$

represents the velocity after the specular reflection of a molecule with incident velocity $\mathbf{v}$. The α_v are the accommodation coefficients which at present are adjustable parameters and have values in the range $0 \leqslant \alpha_v \leqslant 1$. As remarked in Chapter 2, there is a different accommodation coefficient associated with each transfer process: molecular or mass, momentum, and energy. Some typical coefficients of accommodation are tabulated in Tables 5.1 (evaporation), 5.2 (momentum), and 5.3 (thermal). These results derive from older techniques reviewed, for example, by Wachmann[(113)] and Paul[(114)].

Recently some progress has been made in the investigation of molecule–surface interactions by molecular beam techniques.[(115)] Figure 5.1 shows a result found from such experiments. It may be noted that the reflected beam can to a first approximation indeed be considered to consist of a specular and a diffuse part. There has yet to appear, however, a reliable theory which permits the *a priori* calculation of α_v.

In the following sections we shall consider some applications of the classical concepts of accommodation coefficients in the description of transfer processes in the free molecule and transition regions.

TABLE 5.1. TYPICAL DATA FOR THE EVAPORATION COEFFICIENT (FROM PAUL[114])

Material	α_c	Temperature range (°C)
Beryllium	≈1	898–1279
Copper	≈1	913–1193
Iron	≈1	1044–1600
Nickel	≈1	1034–1329
Ammonium chloride	$(3{\cdot}9–29{\cdot}0) \times 10^{-4}$	118–221
Iodine	0·055–1	−21·4 to 70
Mercurous chloride	0·1	97,102
Sodium chloride	0·11–0·23	601–657
Potassium chloride	0·63–1·0	—
Water (liquid)	0·04–>0·25	15–100
Benzene	0·85–0·95	6
Camphor (synthetic)	0·139–0·190	−14·5–5·5
Carbon tetrachloride	≈1	—
Diamyl sebacate	0·50	25, 35
Di-n-butyl phthalate	1	150–350
Ethyl alcohol	0·0241–0·0288	12·4–15·9
Glycerol	Unity, 0·052	18–70
Naphthalene	0·036–0·135	40–70
Tetradeconal	0·68	20

TABLE 5.2. SOME TYPICAL DATA FOR MOMENTUM ACCOMMODATION COEFFICIENTS (FROM PAUL[114])

System	α_m
Air on old shellac or machined brass	1·00
Air on oil	0·895
CO_2 on oil	0·92
Air on glass	0·89
He on oil	0·87
Air on Ag_2O	0·98
He on Ag_2O	1·0

TABLE 5.3. SOME VALUES OF THERMAL ACCOMMODATION COEFFICIENTS (FROM WACHMANN[113])

System	α_t
Air on machined bronze	0·91–0·94
Air on polished cast iron	0·87–0·93
Air on etched aluminium	0·89–0·97
Air on machined aluminium	0·95–0·97
O_2 on bright Pt	0·808
Kr on Pt	0·699
N_2 on Wo	0·35
He on Na	0·090
Ne on K	0·199
He on Wo	0·017

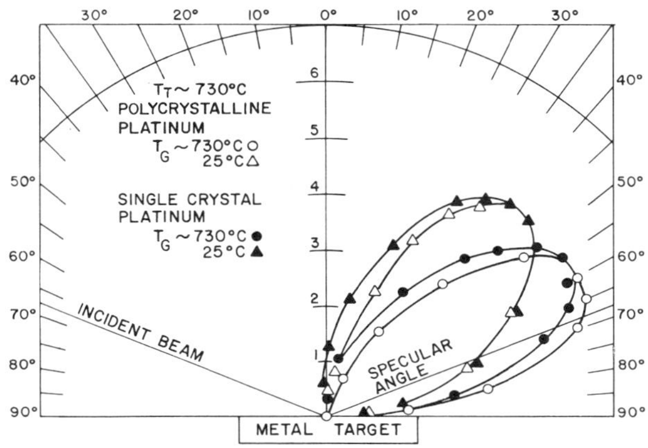

FIG. 5.1. Typical experimental results for the distribution of argon atoms scattered from a polycrystalline and a single crystal platinum target T at 730°C. (From Hinchin and Shepherd.[115])

5.2. Transfer Processes in the Free Molecule Regime

The descriptions by continuum theory of transfer processes from a gas to an aerosol particle given in Chapters 3 and 4 are exact within the restrictions imposed. We turn now to the free molecule regime where within stated restrictions exact descriptions of transfer processes also are possible. However, the accommodation coefficients or other descriptions of the gas–surface interaction must be taken into account so that there exist at present such uncertainties in the free molecule theory.

The free molecule regime for a single aerosol particle in an effectively infinite gas corresponds to the limit $Kn_i \to \infty$. That is, the molecular mean free path of the suspending gas is much greater than the radius of the aerosol particle. By reference to eqn. (2.93) of Chapter 2 it is seen that in this limit, the Boltzmann equation, in the absence of external forces, has the form

$$\mathbf{v} \cdot \nabla f = 0 \tag{5.9}$$

for steady state.

This equation states simply that the gas molecules, on the scale of the aerosol particle, originating at infinity, travel on straightline paths unless intercepted by the particle. Similarly, gas molecules originating from the particle surface travel thence to infinity on straight-line paths, that is without experiencing collisions. Therefore, the particle does not affect the velocity distribution function f^- of molecules which hit the particle. It is clear then that the distribution function f^- at the particle surface may be specified arbitrarily. Also, the distribution function f^+ of molecules leaving the particle surface can be obtained from eqn. (5.6). Therefore, transfer processes to a particle from a gas in any arbitrary state can be described.

An exhaustive account is not presented for transfer processes to particles for all the possible gaseous states. We shall discuss only a few examples which are important in aerosol physics and are illustrative of the type of calculations of transfer processes which can be carried out.

Consider first the molecular (mass) and heat transfer processes in the free molecule region which may be associated with the evaporation or condensation of aerosol particles. Then let us next consider three examples of momentum transfer to a particle—namely, ordinary free molecular drag, thermophoresis, and diffusiophoresis.

As already noted, for the free molecule region, $Kn \geqslant 10$, the simplification appears that the particle is so small that it does not affect the state of impinging gas molecules. Hence the effect of arbitrary states of the gas on the rates of evaporation or condensation can be accounted for quite easily in this region.

A. Molecular and Heat Transfer Rates for Evaporating or Condensing Particles in the Free Molecule Region

Mass Transfer in Binary Gas Mixtures

For this illustration, consider the region of a dilute binary gas mixture surrounding a spherical particle where for generality both components of the mixture composing the particle and the gas are evaporating or condensing. It will be assumed that the velocity of advance or regression of the surface owing to evaporation or condensation is negligible relative to the mean molecular velocity. Also, it will be assumed that the surface properties of the spherical particle have no angular dependence.

In the region of the particle, the state of the gas is specified by assigning number densities n_α^-, temperature T^-, mass velocity $\mathbf{q}$, and gradients ∇n_α, ∇T^-, $\nabla \mathbf{q}$. It will be assumed that the overall gas density is such that all fluxes in the gas may be given by expressions linear in the gradients. A reference coordinate system is taken fixed in the spherical particle. It is seen that a rather general state of the gas has been specified.

With the specification above, the distribution functions f_α^- for the gas surrounding the spherical particle have the form

$$f_\alpha^- = f_\alpha^{(0)-}\{1 - \mathbf{A}_\alpha . \nabla \ln T^- - \mathbf{D}_\alpha . \mathbf{d}_{\alpha\beta} - \bar{\bar{B}}_\alpha : \nabla \mathbf{q}\}, \tag{5.10}$$

where

$$f_\alpha^{(0)-} = n_\alpha^- (\beta_\alpha^-/\pi)^{\frac{3}{2}} \exp\left[-\beta_\alpha^- \{(\mathbf{v}_\alpha - \mathbf{q})^2\}\right], \tag{5.11}$$

$\beta_\alpha^- = (m_\alpha/2kT^-)$, and $\mathbf{v}_i$ is the molecular velocity vector of species i. The functions $\mathbf{A}_\alpha$, $\mathbf{D}_\alpha$, $\mathbf{B}_\alpha$ are defined in Chapter 2 in the discussion of the Chapman–Enskog "normal" solution of the Boltzmann equation (see also ref. (8) for a discussion of these quantities). $\mathbf{q}$ is the mass average velocity defined in Chapter 2, eqn. (2.68). $\mathbf{d}_{\alpha\beta}$ is

$$\mathbf{d}_{\alpha\beta} = \nabla x_\alpha + \frac{n_\alpha n_\beta (m_\beta - m_\alpha)}{n_G \rho_G} \nabla \ln p - \frac{\rho_\alpha \rho_\beta}{p \rho_G} (\mathbf{F}'_\alpha - \mathbf{F}'_\beta), \tag{5.12}$$

where x_α is the mole fraction of component α, p is the scalar pressure, and $\mathbf{F}'_\alpha$ is the external force per unit mass acting on α.

For the molecular (or molar) flux calculations, the distribution functions f_α^+ for the molecules leaving the particle surface are defined by

$$f_\alpha^+ = \alpha_{c\alpha} f_\alpha^{(0)+} + (1 - \alpha_{c\alpha}) f_\alpha^-, \tag{5.13}$$

and $\alpha_{c\alpha}$ is the so-called "condensation coefficient" for component α. The identification of $\alpha_{c\alpha}$ with the condensation coefficient implies that molecules of species α which achieve equilibrium with the surface also undergo condensation. One could distinguish a class of molecules which achieve equilibrium with the surface but do not undergo condensation. However, this class of molecules does not contribute to the net molecular flux and it can be lumped in with f_α^-. In the absence of other data, the coefficients $\alpha_{c\alpha}$ for condensation and evaporation are assumed to be the same. Typical values are listed in Table 5.1.

In eqn. (5.13)

$$f_\alpha^{(0)+} = n_\alpha^+(\beta_\alpha^+/\pi)^{\frac{3}{2}} \exp\left[-\beta_\alpha^+ v_\alpha^2\right],$$
$$\beta_\alpha^+ = (m_\alpha/2kT^+). \tag{5.14}$$

Therefore the total rate of evaporation or condensation $\tilde{\Phi}$ of the spherical particle is given by eqn. (5.2) which is generalized for a droplet composed of a binary mixture. In more concise notation[116]

$$\tilde{\Phi}^* = \sum_{\alpha=A}^{B} \tilde{\Phi}_\alpha^* = \sum_{\alpha=A}^{B} \int_S \mathrm{d}S\mathbf{n} \, . \sum_{\pm} \int_{\pm} \mathbf{v}_\alpha f_\alpha^\pm \, \mathrm{d}\mathbf{v}_\alpha, \tag{5.15†}$$

where

$$\int_{\pm} \equiv \operatorname{sign}(\mathbf{n} \, . \, \mathbf{v}_\alpha) \int_{-\infty}^{\infty} \int_{-\infty}^{\infty} \int_{0}^{\operatorname{sign}(\mathbf{n}\,.\,\mathbf{v}_\alpha)} . \tag{5.16}$$

It is apparent that the calculations can be readily extended to convex bodies of otherwise arbitrary shape.

When the magnitude of the mass average velocity, $|\mathbf{q}|$ is small relative to the mean molecular speed, that is

$$2\beta_\alpha \mathbf{v}_\alpha \, . \, \mathbf{q} \ll 1, \tag{5.17}$$

then a simplification results in the calculation of Φ^*. For this assumption[116]

$$f_\alpha^- = n_\alpha^-(\beta_\alpha^-/\pi)^{\frac{3}{2}} \exp(-\beta^- v^2)\{1 + 2\beta_\alpha^- \mathbf{v}_\alpha . \mathbf{q} - \mathbf{A}_\alpha . \nabla \ln T^- - \mathbf{D}_\alpha . \mathbf{d}_{\alpha\beta} - \bar{\bar{B}}_\alpha : \nabla \mathbf{q}\}, \tag{5.18}$$

where products involving the mass average velocity and the various gradients have been considered to be of second order.

For the approximation of eqn. (5.17), the following expression is obtained from eqns. (5.13), (5.15), and (5.18):

$$\tilde{\Phi}^* = \tilde{\Phi}_A^* + \tilde{\Phi}_B^* = 2\pi R_i^2 \left\{ \sum_{\alpha=A}^{B} \alpha_{c\alpha} \left[n_\alpha^+ \left(\frac{2kT^+}{\pi m_\alpha}\right)^{\frac{1}{2}} - n_\alpha^- \left(\frac{2kT^-}{\pi m_\alpha}\right)^{\frac{1}{2}} \right] \right.$$
$$\left. + \sum_{\alpha=A}^{B} \frac{8}{3} \alpha_{c\alpha} (2\pi m_\alpha k T^-)^{-\frac{1}{2}} \rho_G^- D_{AB} \mathbf{q} \, . \, \nabla c_\alpha^- - \sum_{\alpha=A}^{B} \alpha_{c\alpha} (2/\pi^{\frac{1}{2}}) \frac{n_\alpha^- a_\alpha}{T^-} \mathbf{q} \, . \, \nabla T^- \right\}, \tag{5.19}$$

where $c_\alpha^- = \rho_\alpha^-/\rho_G$ is the mass fraction of α. For the situation where the thermal conductivity k_G of the mixture does not vary appreciably from that of each pure component, the last term on the right may be written[116]

$$\sum \alpha_{c\alpha} \frac{2}{\pi^{\frac{1}{2}}} \frac{n_\alpha^- a_\alpha}{T^-} \mathbf{q} \, . \, \nabla T^- = -\left[\alpha_{CA} n_A^- + \alpha_{CB} n_B^- \left(\frac{m_B}{m_A}\right)^{\frac{1}{2}} \right] \frac{8k_G}{5\pi n_G kT} \frac{1}{\left(\dfrac{2kT}{\pi m_\alpha}\right)^{\frac{1}{2}}} \mathbf{q} . \nabla T^-, \tag{5.20}$$

† To distinguish between the transfer rates in the continuum and free molecule limits, a superscript* is used for the free molecule regime, whereas the superscript c has been used for the continuum results.

which defines the parameter a_α in this case. An order of magnitude calculation reveals that the last two terms on the right-hand side of eqn. (5.19) are of negligible order except for extremely large density and temperature gradients. Contributions of inertial terms have previous been disregarded as of order $|\mathbf{q}|^2$.

An interesting feature of eqn. (5.19) is that there are no contributions to the net flux of first order in $\mathbf{q}$ or the gradients. Thus for small relative $|\mathbf{q}|$ and gradients the evaporation and condensation rates in the free molecule regime are completely specified, as will be seen, for the gas phase by n_α^-, T^-, and the laws of interaction between the gas molecules and the particle surface. The only effect to first order in $\mathbf{q}$ is the indirect one of changing the local environment n_α^- and T^- of the body.

If one of the components of the binary gas system does not undergo condensation or evaporation upon adsorption by the surface an additional simplification results. Actually, this case is one of the most common for aerosol particles in nature where, usually, the particle is in air, an inert gas, which can contain a relatively small amount of a condensable substance such as water. If we let the subscript B indicate the inert component, then

$$\tilde{\Phi}_B^* = 0$$

which results in the specific relation between n_B^+, n_B^-, T^+, and T^-,

$$n_B^+ = n_B^-(T^-/T^+)^{\frac{1}{2}}.$$

In the general case, when the particle is in a binary gas mixture from which both components undergo evaporation or condensation on the particle, then n_α^+ is specified by the thermodynamic relation

$$n_\alpha^+ = n_\alpha^+(T^+, x_{\alpha s}, R_i),$$

where the number density of emitted molecules of species α leaving the surface is some function of the particle surface temperature, surface composition, and particle radius. It is usually assumed that n_α^+ is related to the more familiar vapor pressure p_α through the ideal gas relation:

$$p_\alpha = n_\alpha^+ kT^+.$$

We may consider the condensation and evaporation rates more generally by relaxing the assumptions of eqn. (5.17). In this case, one obtains contributions to $\tilde{\Phi}^*$ of all orders in $\mathbf{q}$. Except for extremely large gradients of the order one might observe in shock waves, the contribution of the gradients can still be neglected. With relaxation of eqn. (5.17), we obtain to all orders in $\mathbf{q}$ the following expression for $\tilde{\Phi}^{*(116)}$:

$$\left.\begin{aligned}\tilde{\Phi}^* = \tilde{\Phi}_A^* + \tilde{\Phi}_B^* = 2\pi R_i^2 \sum_{\alpha=A}^{B} \alpha_{c\alpha} \Bigg[n_\alpha^+ \left(\frac{2kT^+}{\pi m_\alpha}\right)^{\frac{1}{2}} \\ - n_\alpha^- \left(\frac{2kT^-}{\pi m_\alpha}\right)^{\frac{1}{2}} \cdot \tfrac{1}{2}\{(1 + 2\beta_\alpha^- q^2) E_0 + \exp(-\beta_\alpha^- q^2)\}\Bigg],\end{aligned}\right\} \quad (5.21)$$

where

$$E_0 \equiv \int_0^1 d\xi \exp(-\beta_\alpha^- q^2 \xi^2).$$

If B is an inert component in the gas phase, then

$$n_B^+ = n_B^- \left(\frac{T^-}{T^+}\right)^{\frac{1}{2}} \cdot \tfrac{1}{2}\{(1 + 2\beta_\beta^- q^2)\, E_0 + \exp(-\, \beta_\beta^- q^2)\},$$

which follows from eqn. (5.21) on setting $\tilde{\Phi}_B^* = 0$. Equation (5.21) is generalized easily to any number of components ν undergoing evaporation or condensation.

The rates of evaporation or condensation $\tilde{\Phi}_B^*$ and $\tilde{\Phi}_A^*$ are simply related to the rate of change of mass of the particle:

$$\tilde{\Phi}^* = \tilde{\Phi}_A^* m_A + \tilde{\Phi}_B^* m_B = \tfrac{4}{3}\pi \frac{\mathrm{d}}{\mathrm{d}t}(\rho_i R_i^3), \tag{5.22}$$

where m_A and m_B are the molecular masses of species A and B respectively. ρ_i is the mass density of the particle and is itself in the general case a function of the particle radius R_i. It has been assumed implicitly that the equilibrium state of the particle is established virtually instantaneously, so that at any instant the single value ρ_i characterizes the density of the particle.

Equation (5.22) is the starting point for the discussion of the lifetime of aerosol particles. For relatively involatile particles in inert gases, one can make the assumption that $T^+ = T^-$ In the general case, however, the temperature of the particle T^+ must be determined from the equation for the overall heat transfer rate H to the particle.

Heat Transfer to Particles

In the calculation of the heat transfer rate by eqn. (5.4), there are some difficulties. First, heat can be transferred to the particle independently of the surrounding gas molecules through the agency of electromagnetic radiation. If such radiation is present and is important, then it is necessary to add in a term to account for this mode of heat transfer. We can encounter a second difficulty with the calculation of H if the gas molecules possess energy in addition to the translational energy indicated in eqn. (5.4). For example, diatomic and polyatomic gases possess energy of rotation and internal vibration which can contribute to the heat transfer rate to the particle. While progress is being made in the kinetic theory of polyatomic gases, it is not yet possible to calculate on a rigorous basis the heat transfer rate by such molecules to a particle.

With these comments in mind, let us consider the calculation of the total heat transfer rate H with the restriction of eqn. (5.17). It is difficult at present to improve on the following expression for H^*[117].

$$H^* = H_A^* + H_B^* = 4\pi R_i^2 \sum_{\alpha=A}^{B} \alpha_{t\alpha}[\tfrac{1}{4}n_\alpha^+ v_\alpha^+(2kT^+ + \hat{U}_\alpha^+) - \tfrac{1}{4}n_\alpha^- v_\alpha^-(2kT^- + \hat{U}_\alpha^-)] + E_r, \tag{5.23}$$

where $\alpha_{t\alpha}$ is the thermal accommodation coefficient and the assumption is made that the accommodation of translational and internal energies is the same. Some values of $\alpha_{t\alpha}$ are listed in Table 5.2. The $v_\alpha^\pm$ are the mean molecular speeds; $\hat{U}_\alpha^+$ and $\hat{U}_\alpha^-$ represent for each molecule the energy—other than translational energy—leaving and arriving at the particle surface respectively. If the molecules are assumed to equilibrate rapidly at the surface, it can be assumed that[117]:

$$\hat{U}_\alpha^\pm = (c_{v\alpha} - \tfrac{3}{2}k)\,T^\pm \tag{5.24}$$

where $c_{v\alpha}$ is the heat capacity at constant volume per molecule. That is, it is assumed that the energies $\hat{U}_\alpha^\pm$ have their thermodynamic values. In eqn. (5.23) E_r represents the contribution to the heat transfer rate to the particle by electromagnetic radiation.

If one or both gaseous components are inert, i.e. neither condensing nor evaporating, then, according to the approximation of eqn. (5.17),

$$n_\alpha^+ = n_\alpha^- \left(\frac{T^-}{T^+}\right)^{\frac{1}{2}},$$

and the heat transfer rates H_α^* are completely determined in this instance from eqn. (5.23) with specification of n_α^-, T^-, and the thermal properties of the particle.

When evaporation or condensation occurs, eqns. (5.19) and (5.23) are closely coupled for volatile droplets, and simultaneous solution of the two equations is necessary.

If the approximation of eqn. (5.17) is relaxed, the following expression is obtained for the total heat transfer rate H^* for large relative mass average velocities $\mathbf{q}$ in a monatomic suspending gas[116]:

$$\begin{aligned} H^* = H_A^* + H_B^* = 4\pi R_i^2 \sum_{\alpha=A}^{B} \alpha_{t\alpha} & \left\langle n_\alpha^+ kT^+ \left(\frac{2kT^+}{\pi m_\alpha}\right)^{\frac{1}{2}} - n_\alpha^- kT^- \left(\frac{2kT^-}{\pi m_\alpha}\right)^{\frac{1}{2}} \right. \\ & \times \left\{ (\tfrac{1}{2}\beta_\alpha^- q^2 + 1)\,E_0 + \frac{\pi^{\frac{1}{2}}}{4}(\beta_\alpha^- q^2 + \tfrac{5}{2})\,\beta_\alpha^- q^2 \left[E_0\left(\frac{1}{2\beta_\alpha^- q^2} + 1\right) \right.\right. \\ & \left.\left.\left. - \frac{1}{2\beta_\alpha^- q^2}\exp(-\beta_\alpha^- q^2)\right]\right\}\right\rangle. \end{aligned} \tag{5.25}$$

For a particle composed of only one chemical species, the total heat transfer rate H^*, is related in the following manner to various changes of state of the particle:

$$\begin{aligned} \frac{H^*}{4\pi R_i^2} = & \left[-\lambda_v + C_{ps}(T^+ - T_0) + \frac{R_i}{3}\frac{\partial \hat{H}}{\partial R_i} \right] \rho_i \frac{\partial R_i}{\partial t} \\ & + [-\lambda_v + C_{ps}(T^+ - T_0)]\,\frac{R_i}{3}\frac{\partial \rho_i}{\partial t} + \frac{R_i C_{ps}}{3}\rho_i \frac{\partial T^+}{\partial t}. \end{aligned} \tag{5.26}$$

Here it is assumed that the equilibrium termperature distribution of the particle is obtained virtually instantaneously. In this equation, $\partial\hat{H}/\partial R_i$ is the change of enthalpy with particle radius. C_{ps} is the heat capacity of the particle. A reference temperature T_0 corresponding to the initial temperature of the particle can be chosen. λ_v is the heat of vaporization of the particle.

Equations (5.19), (5.22), (5.25), and (5.26) for $\tilde{\Phi}_B^* = 0$ are the starting points for the discussion of the lifetime of single component aerosol particles in the free molecule region.[118] The equations are closely coupled and can be solved numerically to give the particle radius as a function of time. The mass and heat transfer equations can be decoupled, as noted previously, for the evaporation of a relatively non-volatile particle in an inert gas.

From this discussion of the rates of mass and heat transfer in the free molecule region, it is clear that such transfer processes to particles can be described in great generality.

So long as the state of the suspending gas can be specified, the rate of transfer of mass and heat to a particle can be obtained. This statement holds, of course, in turbulent motion of the suspending gas, provided local concentrations, temperatures, and velocities are known as a function of time and space.

B. Momentum Transfer to Particles in the Free Molecule Region

The total rate of momentum transfer to an aerosol particle, i.e. the net force acting on the particle, can be calculated from eqn. (5.3). As with the mass and heat transfer processes just discussed, the calculation of the net force acting on a particle in the free molecule region is particularly simple.

In the free molecule regime, the distribution function f^- of molecules hitting the particle is chosen to describe any arbitrary state of the suspending gas. The distribution function f^+ is specified by the assumption that the molecule–surface interaction can be described by the empirical momentum accommodation coefficient α_m:

$$f^+ = \alpha_m f^{(0)+} + (1 - \alpha_m) f^-(\mathbf{v}'),$$

where the various symbols have been defined previously. Several typical values of α_m are tabulated in Table 5.3. With this information, the rate of momentum transfer, or the force, for an aerosol particle is readily calculated from eqn. (5.3).

If we consider the case of the isothermal drag on a sphere in the free molecule regime, we have for f^-,

$$f^- = n_G \left(\frac{m_G}{2\pi kT}\right)^{\frac{3}{2}} \exp\left\{-\frac{m_G}{2kT}(\mathbf{v} - \mathbf{q})^2\right\}. \tag{5.26}$$

with coordinates centered on the sphere.

For the approximation:

$$\beta \mathbf{v}.\mathbf{q} \ll 1$$

the free molecule drag $\mathscr{D}^*$ has a particularly simple form.[119]

$$\mathscr{D}^* = \tfrac{8}{3} R_i^2 n_G (2\pi m_G kT)^{\frac{1}{2}} [1 + (\pi/8)\alpha_m]\, \mathbf{q}, \tag{5.27}$$

in which the coefficient of $\mathbf{q}$ is recognized as the inverse of the mobility relationship for large molecules in the kinetic theory of gases. While it is true that there are no reliable methods for predicting α_m for aerosol particles and there are few experimental data, the maximum error in the calculation using eqn. (5.27) can only be $\pi/8$ since α_m ranges from 0 to 1.

If the linearization of eqn. (5.26) is not applied, then the particle mobility becomes a function of the relative velocity, and[119]

$$\mathscr{D}^* = \tfrac{8}{3} R_i^2 n_G (2\pi m_G kT)^{\frac{1}{2}} [(1 + \beta q^2)\tilde{E}_0 + \tilde{E}_2 - \beta q^2 \tilde{E}_4 + \alpha_m\{\tfrac{1}{2}\tilde{E}_0 - \tfrac{3}{2}(1 + \tfrac{2}{3}\beta q^2)\tilde{E}_2 - \tfrac{1}{2}\beta q^2 \tilde{E}_4\} + (\pi/6)\,\alpha_m]\, \mathbf{q}, \tag{5.28}$$

where

$$\tilde{E}_n \equiv \int_0^1 d\xi\, \xi^n \exp(-\beta q^2 \xi^2),$$

Phoretic Forces and Gas Non-uniformities

The forces acting on particles in the free molecule regime are coupled directly to non-uniformities in the suspending gas—gradients of concentration, temperature, etc. The free molecule theory for phoretic forces in a non-uniform gas has been treated by several workers. Perhaps best known are the calculations of Waldmann(120) and those of Bakanov and Derjaguin.(121) Let us consider first the problem of the thermal force. When a temperature gradient exists in an otherwise pure monatomic stagnant gas, for example, a thermally induced force acting on the suspended sphere develops. Since, in this case,

$$f^{-} \approx f^{(0)}\left[1 - \frac{2m_G k_G}{5pkT}\left(\frac{m_G v^2}{2kT} - \frac{5}{2}\right)\mathbf{v}.\nabla T\right] \tag{5.29}$$

the thermal force $\mathbf{F}_t^*$ is(120)

$$\mathbf{F}_t^* = -\frac{32}{15}R_i^2(k_G/\bar{v})\mathbf{v}\cdot\nabla T \tag{5.30}$$

where k_G is the thermal conductivity of the gas and ∇T is the temperature gradient in the gas.

If a particle travels with velocity $\mathbf{q}$ through a stagnant gas with an imposed temperature gradient, the thermal force and the frictional resistance balance in steady flow. Therefore in the free molecule region according to eqns. (5.27) and (5.30):

$$\mathbf{q} = \frac{1}{5[1 + (\pi/8)\,\alpha_m]}\frac{k_G}{n_G kT}\nabla T \tag{5.31}$$

The thermal force $\mathbf{F}_t^*$ is associated in part with heat transfer by molecular motion from the gas to the particle. If the particle receives heat from sources other than by such molecular motion, e.g. by absorption of incident electromagnetic radiation, a new force may also develop as a result of the temperature gradient within the particle. The force produced by such radiation is known as the *photophoretic force* and has been discussed in the free molecule region.(122) The photophoretic force $\mathbf{F}_p^*$ in the free molecule region has the approximate form

$$\mathbf{F}_p^* \approx -\pi R_i^3 p\mathbf{I}/6(\tfrac{1}{2}p\bar{v}R + k_p T), \tag{5.32}$$

where $\mathbf{I}$ is the radiation flux through the gas and k_p is the thermal conductivity of the particle.

If there are non-uniformities of composition of a suspending gas mixture, a force may act on suspended particles. This force is commonly termed the *diffusion or diffusiophoretic force.* In this case, the distribution function f_α^- for a component in the gas is given by the approximation

$$f_\alpha^- \approx f^{(0)}\left[1 + \frac{m_\alpha}{kT}\mathbf{v}_\alpha\cdot\mathbf{q}_\alpha\right], \tag{5.33}$$

where $\mathbf{q}_\alpha$ is the mean velocity vector of the molecules α. The diffusion force $\mathbf{F}_d$ acting on a sphere with radius R_i and velocity $\mathbf{q}_i$ follows from eqns. (5.6) and (5.33):

$$\mathbf{F}_\alpha^* \approx -\frac{32}{3}R_i^2\sum_\alpha \frac{p_\alpha}{\bar{v}_\alpha}\left(1 + \frac{\pi}{8}\alpha_{m\alpha}\right)(\mathbf{q}_i - \mathbf{q}_\alpha), \tag{5.34}$$

where $p_\alpha = n_\alpha kT$ is the partial pressure of species α in the gas.

The mean diffusion velocity $\mathbf{W}_\alpha$ of molecules α, relative to the mean molar velocity $\mathbf{w}$ is introduced:

$$\mathbf{W}_\alpha = \mathbf{q}_\alpha - \mathbf{w}, \tag{5.35}$$

where $\mathbf{w} = \sum_\alpha x_\alpha \mathbf{q}_\alpha$, and x_α is the mole fraction of α.

For a binary gas mixture,

$$x_1\mathbf{W}_1 = -x_2\mathbf{W}_2 = -D_{AB}\nabla x_A, \tag{5.36}$$

where D_{AB} is the binary diffusion coefficient.

Two special cases can be noted. If $\mathbf{w} \equiv 0$, a condition commonly termed in a diffusing system, equimolar counterdiffusion, one obtains for $\mathbf{F}_d^*$:

$$\mathbf{F}_d^* = \frac{8}{3} R_i^2 n_G (2\pi kT)^{\frac{1}{2}} D_{AB}\left[m_B^{\frac{1}{2}}\left(1 + \frac{\alpha_{mB}\pi}{8}\right) - m_A^{\frac{1}{2}}\left(1 + \frac{\alpha_{mA}\pi}{8}\right)\right]\nabla x_A \tag{5.37}$$

If $\alpha_{mB} \approx \alpha_{mA}$, one can conclude from eqn. (5.37) that the particle will experience a force tending to move it in the direction of diffusion of the heavy component of the gas mixture.

The special case of diffusion of species A through stagnant species B in a binary system is quite important. For this case we have $\mathbf{q}_B = 0$

and

$$\mathbf{q}_A = \mathbf{w} + \mathbf{W}_A = -\mathbf{W}_A + \mathbf{W}_B = -\frac{D_{AB}}{x_A x_B}\nabla x_A. \tag{5.38}$$

Therefore, for this case, we have for $\mathbf{F}_d^*$:

$$F_d^* = -\frac{8}{3} R_i^2 n_G (2\pi m_A kT)^{\frac{1}{2}}\left(1 + \frac{\alpha_{mA}\pi}{8}\right)\frac{D_{AB}}{x_B}\nabla x_A \tag{5.39}$$

and it is clear that the force is in the direction of the diffusing species A in the gas mixture.

Brownian Motion

In the free molecule region for $Ma_i \ll 1$, when a macroscopic body moves through a continuous fluid, it experiences a resistive force which according to eqn. (3.8) is proportional to its velocity. If the body also is sufficiently small so that it responds to fluctuations in the total rate of the impacts by the gas molecules over its surface, then the body will exhibit an average behavior in accordance with eqn. (5.27), on which is superimposed a random motion caused by the fluctuation in the molecular impacts. This Brownian motion is, of course, one of the distinguishing characteristics of aerosol particles.

The equation of motion of a small particle of mass m_i is therefore a stochastic equation, and is often called the *Langevin equation*.† A form of this equation is

$$\frac{d\mathbf{q}_i}{dt} = -\mathscr{A}^*(\mathbf{q}_i - \mathbf{q}) + \frac{\mathbf{F}_e^*}{m_i} + \frac{\mathbf{F}_{NU}^*}{m_i} + \frac{\mathscr{F}(t)}{m_i}, \tag{5.40}$$

† For further discussion of the questions of particle motion as given by the Langevin equation, see, for example, Rice and Gray[123] or Lebowitz and Rubin.[124]

where

$$\mathscr{A}^* = \frac{1}{B_i^* m_i} = \left\{ m_i \left[\frac{8}{3} R_i^2 n_G (2\pi m_G kT)^{\frac{1}{2}} \left(1 + \frac{\pi \alpha_m}{8}\right)\right]\right\}^{-1} \tag{5.41}$$

is valid for the free molecule region when $Ma_i \ll 1$. As in eqn. (3.21) $\mathbf{F}_e^*$ and $\mathbf{F}_{NU}^*$ respectively are the forces associated with external fields, and from gaseous non-uniformities, $\mathscr{F}(t)$ is the fluctuating force whose time average vanishes. Equation (5.40) may be used for a variety of applications to problems involving aerosol deposition and curvilinear motion. It is the free molecule equivalent to the continuum equation, eqn. (3.29).

C. Summary

From these illustrative examples it is evident that any transfer process by the suspending gas to a particle can be described in a straightforward manner in the free molecule regime. One needs to specify only the distribution f^- of incoming molecules. There remains, of course, the uncertainty in all these calculations of the accommodation coefficient. There are no exact methods for estimating these coefficients, although approximate theories have been proposed (cf. Brundin[(115)]). At present one must regard the accommodation coefficients as parameters in the equation which must either be estimated or, preferably, determined experimentally (see also refs. (113) and (114)).

When the Knudsen number for a single particle is finite, the simple assumptions of the free molecule region no longer hold. In this case, it is necessary to calculate f^- and f^+ from the Boltzmann equation for the suspending gas. This is a difficult task.

Some examples of work on transfer processes in the transition region will now be discussed.

5.3. Transfer Processes in the Transition Regime

As we have seen, transfer processes in the free molecule region where $Kn_i \gtrsim 10$ occur by the collisionless or "free-streaming" motion of gas molecules. On the other hand, in the continuum and slip flow regimes $Kn_i \lesssim 0.25$ the motion of the gas molecules is dominated by collisions and all the transfer processes partake of a "diffusive" character. The region of Knudsen number between these two extremes is commonly referred to as the transition region. In the transition region the molecular motion is difficult to describe analytically. Therefore the theory of the transfer processes in this region has not progressed very rapidly. As well, there are difficulties in obtaining accurate experimental data for transfer processes to particles in the transition regime. For these reasons the study of such processes is still in its early stages relative to other phases of the dynamics of aerosols.

As mentioned previously, there are no exact bounds on the Knudsen number which will delineate the transition region for all transfer processes. Rather arbitrarily, the region $0.25 \leqslant Kn_i \leqslant 10$ is designated here as the transition region. The lower bound, where the simple slip flow description breaks down will usually be different for the various transfer processes.

We have already discussed in Chapter 2 in some detail examples of the analytical methods used for describing dynamical processes in the transition region. Let us recapitulate a few of the important points already given.

If we write the Boltzmann equation in dimensionless form we obtain the result

$$\frac{\partial f^{\oplus}}{\partial t^{\oplus}} + \mathbf{v}^{\oplus} \cdot \nabla^{\oplus} f^{\oplus} + \mathscr{E}^{\oplus} \, . \, \nabla_{\mathbf{v}\oplus} f^{\oplus} = \frac{0.707}{\pi} \frac{R_i}{\lambda_G} J^{\oplus}(f^{\oplus}, f_1^{\oplus}), \tag{2.88}$$

where, for this discussion, the coefficient of the collision term on the right-hand side of this equation is the inverse Knudsen number.

In the free molecule region, $Kn_i \to \infty$, so that

$$\frac{\partial f^{\oplus}}{\partial t^{\oplus}} + \mathbf{v}^{\oplus} \cdot \nabla^{\oplus} f^{\oplus} + \mathscr{E}^{\oplus} \cdot \nabla_{\mathbf{v}^{\oplus}} f^{\oplus} = 0,$$

which describes a collisionless flow of molecules for distances of the order of the particle size. In the transition region, $Kn_i < 10$. Therefore, in order to calculate the rates of transport to a particle in this region in terms of the distribution functions f^+ and f^- it is necessary to solve eqn. (2.88). The collision term on the right-hand side of this equation is non-linear and singular so that a general solution has not been found. Instead various schemes have been used including the three examples of approximations described in Chapter 2.

In the description of the free molecule region exact results are possible as we have noted in Section 5.2. The free molecule results are regarded as a "rock of faith". In the transition region, however, for transfer processes to a particle the results obtained theoretically to date are not exact. For this reason and also because a detailed exposition of the theoretical calculations in the transition region is outside the scope of this book, we shall simply summarize what we believe to be at present the most satisfactory equations for the various transfer processes to aerosol particles in the transition region. The form of these equations is inferred from both experimental and theoretical evidence. In view of the current inexact state of the theory of the transition region already noted, this approach should be the most generally useful. In the future, as more accurate experimental data become available and the theoretical description is perfected, the exposition which follows will undoubtedly be regarded as only a first approximation. However, in the interest of progress in this field of study, the present compromise is believed to be a satisfactory one. We proceed thus in this manner to give for the transition region the approximate equations for mass, heat, and momentum transfer to a single isolated spherical aerosol particle.

A. Mass and Heat Transfer in the Transition Region

There have been several experimental studies of the slow, radially directed evaporation of molecules from isolated drops for $Ma_i \ll 1$ in the transition region.[125–127] In only one of these studies[125] was the transition region investigated in detail. These experimental results, together with recent theoretical work,[128] indicate that the total radial molecular transfer to or from a single component A particle in a gas B has the approximate form

$$\tilde{\Phi}_A \approx \tilde{\Phi}_A^*/(1 + \alpha_c \tilde{\mathscr{C}} R_i/\lambda_B), \tag{5.42}$$

where λ_B is the mean free path for the suspending gas molecules defined by the equation for viscosity of component B

$$\mu_{GB} = 0.499 \rho_B \bar{v}_B \lambda_B, \tag{5.43}$$

$\tilde{\Phi}_A^*$ is the free molecule transfer rate given by eqn. (5.22), α_c is the condensation coefficient, and $\tilde{\mathscr{C}}$ is a constant. There is no accurate information on the range of validity of eqn. (5.42). We are guided in this case by analogous experimental results for heat transfer from a sphere[129, 130] in proposing the following approximate range of validity of eqn. (5.42): $0.25 \leqslant Kn_i \leqslant \infty$. The linear relation predicted between $\tilde{\Phi}_A^*/\tilde{\Phi}_A$ and R_i/λ_B by eqn. (5.42) has been found by Monchick and Reiss.[125] Their experimental results are shown in Fig. 5.2.

The evaporation rate of small particles has been studied theoretically[128] in the transition region using model equations for mixtures similar to the BGK model for a single component gas (Section 2.2G). In this theoretical study, the model equations were solved for $Ma_i \ll 1$ for the distribution functions by the method of Knudsen iteration discussed in Chapter 2. The first iterative solution yielded an equation of the form

$$\tilde{\Phi}_A = \tilde{\Phi}_A^*(1 - \alpha_c \tilde{\mathscr{C}} R_i/\lambda_B), \tag{5.44}$$

which is recognized as the first term in the expansion of the empirical relation, eqn. (5.42), for small Kn_i^{-1}. For $Ma \gg 1$ a relative particle–gas velocity has been found to have no effect on the evaporation rate.[128]

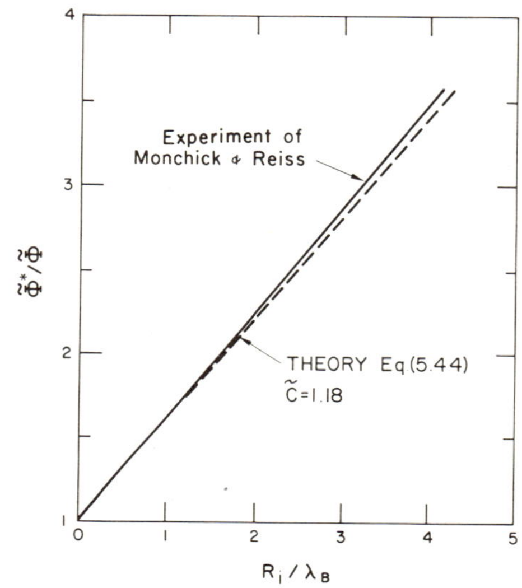

FIG. 5.2. Comparison between theory of the transition regime and data of Monchick and Reiss[125] for evaporation of spheres of diamyl sebacate.

For the special case of evaporation of species (A) into a suspending gas of species (B) such that $n_A \ll n_B$, the parameter $\tilde{\mathscr{C}}$ has been calculated[128] and has been found to be

$$\tilde{\mathscr{C}} = 0.807(1/\mu_B)^{\frac{1}{2}} (d_{AB}/d_B)^2 \tilde{\mathscr{I}}, \tag{5.45}$$

where μ_B is the reduced molecular $\mu_B = m_B/(m_A + m_B)$, d_A and d_B are the molecular diameters of species A and B respectively, and $d_{AB} = \frac{1}{2}(d_A + d_B)$. $\tilde{\mathscr{I}}$ is an integral function of μ_B which must be determined numerically. In Table 5.4 values of $\tilde{\mathscr{I}}$ as a function of μ_B are given. In Table 5.5 values of $\tilde{\mathscr{C}}$ are given for certain special cases. As an example, the theoretical value of $\tilde{\mathscr{C}}$, 1.18, calculated for diamyl sebecate droplets evaporating in air is in close agreement with an experimental value of $\tilde{\mathscr{C}} = 1.2$ obtained by Monchick and Reiss.[125] However, such close agreement is likely to be fortuitous. The molecular parameter d_A, for diamyl sebecate is not known exactly. Also, the value of $\tilde{\mathscr{C}}$ attributed to Monchick and Reiss is only an estimation. In addition, the collisional model employed in the derivation of eqn. (5.45) is known to become highly inaccurate as $\mu_B \to 0$. In Fig. 5.2 the apparent agreement between the theory giving $\tilde{\mathscr{C}}$ and the experimental results of Monchick and Reiss[125] is indicated.

TABLE 5.4. RESULTS OF NUMERICAL CALCULATIONS OF $\tilde{\mathscr{I}}$ AS A FUNCTION OF REDUCED MASS (FROM BROCK[128])

$\underline{\mu}_B$	$\mathscr{I}$
1·000	0.295
0.900	0.269
0.800	0.245
0.700	0.222
0.600	0.201
0.500	0.180
0.400	0.162
0.300	0.144
0.200	0.130
0.100	0.115

TABLE 5.5. $\mathscr{C}$, EQN. (5.45), FOR EVAPORATION OF PARTICLE OF SPECIES A INTO A GAS CONTAINING ONLY SPECIES B (FROM BROCK[128])

(1) $\underline{\mu}_B = 1.000$; $d_A/d_B = 0.000$	0.060
(2) $\underline{\mu}_B = \frac{1}{2}$; $d_A/d_B = 1.000$	0.206
(3) Water (1), air (2); $\underline{\mu}_B = 0.616$; $d_A/d_B = 1.235$	0.264
(4) n-Octane (1), air (2); $\underline{\mu}_B = 0.21$; $d_A/d_B = 2.8$ (estimated)	0.46
(5) Diamyl sebacate (1), air (2); $\underline{\mu}_B = 0.08$, $d_A/d_B = 2.8$ (estimated)	1.18

Semi-empirical Analyses of Diffusional Processes

Before closing this discussion of the mass transfer rate in the transition region, two interpolation schemes which have been proposed for the entire region of Kn_i should be noted. One of these, which we term the "mean free path" or the "dividing-sphere" method, has been proposed some years ago by Fuchs[90] for estimating the evaporation rate or condensation rate of small particles. Fuchs, in discussing the evaporation of spherical drops, introduced a distance Δ from the sphere such that for distances $r > \Delta$ the mass transfer process follows the continuum equations, and for $r < \Delta$ the mass transfer is by the free molecule mechanism. The principal objections to this interpolation scheme and to others similar to it are that the distance Δ is unknown and there is no systematic way to calculate it. Also, this mean free path method yields an incorrect analytical form for $\tilde{\Phi}$, inasmuch as for $Kn_i \gg 1$ one obtains from this method

$$\tilde{\Phi}/\tilde{\Phi}^* = 1 - \tilde{a}(R_i/\lambda_B)^2, \tag{5.46}$$

where $\tilde{a}$ is a constant. That is, the method yields no first order term in R_i/λ_B for small R_i/λ_B. Yet the non-intuitive theory[128] demonstrates the existence of such a first order term. In the treatment of heat transfer from an isolated sphere and in sphere drag, the same objections to the mean free path method arise as given here for mass transfer from an isolated sphere.

The other interpolation scheme for the entire region of Kn_i, $0 \leqslant Kn_i \leqslant \infty$, was introduced by Shermann,[131] who noted that many non-continuum transfer processes to bodies of various geometries could be described by an interpolation of the form

$$\mathscr{p} = \mathscr{p}^*/(1 + \mathscr{p}^*/\mathscr{p}^c) \tag{5.47}$$

where p is the transfer property and p^* and p^c are the values of this function in the free molecule and continuum limits respectively. Moment solutions of the Boltzmann equation for heat transfer from a sphere[132] and rotational drag of a sphere[133] give exactly equations which agree with eqn. (5.47). However, for these problems, eqn. (5.47) is known to be only approximate.

If we apply Sherman's relation, eqn. (5.47), to find the mass transfer rate $\tilde{\Phi}_A$, of component A from a sphere to a gas containing A and B we get

$$\tilde{\Phi}_A = \tilde{\Phi}_A^*/(1 + \alpha_c \tilde{\mathscr{C}}' R_i/\lambda_B), \tag{5.48}$$

where

$$\tilde{\mathscr{C}}' = \bar{v}_A \lambda_B / 4D_{AB}. \tag{5.49}$$

Here D_{AB} is the mutual diffusion coefficient of the gas components, and $\bar{v}_A$ is the mean molecular speed of component A. This value of $\tilde{\mathscr{C}}'$ appears to be much too small to explain, for example, the experimental results of Monchick and Reiss for diamyl sebecate.[125] Hence eqn. (5.47) cannot be used with confidence, though for the transition region it does possess what appears to be the proper analytical form (cf. Fig. 5.2).

The limited amount of experimental and theoretical work cited at the beginning of this discussion of mass transfer suggests that the general form of eqn. (5.42) for evaporation or condensation of single particles in the transition region is probably approximately correct. However, the experimental evidence exists only for relatively non-volatile particles. The theoretical results for this problem have proceeded from models of the Boltzmann equation which are not entirely satisfactory approximations (cf. ref. (128), (134) and (135)).

From these facts one may conclude that much more remains to be done on the theory of evaporation and condensation of particles in the transition region.

Heat Transfer in the Transition Region

Experiment[129, 130] and theory[132] have shown that the radial heat transfer from a sphere for $Ma_i \ll 1$ can be described by an equation very similar in form to the mass transfer equation

$$H \approx H^*/(1 + \alpha_t \tilde{\mathscr{D}} R_i/\lambda_G), \tag{5.50}$$

where H^* is the free molecule heat transfer rate given by eqn. (5.19). $\tilde{\mathscr{D}}$ is a parameter depending on the nature of the gas surrounding the particle. The mean free path λ_G is defined in eqn. (5.43). α_t is the thermal accommodation coefficient. We set arbitrarily the limits of validity of eqn. (5.50) as $0.25 \lesssim Kn_i \leqslant \infty$.

There is ample evidence[131] that relative particle–gas motion does not influence the heat transfer rate provided $Ma_i < 1$. For $Ma \gtrsim 1$ the heat transfer rate is increased by the relative motion. There is no theory as yet to describe this increase for the transition region.

Heat transfer from a sphere has been studied[24] using the method of Knudsen iteration and the BGK model of the Boltzmann equation. For monatomic gases the first iterate solution for perfect accommodation gives[136] the result

$$H/H^* = 1 - (7\pi/144)\, Kn_i, \tag{5.51}$$

which is seen to be the first term in the expansion expression of eqn. (5.50) with $\alpha_t = 1.0$. Therefore $\tilde{\mathscr{D}} = 7\pi/144$. A similar study[132] using the moment method gives a value of

$\tilde{\mathscr{D}} = 4/15$. Moment methods are known to be unreliable for large Knudsen numbers.[137] Therefore we favor the value $\tilde{\mathscr{D}} = 7\pi/144$ for monatomic molecules over the indicated range of Knudsen number, $0.25 \lesssim Kn_i \leqslant \infty$. From the proposed free molecule results for heat transfer theory for polyatomic molecules (eqn. (5.23)), it may be inferred that for diatomic molecules the coefficient $\tilde{\mathscr{D}}$ should be increased by a factor of $\frac{3}{2}$.[132] Thus for diatomic molecules, $\tilde{\mathscr{D}} = 7\pi/96$. Generalizing this approximation for any polyatomic gas,

$$\tilde{\mathscr{D}} \approx \frac{7\pi}{144}\left(\frac{C_v + k/2}{2k}\right). \tag{5.52}$$

In Fig. 5.3, the linear relation predicted by the inverse of eqn. (5.50) is shown to be exhibited by the experimental data of Harbour[130] for heat transfer from an isolated sphere immersed in air. The experimental data are fit in the region $0.25 \lesssim Kn_i \leqslant \infty$ by a straight line which yields a value of $\tilde{\mathscr{D}} = 0.256$. The theory, based on the BGK model, eqn. (5.52) predicts a value of $\tilde{\mathscr{D}} = 0.24$ for air. At small Kn_i, Harbour's data do not approach the free

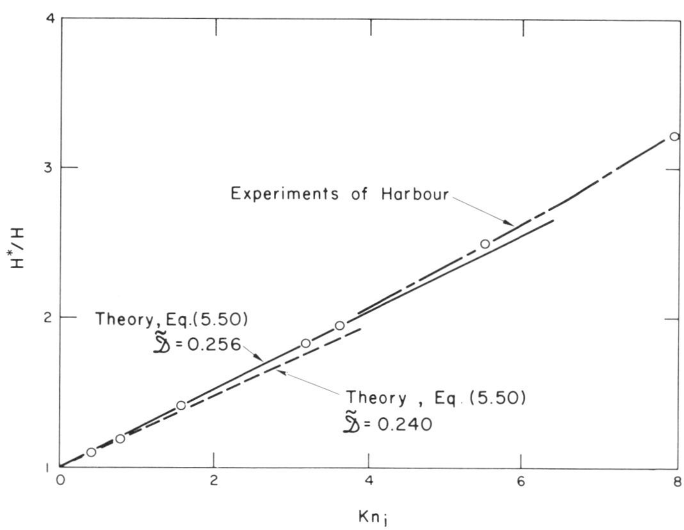

FIG. 5.3. Comparison between theory of the transition region and heat transfer data of Harbour[130].

molecular values sufficiently rapidly so that for more accurate experimental data at small Kn_i, the agreement between theory and experiment may be even better than this comparison would indicate. For $R_i/\lambda_G \gtrsim 4$ from Fig. 5.3 it is clear that eqn. (5.50) becomes invalid as the slip flow region is entered. The only additional data at this time for heat transfer from an isolated sphere are from the study of Takao,[129] who investigated also the heat transfer into air from a sphere.

B. Momentum Transfer in the Transition Region

Isothermal Drag Force

From the experimental standpoint, the isothermal drag of a sphere has been studied in non-continuum regions more intensively than any other phenomenon. Millikan[50] and subsequent workers, e.g. refs. (138), (139) and (140), have established the following empirical interpolation formula for the isothermal drag force on an isolated sphere for $0 \leqslant Kn_i \leqslant \infty$:

$$\mathscr{D} = 6\pi\mu_G R_i \mathbf{q}/\{1 + Kn_i[\hat{A} + \hat{B}\exp(-\hat{C}/Kn_i)]\}, \tag{5.53}$$

which is presumed to be valid for $Ma_i \ll 1$. (See also p. 58).

The constants $\hat{A}$ and $\hat{B}$ are fixed to some extent by the requirement that eqn. (5.53) agree in the limit $Kn_i \to \infty$ with the exact free molecule relation, eqn. (5.27) for the isothermal drag force. That is,

$$\hat{A} + \hat{B} = 2.2/(1 + \pi\alpha_m/8). \tag{5.54a}$$

In addition, for small Kn, the value of $\hat{A}$ is given by the slip-flow theory (cf. Chapter 6) by the equation

$$\hat{A} = 1.00(2 - \alpha_m)/\alpha_m, \tag{5.54b}$$

according to the careful, but limited, experimental work by Millikan and co-workers.[50] There is a difficulty with eqns. (5.54a) and (5.54b) which leads one to believe that eqn. (5.54b) is probably not generally correct insofar as the indicated dependence on α_m is concerned. This statement is based on the fact that $\hat{B}$ calculated from eqns. (5.54a) and (5.54b) has a value of zero when $\alpha_m \simeq 0.73$. This prediction seems to be contradicted by experimental measurements[138] for α_m as low as 0.78 which show no substantial decline in the value of $\hat{B}$. From this apparent contradiction one concludes that the eqn. (5.54b) is not correct, insofar as its dependence on α_m is concerned, for values of α_m very different from unity. Also, eqn. (5.53) probably does not have the proper analytical form, although the gross features of the isothermal drag force seem to be described adequately by eqn. (5.53), regarded as a curve-fitting equation. Further experimental work involving particle–gas systems with small accommodation coefficients appears necessary to resolve the questions implied above.

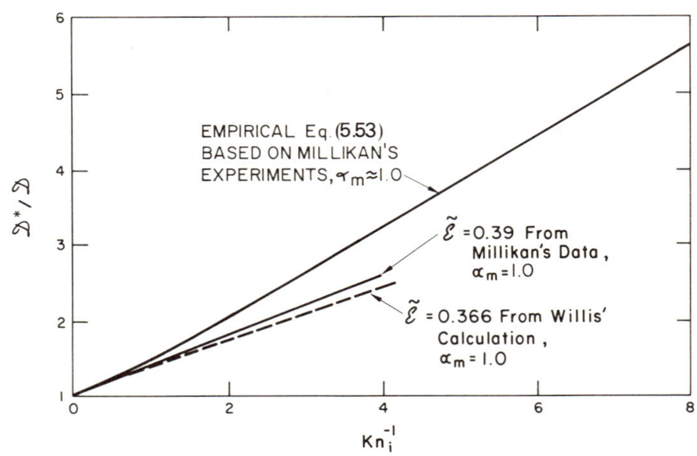

FIG. 5.4. Aerodynamic drag as predicted by theory with Millikan's empirical equation.

Recently Willis[137] has used the method of Knudsen iteration and the BGK model (cf. Section 2.2G) to calculate the isothermal drag force of a sphere with $\alpha_m = 1.0$. Willis found

$$\mathscr{D} = \mathscr{D}^*[1 - \tilde{\mathscr{E}} Kn_i^{-1}], \tag{5.55}$$

where $\mathscr{D}^*$ is the free molecule drag force given by eqn. (5.27). Willis found that the constant $\tilde{\mathscr{E}} = 0.366$. Earlier, a polynomial expansion technique had been employed[141] and an equation similar in form to eqn. (5.55) had been obtained valid for $Ma_i \lesssim 1$. However, the coefficient of Kn_i^{-1} in this work for $Ma_i \ll 1$ was 0.278 and the discrepancy with eqns. (5.55) has been attributed[137] to inadequacies in the polynomial expansion method.

The coefficient of Kn_i^{-1} in eqn. (5.55) $\tilde{\mathscr{E}}$ is related in the following manner to the constants of Millikan's empirical equation, eqn. (5.53), for $\alpha_m = 1$:

$$\tilde{\mathscr{E}} = \frac{1 - \hat{C}\hat{B}}{\hat{A} + \hat{B}} = 0.366. \tag{5.56}$$

The calculation of the separate constants $\hat{C}$ and $\hat{B}$ cannot be performed with confidence from eqns. (5.54a), (5.54b), and (5.56) owing to the difficulty noted in the definition of $\hat{B}$.

For comparison, Millikan's empirical curve given by eqn. (5.53) is shown with calculations based on Willis' theory in Fig. 5.4. With $\tilde{\mathscr{E}} = 0.366$, $\alpha_m = 1.0$, the transition theory agrees satisfactorily with experiment down to $Kn_i \approx 1.4$.

The Thermal Force

In the free molecule region the thermal force was easily understood as resulting from the fact that gas molecules in an imposed temperature gradient will impart more momentum to a suspended particle in the direction opposite to the temperature gradient than in the direction of the gradient. The result of this difference is a force producing particle motion directed toward lower temperatures. In the transition region as collisions of gas molecules become important, the collisions have the effect of reducing the thermal force imparted by the gas molecules.

Experimental studies of the thermal force in the transition region using the Millikan cell technique[140, 142] have found the following empirical relationship for the thermal force $\mathbf{F}_t$:

$$\mathbf{F}_t = \mathbf{F}_t^* \exp[-\hat{\tau}/Kn_i] \tag{5.57}$$

valid over the range $0 \leqslant (R_i/\lambda_G)^{-1} \lesssim 5$ as indicated in Fig. 5.5. $\hat{\tau}$ is a parameter for a particular aerosol system and is independent of R_i/λ_G. Through an integral iteration technique working with the BGK model of the Boltzmann equation[28] the value of $\hat{\tau}$ has been found for monatomic molecules:

$$\hat{\tau} \approx 0.06 + 0.09\alpha_m + 0.28\alpha_m(1 - \alpha_t k_G/2k_p), \tag{5.58}$$

where k_G is the thermal conductivity of the gas and k_p the thermal conductivity of the particle. Approximately, for polyatomic molecules, $\hat{\tau}$ is given also by eqn. (5.58). Figure 5.6 gives a comparison with experiment[140, 142–144] of values of $\hat{\tau}$ according to eqn. (5.58). Table 5.6 gives various experimental values of $\hat{\tau}$; in this table they are calculated from isothermal drag measurements from eqn. (5.54). The coefficient α_t is generally not known for aerosols. For monatomic gases at room temperature it seems safe to assume $\alpha_t \approx \alpha_m$.

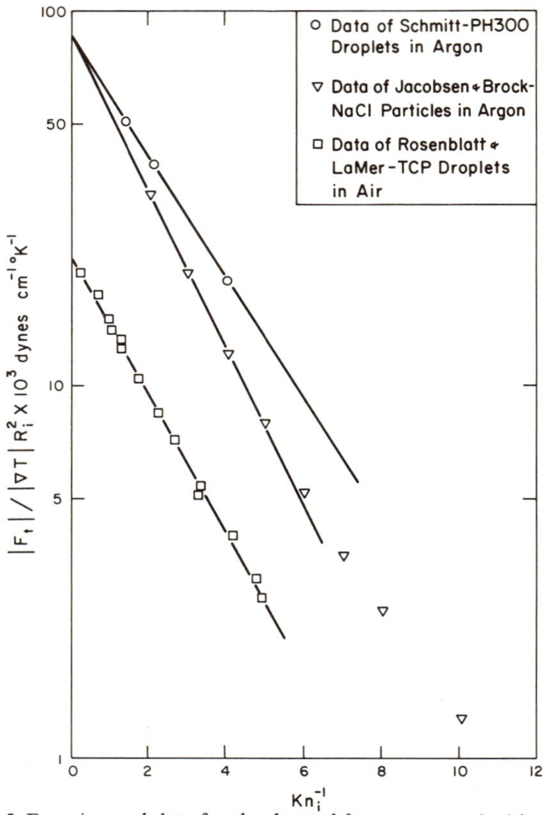

FIG. 5.5. Experimental data for the thermal force compared with eqn. (5.57).

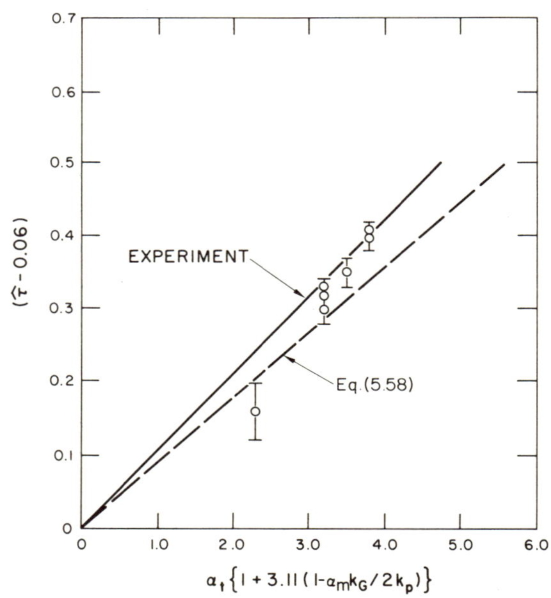

FIG. 5.6. Experiment and theory for a relationship of $\hat{\tau}$. (From Brock.[28])

Except for gases with high thermal conductivity such as hydrogen, the term $k_G/2k_p$ in eqn. (5.58) is generally not of much importance. This result is corroborated by experiment. The accommodation coefficient α_m is seen to be of greatest importance in determining the magnitude of $\hat{\tau}$.

The calculation of $\hat{\tau}$ was based on the BKG model which is known to be inaccurate in describing heat and momentum transfer.[137] Hence eqn. (5.58) represents only an approximation which, however, should be correct as to the general functional relation of the

TABLE 5.6. VALUES OF $\hat{\tau}$ IN EQN. (5.57) FROM EXPERIMENTAL THERMAL FORCE FROM BROCK[28])

Reference	Aerosol	Gas	$\hat{\tau}$	α_m
140	M300	A	0.39 ± 0.01	0.80
140	M300	N_2	0.38 ± 0.01	0.81
140	M300	CO_2	0.36 ± 0.02	0.80
140	M300	H_2	0.22 ± 0.04	0.79
144	TCP	Air	0.41 ± 0.02	0.88
142	NaCl (amor.)	A	0.47 ± 0.01	0.91
143	NaCl (cryst.)	Air	0.46 ± 0.02	(0.91)

parameters which enter into $\hat{\tau}$. The available experimental data for monatomic gases appear to be fit most accurately by the following expression for $\hat{\tau}$:

$$\hat{\tau} = 0.09 + 0.12\,\alpha_m + 0.28\,\alpha_m\,[1 - \alpha_t k_G/2k_p]. \qquad (5.59)$$

This equation differs from (5.58) only in the values of the first two numerical constants which occur.

Equations (5.57) and (5.59) provide a good description of the thermal force for an aerosol particle in an effectively infinite, motionless gas. Provided there are no large velocity gradients G_q, that is

$$\left(\frac{G_q R_i}{\bar{v}_G}\right) \ll 1, \qquad (5.60)$$

the total force in the transition region **F** acting on a particle in a flowing gas with a known local temperature gradient should be given by the sum of the separate isothermal drag force, eqn. (5.53), and the thermal force, eqn. (5.57),

$$\mathbf{F} = \mathscr{D} + \mathbf{F}_t. \qquad (5.61)$$

A particle experiencing a thermal force may be regarded as being in an effectively infinite gas provided that the distance L of the particle from a boundary or other surface is such that

$$\lambda_G/L \to 0. \qquad (5.62)$$

In the instance of deposition of particles on surfaces in a device such as the thermal precipitator, the particle spends only a very brief time in which the condition of eqn. (5.62) does not hold. Therefore, eqn. (5.61) should describe adequately the deposition of particles on boundary surfaces.

In the transition region there is no general theory for the photophoretic force, which arises on heating of the particles by incident radiation. Photophoresis is a phenomenon which has not yet been found to be of great importance in either practical applications or natural phenomena (see, for example, Hidy and Brock[122]).

The Diffusion Force

The explanation for the diffusion force in the free molecule region is very simple, as is evident from the discussion of Section 2.2. In the free molecule region, each component of a diffusing gas mixture is moving independently of the other components with, in general, a mean velocity $\mathbf{q}_\alpha$. Each species, therefore, produces an individual free molecule drag force on the suspended particle. The vector sum of these separate drag forces is the diffusion force.

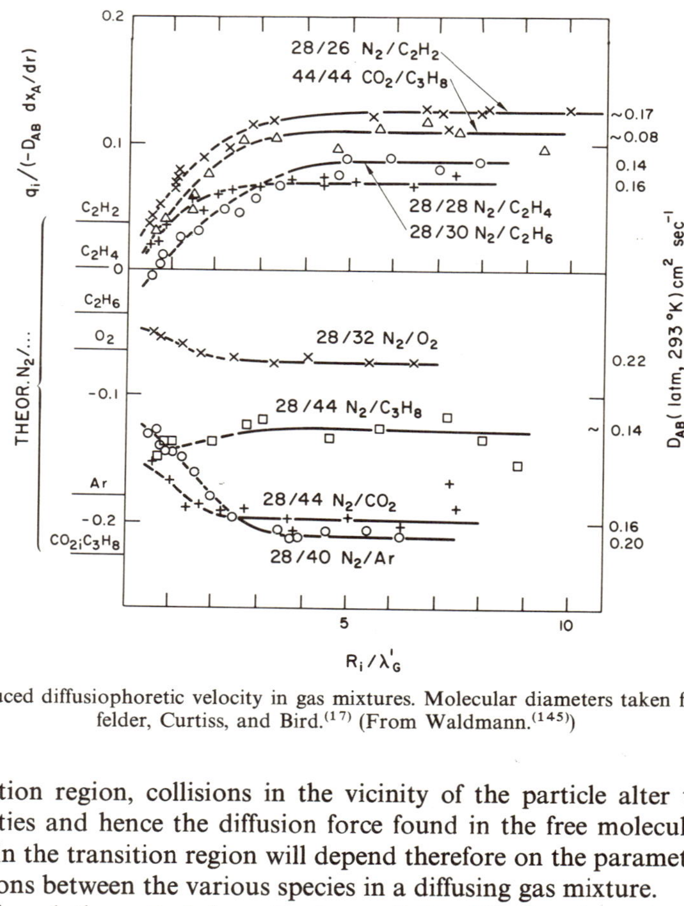

FIG. 5.7. Reduced diffusiophoretic velocity in gas mixtures. Molecular diameters taken from Hirschfelder, Curtiss, and Bird.[17] (From Waldmann.[145])

In the transition region, collisions in the vicinity of the particle alter the "normal" diffusion velocities and hence the diffusion force found in the free molecule region. The diffusion force in the transition region will depend therefore on the parameters characterizing the collisions between the various species in a diffusing gas mixture.

Experimental and theoretical investigations of the diffusion force in the transition region have been carried out for a particle in a diffusing binary gas mixture for two special cases: (1) equimolar counter diffusion (ECD), (2) diffusion of one species (A) through a second stagnant species (SD).[145, 146, 147]

The experimental results of Waldmann[145, 146] are shown for ECD in Fig. 5.7 and for SD in Fig. 5.8. The theory of the diffusion force for ECD and SD has been developed[147]

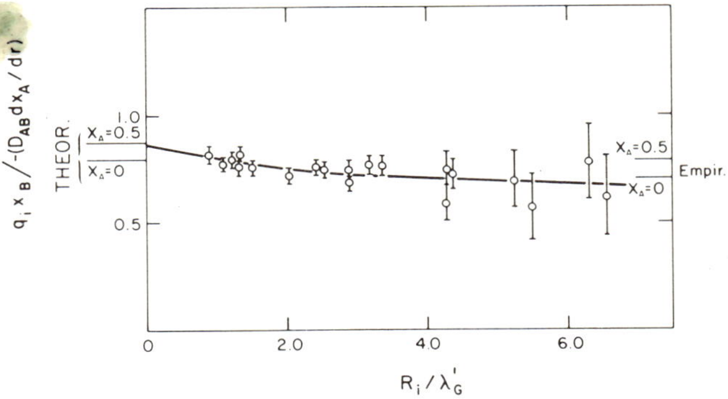

FIG. 5.8. Reduced diffusiophoretic velocity in gas mixtures for the case of gaseous species A diffusing through stagnant gas of species B. (From Waldmann and Schmitt.[146])

using models of the Boltzmann equation and the method of Knudsen iteration (Section 2.2). In this study the diffusion force was investigated for binary gas mixtures in which one of the components is very dilute. This restriction permitted a more accurate accounting of the intermolecular collisions. For ECD, the diffusion force $\mathbf{F}_d$ for large Kn_i was found to have the form

$$\mathbf{F}_d = \frac{8}{3} R_i^2 n_G (2\pi kT)^{\frac{1}{2}} \left(1 + \frac{\pi}{8}\right) D_{AB} (\nabla x_A)_\infty \times \left\{ (m_B^{\frac{1}{2}} - m_A^{\frac{1}{2}}) + \left[\frac{d_A^2 m_A^{\frac{1}{2}}}{d_{AB}^2} - \left(\frac{2 m_A m_B}{m_A + m_B} \right)^{\frac{1}{2}} \right] \times 0.311\, R_i/\lambda'_G \right\}. \tag{5.63}$$

for $n_A \gg n_B$.
For SD the diffusion force for large Kn_i was

$$\mathbf{F}_d = -\frac{8}{3} R_i^2 n_G (2\pi m_A kT)^{\frac{1}{2}} \left(1 + \frac{\pi}{8}\right) \frac{D_{AB}}{x_B} (\nabla x_A)_\infty \times [1 - 0.071 (2m_B/(m_A + m_B))^{\frac{1}{2}} R_i/\lambda'_G], \tag{5.64}$$

where $n_A \ll n_B$. In both eqns. (5.63) and (5.64) the following definition has been used for the mean free path λ'_G:

$$\lambda'_G = 1/\sqrt{2}\pi n_G d_{AB}^2, \tag{5.65}$$

where $d_{AB} = \frac{1}{2}(d_A + d_B)$, and d_A and d_B are the effective molecular diameters of species A and B respectively. These equations have been derived under the assumption of perfect momentum accommodation, i.e. $\alpha_{mA} = \alpha_{mB} = 1.0$. The range of validity of the linear relations indicated in eqns. (5.63) and (5.64) appears to be $0.5 \lesssim Kn_i \leqslant \infty$. For the majority of the systems studied it would appear that in this indicated range of Knudsen number the developed theory explains in an approximate manner the experimental data which appear to be difficult to obtain with accuracy.[145, 146] Table 5.7 gives a comparison between this transition region theory[147] and experiment.[145, 146]

The existence of a mode of aerosol particle motion termed "photodiffusiophoresis" has been pointed out recently.[148] This motion arises from molecular diffusion velocities produced by electromagnetic radiation incident on a particle. The electromagnetic radi-

TABLE 5.7. A COMPARISON OF THE PROPOSED THEORY WITH EXPERIMENT FOR THE DIFFUSION FORCE

Gas pair A/B	Mass ratio, m_A/m_B	Slope of curve at $Kn_i = 0$ established from Figs. 5.7 and 5.8	Slope of curves at $Kn_i = 0$ from theory, eqn. (5.63)
Equimolar counterdiffusion			
N_2/O_2	0.875	−0.011	−0.014
N_2/C_2H_4	1.000	+0.039	+0.041
N_2/C_2H_6	0.933	+0.060	+0.059
N_2/C_2H_2	1.078	+0.040	+0.035
CO_2/C_3H_8	1.000	+0.050	+0.068
N_2/A	0.700	−0.010	+0.003
CO_2/A	1.100	—	−0.058
Diffusion of species A through stagnant species B			Eqn. (5.64)
H_2O/N_2	0.642	−0.060	−0.063

ation produces the diffusion velocities by altering the chemical or physical equilibrium between an aerosol particle and a surrounding gas mixture. It is not known as yet what importance this effect has in either practical applications or natural phenomena.

C. Summary

TABLE 5.8. SUMMARY OF TRANSFER EQUATIONS FOR AN ISOLATED SPHERICAL PARTICLE IN THE TRANSITION REGION

Transfer process	Equation	Restrictions	Equation reference
Radial molecular (mass) transfer of species A from an isolated spherical particle ($\sec^{-1}$)	$\tilde{\Phi}_A = \dfrac{2\pi R_i^2 \alpha_{CA}\left[n_A^+\left(\dfrac{2kT^+}{\pi m_A}\right)^{\frac{1}{2}} - n_A^-\left(\dfrac{2kT}{\pi m_A}\right)^{\frac{1}{2}}\right]}{1 + \alpha_{CA}\tilde{\mathscr{C}} R_i/\lambda_B}$	$0.25 \lesssim \lambda_B/R_i \leqslant \infty$ $Ma_i \ll 1$	(5.42)
Radial heat transfer H from an isolated spherical particle	$H = \dfrac{4\pi R_i^2 \alpha_t n^- k\left(\dfrac{2kT^-}{\pi m_G}\right)^{\frac{1}{2}}[T^+ - T^-]}{1 + \alpha_t \tilde{\mathscr{D}} R_i/\lambda_G}$	Stable particle. Single component. $0.25 \lesssim \lambda_G/R_i \leqslant \infty$ $Ma_i < 1$	(5.50)
Isothermal drag force	$\mathscr{D} = 6\pi\mu_G R_i \mathbf{q}/\{1 + (\lambda_G/R_i)[\hat{A} + \hat{B}\exp(-\hat{C}R_i/\lambda_G)]\}$	$Re_i < 1$ $Ma \ll 1$ $0 \leqslant \lambda_G/R_i \leqslant \infty$	(5.53)
Thermal force	$\mathbf{F}_t = -\frac{32}{15}R_i^2(k_G/\bar{v}_G)(\nabla T)\ \exp[-\hat{\tau}R_i/\lambda_G]$	$0.25 \lesssim \lambda_G/R_i \leqslant \infty$ $Ma_i \ll 1$	(5.57)
Diffusion force (a) Equimolar counter-diffusion	$\mathbf{F}_d = \dfrac{8}{3}R_i^2 n_G(2\pi kT)^{\frac{1}{2}}\left(1 + \dfrac{\pi}{8}\right)D_{AB}(\nabla x_A)\left\{m_B^{\frac{1}{2}} - m_A^{\frac{1}{2}} + \left[\dfrac{d_A^2 m_A^{\frac{1}{2}}}{d_{AB}^2} - \left(\dfrac{2m_A m_B}{m_A + m_B}\right)^{\frac{1}{2}}\right]0.311\ R_i/\lambda'_G\right\}$	Only diffusive motion of gas. Binary gas. $n_A \gg n_B$ $\alpha_{mA} = \alpha_{mB} = 1.0$ $0.5 \lesssim \lambda'_G/R_i \leqslant \infty$	(5.63)
(b) Diffusion of component A through stagnant component B	$\mathbf{F}_d = -\dfrac{8}{3}R_i^2 n_G(2\pi m_A kT)^{\frac{1}{2}}\left(1 + \dfrac{\pi}{8}\right)\dfrac{D_{AB}}{x_B}(\nabla x_A)_\infty\Big\} \times\left[1 - 0.071\left(\dfrac{2m_B}{m_A + m_B}\right)^{\frac{1}{2}} R_i/\lambda'_G\right]$	Only diffusive motion of gas. Binary gas $n_A \ll n_B$ $\alpha_{mA} = \alpha_{mB} = 1.0$ $0.5 \lesssim \lambda'_G/R_i \leqslant \infty$	(5.64)

In this section we have given a few examples of methods for describing the transfer processes to aerosol particles in the transition region. The descriptions which we have outlined of the various transfer processes appear to be the most complete available at present from the theory of rarified gas dynamics. In Table 5.8 the various formulae for the transfer processes to single aerosol particles are summarized. Undoubtedly, in the future these descriptions will have to be modified as improved theories and extensive experimental data become available.

CHAPTER 6

TRANSFER PROCESSES TO AN AEROSOL PARTICLE IN THE SLIP FLOW REGION

IN CHAPTER 5 we have discussed the transfer processes to an aerosol particle in the regime $0.25 \lesssim Kn_i \leqslant \infty$, where there is a large departure from continuum conditions. As the Knudsen number becomes small, $\ll 1$, it might be hoped that the analytical difficulties found in the description of transfer processes in the transition region could be avoided by "patching up" the relatively simple continuum description, known to be valid for $Kn_i \to 0$.

This "patching up" procedure for $Kn_i \ll 1$, known as the slip flow procedure, was suggested notably by Maxwell.[89] He proposed that the non-continuum effects for small Knudsen number could be accounted for in the boundary conditions of the continuum transport equations, the Navier–Stokes equations. That is, in the slip flow procedure the transport processes in the gas are described by the continuum equations everywhere but in a very thin layer of gas next to a boundary surface, where non-continuum effects are assumed to be present.

In the continuum description through the Navier–Stokes equation, the boundary conditions assume, for example, that the tangential fluid velocity and the fluid temperature at the surface are equal respectively to the tangential velocity of the surface and temperature at the surface. The boundary conditions for slip flow do not assume such equalities, but instead allow for a slip of the gas relative to the surface, and a temperature "jump" or temperature difference between the gas and the surface with which it is in contact.

Consider then the application of the slip flow procedure to the classical problem of the drag on a sphere produced by the creeping, relative motion of a suspending gas when $Kn_i > 0$. In this case one solves the Navier–Stokes equations for the mean velocity components of the gas according to Stokes' procedure, but allows for a "slip" of the gas past the surface. With these derived expressions for the mean velocity components the pressure tensor, and hence the drag force on the sphere can be obtained. The assumption is that by allowing for the velocity "slip" for $Kn > 0$ the non-continuum effects are correctly accounted for. Actually, the equation derived for the drag force on a particle as estimated by the slip flow procedure just described is strictly correct only in a limiting sense as $Kn_i \to 0$.

More generally, we shall see that the slip flow procedure does not yield in all cases analytical expressions for transfer processes to aerosol particles. In some instances this method appears to give correct results only for limitingly small Kn_i, as for the particle drag problem just discussed. However, in the slip flow regime, where $0 < Kn_i \lesssim 0.25$, expressions for transfer rates as determined from experiment can be derived from theoretical considerations in such a way that only the parameters introduced in the slip flow analysis appear. These parameters, in principle, can be determined independently on theoretical grounds, and are presumed to be rather general in applicability. For this reason the slip flow procedure is favored over other more intuitive interpolation procedures such as mean free path methods [e.g. refs. (90), (133), (149)], which yield analytical forms that have to be verified by experiment. Furthermore, the parameters arising from such intuitive calculations cannot be estimated theoretically, but also must be determined empirically from the experiments. Preferable, of course, to both the slip flow procedures and various interpolation schemes would be descriptions of transfer processes in the regime, $0 < Kn_i \lesssim 0.25$, obtained from solutions of the Boltzmann equation. Unfortunately, such solutions are not yet available at this time.

The purpose of this chapter, then, is not to present the slip flow procedure as a useful technique of complete generality for describing various transfer processes to aerosol particles at small Knudsen numbers. Our purpose, instead, is to point out the limitation of this procedure and for specific illustrative transfer processes, to give formulae, based on slip flow analyses, which permit approximate quantitative description of these transfer processes in the region, $0 \leqslant Kn_i \lesssim 0.25$.

We begin with a more detailed discussion of the nature of the slip flow procedure as applied to transfer processes involving spheres. Then illustrations of results from application of the slip flow procedure are presented and interpreted. We discuss radial molecular (mass) and heat transfer from a spherical particle in a motionless gas. Subsequently, three examples of momentum transfer are considered, which are the frictional drag in an isothermal medium, the thermal force, and the diffusion force. In all these examples, comparison with experiment is made. From such comparisons, equations are developed which together with the equations for the free molecule and transition regime, given in Chapter 5 for the same problems, permit approximate description of these transfer processes over the complete range of Kn_i.

6.1. The Slip Flow Procedure

Simply stated, the slip flow procedure consists in solving the Navier–Stokes equations (Section 2.3) with boundary conditions, the slip flow boundary conditions, which account for the non-continuum effects. No rigorous justification for this procedure has been offered, and our discussion will, therefore, reflect the intuitive basis of this classical method.

In Chapter 2 we reviewed the derivation of the Navier–Stokes equations for a dilute gas from a class of solutions to the Boltzmann equation. There it was shown that the Navier–Stokes equation resulted from the second approximation in a perturbation solution of the Boltzmann equation for the molecular velocity distribution function. This solution is expected to apply best for small departures from continuum conditions. The nature of this solution is special in that the distribution function f depends on no parameters other than the values of the number density n_G, mean velocity $\mathbf{q}$, and temperature T of the gas. In this class of special solution, known as the "normal" solution, time and spatial variables do not appear explicitly but only implicitly in the functions $n_G(\mathbf{x}, t)$, $\mathbf{q}(\mathbf{x}, t)$, and $T(\mathbf{x}, t)$, and their time and space derivatives. Then in the analytical description of any physical system with solutions of the Navier–Stokes equations, the time and spatial variables are introduced explictly only in the initial and boundary conditions for the system. This is in sharp contrast to the description of physical systems in the transition region of Knudsen numbers, where the time and spatial variables appear explicitly in f, and hence in the description of the system. Such a difference simply reflects the fact that in the transition region, the direct influence of the boundary surface for large Kn_i is present for a considerable distance in the gas away from the boundary surface. Therefore, one suspects that the Navier–Stokes equations together with the slip flow boundary conditions probably will furnish a satisfactory description of the system for $0 < Kn_i \lesssim 0.25$ where the influence of the surface conditions is confined to a very small layer of gas adjacent to the surface.

The "normal" solution procedure, or procedures similar to it, which yield in a second approximation the Navier–Stokes equations, can be extended to the third and higher order approximation to give equations called the Burnett equations,(150) or Grad's thirteen moment equations.(151) These higher order approximations appear to be of questionable utility. No rigorous theoretical reasoning can be advanced for this statement but insofar as agreement with experiment goes, the use of the Navier–Stokes equations have been found time and again to be adequate for the slip flow region.(152) We shall therefore adopt past practice in gas dynamic theory, and simply discuss the slip flow procedure without referring to these higher approximations except in one illustrative instance.

A. Application of the Navier–Stokes Equations

As a starting point, let us write once again the Navier–Stokes equations for an incompressible binary gas mixture, as reviewed in Chapters 3 and 4:
Conservation of mass:

α species:

$$\frac{\partial \rho_A}{\partial t} + \mathbf{q} \cdot \nabla \rho_\alpha = D_{\alpha\beta} \nabla^2 \rho_\alpha.$$

Total gas:

$$\frac{\partial \rho_G}{\partial t} + \mathbf{q} \cdot \nabla \rho_G = 0,$$

where

$$\rho_G = \sum_\alpha \rho_\alpha.$$

Conservation of momentum:

$$\frac{\partial \mathbf{q}}{\partial t} + \mathbf{q} \cdot \nabla \mathbf{q} = -\frac{1}{\rho_G} \nabla p + \mu_G \nabla^2 \mathbf{q}.$$

Conservation of energy (single gaseous component):

$$\frac{\partial T}{\partial t} + \mathbf{q} \cdot \nabla T = \frac{k_G}{\rho_G c_p} \nabla^2 T.$$

Then for any particular system that is to be analyzed by the slip flow procedure, the problem is to solve these appropriate conservation equations with the slip flow boundary condition introduced below.

Boundary Conditions

In the continuum analysis outlined in Chapters 3 and 4, for example, it is assumed that the following boundary conditions hold:

(a) tangential gas velocity = velocity of the surface, or

$$q(\text{tang.}) = U_s;$$

(b) gas temperature = surface temperature, or

$$T_G = T_s = T;$$

(c) number density of gaseous species n_α = number density of α at equilibrium with the surface, or

$$n_\alpha = n_{\alpha s}.$$

Furthermore, the assumption usually is made that the tangential gradients of n_α and T_G at the surface do not produce mean motion in the gas. Now when $Kn_i > 0$, it has been observed experimentally that all these boundary assumptions are incorrect. The gas is found to slip or slide over the surface and its temperature and concentrations at the surface are found not to be the same as the surface temperature and concentrations.(117) Furthermore, the tangential gradients of n_α and T_G at the surface are found to produce a mean motion of the gas.(117) The logical conclusion of these observations is to replace the assumed boundary conditions listed above with the more appropriate non-continuum boundary conditions. Of course, it will remain to be seen whether or not this procedure gives descriptions of transfer processes in the non-continuum regions, which are more consistent with experimental results.

The slip velocity and temperature jump boundary conditions can be derived from simple argument, as was done many years ago by Maxwell.(89) Maxwell's arguments are discussed fully in Kennard's book.(117) Modern theoretical treatments based upon kinetic theory give expressions for the boundary condition for slip flow which are of the same form as Maxwell's conditions, and usually differ only by a relatively small numerical factor.

If we postulate the existence of a slip velocity, and temperature and (gas) concentration differences at a surface, the following slip flow boundary conditions can be derived for a spherical aerosol particle in spherical coordinates (r, θ, ϕ),(145) assuming the Navier–Stokes equations remain valid up to the surface.

For the tangential component of velocity, the θ component in these problems,

$$q_\theta = c_m \lambda_G \left[\left(\frac{\partial q_\theta}{\partial r} - \frac{q_\theta}{r} + \frac{1}{r}\frac{\partial q_r}{\partial \theta}\right)\right] + c_{tm}\lambda_G \frac{1}{r}\frac{\partial T}{\partial \theta} + c_{dm}\lambda_G \frac{1}{r}\frac{\partial x_\alpha}{\partial \theta} \tag{6.1}$$

at the surface of the sphere. In this and the following equations λ_G is defined by the relation

$$\mu_G = 0.499 \rho_G \bar{v}_G \lambda_G.$$

The *isothermal slip coefficient* c_m is

$$c_m = \mathscr{P}_m \left(\frac{2 - \alpha_m}{\alpha_m}\right). \tag{6.2}$$

$\mathscr{P}_m$ is a numerical constant which, with perfect accommodation for any law of interaction of neutral monatomic molecules, has the following bounds:(153)

$$1.00 \leqslant \mathscr{P}_m \leqslant 1.274. \tag{6.3}$$

c_{tm} is the "*thermal creep*" *coefficient* having the classical value

$$c_{tm} \simeq \frac{3}{4}\frac{\mu_G}{\rho_G T_G \lambda_G} \tag{6.4}$$

c_{dm} is the coefficient of the tangential concentration gradient written for a binary gas mixture. The form of this term will be discussed later in the section dealing with the diffusion force.

Equation (6.1), of course, is the usual expression for the slip velocity, expressed here in a spherical coordinate system.

The "temperature jump" boundary condition can also be given in a spherical coordinate system:

$$T_G - T_s = c_t \lambda_G \frac{\partial T_G}{\partial r} + c_{mt} \lambda_G \frac{2 \partial q_r}{\partial r}, \tag{6.5}$$

where c_t is the *temperature jump coefficient* which has the classical form[117]

$$c_t = \mathscr{P}_t \frac{(2 - \alpha_t)}{\alpha_t}. \tag{6.6}$$

$\mathscr{P}_t$ is a numerical constant which has for perfect accommodation the known bounds for any law of interaction of neutral monatomic molecules:[153]

$$1.875 \leqslant \mathscr{P}_t \leqslant 2.48. \tag{6.7}$$

For polyatomic molecules, one has the semi-macroscopic result[117]

$$\mathscr{P}_t \simeq \frac{2}{\gamma_G + 1} \frac{k_G}{\mu_G C_v}. \tag{6.8}$$

γ_G is the ratio of heat capacity at constant pressure to that at constant volume. The ratio $\mu_G C_v / k_G$ is the Prandtl number (usually defined in terms of C_p). c_{mt} has the definition for monatomic molecules

$$c_{mt} \approx \frac{\mu_G}{4 n_G k_G \lambda_G}. \tag{6.9}$$

If there is a gradient in gas concentration normal to a surface, one find also a concentration difference condition similar to the temperature jump condition

$$n_\alpha - n_\alpha^c = c_{d\alpha} \lambda_G \frac{\partial n_\alpha}{\partial r}, \tag{6.10}$$

where n_α^c is the concentration of α at the surface expected under the continuum conditions, and n_α is the actual number concentration of species α. For a binary system, c_{dA} can be obtained by arguments analogous to those used in deriving the classical value for the temperature jump.[117] One finds through this procedure

$$c_{dA} = \frac{2 - \alpha_{cA}}{\alpha_{cA}} \left(\frac{2 D_{AB}}{\bar{v}_A \lambda_G} \right). \tag{6.11}$$

α_{cA}, the condensation–evaporation coefficient, is related to the condensation–evaporation coefficient employed in the free molecule and transition regions. They are, however, not precisely the same coefficients, although we shall not distinguish between them in this treatment.

The set of slip flow boundary conditions is now complete for the purposes of our discussion. This set, eqns. (6.1), (6.5), and (6.10), includes only first order linear gradients of n_α, $\mathbf{q}$, and T_G, which are consistent with the application of the Navier–Stokes equations.

Boundary conditions involving higher order and non-linear gradients of n_α, $\mathbf{q}$, and T_G have been obtained by Grad[151] for use with his thirteen moment equations. The use of such higher order equations and boundary conditions will be mentioned briefly in the section on the thermal force.

6.2. Illustrations of the Slip Flow Procedure

As remarked in the above sections, the slip flow regime covers that region of Knudsen number where one may apply the continuum equations of mass, momentum, and energy and account approximately for the non-continuum effects through the introduction of slip flow boundary conditions. This procedure, however, is found to fail as the Knudsen number increases. If, however, the slip flow procedure does give correct results for small Knudsen numbers, then there is the possibility of developing analytical relations based on experimental observations, much as we have done for the transition region in Chapter 5. Thus our viewpoint is that the region $0 < Kn_i \lesssim 0.25$, which we have designated as the slip flow region, may be described using the slip flow theory and experimental observation. We can illustrate this viewpoint by considering some physical systems which correspond in part to the systems discussed in Chapter 5. In the exposition which follows, the slip flow procedure is applied to treat (a) the evaporation or condensation of a particle in a stagnant gas, (b) heat transfer from a particle in a stagnant gas, and (c) momentum transfer to a particle. In this last discussion isothermal drag, the thermal force, and the diffusion force are included.

A. Mass Transfer to an Aerosol Particle in a Stagnant Gas

With the slip flow procedure let us investigate the quasi-steady mass transfer of an aerosol particle composed of species A in a stagnant dilute binary gas mixture composed of species A in dilution and an inert gas which we designate as component B. This essentially represents the process of evaporation or condensation of vapor on the particle. Evaporation or condensation is assumed to be very slow so that the heat transfer process can be ignored and the entire physical system can be regarded as isothermal.

For this system, the continuum analysis is quite simple and straightforward, as noted in Chapter 4. From eqn. (4.41) we have for the total rate of transfer $\tilde{\Phi}_A$ of component A in the gas, given in units of molecules $\sec^{-1}$:

$$\tilde{\Phi}_A = -D_{AB}\frac{dn_A}{dr}4\pi r^2, \tag{6.12}$$

where D_{AB} is the mutual diffusion coefficient and r is the distance from the center of the spherical particle.

In proceeding with a continuum description, we simply integrate eqn. (4.38) using the boundary conditions

$$n_A = n_A^c(R_i), \quad r = R_i; \tag{6.13a}$$

$$n_A = n_A(\infty), \quad r = \infty, \tag{6.13b}$$

where n_A is the equilibrium concentration of species A at the particle surface and $n_A(\infty)$ is the (constant) concentration of species A far removed from the particle. The expression for the total evaporation or condensation rate Φ_A is then, in the continuum approximation,

$$\tilde{\Phi}_A/4\pi R_i^2 = D_{AB}[n_A^c(R_i) - n_A(\infty)]\, R_i^{-1}. \qquad (6.14)\dagger$$

This result is essentially that given in eqn. (4.42). Of course, this expression is not valid as $R_i \to 0$ inasmuch as the local flux $\tilde{\Phi}_A/4\pi R_i^2$ will remain finite in this limit, but the right-hand side of eqn. (6.14) becomes infinite. As $R_i \to 0$ we must certainly pass into non-continuum regimes where the continuum analysis as outlined will be incorrect.

To extend the continuum results to small, but finite Knudsen numbers, the continuum boundary condition [eqn. (6.13a)] is replaced by the slip flow boundary condition

$$n_A(R_i) - n_A^c(R_i) = c_{dA}\lambda_G \frac{dn_A}{dr}. \qquad (6.15)$$

Upon integration of eqn. (4.38) and application of boundary conditions, eqns. (6.15) and (6.13b), we find the following expression for $\tilde{\Phi}_A^s$

$$\tilde{\Phi}_A^s/4\pi R_i^2 = \frac{D_{AB}[n_A^c(R_i) - n_A(\infty)]\, R_i^{-1}}{1 + c_{dA}\lambda_G/R_i}. \qquad (6.16)$$

Equation (6.16), of course, remains finite as $R_i \to 0$.

As yet there have been no rigorous theoretical estimates of the coefficient c_{dA} except for the special case of equimolar counterdiffusion of two gaseous species having the same molecular weight.(154) For more general knowledge of c_{dA}, eqn. (6.11) must be used. One expects that this equation will give at least a first estimation of the value of c_{dA} inasmuch as the classical derivation of the velocity slip coefficient(117) gives a value for c_m which is in approximate agreement with experiment, and with more rigorous calculations.

For purposes of comparison of eqn. (6.16) with experiment, the most accurate experimental data for small Kn_i for the radial evaporation of a spherical particle in a stagnant gas appear to be those of Birks and Bradley.(126) These investigators studied the evaporation in air at low pressure of dibutyl phthalate droplets ($R_i \approx 0.5$ mm) suspended by quartz fibers from a microbalance. In Fig. 6.1 their data are plotted as a function of λ_G/R_i. As predicted by eqn. (6.16), $\tilde{\Phi}_A^c/\tilde{\Phi}_A^s$ is approximately linear in Kn_i, at least for small Kn_i. The scatter of the data is rather large so that the exact slope of $Kn_i = 0$ cannot be determined. A best fit of the data for small Kn_i indicates the experimental value $c_{dA} \approx 1.2$.

Birks and Bradley also measured c_{dA} by evaporating a flat surface of dibutyl phthalate under high vacuum conditions.(126) They found $\alpha_{cA} \simeq 0.69$. With this value for α_{cA} and their values for D_{AB} and $\bar{v}_A$, we calculate from eqn. (6.11) the classical value of $c_{dA} \approx 1.5$. The agreement between these two estimates is not too good, but in view of experimental and theoretical uncertainties such a difference may be anticipated.

One concludes on the basis of present knowledge of this particular problem that eqns. (6.11) and (6.16) probably provide a reasonable method for estimating $\tilde{\Phi}_A$ for small Kn_i, i.e. $0 \leqslant Kn_i \lesssim 0.25$. In the experimental example just studied, the 25 per cent difference of the classical value of c_{dA}, eqn. (6.11), and the experimental value actually results in a maximum error in $\tilde{\Phi}_A^s/\tilde{\Phi}_A^c$ at $Kn_i = 0.25$ of only 5 per cent. For larger Kn_i one should certainly use the theory for the transition region.

† Here, for example, if p_A is the vapor pressure of the particle at temperature T

$$n_A^c = p_A/kT.$$

The data of Birks and Bradley scatter badly for larger Kn_i, but one suspects from the analogous heat transfer problem, that the linear relation predicted by eqn. (6.16) for $\tilde{\Phi}^s_A/\tilde{\Phi}^c_A$ as a function of Kn_i breaks down somewhere around $Kn_i \approx 0.25$. Therefore, in the absence of better information, the upper limit $Kn_i = 0.25$ is recommended for application of the slip flow method to evaporation or to condensation.

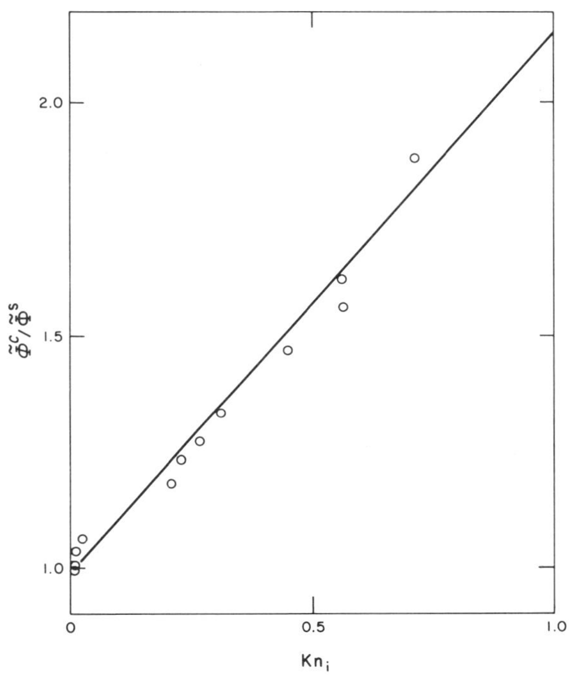

FIG. 6.1. Comparison between slip flow theory and experiments of Birks and Bradley[126] for the evaporation of phthalate droplets.

B. Heat Transfer from an Aerosol Particle to a Stagnant Gas

Next let us turn to the problem of heat transfer from a spherical particle in a stagnant gas. The description and results are closely analogous to the preceding example for nearly isothermal evaporation or condensation of a particle.

Consider a sphere suspended in a stagnant gas. The assumption is made, as in the previous example, that the process of heat transfer may be regarded as being quasi-stationary. From eqn. (4.41) we have for the total conduction heat flux H^c,

$$H^c = -\, k_G \frac{\mathrm{d}T_G}{\mathrm{d}r} 4\pi r^2. \tag{6.17}$$

If, in addition, heat is transferred to the particle by radiation, a term to account for this effect must be included in eqn. (6.17).

In the slip flow procedure the temperature jump condition, eqn. (6.5), and the specification of the temperature at large distances from the particle provide the boundary conditions for this problem. These conditions are:

$$T_G = T_G(\infty), \quad r \to \infty, \tag{6.18a}$$

$$T_G - T_s = c_t \lambda_G \left(\frac{\mathrm{d}T_G}{\mathrm{d}r}\right)_{r=R_i}. \tag{6.18b}$$

With the integration of eqn. (4.38) using the boundary conditions, eqn. (6.18a) and (6.18b), we obtain for the total heat transfer rate, H^s

$$H^s = \frac{4\pi k_G R_i [T_s - T_G(\infty)]}{1 + c_t \lambda_G / R_i}. \tag{6.19}$$

As expected from the discussion in Section 4.1, this equation is quite similar to the equation for the radial evaporation or condensation of a particle in a stagnant gas, eqn. (6.16). These two equations, together with eqns. (5.42) and (5.50), are the starting point for the analysis of the lifetime of evaporating aerosol particles in the slip flow region.

In Fig. 6.2 the linear relation predicted by eqn. (6.19) for H^c/H^s as a function of Kn_i is shown with the experimental data of Takao[129] and Harbour[130] for radial heat transfer

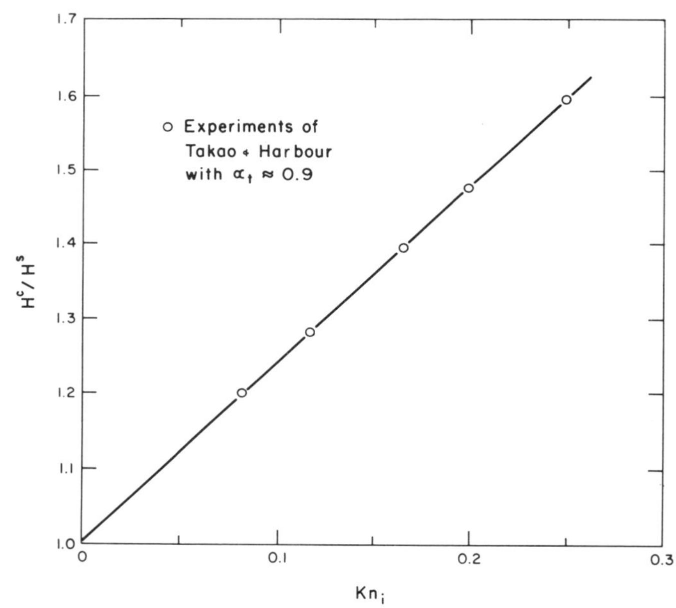

FIG. 6.2. Measurements of heat transfer to spheres in the slip flow regime by Takao[129] and Harbour[130] Theoretical line is drawn for comparison.

from a sphere in air. Under conditions where $Kn_i \lesssim 1$, Harbour's and Takao's data coincide almost exactly, even though Harbour reports a value for $\alpha_t = 0.9$, whereas Takao reports a value for $\alpha_t \approx 1.0$. However, Lord and Harbour[133] have pointed out that Takao's data show considerable scatter at large Kn_i which indicates an uncertainty in the value for α_t reported by Takao. We therefore accept the value 0.9 for α_t for both sets of data. The experimental value for c_t determined from these data is $c_{t(\text{expt})} \approx 2.3$. The classical value for c_t for air with $\alpha_t = 0.9$ from eqn. (6.8) is $c_t = 2.0$. Yet a more rigorous calculation of c_t [cf. Ref. (7)] shows a value, for monatomic gases with $\alpha_t = 1.0$, $c_t = 2.16$. This is higher than the classical value of 1.87 as calculated for $\alpha_t = 1.0$. If the classical value for c_t for air is increased in this proportion, one gets for air $c_t \approx 2.3$, for $\alpha_t = 0.9$ which, of course, is in good agreement with the experimental value. There is reason then to expect that if the calculation of c_t were refined, reasonable agreement between theory and experiment could be achieved for the region $0 \leqslant Kn_i \lesssim 0.25$.

The problem of heat transfer has been analyzed by means of the moment method[132] (cf. Chapter 2). The moment method gave the following expression for the heat transfer from a sphere:

$$H^s = H^c/(1 + \tilde{B}\lambda_G/R_i), \tag{6.20}$$

supposedly valid (but now known not to be so) over the region $0 \leqslant Kn_i \leqslant \infty$. The moment method has been found to be incorrect for large Kn_i.[137] In eqn. (6.20), $\tilde{B}$ for monatomic molecules has the value 15/4 for $\alpha_t = 1.0$, and the value 5/2 for polyatomic gases for $\alpha_t = 1.0$. The value 5/2 is not too different from the experimental value of 2.3 obtained from the data of Takao. However, the value 15/4 for monatomic molecules seems too large if we accept the slip flow procedure and the bounds on c_t given above. In fact, comparison of this moment calculation with the heat transfer data of Harbour[130] indicates a discrepancy.

Some success in analysis of this problem has been achieved using mean free path methods[130] although it does not appear that there are enough experimental data available as yet for any real claim of validity for these more intuitive methods. It is possible that the apparent successes of these methods may be fortuitous.

In summary, for the problem of radial heat transfer from a sphere in a stagnant gas in the region $0 \leqslant Kn_i \lesssim 0.25$, the slip flow procedure apparently gives a correct functional form for the transfer rate, as given by eqn. (6.19). Over the slip flow range of Knudsen number the classical value of c_t calculated from eqn. (6.8), but corrected by more rigorous calculation,[152] gives reasonably good agreement with available experimental data.

C. Momentum Transfer to an Aerosol Particle in the Slip Flow Region

Modification of momentum transfer to spheres in the presence of slippage of gas has been recognized for some time. One of the first attempts to construct a solution for the creeping flow of an isothermal gas past a sphere with slip was undertaken by Basset in the mid-nineteenth century. More recently, Derjaguin and coworkers[155, 156] have reported an interesting calculation for the thermal force and the diffusiophoretic force in the slip flow regime by applying intuitive arguments based on the theory of irreversible thermodynamics. Because Waldmann[157] has found that there are difficulties and uncertainties in specifying proper boundary conditions in calculations using irreversible thermodynamics, we arbitrarily have chosen to rely on the classical slip flow procedure with the Navier–Stokes equations. In the succeeding discussion this viewpoint is explored, bearing in mind that other equally suitable methods may evolve from the theories of Derjaguin *et al.*[155, 156]

The Isothermal Drag Force

Basset's[53] calculation of the isothermal drag force on a sphere with a velocity slip condition at the surface represents perhaps one of the earliest applications of the slip flow procedure. Even if this classical problem is rather well known, the results obtained therein should serve as a caution in the application of the slip flow procedure. Here is a good example of a case where the expectation is not always realized that from the slip flow analysis a correct analytical form will result which is uniformly valid (for the range $0 \leqslant Kn_i \lesssim 0.25$).

In order to obtain the drag force by the slip flow procedure, we proceed in the manner outlined above. The equation of motion, eqn. (3.1b), is solved for the regime $Re_i \ll 1$ with the velocity slip boundary condition, eqn. (6.1). The result of this procedure is the familiar equation:

$$\mathscr{D}^s = 6\pi\mu_G R_i \mathbf{q}\left[\frac{1 + 2c_m\lambda_G/R_i}{1 + 3c_m\lambda_G/R_i}\right] \tag{6.21}$$

with the coordinates centered on the sphere.

Despite the relative success in applying the slip flow procedure to heat and mass transfer up to $Kn_i \approx 0.25$, it has been found that eqn. (6.21) is correct in form only for $Kn_i \to 0$.

For purposes of comparison with experiment, we can write in place of the drag force, the velocity of fall of particle v_s^s in terms of the velocity of fall v_s^c, the particle would have under continuum conditions. That is

$$\frac{v_s^s}{v_s^c} = \left[\frac{1 + 3c_m Kn_i}{1 + 2c_m Kn_i}\right]. \tag{6.22}$$

This ratio has been measured experimentally by many investigators [cf. ref. (117)]. For different particle–gas systems, similar results have been obtained and verified independently. Therefore, we give a single example of an experimental determination(142) which is taken over much of the slip flow regime $0 \leqslant Kn_i \lesssim 0.25$. In Fig. 6.3, Basset's equation in the form of eqn. (6.22) is shown in comparison with some observed experimental results. It is readily

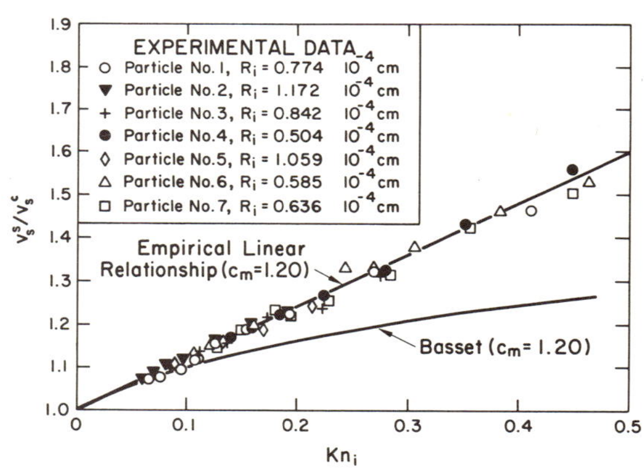

FIG. 6.3. Comparison between Basset's theory and experiment for the gravitational settling velocity of a single sphere. (From Jacobsen and Brock.(142))

seen from this diagram that Basset's expression is correct only in the sense of the limit $Kn_i \to 0$. The observed analytical form in this case for v_s^s/v_s^c is linear in Kn_i and all other experimental data show this same result.

In this rather important example the slip flow procedure has failed to give the correct analytical form over the region $0 \leqslant Kn_i \lesssim 0.25$. It is perhaps tempting to rationalize this difficulty by requiring that we linearize eqn. (2.21) in view of the linear nature of the boundary conditions. However, this is not generally acceptable because the linearization of the previous two examples for mass and heat transfer would in those instances lead to incorrect analytical forms over the region $0 \leqslant Kn_i \lesssim 0.25$. Equation (2.21) obtained by the slip flow procedure illustrates the danger in unrestricted application of this method. One cannot generally apply the slip flow procedure and claim, without experimental verification, that the results will be valid outside of a region $Kn_i \to 0$.

The Thermal Force

For $Kn_i \lesssim 0.25$ the calculation of the thermal force by eqn. (5.57) is difficult. In fact in the past the attempt at a molecular description has been abandoned entirely. Instead one assumes in this region that the system is hydrodynamic in nature and applies the macroscopic conservation equations with slip flow boundary conditions to calculate the thermal force. This is the usual slip flow procedure.

For calculation of the thermal force the slip flow procedure was first applied by Epstein[158] who failed to use the full set of slip flow boundary conditions. The following equation was obtained:

$$\mathbf{F}_t^s = -9\pi R_i^2 \frac{\mu_G}{\rho_G T_G} [k_G/(2k_G + k_p)] \nabla T_G. \tag{6.23}$$

The full set of first order slip flow boundary conditions, the first two terms on the right-hand side of eqn. (6.1) and (6.5) for this problem, was applied in deriving the following equation:[159]

$$\mathbf{F}_t^s = -\frac{12\pi\mu_G R_i^2 (c_{tm} Kn_i)(k_G/k_p + c_t Kn_i)}{(1 + 3c_m Kn_i)(1 + 2k_G/k_p + 2c_t Kn_i)} \nabla T_G. \tag{6.24}$$

Equation (6.24) was found[159] to agree well with the experimental data of Rosenblatt and LaMer[144] for tricresyl phosphate aerosols and the data of Schmitt[140] for silicon oil aerosols in the range $0 \leqslant Kn_i \lesssim 0.2$. Equation (6.23) was found to deviate considerably (20 per cent at $Kn_i = 0.1$, 50 per cent at $Kn_i = 0.2$) from the experimental results for those

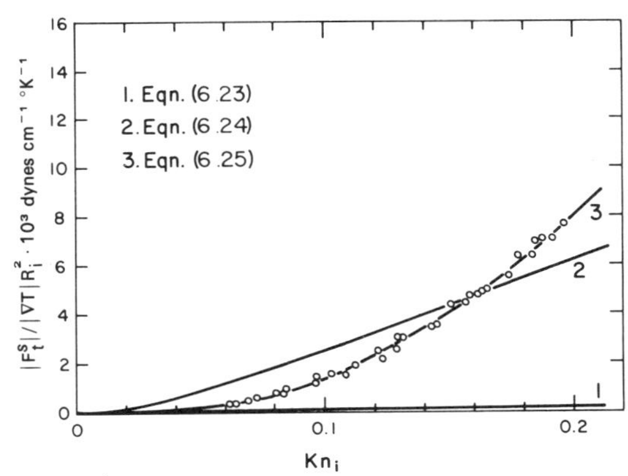

FIG. 6.4. Theory and experiments for the thermal force in the slip flow region. (From Jacobsen and Brock.[142])

aerosols.[159] On the basis of comparison with existing experimental data, eqn. (6.24) is apparently a useful equation for the *a priori* calculation of the thermal force from theoretical parameters for liquid aerosols with usual thermal conductivities in the range of $0 \leqslant Kn_i \lesssim 0.2$.

Experimental measurements of a high thermal conductivity aerosol, however, revealed that eqn. (6.24) failed to represent adequately the data (Fig. 6.4). This failure was attributed to the inadequacy of a first order slip flow analysis in accounting for the large departure from local equilibrium produced in the vicinity of a high thermal conductivity particle. Accordingly, the following equation was derived for the thermal force using the flux expressions obtained from the third approximation to the solution of the Boltzmann equation.[159]

$$\mathbf{F}_t^s = -12\pi\mu_G R_i^2 \times \frac{(c_{tm}Kn_i)\left[(k_G/k_p + c_t Kn_i)(1 + 1.33a_3c_mKn_i) - 1.33a_3c_mKn_i\right]\nabla T_G}{(1 + 3c_mKn_i)(1 + 2k_G/k_p + 2c_tKn_i)}, \tag{6.25}$$

where a_3 is a second order constant.

This equation presumably can be made to fit the experimental data for aerosols with high thermal conductivity. Also it can be made to fit the existent data for liquid aerosols. However, eqn. (6.25) is very sensitive to the values selected for the slip coefficients. Therefore, with present knowledge of the slip coefficients, eqn. (6.25) is not recommended for the *a priori* calculation of the thermal force: it is useful as a curve-fitting equation for high thermal conductivity aerosols.

The Diffusion Force

The slip flow analysis for the diffusion force is closely analogous to the problem of the thermal force just discussed. For the diffusion force, however, the classical value of the coefficient c_{dm} appears to be incorrect,[145] although so far as is known at present, the classical value of c_{tm} is approximately correct.

We consider the force acting on an aerosol particle in an effectively infinite binary gas mixture undergoing either of two types of diffusion: (a) equimolar counterdiffusion (ECD), and (b) diffusion of species A through a stagnant species B (SD). We assume isothermal conditions so that we need only consider the equation of conservation of mass [eqn. (2.118)] and momentum [eqn. (2.119)]. For the boundary conditions, we apply the diffusional slip condition, eqn. (6.1). Waldmann's[145] derivation is followed in ignoring the velocity slip condition. Later a correlation based in part on this assumption is presented without complete analytical justification.

The boundary conditions, eqn. (6.1), are rewritten in terms of the gaseous average molar velocity $\mathbf{w}$ of an isothermal binary mixture:

$$q_\theta = \frac{\sigma_{AB}D_{AB}}{R_i}\frac{\partial x_A}{\partial \theta}, \quad r = R_i, \tag{6.26}$$

where

$$\mathbf{q} = \mathbf{w} + \frac{1}{m_A + m_B}\sum_{\alpha=A}^{B} m_\alpha x_\alpha(\mathbf{q}_\alpha - \mathbf{w}) \tag{6.27}$$

defines the average molecular velocity **w** in terms of the mass average velocity **q**. Equation (6.26) is the defining equation for the coefficient σ_{AB}. With this boundary condition, and the equation of motion eqn. (3.1b) for $Re_i \ll 1$, we obtain as the diffusion force for ECD:

$$\mathbf{F}_d = -6\pi\mu_G R_i \sigma_{AB} D_{AB} (\nabla x_A)_\infty. \tag{6.28}$$

Here, of course, the velocity slip has been neglected. For SD, one obtains, by the same method

$$\mathbf{F}_d = -6\pi\mu_G R_i (1 + \sigma_{AB} x_B) \frac{D_{AB}}{x_B} (\nabla x_A)_\infty. \tag{6.29}$$

Waldmann and Schmitt[(146)] simply correlate all their data through an empirical expression for σ_{AB}, which is

$$\sigma_{AB} = 0.95 \frac{m_A - m_B}{m_A + m_B} - 1.05 \frac{d_A - d_B}{d_A + d_B}, \tag{6.30}$$

where d_α = molecular diameter of species α. This expression and eqns. (6.28) and (6.29) apparently reproduce all the existing accurate data for the diffusion force for the slip flow region $0 \leqslant Kn_i \lesssim 0.25$ as indicated in Table 6.1.

Of course this leaves a very unsatisfactory status for the diffusion force in the slip flow region. What is needed is a rigorous calculation of c_{dm} for the diffusion slip condition which will explain these experimental results and be of more general validity than the empirical eqn. (6.30).

TABLE 6.1. TEST OF WALDMANN AND SCHMITT'S EMPIRICAL FORMULA [Ref. (146)]

$$\sigma_{AB} = 0.95 \frac{m_A - m_B}{m_A + m_B} - 1.05 \frac{d_A - d_B}{d_A + d_B}$$

Gas	m	d	$\frac{m_A - m_B}{m_A + m_B}$	$\frac{d_A - d_B}{d_A + d_B}$	σ_{AB} (calc.)	σ_{AB} (exp.)
N_2^a	28	3.68				
H_2	2	2.97	0.865	−0.167	0.65	0.90
C_2H_2	26	4.22	0.037	0.069	0.11	0.13
C_2H_4	28	4.23	0	0.070	0.073	0.073 adj[b]
C_2H_6	30	4.42	−0.035	0.092	0.063	0.085
O_2	32	3.43	−0.067	−0.035	−0.10	−0.10 adj[b]
A	40	3.42	−0.176	−0.037	−0.21	−0.22
CO_2	44	4.00	−0.222	0.039	−0.17	−0.20
C_3H_8	44	5.06	−0.222	0.158	−0.04	−0.13
CO_2^a						
C_3H_8			0	0.117	0.12	0.11

[a] Component A.
[b] These two experimental values for σ_{AB} were used by Waldmann and Schmitt to determine the constants 0.95 and 1.05 in their formula.

6.3. Summary

The examples of the method of slip flow analysis discussed in this chapter are not exhaustive. They do represent, however, examples for which there is at least some body of experimental evidence for verification of the theory. From the comparisons given, the difficulties in the use of the slip flow analysis, hopefully, are made apparent.

For example, it does not follow that the consideration of the evaporation or condensation or heat transfer from a particle in relative motion to a suspended gas will be correctly described by the slip flow procedure in the region $0 \leqslant Kn_i \leqslant 0.25$. We would have no assurance (cf. the isothermal drag problem) that the slip flow analysis of this problem would yield a correct analytical form.

In general, the best one can hope to achieve with the slip flow procedure is to describe correctly a particular transfer process to an aerosol particle in the limit of very small Kn_i. If the experimental data are available and the approximate analytical forms of the transfer equations can be determined for the region $0 \leqslant Kn_i \lesssim 0.25$, then these forms usually will be characterized by the slip flow coefficients. The advantage here is that these coefficients can in principle be calculated accurately and in a general way from theoretical considerations.

TABLE 6.2. SUMMARY OF TRANSFER EQUATIONS FOR AN ISOLATED SPHERE IN THE SLIP FLOW REGIME

Transfer process	Equation	Restrictions	Equation Reference
Radial mass transfer from sphere species A through stagnant inert gas of species B	$\tilde{\Phi}^s_A = \tilde{\Phi}^c_A/(1 + c_{dA}\lambda_G/R_i)$	$0 \leqslant Kn_i \lesssim 0.25$ $Ma_i \to 0$	(6.16)
Radial heat transfer from sphere through stagnant gas	$H^s = H^c/(1 + c_t\lambda_G/R_i)$	$0 \leqslant Kn_i \lesssim 0.25$ $Ma_i \to 0$	(6.19)
Isothermal drag force on sphere	See Table 5.8	See Table 5.8	(5.53)
Thermal force on sphere in effectively infinite gas	$\mathbf{F}^s_t = -12\pi\mu_G R_i^2 c_{tm} Kn_i \nabla T_G \times [(k_G/k_p + c_t Kn_i)(1 + \frac{4}{3}a_3 c_m Kn_i) - \frac{4}{3}a_3 c_m Kn_i]/(1 + 3c_m Kn_i)(1 + 2k_G/k_p + 2c_t Kn_i)$	$0 \leqslant Kn_i \lesssim 0.25$ Recommended only as curve fitting equation	(6.25)
Diffusion force on sphere in effectively infinite gas		Binary gas only Coefficient σ_{AB} is empirical $0 \leqslant Kn_i \lesssim 0.25$	(6.28)
(a) Equimolar counter-diffusion	$\mathbf{F}^s_d = -6\pi\mu_G R_i \sigma_{AB}(\nabla x_A)_\infty$		
(b) Diffusion of A through stagnant B	$\mathbf{F}^s_d = -6\pi\mu_G R_i(1 + \sigma_{AB}x_B)(D_{AB}/x_B)(\nabla x_A)_\infty$		

In Table 6.2 the equations for the examples discussed in this chapter are summarized. For the mass and heat transfer examples it is clear that more restrictions are introduced in the slip flow region than in the transition region where relative particle–gas motion does not affect the mass or heat transfer rates so long as $Ma_i \ll 1$.

The slip flow procedure outlined here is rather limited. As already indicated, it cannot be relied on to yield *a priori* a correct description of an arbitrary system. It is, however, at a level above the interpolation schemes, such as the mean free path methods which yield equations whose analytical forms are of questionable validity, and which contain constants which must be determined empirically. The slip flow coefficients, at least, have the possibility of being determined theoretically without recourse to experiment. Much to be preferred over all those methods would be the solution of the Boltzmann equation for the slip flow regime, and the calculation of the various transfer processes by eqns. (5.2)–(5.4). One method which might be very promising in the slip flow regime would be the moment method which, however, is very difficult to apply in the absence of spherical symmetry such as found in the radial heat transfer problems. As our survey of the slip flow region shows, there is much work to be done in describing analytically transfer processes to aerosol particles in the important regime $0 \leqslant Kn_i \lesssim 0.25$ to the same degree of accuracy as the extremes of $Kn_i \to 0$ and $Kn_i \to \infty$.

CHAPTER 7

DIFFUSION AND DISPERSION OF AEROSOL PARTICLES

SINCE the work of Robert Brown early in the nineteenth century, it has been known that colloidal particles can disperse themselves through their suspending medium by thermal agitation in a manner analogous to the diffusion of gas molecules. The applicability of a Fick's equations for the diffusion of tiny particles in a fluid generally has been accepted after the work of Einstein(160) and others in the early 1900's.

In applying the theory of Brownian movement to diffusion of aerosols, it is useful to consider first the general properties of diffusional processes in the light of the mathematics of probability. By means of this introduction, a reasonably rigorous framework for the structure of the theory of aerosol diffusion may be established and developed.

7.1. General Properties of the Diffusion Equations

Diffusion of particles in one direction x of a stagnant fluid medium traditionally has been described mathematically by the parabolic partial differential equation

$$\frac{\partial N}{\partial t} = D_i \frac{\partial}{\partial x}\left(\frac{\partial N}{\partial x}\right), \tag{7.1a}$$

called *Fick's second law*. Here the term on the left is the rate of change with time of the concentration of particles N, and the term on the right represents the divergence of the mass flux $D_i(\partial N/\partial x)$, where D_i is the diffusivity of the particle. Equation (7.1a) has been derived in an elegant way by Chandrasekhar(5) (see also Wang and Uhlenbeck(161)) by application of a Markoff "random walk" process to the random motion of particles. This relation can, of course, be generalized to include all three spatial dimensions. In vector form eqn. (7.1a) reads

$$\frac{\partial N}{\partial t} = D_i \nabla^2 N. \tag{7.1b}$$

This relation may also be formulated from the kinetic theory of gases and from irreversible thermodynamics. An equivalent derivation of the diffusion equations stems from considerations of the mathematical development of the theory of random process.

A. Some Concepts of Random Processes

A random process $x(t)$ generally means a process in which the dependent variable x does not depend in a definite way on the independent variable t (equal to time). A random process then can be described the following set of probability distributions.(161)

$\tilde{W}_1(x, t)\,dx$ = probability of finding x in the range $(x, x + dx)$ at time t;
$\tilde{W}_2(x_1, t_1; x_2, t_2)\,dx_1 dx_2$ = joint probability of finding x in the range $(x_1, x_1$ $dx_1)$ at time t_1, and in the range $(x_2, x_2 + dx_2)$ at time t_2; (7.2)
$\tilde{W}_3(x_1, t_1; x_2, t_2; x_3, t_3)\,dx_1 dx_2 dx_3$ = joint probability of finding the three values of x in the ranges $x_1 + dx_1, x_2 + dx_2, x_3 + dx_3$ at times t_1, t_2, t_3, etc.

The set of relations in (7.2) must follow the conditions:

(a) $\tilde{W}_n \geqslant 0$;
(b) $\tilde{W}_n(x_1, t_1; x_2, t_2; \ldots x_n, t_n)$ is a symmetric function in the variables $x_1, t_1, \ldots, x_n, t_n$.
(c) $\tilde{W}_k(x_1, t_1; \ldots x_k, t_k) = \int \ldots \int dx_{k+1} \ldots dx_n \tilde{W}_n(x_1, t_1; \ldots x_n, t_n)$; (7.3)

because each function $\tilde{W}_n$ implies all the previous $\tilde{W}_k$ where $k < n$. The sequence of functions in (7.2) thus constitutes a hierarchy (7.3) for specifying the process in successively more detail.

In problems involving the Brownian motion of particles, the simplification is made that the processes are stationary in time. This means that the physical mechanism which causes a disturbance or fluctuation from the average behavior of the system does not change with time. In other words, a shift in the t coordinate will not affect the functions $\tilde{W}_n$, and consequently (7.2) becomes:

$\tilde{W}_1(x)\,dx$ = probability of finding x between x and $x + dx$,
$\tilde{W}_2(x_1; x_2, t)\,dx_1 dx_2$ = joint probability of finding a pair of values of x in the ranges dx_1, dx_2 in the time interval $\Delta t = t(=t_2 - t_1)$, etc.

Markoff Processes

Random processes have been classified[161] in terms of the set of probability distributions in (7.2). For example, a purely random process, in which no correlation exists between values of x is given by

$$\tilde{W}_2(x_1, t_1; x_2, t_2) = \tilde{W}_1(x_1, t_1) \,.\, \tilde{W}_1(x_2, t_2),$$

and so on for higher $\tilde{W}_n$.

Thus, all the information about such a process is contained in the distribution function $\tilde{W}_1$.

The Markoff process constitutes the next more complicated case of a random process which is specified completely by the function $\tilde{W}_2$. This kind of random process has been defined, for example, by Wang and Uhlenbeck[161] in terms of a conditional or transition probability $\tilde{P}_2(x_1|x_2,t)dx_2$. This function means that given x_1, x is found in the range $x_2, x_2 + dx_2$ at time $\Delta t(=t_2 - t_1)$ later. Then, for a stationary process, $\tilde{W}_2$ becomes

$$\tilde{W}_2(x_1; x_2, t) = \tilde{W}_1(x_1)\,\tilde{P}_2(x_1|x_2, t), \tag{7.4}$$

where $\tilde{P}_2$ fulfills the restrictions from (7.3):

$$\tilde{P}_2(x_1|x_2, t) \geqslant 0, \tag{7.5a}$$

$$\int_{-\infty}^{\infty} dx_2 \tilde{P}_2(x_1|x_2, t) = 1, \tag{7.5b}$$

$$\tilde{W}_1(x_2) = \int_{-\infty}^{\infty} \tilde{W}_1(x_1)\,\tilde{P}_2(x_1|x_2, t)\,dx_1. \tag{7.5c}$$

For problems in Brownian motion[161] one also requires that

$$\lim_{\Delta t \to \infty} \tilde{P}_2(x_1|x_2, t) = \tilde{W}_1(x_2). \tag{7.6}$$

We then note that as $\Delta t \to \infty$, the Markoff process tends to be a purely random process. The Markoff process now can be identified more precisely as a process for which the transitional probability that x lies in the interval $(x_n, x_n + dx_n)$ at time t_n, given that x is $x_1, x_2, \ldots, x_{n-1}$ at times $t_1, t_2, \ldots, t_{n-1}$ $(t_n > t_{n-1} \ldots t_2 > t_1)$ depends only on x_n, t_n and on x at previous time t_{n-1}. Mathematically this reads

$$\tilde{P}_n(x_1, t_1, x_2, t_2 \ldots x_{n-1}, t_{n-1}|x_n, t_n) = \tilde{P}_2(x_{n-1}, t_{n-1}|x_n, t_n). \tag{7.7}$$

Thus all $\tilde{W}_n$ for $n > 2$ can be determined when $\tilde{W}_2$ is known. We note, however, that $\tilde{P}_2$ is not an arbitrary function of its variables. Besides the relations in (7.5) and (7.6), $\tilde{P}_2$ has the restriction

$$\tilde{P}_2(x_2|x_2, t) = \int_{-\infty}^{\infty} dx \tilde{P}_2(x_1|x, t') \tilde{P}_2(x|x_2, t - t'), \tag{7.8}$$

for all values of t' between zero and t. This is the essential equation for the theory,[161] and has been named for M. V. Smoluchowski.

Discrete Jumps

Let us consider the problem of a particle jumping in discrete steps in time Δt, and in one dimension Δx by a Markoff process. The solution to this problem will require specifying the transitional probability $P(i|j, s\Delta t)$ when $P(n|m, \Delta t)$ only is known. P here is the analog of the continuous distribution $\tilde{P}(x_1|x_2, t)$, where $n|m$ specify the discrete changes in x_1 and x_2 for discrete time steps, $s\Delta t$ where $s = 1, 2, 3, \ldots$. Let us drop the use of $s\Delta t$ and write $P_{i,j}$ for $P(i|j, \Delta t)$ noting that $P_{i,j}$ is the basic probability of i given by the "physical cause" for the random process.

To find $P(i|j, s)$ it is possible, in principle, to use the discrete form of eqn. (7.8):

$$P(i|j, s) = \sum_k P(i|k, s-1) P_{k,j}. \tag{7.9a}$$

Use of eqn. (7.9a) becomes impractical for large values of s.

Some properties of the function $P_{i,j}$ may be used to give an alternate form for determining $P(i|j, s)$. Recalling that

$$\sum_j P_{k,j} = 1 \quad \text{or} \quad P_{k,k} + \sum_j{}' P_{k,j} = 1,$$

the prime noting that the value $k = j$ is omitted in the summation, equation (7.9a) may be rewritten as

$$P(i|j, s) - P(i|j, s-1) = -P(i|j, s-1) \sum_k{}' P_{j,k} + \sum_k{}' P(i|k, s-1) P_{k,j}. \tag{7.9b}$$

Thus the rate of change of $P(i|j, s)$ with increments of time s equals the gains in P from transitions from k to j less the losses in P attributed to transitions of j to all possible k. Wang and Uhlenbeck[161] have pointed out that eqn. (7.9b) is completely analogous to the Boltzmann equation, eqn. (2.38). To obtain an expression for $P(i|j, s)$ one solves eqn.

(7.9b) for an initial distribution, which introduces the index i since

$$P(i|j, 0) = \delta(i, j), \tag{7.10}$$

where δ is the Dirac delta function. In some cases the dependent variable k can change only by ± 1 in a step. This means that $P_{k,j} = 0$ except for $j - k, k = \pm 1$, and eqns. (7.9a) and (7.9b) reduce to a simpler form.

If we apply these principles to an example of random walk along a straight line x we can arrive at a specific solution for $P(i|j, s)$. In this simple case, the step along x and Δx for each time interval $s\Delta t$. Then $P(i|j, s)$ refers to the probability of a particle being at position $j - x$ at $s - t$ if at $s = 0$, the particle is located at $i\Delta t$. The basic transition probability is

$$P_{k,j} = \tfrac{1}{2}\delta(j, k-1) + \tfrac{1}{2}\delta(j, k+1),$$

and eqn. (7.9) becomes the difference equation

$$P(i|j, s) = \tfrac{1}{2}P(i|j+1, s-1) + \tfrac{1}{2}P(i|j-1, s-1), \tag{7.11}$$

which can be solved for the initial condition (7.10). The solution is[161]

$$P(i|j, s) = (\tfrac{1}{2})^s s! \Big/ \left(\frac{i-j+s}{2}\right)! \left(\frac{i-j-s}{2}\right)! \tag{7.12}$$

Using eqn. (7.5b) we then find that

$$W(j) = \sum_i W(i) \,.\, P(i|j, s),$$

where $W(i)$ is a constant. Here $W(j)$ is the jth distribution function, giving a relation analogous to eqn. (7.5c).

Suppose we consider the limiting case where in the example above, the diffusional steps become continuous. That is, consider the limit of eqn. (7.11) when

$$i\Delta x = x \quad \text{and} \quad a\Delta t = t.$$

Rewriting eqn. (7.11) as

$$P(i|j, s) - P(i|j, s-1) = \tfrac{1}{2}[P(i|j+1, s-1) - 2P(i|j, s-1) + P(i|j-1, s-1)],$$

and taking the limit yields

$$\frac{\partial \tilde{P}}{\partial t} = D\left(\frac{\partial^2 \tilde{P}}{\partial x^2}\right), \tag{7.13}$$

where $D = \lim_{\Delta t \to 0} \Delta x^2/2\Delta t$, which is named the *Einstein diffusivity*. Equation (7.13) is equivalent to the well known diffusion equation, eqn. (7.1).

Within the continuum limit, the determination of the probability distribution reduces to the problem of finding a solution to the parabolic partial differential equation into which Smoluchowski's equation has degenerated. For the case where the initial condition is

$$\tilde{P}(\chi_0|\chi_0, 0) = \delta(x - x_0) \quad \text{at} \quad t = 0,$$

the solution to eqn. (7.13) reads

$$\tilde{P}(x_0|x, t) = (4\pi D t)^{-\frac{1}{2}} \exp\left[-(x - x_0)^2/4Dt\right]. \tag{7.14}$$

This is exactly the continuum limit of the solution for the discrete case eqn. (7.12). Thus for the restriction that in the time Δt, i can only change by ± 1 the Fickian diffusion equation is obtained in the continuum limit. This is equivalent to *assuming* in addition that the Markoff process is a special one involving a Gaussian function for the probability distributions [like eqn. (7.14)].

B. Diffusion by Discontinuous Jumps

It is possible to construct more general "diffusion" equations than the parabolic form by considering a discrete process that involves certain non-isotropic properties. This more general case has been discussed by Goldstein[(162)] (see also Michelson[(163)]).

At some initial time $t = 0$, let us consider again a stochastic process in which a large number of non-interacting particles start from an origin and travel with a uniform jump velocity v_p along a line x in a sequence of time intervals Δt. At the beginning of displacement, half the particles jump one incremental step in either direction from the origin. Thereafter, and at the end of each successive step in time interval Δt, each particle sets out on a new particle path with a speed v_p. The probability functions for such a model are defined as follows:

1. $\hat{P} = \hat{P}(n|n + 1|n + 2)$ denotes the probability that a particle will move from its present position one jump to the right and continue towards the same direction during the succeeding increment of time;
2. $\hat{Q} = \hat{P}(n|n + 1|n)$ refers to the probability that a particle will start from its initial position, move one step to the right, and then return to its initial position;
3. $\hat{P}' = \hat{P}(n|n - 1|n - 2)$ equals the probability that a particle will start from its initial position, travel one jump to the left, and then continue to the left in the next instant of time;
4. $Q' = \hat{P}(n|n - 1|n)$ equals the probability that a particle will begin from its present position, move one jump to the left, and then return to its initial position.

In defining this type of random walk process with probabilities $\hat{P}, \hat{P}', \hat{Q}$, and $\hat{Q}'$, it is assumed implicitly that the particles never stand still, and the particles move only a finite distance in a finite interval of time. Then, the constraints

$$\left.\begin{aligned} \hat{P} + \hat{Q} &= 1, \\ \hat{P}' + \hat{Q}' &= 1, \end{aligned}\right\} \tag{7.15}$$

must be true. Furthermore, the constants

$$\left.\begin{aligned} C &\equiv \hat{P} - \hat{Q}, \\ K &\equiv \hat{P} - \hat{P}', \end{aligned}\right\} \tag{7.16}$$

can be defined, which are related either to the *process*, or to the *system* in which the diffusional process occurs. Hence, the term C has physical significance in that it deals with a probability of a systematic drift in the diffusional process, or to the degree of isotropy in the jump process. When C equals zero, the particles, after one jump, have equal likelihood of proceeding forward or returning in the reverse direction. On the other hand, the constant K has to be related to the system, and if $K = 0$, the act of traveling to the right or to the left with succeeding jumps is equally probable.

The statistics of the distribution of particles can be estimated in terms of the probability distribution function $W(m, n)$, which represents the fractional number of particles at a fixed distance mx to the right or the left of the origin in a fixed time nt. The function W

then may be divided into $W^+(m, n)$, the fraction of particles moving to the right, having arrived at $m\,\Delta x$ from the left, and $W^-(m, n)$, the fraction of particles moving to the left, from the right at $m\,\Delta x$ steps. Then,

$$W(m + \mathrm{i}, n + j) = W^+(m + i, n + j) + W^-(m + i, n + j), \tag{7.17}$$

or, abbreviating the notation,

$$W_{i,j} = W^+_{i,j} + W^-_{i,j},$$

where i denotes the time index (always +) and j refers to the position index (± or 0). Similarly, conservation of matter requires that

$$W_{i,j} = W^+_{i+1,j+1} + W^-_{i+1,j-1}, \tag{7.18}$$

indicating that the probability density of particles $W_{i,j}$ at position $m\,\Delta x$, time $n\,\Delta t$, can be given in the first term on the right by particles coming from a unit distance on the left one step to the right in one time interval after j, plus, in the second term on the right, the particles arriving from the left in time step preceeding j.

Hence, from eqns. (7.17) and (7.18) the flux of particles in time Δt to a particular point in space equals

$$W_{i+1,j} - W_{i,j} = W^+_{i+1,j} - W^+_{i+1,j+1} + W^-_{i+1,j} - W^-_{i+1,j-1}. \tag{7.19}$$

Using eqns. (7.16)–(7.19), we find that

$$W(m + 1, n) = \frac{1 + C}{2}[W(m, n - 1) + W(m, n + 1)] - CW(m - 1, n) - K[W(m, n - 1) - W(m - 1, n)]. \tag{7.20}$$

To pass from the finite difference scheme to a continuous form, let

$$m\Delta x \to x\ , \quad (n + 1)\,\Delta t \to t\,; \qquad \Delta x \to \mathrm{d}x, \quad \Delta t \to \mathrm{d}t\,;$$

and eqn. (7.10) is expanded in a Taylor series about t and x. Terms with the same coefficients are collected, and we find an equation of the form

$$\mathrm{d}t\frac{\partial \tilde{W}}{\partial t}(C - 1 - K) - (K\,\mathrm{d}x)\frac{\partial \tilde{W}}{\partial x} + \frac{(\mathrm{d}t)^2}{2}\frac{\partial^2 \tilde{W}}{\partial t^2}[K - 1 - C] + \frac{(\mathrm{d}x)^2}{2}\frac{\partial^2 W}{\partial x^2}[1 + C - K] + \frac{(\mathrm{d}t)^3}{3!}\frac{\partial^3 W}{\partial t^3}[C - 1 - K] - \frac{K(\mathrm{d}x)^3}{3!}\frac{\partial^3 W}{\partial x^3} + 0[(\mathrm{d}x)^4, (\mathrm{d}t)^4] = 0, \tag{7.21}$$

where the last term indicates a continuation of the series in terms of fourth order and higher.

The Generalized Diffusion Equation

A general form of a "diffusion" equation now can be formed from eqn. (7.21) by considering the limits:

$$\lim_{dt\to 0} \frac{K}{dt} = \lim_{dt\to 0} \frac{\hat{P} - \hat{P}'}{dt} \to \text{const.}, B;$$

$$\lim_{dt\to 0} \frac{1 - C}{dt} = \lim_{dt\to 0} \frac{1 - \hat{P} + \hat{Q}}{dt} \to \text{const.}, A;$$

$$\lim_{dt\to 0} \frac{dx}{dt} \to v_p.$$

Dropping terms of third and higher order, this limiting result reads

$$\frac{\partial^2 \tilde{W}}{\partial t^2} + (A + B)\frac{\partial \tilde{W}}{\partial t} + Bv_p \frac{\partial \tilde{W}}{\partial x} + v_p^2 \frac{\partial^2 W}{\partial x^2} = 0, \tag{7.23}$$

which represents a relation analogous to the well known telegrapher's equation. This is a hyperbolic equation in contrast to the parabolic form of eqn. (7.1). Therefore, the mathematical solutions of eqn. (7.23) generally may be expected to have properties differing from those of eqn. (7.1).

Equation (7.23), of course, mathematically contains Fick's diffusion equation, provided that the first term and the next to the last term on the left side are absent. This implies that in passing to the limit as $dt \to 0$, and dropping terms of third and higher order,

$$\begin{matrix} K = 0, \\ C = 0, \end{matrix} \quad \lim_{dt\to 0} \frac{(dx)^2}{2dt} \to \text{const.}, \tag{7.24}$$

equal to the Einstein diffusivity D, and

$$\lim_{dt\to 0} \frac{dx}{dt} \to v_p.$$

Under other restrictions, there exist quite different forms of eqn. (7.23). For example, if eqn. (7.21) is divided by dt and the limit $dt \to 0$ is taken,

$$\frac{1}{v_p} \frac{1 - C + K}{K} \frac{\partial \tilde{W}}{\partial t} + \frac{\partial \tilde{W}}{\partial x} = 0 \tag{7.25}$$

provided that $dx \to 0$ as $dt \to 0$. This relation corresponds to a (hyperbolic) half wave equation describing the uniform propagation of a disturbance in one direction, where

$$\lim_{dt\to 0} v_p \left(\frac{K}{1 - C + K}\right) \to \text{const.}$$

equals the propagation velocity. On the other hand, if C = unity, for perfect correlation between jumps, and $K = 0$, we obtain the well known wave equation

$$v_p^2 \frac{\partial^2 \tilde{W}}{\partial t^2} = \frac{\partial^2 \tilde{W}}{\partial x^2} \tag{7.26}$$

in which v_p now denotes the speed of waves propagating in both directions along x.

A third case of interest crops up if the system is isotropic, or if C is assumed to be zero. Then if

$$\lim_{dt \to 0} \frac{(dx)^2}{2dt} \to \text{const.}, D$$

and

$$\lim_{dt \to 0} \frac{dx}{dt} \to v_p,$$

we find that

$$\frac{\partial \tilde{W}}{\partial t}(1 + K) + v_p K \frac{\partial \tilde{W}}{\partial x} = D(1 - K)\frac{\partial^2 \tilde{W}}{\partial x^2}. \tag{7.27}$$

This expression is analogous to the customary equation written for convective diffusion. It is also similar in form to the well known Fokker–Planck equation which may be derived from considerations of a Gaussian Markoff process (Wang and Uhlenbeck[161]). However, v_p now denotes a drift velocity, and not the mass average velocity of the suspending medium.

Thus, depending on the nature of mathematical assumptions, combined with the requirements of isotropy in the diffusional process and in the system in which diffusion occurs, different kinds of expressions describing a diffusional process can be obtained. A reasonably general form of the "diffusion" equation, eqn. (7.23), can be found involving only the assumptions that terms of third and higher order are disregarded, that particles must always move, with finite speed, and that the processes of discontinuous jumps of particles, and of course the probability density for $\tilde{W}$ pass to a continuous function in the continuum limit.

C. Diffusion of Particles by Brownian Motion

To apply the ideas of probability theory to the diffusion of aerosols in Brownian motion, one considers the statistical behavior of a system containing very large numbers of particles $\mathcal{N}$. The number of particles in the range $\mathbf{r}, \mathbf{r} + d\mathbf{r}$, at a given time t is

$$N = \mathcal{N}\tilde{W}_1(\mathbf{r}, t). \tag{7.28}$$

The position vector $\mathbf{r}$ is given in magnitude by

$$|\mathbf{r}| = (x^2 + y^2 + z^2)^{\frac{1}{2}}.$$

Hence, generalizing eqn. (7.5c) to three dimensions, with eqn. (7.28),

$$N(\mathbf{r}, \Delta t) = \int_{-\infty}^{\infty} N(\mathbf{r} - \Delta\mathbf{r}, t)\, \tilde{P}_2(\mathbf{r} - \Delta\mathbf{r}, t; \Delta t, \Delta\mathbf{r})\, d(\Delta\mathbf{r}), \tag{7.5d}$$

where

$$\Delta t = t_2 - t_1 \quad \text{and} \quad \Delta\mathbf{r} = \mathbf{r}_2 - \mathbf{r}_1.$$

If eqn. (7.5d) is expanded in a Taylor series about $d\Delta t$ and the integrations are carried out term by term, one finds that the expression,

$$\frac{\partial N}{\partial t} = D\nabla^2 N, \tag{7.1b}$$

is obtained provided that terms of order $(\Delta x)^3$ are neglected, and we assume that

$$\tilde{P}_2(\mathbf{r} - \Delta\mathbf{r}; \Delta t, \Delta\mathbf{r}) = \tilde{P}_{2\Delta x} \cdot \tilde{P}_{2\Delta y} \cdot \tilde{P}_{2\Delta z}$$

to remove the off-diagonal terms in the tensor resulting from the Taylor series expansion. The Einstein diffusivity then has the symmetrical form of

$$D = \lim_{\substack{\Delta t \to 0 \\ \Delta x \to 0}} \frac{1}{2} \frac{[\overline{\Delta x^2} + \overline{\Delta y^2} + \overline{\Delta z^2}]}{\Delta t}$$

where the over bars refer to displacement of particles averaged over an ensemble of many realizations of trajectories.

In summary, then, we assume that the Brownian diffusion of aerosol particles is a Gaussian random process of the Markoff type. The time scale for transition Δt is small compared to changes in the entire system, but long enough to provide for particle movement in which the particle speed is uncorrelated with position. The system and the process are isotropic, and no ambiguities develop in the mathematical process of taking the limits to form the continuum equations.

Evaluation of the Einstein Diffusivity

To obtain a specific relation for the diffusivity, the dynamics of the individual particles is introduced. In the classical treatment, a form analogous to eqn. (3.29), called the *Langevin equation*, is used to give the acceleration of a particle:

$$\frac{dv_{ix}}{dt} = \mathscr{A} v_{ix} + \mathscr{F}(t), \tag{7.29†}$$

where v_{ix} represents the particle velocity in the x direction. $\mathscr{F}(t)$ denotes a function accounting only for the acceleration associated with thermal motion of the suspending fluid. Now eqn. (7.29) is used to find a relation for the product of v_{ix} and a distance of travel Δx for the particle. Multiplying eqn. (7.29) by Δx integrating and solving for the product[164]

$$v_{ix}\Delta x = (v_{ix}\Delta x)_0 \, e^{-\mathscr{A}t} + e^{-\mathscr{A}t} \int_0^t v_{ix}^2 e^{\mathscr{A}\gamma} d\gamma + e^{-\mathscr{A}t} \int_0^t \mathscr{F} \Delta x \, e^{\mathscr{A}\gamma} d\gamma. \tag{7.30}$$

Here the subscript 0 refers to the initial value.

When eqn. (7.30) is averaged over an ensemble of all diffusing species, we find that

$$\overline{v_{ix}\Delta x} = \frac{\overline{v_{ix}^2}}{\mathscr{A}} (1 - e^{-\mathscr{A}t}), \tag{7.31}$$

assuming that $\mathscr{F}(t)$ is a random function, or $\mathscr{F}(t) = 0$, and t is taken long enough for a given Δx so that there remains no correlation between $\mathscr{F}(t)$ and Δx. However,

$$\overline{v_{ix}\Delta x} = \frac{\overline{\Delta x \, d\Delta x}}{dt} = \frac{1}{2} \frac{d\overline{\Delta x^2}}{dt}.$$

† In this case the medium is stagnant so that $q_x = 0$. The perturbation function $F_e'(t)$ is also assumed to be zero, or at least that it is a random function that can be incorporated into $\mathscr{F}(t)$ [see also eqn. (5.40)].

So by integrating eqn. (7.31), the mean square displacement for a large number of particles reads

$$\frac{\overline{\Delta x^2}}{2} = \frac{\overline{v_{ix}^2}\,\Delta t}{\mathscr{A}} + \frac{\overline{v_{ix}^2}}{\mathscr{A}^2}(\mathrm{e}^{-\mathscr{A}t} - 1), \tag{7.23}$$

or for times much greater than $\mathscr{A}^{-1}$, the relaxation time of the particle,

$$\frac{\overline{\Delta x^2}}{2} = \frac{\overline{v_{ix}^2}\,\Delta t}{\mathscr{A}}. \tag{7.33}$$

If the gas–particle system is at equilibrium, it is plausible to assume that there exists an equipartition of kinetic energy, or

$$m_i\overline{v_{ix}^2}/2 = kT/2, \tag{7.34}$$

where k is Boltzmann's constant. This relation defines an average velocity for Brownian motion of

$$\bar{c}_i = \left(\overline{v_{ix}^2}\right)^{\frac{1}{2}} = \left(\frac{kT}{m_i}\right)^{\frac{1}{2}}. \tag{7.34a}$$

From eqns. (7.33) and (7.34), the Einstein diffusivity must be

$$D_i = \frac{\overline{\Delta x^2}}{2\Delta t} = B_i kT. \tag{7.35a}$$

The term B_i is the mobility of the particle, defined by eqn. (3.9), or by eqn. (5.41), for example. Then

$$D_i = B_i kT = \frac{kT\mathscr{A}^{-1}}{m_i} = \bar{c}_i^2\mathscr{A}^{-1}. \tag{7.35b}$$

The relation B_ikT bears the name of the *Stokes–Einstein diffusivity*.

Making an analogy to simple kinetic theory of gases, the apparent mean free path of a particle λ_i can be defined as

$$\lambda_i = \bar{v}_i\mathscr{A}^{-1}, \tag{7.36}$$

where the average "thermal" speed of the particle is given by

$$\bar{v}_i = \frac{2\sqrt{2}}{\sqrt{\pi}}\bar{c}_i = \left(\frac{8kT}{\pi m_i}\right)^{\frac{1}{2}}. \tag{7.36a}$$

Hence

$$D_i = \frac{\pi}{8}\bar{v}_i\lambda_i. \tag{7.35b}$$

Some examples of the characteristic numbers for diffusing aerosols from eqns. (7.35) and (7.36) are given in Table 7.1.

Although discrepancies between observed diffusivities of aerosols and those calculated from eqn. (7.35) with eqn. (3.9a) have been reported by some workers, very careful experimentation has indicated that values of D_i as calculated by these relations in uniform, stagnant gases are well within the error of the experiments.

TABLE 7.1. SOME CHARACTERISTIC TRANSPORT PROPERTIES OF AEROSOL PARTICLES OF UNIT DENSITY IN AIR AT 1 ATM PRESSURE AND 20°C (CGS UNITS)

Particle radius	B_i	D_i	$\mathscr{A}$	$\bar{v}_i$	$\lambda_i = \bar{v}_i \mathscr{A}^{-1}$
1×10^{-3}	2.94×10^5	1.19×10^{-8}	8.13×10^2	4.96×10^{-3}	6.11×10^{-6}
5×10^{-4}	5.96×10^5	2.41×10^{-8}	3.24×10^3	1.41×10^{-2}	4.34×10^{-6}
1×10^{-4}	3.17×10^6	1.28×10^{-7}	7.59×10^4	0.157	2.07×10^{-6}
5×10^{-5}	6.71×10^6	2.71×10^{-7}	2.85×10^5	0.444	1.54×10^{-6}
1×10^{-5}	5.38×10^7	2.17×10^{-6}	4.44×10^6	4.97	1.12×10^{-6}
5×10^{-6}	1.64×10^8	6.63×10^{-6}	11.7×10^6	14.9	1.20×10^{-6}
1×10^{-6}	3.26×10^9	1.32×10^{-4}	7.35×10^7	157	2.14×10^{-6}
5×10^{-7}	1.26×10^{10}	5.09×10^{-4}	1.52×10^8	443	2.91×10^{-6}
1×10^{-7}	3.08×10^{11}	1.25×10^{-2}	7.79×10^8	4970	6.39×10^{-6}

D. Diffusion in the Presence of an External Force Field

In the above development it was assumed that $\tilde{P}_2 = \tilde{P}_2(\Delta\mathbf{r}, \Delta t)$ only. However, when gradients in properties persist in the suspending medium, or external forces act on the particles, at best the transitional probability $\tilde{P}_2$ will have a more general form, depending on location in space, or $\tilde{P}_2 = \tilde{P}_2(\mathbf{r}, \Delta\mathbf{r}, \Delta t)$. Potentially more serious is the possibility that such complications might invalidate the use of the Fickian laws of diffusion as discussed above. Under certain conditions, however, it seems possible, at least in practice, to use the ordinary diffusion equation of parabolic form for these cases.

In one special case, one may apply the Gaussian Markoff process[5] to diffusion of aerosols in a stagnant medium (or a uniformly translating system under the influence of an external field of force). This example involves the simplifying assumption that P_2 is an unsymmetrical function of $\Delta\mathbf{r}$ and Δt, but there is no explicit dependence on $\mathbf{r}$. In the one dimensional case, then, the limit of a Taylor expansion of eqn. (7.5d) around $d(\Delta t)$ yields[7]

$$\Delta t \frac{\partial N}{\partial t} = -\frac{\partial}{\partial x}\left[N \int_{-\infty}^{\infty} \tilde{P}_2 \Delta x \, d\Delta x\right] + \frac{1}{2}\frac{\partial^2}{\partial x^2}\left[N \int_{-\infty}^{\infty} \tilde{P}_2 (\Delta x)^2 \, d(\Delta x)\right] + 0(\Delta x^3). \qquad (7.37)$$

The integral in the last term on the right equals $\overline{\Delta x^2}$, but the integral in the first term on the right is $\overline{\Delta x}$ and not zero as for the case of a symmetrical form of $\tilde{P}_2$. Hence, in analogy to eqn. (7.1b), and neglecting terms of order higher than Δx^2, eqn. (7.37) becomes

$$\frac{\partial N}{\partial t} = \frac{\overline{\Delta x^2}}{2\Delta t}\left(\frac{\partial^2 N}{\partial \chi^2}\right) - \frac{\overline{\Delta x}}{\Delta t}\left(\frac{\partial N}{\partial x}\right), \qquad (7.38)$$

where $\overline{\Delta x}/\Delta t$ is identified as average drift velocity q_{ix} of particles relative to any uniform formulation of the entire gas–particle system. Equation (7.38) has a form like the Fokker–Planck equation. It may be generalized to encompass three spatial dimensions[5] as

$$\frac{\partial N}{\partial t} + \nabla \, \mathbf{q}_i N = \nabla \, . \, D_i \nabla N. \qquad (7.39)$$

Chandrasekhar[5] has remarked that this expression is sometimes called the *Smoluchowski equation.* This generalization appears to be rigorous within the theory of Markoff process when

$$\mathbf{q}_i = -m_i B_i \nabla \Phi,$$

where Φ is the potential of the external field of force.

Equation (7.39) represents the extension of the classical theory of diffusion by Brownian motion to a case where a drift in particles may result from an external force field. The term containing the drift velocity $\mathbf{q}_i$ is analogous to the convective diffusion term appearing in the equations for diffusion of gases in a flowing medium.

Mobility and the Diffusivity of Particles in a Force Field

When a field of force exerts its influence on particles in Brownian motion, the equation for the particle dynamics contains a term added to the Langevin equation. In the one dimensional case,

$$\frac{\mathrm{d}v_{ix}}{\mathrm{d}t} + \mathscr{A} v_{ix} = \mathscr{F}(t) + F_e'(x), \tag{7.40}$$

where $F_e'(x)$ is the external force per unit mass. If this relation is integrated over t to $t + \Delta t$, for a small interval of x, Δx, and averaged over many particles, a specific relation for the average particle velocity develops, as before:[164]

$$v_{ix} - v_{i0} + \mathscr{A}\Delta x \approx F_e'\Delta t.$$

But $v_{ix} = \mathrm{d}x/\mathrm{d}t$. Therefore, integrating again, we find that

$$\Delta x = \mathrm{e}^{-\mathscr{A}t}\int_0^t v_{i0}\,\mathrm{e}^{\mathscr{A}\gamma}\,\mathrm{d}\gamma + \mathrm{e}^{-\mathscr{A}t}\int_0^t F_e'\gamma\,\mathrm{e}^{\mathscr{A}\gamma}\,\mathrm{d}\gamma = \frac{v_{i0}}{\mathscr{A}}(1 - \mathrm{e}^{-\mathscr{A}\Delta t}) + \frac{F_e'\Delta t}{\mathscr{A}} - \frac{F_e'}{\mathscr{A}^2}(1 - \mathrm{e}^{-\mathscr{A}\Delta t}).$$

As before, it is assumed that $\Delta t \gg \mathscr{A}^{-1}$, and integrating again,

$$\overline{\Delta x} \approx \frac{v_{i0}}{\mathscr{A}} + \frac{F_e'\Delta t}{\mathscr{A}} - \frac{F_e'}{\mathscr{A}^2} \tag{7.41}$$

But the first and the third term on the right-hand side have a value much less than the middle right-hand term, so by choosing the time interval for integration suitably,

$$q_{ix} = \frac{\overline{\Delta x}}{\Delta t} \approx \frac{F_e'}{\mathscr{A}}. \tag{7.42}$$

We recognize this expression immediately as the relation between the particle velocity, a force and the mobility B_i.

Using eqn. (7.42) it is possible to deduce that the Stokes–Einstein relation follows from eqn. (7.40) with certain restrictions. The argument of Friedlander[164] proceeds by substituting the mobility relation into the force balance, the Langevin equation, giving a form that is the same as the force-free case:

$$\frac{\mathrm{d}(v_{ix} - q_{ix})}{\mathrm{d}t} + \mathscr{A}(v_{ix} - q_{ix}) = \mathscr{F}(t). \tag{7.43}$$

Now, instead of v_{ix} having a random or (symmetric) Gaussian distribution, it is assumed that the quantity $(v_{ix} - q_{ix})$ has such a distribution. Then, with a Galilean transformation of coordinates,

$$x' = x - q_{ix}t.$$

Equation (7.43) can be rewritten as

$$\frac{\mathrm{d}[\Delta x'(v_{ix} - q_{ix})]}{\mathrm{d}t} = \mathscr{A}\Delta x'(v_{ix} - q_{ix}) = \mathscr{F}\Delta x' + (v_{ix} - q_{ix})^2. \tag{7.44}$$

Averaging over a large number of particles,

$$\overline{\mathscr{A}\Delta x'(v_{ix} - q_{ix})} = \frac{\mathscr{A}}{2}\frac{\mathrm{d}\overline{\Delta x'^2}}{dt} = \overline{(v_{ix} - q_{ix})^2}. \tag{7.45}$$

Now it is assumed that the principle of equipartition applies to $\overline{(v_{ix} - q_{ix})^2}$. Therefore, noting that

$$\overline{\Delta x'^2} = \overline{\Delta x^2} - 2\Delta x q_{ix}\,\Delta t + \overline{q_{ix}^2}\,\Delta t^2,$$

the Stokes–Einstein relation is recovered, provided that the force field is weak enough to insure that $\overline{q_{ix}^2}\Delta t \to 0$ as $t \to 0$. In other words,

$$\frac{1}{2}\frac{\overline{\Delta x^2}}{\Delta t} = \lim_{\Delta t \to 0}\,[B_i kT - \overline{q_{ix}^2}\Delta t] \approx B_i kT. \tag{7.46}$$

7.2. Deposition of Aerosols by Diffusion

A. Diffusion in a Stagnant Medium

Once the applicability and limitations of the theory for diffusion of particles by Brownian motion has been established, a wide variety of problems of practical interest may be examined. Often one deals with situations where aerosols are deposited on a surface as a result of Brownian diffusion towards the surface. The surface on which the particles alight is generally considered a perfect absorber. Therefore in the absence of gas flow, the theoretical rates of "absorption" of particles on a surface may be calculated, for example, from the diffusion equation, eqns. (7.1) or (7.1a) with the boundary condition that the particle concentration N equal zero right at the surface in question. Strictly speaking, this condition applies to a region a distance of one particle radius away from the surface. However, the bulk of the early calculations for deposition by diffusion use the simpler boundary condition. This, of course, is equivalent to assuming the particles are volumeless points suspended in the gas. It is analogous to neglecting the interception effect in treatment of inertial deposition (p. 79).

Solutions to the parabolic diffusion equation with the boundary condition of an absorbing surface are well known. The general mathematical properties of solutions to such boundary value problems are discussed by Sneddon,[165] for example. A comprehensive list of solutions for various geometries of surfaces are contained in texts such as Carslaw and Jaeger[87] and Crank[166].

To illustrate the nature of different solutions for diffusive deposition surfaces several special examples are outlined below. In all of the cases, the radius of the aerosol particles is assumed to be vanishingly small.

A1. *Diffusion to a Vertical Wall*

Perhaps the simplest case of diffusional deposition of aerosols deals with a plain, vertical wall at $x = x_0$ in contact with a large volume of stagnant gas containing aerosols, whose initial concentration N_∞ is spatially uniform. The one-dimensional diffusion equation, eqn. (7.1), is applicable to this case, with

$$N(x, 0) = N_\infty, \quad x > x_0;$$
$$N(x_0, t) = 0, \quad t > 0.$$

For these requirements, the distribution of particles $N(x, t)$ then is given by the solution to eqn. (7.1) as

$$N(x, t) = N_\infty \operatorname{erf}\left[\frac{x - x_0}{\sqrt{(4D_i t)}}\right], \tag{7.47}$$

where

$$\operatorname{erf} \xi = \frac{2}{\sqrt{\pi}} \int_0^\xi e^{-\gamma^2} \, d\gamma.$$

The rate of transfer or deposition of particles to a unit area of surface often has more interest in practice than the actual distribution of particles. The transfer rate j_i per unit surface area can be estimated from the expression

$$j_i = D_i \left(\frac{\partial N}{\partial x}\right)_{x = x_0}, \tag{7.48}$$

or with (7.47),

$$j_i = N_\infty \left(\frac{D_i}{\pi t}\right)^{\frac{1}{2}}, \tag{7.49}$$

In an internal of time $t_1 - t_0$ the number of particles reaching the wall per unit area then reads

$$\int_0^t j_i \, dt = 2N_\infty \left[\frac{D_i(t_1 - t_0)}{\pi}\right]^{\frac{1}{2}}. \tag{7.50}$$

A2. *Diffusion to Parallel Vertical Walls*

For examining clouds of particles in a closed box, such as in a chamber for an ultra-microscope, it is important to know the rate of depletion of particles in the aerosol cloud by diffusion to the walls. Suppose the cavity holding the cloud is long and narrow so that only diffusion to the vertical walls is of interest. If the origin of coordinates is located on one wall, the diffusion rate can be estimated from eqn. (7.1) with the boundary conditions

$$N(x, 0) = N_\infty, \quad 0 < x < h; \quad N(x, 0) = 0, \quad x < 0 \quad \text{and} \quad x > \text{h}.$$
$$N(0, t) = 0; N(h, t) = 0 \quad t > 0,$$

where h denotes the spacing between the walls. The solution to eqn. (7.1) for these conditions reads

$$N(x, t) = \frac{4N_\infty}{\pi} \sum_{i=1}^{\infty} \frac{1}{2i-1} \sin\left[(2i-1)\frac{\pi x}{h}\right] \times \exp\left[-\frac{(2i-1)^2 \pi^2 D_i t}{h^2}\right]. \tag{7.51}$$

Using eqn. (7.48) we find that the total rate of deposition on the walls per unit surface area is given by:

$$j_{i,x=0} + j_{i,x=h} = 2D_i \left(\frac{\partial N}{\partial x}\right)_{x=0} = \frac{8D_i N_\infty}{h} \sum_{i=1}^{\infty} \exp\left[-\frac{(2i-1)^2 \pi D_i t}{h^2}\right]. \tag{7.52}$$

A3. *Decay of Aerosol Concentration inside Spherical and Cylindrical Vessels*

In storing aerosol clouds, or in observing their properties, one often contains them in spherical or cylindrical vessels. Therefore, as in A2, one needs to know how the concentration of particles changes in these containers in time by diffusional deposition on the walls. To deal with the case of a spherical chamber of radius a we consider a spherical geometry with origins of coordinates at the center of the vessel. No dependence on the azimuthal or polar angle is considered. Then eqn. (7.1) may be rewritten as

$$\frac{\partial N}{\partial t} = D_i \left[\frac{1}{r^2}\frac{\partial}{\partial r}\left(r^2 \frac{\partial N}{\partial r}\right)\right], \tag{7.53}$$

where r is the radial vector taken with its origin at the center of the sphere. Boundary conditions for an aerosol cloud initially distributed uniformly in space are

$$\left.\begin{aligned} N(r, 0) &= N_\infty, \quad r < a; \\ N(a, t) &= 0, \quad t > 0. \end{aligned}\right\} \tag{7.54}$$

And the solution of eqn. (7.53) for these conditions reads[87]

$$N(r, t) = \frac{2aN_\infty}{\pi r} \sum_{n=1}^{\infty} \frac{(-1)^n}{n} \sin\frac{n\pi r}{a} \exp\left(-\frac{n^2 \pi^2 D_i t}{a^2}\right). \tag{7.55}$$

To calculate the total rate of removal of particles on the walls, we apply the relation analogous to eqn. (7.48),

$$\tilde{\Phi}_i = 4\pi D_i \left(r^2 \frac{\partial N}{\partial r}\right)_{r=a} \tag{7.56}$$

Using eqn. (7.55) we find that the flux of particles striking the entire inner surface of the sphere is

$$\tilde{\Phi}_i = 8aD_i N_\infty \sum_{n=1}^{\infty} \frac{(-1)^n}{n} \exp\left(-\frac{n^2 \pi^2 D_i t}{a^2}\right) \tag{7.57}$$

Similarly, the concentration for an aerosol diffusing to the walls of an infinitely long cylindrical vessel of radius a can be derived from the diffusion equation written in cylindrical coordinates

$$\frac{\partial N}{\partial t} = D_i\left[\frac{1}{r}\frac{\partial}{\partial r}\left(r\frac{\partial N}{\partial r}\right)\right], \tag{7.58}$$

with boundary conditions

$$N(r,0) = N_\infty, \quad r < a;$$
$$N(a,t) = 0, \quad t > 0,$$

where r is the radial distance measured from the axis of the cylinder. The solution in this case is

$$N(r,t) = \frac{2N_\infty}{a}\sum_{n=1}^{\infty} \exp(-D_i\tilde{\alpha}_n^2 t)\frac{J_0(r\tilde{\alpha}_n)}{\tilde{\alpha}_n J_1(a\tilde{\alpha}_n)}, \tag{7.59}$$

where $\tilde{\alpha}_n$ are the positive roots of the Bessel Function

$$J_0(\tilde{\alpha}_n a) = 0.$$

The total flux of particles per unit length of the cylinder may be determined from

$$\tilde{\Phi}_i = 2\pi D_i\left(r\frac{\partial N}{\partial r}\right)_{r=a}; \tag{7.60}$$

with this relation and eqn. (7.59) we find that

$$\tilde{\Phi}_i = 4\pi N_\infty \sum_{n=1}^{\infty}(-1)^n \exp(-D_i\tilde{\alpha}_n t). \tag{7.61}$$

A4. *Diffusion to Outer Surfaces of Spheres or Cylinders*

The problem of diffusion of particles to the outer surfaces of spherical and cylindrical bodies finds application to several areas of aerosol physics including the theories of coagulation and filtration. Diffusive deposition may also introduce important modifications in the collection efficiencies of these bodies if the aerosol cloud flows by the outer surface. Therefore, this topic has special importance.

For spherical bodies of radius a, the rate of diffusion to the surface can be found using a solution to eqn. (7.53) with the boundary conditions

$$N(r,0) = N_\infty, \quad r > a;$$
$$N(a,t) = 0, \quad t > 0. \tag{7.62}†$$

Thus, solving for $N(r, t)$, we find that

$$N(r,t) = N_\infty\left[1 - \frac{a}{r} + \frac{2a}{2\sqrt{\pi}}\int_0^{(r-a)/2D_i)^{1/2}} \exp(-\gamma^2)\,d\gamma\right]. \tag{7.63}$$

† We note here that if the ratio of the mean free path of the gas to the curvature is finite, account has to be taken of non-continuum effects at the surface (see also chapters 5 and 6).

The flux of particles to the entire surface then becomes

$$\tilde{\Phi}_i = 4\pi D_i \left(r^2 \frac{\partial N}{\partial r}\right)_{r=a} = 4\pi D_i a N_\infty \left[1 + \frac{a}{\sqrt{(\pi D_i t)}}\right]. \tag{7.64}$$

In an interval of time Δt, the number deposited on the outside of the spherical collector is

$$\int_0^{\Delta t} \tilde{\Phi}_i \, \mathrm{d}t = 4\pi a D_i N_\infty \left[\Delta t + \frac{2a\sqrt{(\Delta t)}}{\sqrt{(\pi D_i)}}\right]. \tag{7.65}$$

If a characteristic diffusion time defined as a^2/D_i is much larger than the time interval, $a^2/D_i \gg \Delta t$, then the number of particles deposited per unit area on the surface of the sphere is the same as that for a unit area of flat wall, eqn. (7.50). On the other hand, the circumstances when $a^2/D_i \ll \Delta t$, leads to a steady concentration profile around the sphere

$$N = N_\infty \left(1 - \frac{a}{r}\right), \tag{7.66}$$

yielding a steady flux to the spherical surface

$$\tilde{\Phi}_i = 4\pi D_i a N_\infty. \tag{7.67}$$

Equation (7.67) is closely linked to the classical theory of coagulation, as discussed in Chapter 10.

In the case of deposition to the outer surface of an infinitely long cylinder, the solution to eqn. (7.58) has a rather complicated form, which reads[167]

$$N(r, t) = -\frac{2N_\infty}{\pi} \int_0^\infty \mathrm{e}^{-D_i \gamma^2 t} \frac{J_0(\gamma r)\, Y_0(\gamma a) - J_0(\gamma a)\, Y_0(\gamma r)}{J_0^2(a\gamma) + Y_0^2(a\gamma)} \frac{\mathrm{d}\gamma}{\gamma}. \tag{7.68}$$

The spatial derivative of this expression evaluated at $r = a$ multiplied by $2\pi a D_i$, provides the equation for the rate of diffusive deposition per unit length of cylinder:

$$\tilde{\Phi}_i = \frac{8N_\infty D_i}{\pi} \int_0^\infty \mathrm{e}^{-D_i \gamma^2 t} \frac{\mathrm{d}\gamma}{\gamma} [J_0^2(a\gamma) + Y_0^2(a\gamma)]. \tag{7.69}$$

For short times,

$$\tilde{\Phi}_i \approx 2\pi D_i N_\infty [(\pi \underline{\tau})^{-\frac{1}{2}} + \tfrac{1}{2} - \tfrac{1}{4}(\underline{\tau}/\pi)^{\frac{1}{2}} + (1/8)\, \underline{\tau} \ldots \text{(7.69a)}$$

For long times,

$$\tilde{\Phi}_i \approx 2\pi D_i N_\infty \{[1/\ln(4\underline{\tau})] - 2\tilde{\gamma} - [\tilde{\gamma}/(4 \ln(4\underline{\tau}) - 2)^2] \ldots\}, \tag{7.69b}$$

where

$$\underline{\tau} = D_i t^2/a^2 \quad \text{and} \quad \tilde{\gamma} = 0.577.$$

B. Diffusion in a Gravity Field

Particles tend to settle out towards a horizontal surface under the action of gravity. If no Brownian movement occurs in an aerosol cloud, the top of the cloud would sink towards the bottom of a containing vessel, maintaining a sharp boundary. However, with Brownian diffusion, sharp boundaries are smeared out.

A classical problem, discussed by Chandrasekhar,[5] considers particle settling with Brownian motion in terms of Smoluchowski's equation' Since the external force is given by

$$F_e = \frac{4}{3}\pi R_i^3(\rho_i - \rho_G)\,g,$$

the drift speed

$$v_s = \frac{F_e}{B_i} = \frac{2R_i^2(\rho_i - \rho_G)\,g}{9\mu_G}, \tag{3.14a}$$

the terminal settling velocity of a Stokesian particle. Then we use eqn. (7.39) in one-dimensional form for the vertical direction with the boundary conditions:

(1) bottom surface at $z = 0$, $\dfrac{\partial N}{\partial t} = -\tilde{\Phi}_{iz} = -D_i\dfrac{\partial N}{\partial z} + v_s N = 0, \quad t > 0.$

(2) $z = z_0$, $N_0 = \int_{z_0-\Delta z}^{z_0+\Delta z} N(z, 0)\,dz$, where Δz is an infinitesimal distance away from z_0.

Chandrasekhar[5] found the solution

$$N(z,t) = \frac{N_0}{2(\pi D_i t)^{\frac{1}{2}}}\left\{\exp\left[-\frac{(z-z_0)^2}{4D_i t}\right] + \exp\left[-\frac{(z+z_0)^2}{4D_i t}\right]\right\}$$
$$\times \exp\left[-\frac{v_s}{2D_i}(z-z_0) - \frac{v_s^2 t}{4D_i}\right] + \frac{N_0 v_s}{D_i\sqrt{\pi}}\,e^{-v_s z/D_i}\int_{\frac{z+z_0+v_s t}{2(D_i t)^{\frac{1}{2}}}}^{\infty} \exp(-\gamma^2)\,d\gamma \tag{7.70}$$

The distribution of particles for a particular value of z_0, based on eqn. (7.70), at different times is sketched in Fig. 7.1. We observe that for the case where a number of particles is distributed uniformly in the plane $z = z_0$ initially, Brownian diffusion acts very rapidly to spread the particles as in the no field case, as indicated by curves 1 and 2. However, gravity also exerts its influence quickly, causing a displacement in the location of maximum number of particles downward with a velocity v_s (curves 3, 4, and 5). Simultaneously, the region of maximum concentration tends to flatten out with the spreading by the random motion of particles.

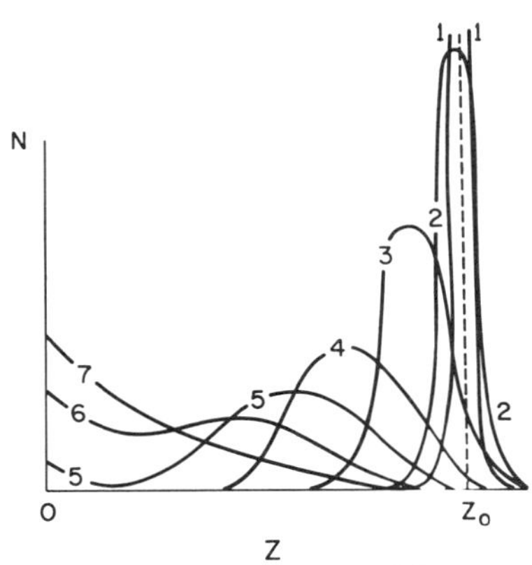

FIG. 7.1. Sedimentation of an aerosol cloud in which particles are undergoing Brownian motion. (After Chandrasekhar.[5])

When the probability of finding particles at the bottom of the vessel becomes sufficiently large, the curves of particle concentration begin to curve upwards near $z = 0$, as drawn in curves 5 and 6, as a result of reflection of particles from the bottom. Finally, as $t \to \infty$, the system seeks an equilibrium distribution of particles given by

$$\frac{N}{N_0}(z, \infty) = \frac{v_s}{D_i} \exp(-v_s z/D_i). \tag{7.71}$$

Chandrasekhar[5] has noted the interesting fact that since

$$\frac{v_s}{D_i} = \frac{m\mathscr{g}}{2kT}\left(1 - \frac{\rho_G}{\rho_i}\right),$$

the equilibrium distribution, eqn. (7.71), corresponds to the law of the ideal, isothermal model of an atmosphere.

In connection with aerosol microscopy, Fuchs[2] has remarked that eqn. (7.70) may be applied to determine values of the particle diffusivity D_i. Measurements may easily be made of the time for the image of a particle to pass between two horizontal lines in a microscope eyepiece. The probability that a particle will reach a horizontal line at plane $z = 0$ from a given height z_0 in the time interval $t, t + dt$ reads

$$\tilde{W}_1(z, t)\,\mathrm{d}t = \frac{z_0}{(4\pi D_i t^3)^{\frac{1}{2}}} \exp\left[-\frac{(z_0 - v_s t)^2}{4D_i t}\ \mathrm{d}t.\right] \tag{7.72}$$

Hence, the mean time of passage to the plane $z = 0$ is

$$\bar{t} = \int_0^\infty t W_1\,\mathrm{d}t = \frac{z_0}{\sqrt{(4\pi D_i)}} \int_0^\infty t^{-\frac{1}{2}} \exp\left[-(z_0 - v_s t)^2/4D_i t\right] \mathrm{d}t = \frac{z_0}{v_s}, \tag{7.73}$$

which is precisely the value expected for the case where no Brownian motion exists. The mean of t^{-1} has the value

$$\overline{\left(\frac{1}{t}\right)} = \frac{z_0}{\sqrt{4\pi D_i}} \int_0^\infty t^{-\frac{5}{2}} \exp\left[-(z_0 - v_s t)^2/4D_i t\right] \mathrm{d}t = \frac{2D_i}{z_0^2} + \frac{v_s}{z_0}. \tag{7.74}$$

Therefore, from eqns. (7.74) and (7.75),

$$D_i = \frac{z_0^2}{2}\left[\overline{\left(\frac{1}{t}\right)} - \frac{1}{\bar{t}}\right]. \tag{7.75}$$

Thus the average value of the time t required to pass between two lines in an eyepiece, with the average of t^{-1}, can yield values of D_i.

If $v_s = 0$, or no settling occurs, a simpler formula than (7.75) has been derived by Fuchs:[2]

$$\overline{\left(\frac{1}{t}\right)} = \frac{2D_i}{z_0^2}. \tag{7.76}$$

An interesting conclusion comes from the analysis of the problem of diffusion to parallel walls when settling is included. Suppose the vertical walls for case A1 above are reoriented horizontally. If the dimensionless variables $x' = x/h$, and $t' = tv_s/h$ are introduced into eqn. (7.51), the expression for the distribution of aerosol concentration has the form

$$N = N_\infty \sum_{i=1}^{\infty} \frac{2}{\pi i}\left[\frac{1-(-1)^i \exp(-2\hat{S}^2)}{1+4\hat{S}^4/\pi^2 i^2}\right] \sin \pi i x' \exp\left\{\frac{2x' - [1+(\pi^2 i^2/4\hat{S}^4)]\, t'}{\hat{S}^2}\right\}, \tag{7.77}$$

where $\hat{S}$ denotes the ratio $(v_s d/4D)^{\frac{1}{2}}$. From this relation, Fuchs[2] has demonstrated, as indicated in Fig. 7.2, that clouds of highly dispersed aerosols cannot maintain sharp upper boundaries in a long box. Even under ideal conditions of spatial uniformity of concentration initially, with no convection in the vessel, the cloud tends to become highly diffuse.

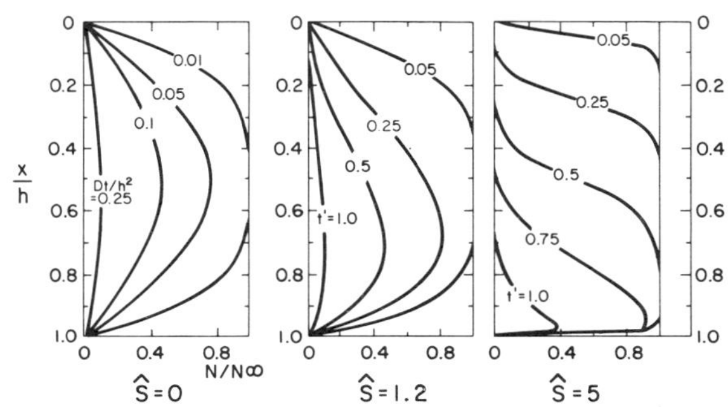

FIG. 7.2. Distribution of aerosol particles between horizontal absorbing plates in a gravity field. (After Fuchs.[2])

7.3. Laminar Convection and Diffusion

A. Diffusion of Particles in a Slow, Uniform Velocity Field

In many cases, particles may diffuse in a system which effectively contains no boundaries. In these circumstances, particles can spread laterally as they are swept downstream by a gas flowing at a constant speed. Several simple cases of dispersion from localized sources have been worked out. Three classical problems will be considered here as illustrations of the nature of these kinds of unbounded solutions.

Suppose first a stream of particles of strength N_0 is ejected instantaneously from a point source. A uniform stream velocity U moves past the source in the x direction. Using the Galilean transformation $x' = (x - Ut)$, we can solve eqn. (7.1b) in terms of these conditions for $N(x, y, z, t)$. From known solutions[87] of the heat conduction equations,

$$N(x, y, z, t) = \frac{N_0}{(4\pi D_i t)^{\frac{3}{2}}} \exp\{-[(x - Ut)^2 + y^2 + z^2]/4D_i t\}. \tag{7.78†}$$

Similarly, the solution for infinitely long line source of particles oriented along the y coordinate and through the point $(x = 0, z = 0)$ may be found:

$$N(x, z, t) = \frac{N_0'}{(4\pi D_i t)} \exp\{[-(x - Ut)^2 + z^2]/4D_i t\}, \tag{7.79}$$

† This solution may also be obtained by assuming that $\boldsymbol{F}_e/B_i = \mathbf{q}_i$ in Smoluchowski's equation [eqn. (7.39)].

where N'_0 denotes the number of particles emitted per unit length of the line source.

In contrast to cases in which a blob of aerosol is injected into a stagnant volume of air, a continuous source must yield a steady state distribution of particles. Consider, for example, the line source again oriented along the y axis and through the point ($x = 0$, $z = 0$), but assume now that N'_0 represents the (constant) number of particles emitted per unit time per unit length of line source. The complete solution of eqn. (7.1b) for emission of particles into the medium is

$$N(x, z, t) = \frac{N'_0}{4\pi D_i} \int_{\hat{r}^2/4\pi t}^{\infty} \gamma^{-1} e^{-\gamma} d\gamma, \tag{7.80}$$

where $\gamma = \hat{r}^2/4\pi t'$ and $\hat{r} = \sqrt{(x^2 + z^2)}$.

The steady state solution may be obtained from this expression by considering the limit as $t \to \infty$, or

$$\underset{t \to \infty}{N(x, z, t)} = \frac{N'_0}{4\pi D_i} \left(\ln \frac{4D_i t}{\hat{r}^2} - 0.5772 \right). \tag{7.81}$$

B. Diffusion of Particles in Laminar Shearing Flows

The situation where particles diffuse in an infinite, uniformly flowing medium is an ideal that rarely can be realized in practical conditions. In most cases, one deals with situations where diffusion of particles occurs in a gas undergoing shear. The most important examples of such cases are those involving diffusion to absorbing surfaces over which the gas moves. Because of the influence of friction near the surface of the body, the gas near the solid boundary must develop gradients in velocity (and consequently shear) even though the fluid may be flowing uniformly far from the boundary. Under these conditions the diffusion to the surface occurs both through Brownian diffusion and convective processes. Convection involves the transport of particles by the fluid itself towards the boundary.

In Smoluchowski's equation [eqn. (7.39)] we found that the random walk model for Brownian diffusion could account for a drift of particles. The speed of drift $\mathbf{q}_i$, which was related to external forces referred to a *particle velocity* relative to the motion of the gas–particle system. For the diffusion of gases in shearing flow, the convective term involves the mass average fluid velocity only. As a first approximation, the continuum equations for heat and mass transfer have been applied to the convective diffusion of aerosol particles. The use of these equations involves the assumptions that the suspended particles are points of the nearly same mass as the carrying fluid so that their particle velocity $\mathbf{q}_i$ equals the mass average fluid velocity $\mathbf{q}$. Using Eq. (7.39) again with $\mathbf{q}$ in place of $\mathbf{q}_i$, we can express the change in the concentration of particles at a fixed point in terms of the total mass flux of particles, the number passing through a unit area of fluid, using Brownian diffusion and the convective terms.

Suppose we assume that the particle–gas suspension is a quasi-binary mixture. For an isothermal system subjected to no significant external forces, the total mass flux $\mathbf{j}_i$ is given by two terms

$$\mathbf{j}_i = \mathbf{q}N - D_i \nabla N, \tag{7.82}$$

where the first term on the left is the flux of particles associated with convective motion of the fluid, and the second term on the right denotes the flux of particles by Brownian motion of the particles themselves. Now in any arbitrary volume of fluid V_G, the number of particles passing through the volume's surface S is given by the expression

$$\tilde{\Phi}_i = -\oint_S \mathbf{j}_i \, \mathrm{d}S,$$

where the integral is taken over the entire closed surface S. Then the change in concentration of particles in the volume V_G

$$\int \frac{\partial N}{\partial t} \mathrm{d}V_G$$

must equal $\tilde{\Phi}_i$ or

$$\int_{V_G} \frac{\partial N}{\partial t} \mathrm{d}V_G = -\int_S \mathbf{j}_i \, \mathrm{d}S. \tag{7.83}$$

Using Gauss' theorem, the surface integral may be transformed to a volume integral or

$$\int_{V_G} \frac{\partial N}{\partial t} \mathrm{d}V_G = -\int_{V_G} \nabla . \mathbf{j} \, \mathrm{d}V_G. \tag{7.84}$$

Since eqn. (7.84) applies to an arbitrary volume, then

$$\frac{\partial N}{\partial t} = -\nabla . \mathbf{j} = \nabla . (D_i \nabla N - \mathbf{q}N). \tag{7.85}$$

If we assume farther that the fluid is incompressible, and the diffusion coefficient remains constant, then eqn. (7.85) becomes

$$\frac{\partial N}{\partial t} + \mathbf{q} . \nabla N = D_i \nabla^2 N, \tag{7.86a}$$

which is directly analogous to Smoluchowski's relation, eqn. (7.39). Now, however, the second term on the left is the convective diffusion term as conventionally defined. In cartesian coordinates eqn. (7.86) has the form

$$\frac{\partial N}{\partial t} + q_x \frac{\partial N}{\partial x} + q_y \frac{\partial N}{\partial y} + q_z \frac{\partial N}{\partial z} = D_i \left(\frac{\partial^2 N}{\partial x^2} + \frac{\partial^2 N}{\partial y^2} + \frac{\partial^2 N}{\partial z^2} \right), \tag{7.86b}$$

where q_x, q_y, q_z correspond to the components of the velocity vector $\mathbf{q}$ in the x, y, and z directions.

The extent to which eqns. (7.86a) and (7.86b) may apply to the behavior of aerosols depends on satisfying the many assumptions involved in deriving these expressions. These assumptions as applied to gas diffusion are treated in some detail in Merk's[83] review. The extent to which the equations may be used for calculating aerosol behavior depends to a large degree on the departure of the average particle velocity from the fluid velocity (see also the discussion in Section 3.2).

The complete formulation of a problem in convective diffusion that involves eqn. (7.86) requires knowledge about the velocity field of the fluid as well as specification of the appropriate mathematical boundary conditions. Even in the cases of the simplest geometries, eqn. (7.86) is difficult to solve for the unsteady case, in contrast to the simpler equation (7.1).

In many problems, only the steady state solution is of interest, where the term

$$\frac{\partial N}{\partial t} = 0.$$

Generally, one can specify either the value of the particle concentration $N(x, y, z)$, or the gradient in particle concentration at geometrical boundaries. For aerosol diffusion problems, one usually assumes as in the previous section that the solid surface is a perfect absorber of particles, or at the boundary, $N = 0$. In addition, it is customarily assumed that at an infinite distance from the surface the concentration remains uniformly distributed and constant N_∞.

In many practical cases aerosol particles are small enough to be carried along in steady flow as apparent fluid particles. In this same range, Brownian motion plays a vital role in deposition of particles on surfaces. Therefore, eqn. (7.86) should be applicable as an approximate model for aerosol behavior. Once the validity of these relations is assumed, one can tap the vast resources of literature dealing with solutions for heat and mass transfer in continuous media.

The Péclét Number

Before going into some features of solutions to simplified forms of eqn. (7.86), it is of interest to examine this expression in dimensionless form. Let us scale all terms in eqn. (7.86b) by a characteristic length L and a characteristic convective velocity q_∞. The scaling will be referred to spatial coordinates, $\mathbf{x}^\oplus = \mathbf{x}/L$, velocities $q^\oplus = q/q_\infty$, and concentrations $N^\oplus = N/N_\infty$. For the steady state case, the dimensionless equation for diffusion is[54]

$$q_x^\oplus \frac{\partial N^\oplus}{\partial x^\oplus} + q_y^\oplus \frac{\partial N^\oplus}{\partial y^\oplus} + q_z^\oplus \frac{\partial N^\oplus}{\partial z^\oplus} = \frac{1}{Pe_i}\left(\frac{\partial^2 N^\oplus}{\partial x^{\oplus 2}} + \frac{\partial^2 N^\oplus}{\partial y^{\oplus 2}} + \frac{\partial^2 N^\oplus}{\partial x^{\oplus 2}}\right), \tag{7.87}$$

The important dimensionless ratio Pe_i is the *Péclét number*, $q_\infty L/D_i$. This parameter plays the same part in the equation for convective diffusion as the Reynolds number in the equations of motion for a viscous fluid (the Navier–Stokes equations in Chapter 3). From the discussion in Section 4.3 and from the examination of eqn. (7.87), we note that if terms involving gradients in $N^\oplus$ are the same order, the Péclét number gives a criterion for the controlling mechanism for particle diffusion. Thus, if $Pe_i \gg 1$ the right-hand side of eqn. (7.87) is small, and the fluid motion will dominate the dispersion of particles. However, if $Pe_i \ll 1$, thermal agitation will control diffusion. When $Pe_i \approx 1$, both processes become of about equal importance. As seen in the next section, near boundaries, the gradients in dimensionless concentration $N^\oplus$ tend to become very large in size regardless of conditions far from the boundary. Under these conditions the terms on the right-hand side of eqn. (7.87) have to be retained locally even if $Pe_i \gg 1$.

The Péclét number for diffusion of aerosol particles may be identified with the particle Reynolds number Re_i and a Schmidt number Sc_i. That is,

$$Pe_i = Sc_i \, . \, Re_i = \left(\frac{\nu_G}{D_i}\right)\left(\frac{q_\infty R_i}{\nu_G}\right). \tag{7.88}$$

In aerosol systems, the Reynolds number for particles Re_i often is the order of 0.10 or smaller. The kinematic viscosity of gases is about 0.1 $cm^2 sec^{-1}$, which typical values of D_i for particles small enough for Brownian motion to be important is ~ 10^{-4} $cm^2 sec^{-1}$. Therefore, in systems of aerosols, the Péclét number is always relatively large, ~ 10^2 or higher, and we must be concerned with the behavior of solutions to eqn. (7.87) in the limiting cases where $Pe_i \to \infty$. That is, under virtually all conditions encountered in practice, convective diffusion should control the spatial distribution of aerosol particles in a gas, except very near boundaries. This is in contrast to cases of evaporation and condensation and spheres when frequently $Pe \ll 1$ (Section 4.3).

Boundary Layer Theory

Since the diffusion of aerosols occurs under conditions of very high Péclét number, the second order terms in eqn. (7.86) can be disregarded for a fluid of infinite extent. When steady flow of gas exists, eqn. (7.86) reduces to

$$\mathbf{q} \cdot \nabla N = 0 \tag{7.88a}$$

having a particular solution

$$N = \text{const.} = N_\infty, \tag{7.89}$$

which fulfills the condition of constant concentration far from surfaces.

The solution to eqn. (7.86), corresponding to (7.89), cannot apply near surfaces because it does not satisfy the condition that $N = 0$ at the boundary. Therefore, near the absorbing surface a thin layer must exist where N changes very rapidly from N_∞ to 0. In this zone, the gradients in N must become very large so that the second order terms accounting for thermal agitation become the same order as the convective terms. Therefore, the terms involving Brownian diffusion must be retained even though the diffusion coefficient is quite small.

Hence, at large values of Pe_i, the fluid containing aerosols can be divided crudely into two regions, the first of constant concentration far from boundaries, and the second a zone of large changes in concentration very near the surface. This kind of division exactly corresponds to the well known aerodynamic treatment of viscous flow over bodies at high Reynolds number.[67] The thin layer of fluid wherein the Brownian diffusion processes become important is analogous to the region of the momentum boundary layer, in which viscosity plays a primary role in transport of momentum. Thus, the second region near an absorbing surface is called the diffusion boundary layer, which has a thickness of δ_D.† The relationship of the diffusion boundary layer to the momentum boundary layer is shown schematically in Fig. 7.3.

The thickness of the diffusion layer may be used to scale eqn. (7.86) instead of the arbitrary length L. Suppose we consider the case where the suspension flows over a flat plate. Let x denote the distance downstream (in the x direction) from the leading edge of the plate (Fig. 7.3). The coordinate z will refer to the direction normal to the plate and into the fluid. The plate is infinitely wide, and no gradients of N exist along a direction y parallel to the leading edge. We then find that

$$\frac{\partial N}{\partial x} \approx \frac{N}{x}, \quad \frac{\partial N}{\partial z} \approx \frac{N}{\delta_D}, \quad \frac{\partial^2 N}{\partial x^2} \approx \frac{N}{x}, \quad \frac{\partial^2 N}{\partial z^2} \approx \frac{N}{\delta_D^2},$$

† In particular, δ_D is defined as the distance from a surface, normal to the surface, where $N = 0.999\, N_\infty$.

or

$$\frac{\partial^2 N}{\partial z^2} \gg \frac{\partial^2 N}{\partial x^2}$$

since $x \gg \delta_D$ generally.

Therefore the equation describing convective diffusion reads for conditions of steady state

$$q_x \frac{\partial N}{\partial x} + q_z \frac{\partial N}{\partial z} = D_i \frac{\partial^2 N}{\partial z^2} \tag{7.90}$$

with boundary conditions

$$\left.\begin{aligned} N &= 0, & z &= 0; \\ N &= N_\infty, & z &\to \infty. \end{aligned}\right\} \tag{7.91}$$

Using the results of theory for laminar boundary layers,[67] it is possible to estimate the magnitude of thickness δ_D for high values of *Pe*. The thickness of the momentum boundary layer δ_M is proportional to $\nu_G^{\frac{1}{2}}$. But D_i is analogous to ν_G, and about 10^3 times smaller.

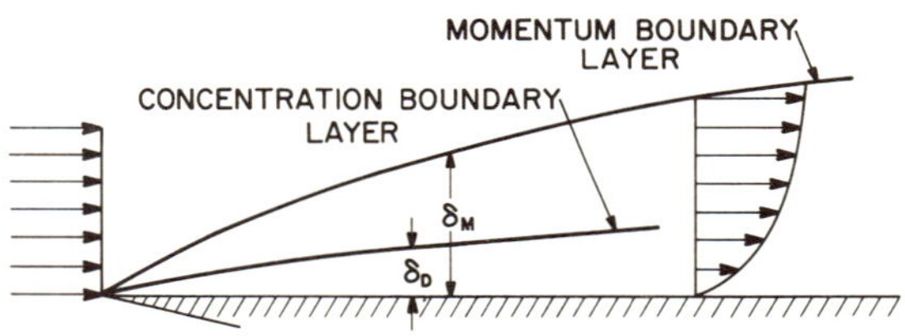

FIG. 7.3. Schematic diagram of the development of momentum and concentration boundary layers over a flat plate at high values of the Péclét number.

Therefore one expects that the thickness of the diffusion layer will be much less than δ_M for Pe_i large. This allows us to approximate the vertical velocity component q_z by $\nu_G z^2/\delta_M^3$. Thus

$$q_z \frac{\partial N}{\partial z} \approx q_z \frac{N}{\delta_D} \approx \frac{\nu_G \delta_D N}{\delta_M^3}.$$

From the continuity equation for an incompressible fluid, $\nabla . \mathbf{q} = 0$,

$$\frac{\partial q_x}{\partial x} \approx q_\infty \frac{\delta_D}{\delta_M x} \approx \frac{\partial q_z}{\partial z} . \frac{q_z}{\delta_D}$$

or

$$q_x \frac{\partial N}{\partial x} \approx q_\infty \frac{\delta_D}{\delta_M} \frac{N}{x} \approx q_z \frac{N}{\delta_D} \approx q_z \frac{\partial N}{\partial z}.$$

Bearing in mind that for $z > \delta_D$, the term on the right-hand side of eqn. (7.90) is negligible while at $z = \delta_D$, it becomes the same order as the convective terms because $D_i \partial^2 N/\partial z^2 \approx$

$D_i N/\delta_D^2$, then at $z = \delta_D$,

$$v_G \delta_D \frac{N}{\delta_M^3} \approx \frac{D_i N}{\delta_D^2}. \tag{7.92}$$

Thus

$$\delta_D \approx \left(\frac{D_i}{v_G}\right)^{\frac{1}{3}} \delta_M = \frac{\delta_M}{(Sc_i)^{\frac{1}{3}}}. \tag{7.93}$$

Since $Sc_i \simeq 10^3$ or larger for aerosol systems, the thickness of the diffusion layer has to be about 0.10 of the thickness of momentum layer in such media. That is, the velocities of the fluid are about 10 per cent of the free stream value q_∞ at the outer portion of the diffusion layer.

C. Solutions for Convective Diffusion in Laminar Flows

If it is assumed that no coupling exists between the equation of motion of the fluid and the diffusion of particles, solutions may be obtained to eqn. (7.90) once the velocity field is known analytically. Other than affecting the value of the viscosity, the only coupling between the motion of the fluid and the diffusion equation that is possible lies in the continuity relation. In particular this restriction implies that the vertical velocity of the total mass of fluid is unaffected by the diffusional flux of particles moving randomly towards the surface. This condition appears to be somewhat less restrictive than cases where heated surfaces (and heat transfer) are involved.

The Flat Plate

Approximate solutions for the velocity field for laminar flow over a flat plate are well known.[67] These results may be used with eqns. (7.90) and (7.91). The solution for the concentration field given by Levich[54] in this case reads

$$N(x, z) = \frac{N_\infty \int_0^{\frac{1}{2}\sqrt{[(q_\infty z^2)/(v_G x)]}} \exp(-0.22\, Sc_i \gamma^3)\, d\gamma}{\int_0^\infty \exp(-0.22\, Sc_i \gamma^3)\, d\gamma}. \tag{7.94}$$

The integral in the denominator of eqn. (7.94) can be expressed in terms of a gamma function

$$\int_0^\infty \exp(-0.22\, Sc_i \gamma^3)\, d\gamma = \frac{\frac{1}{3}\Gamma(\frac{1}{3})}{(0.22\, Sc_i)^{\frac{1}{3}}} \approx \frac{0.89}{(0.22\, Sc_i)^{\frac{1}{3}}}. \tag{7.95}$$

Hence, eqn. (7.94) becomes

$$N(x, z) = \frac{(0.22\, Sc_i)^{\frac{1}{3}} N_\infty}{0.89} \int_0^{\frac{1}{2}\sqrt{[(q_\infty z^2)/v_G x)]}} \exp(-0.22\, Sc_i \gamma^3)\, d\gamma. \tag{7.96}$$

The diffusional flux to the plate then will have the form:

$$j_i = D_i \left(\frac{\partial N}{\partial z}\right)_{z=0} = 0.34 D_i N_\infty \left(\frac{q_\infty}{v_G x}\right)^{\frac{1}{2}} \left(\frac{v_G}{D_i}\right)^{\frac{1}{3}}. \tag{7.97}$$

Using the definition for the thickness of the diffusion boundary layer, and eqn. (7.94), we find

$$\delta_D \approx 3\, Sc_i^{-\frac{1}{3}} \left(\frac{v_G x}{q_\infty}\right)^{\frac{1}{2}} \tag{7.98}$$

or, from eqn. (7.97),

$$\delta_D = \frac{D_i N_\infty}{j_i} \approx 0.6\, Sc_i^{-\frac{1}{3}}\, \delta_M. \tag{7.99}$$

This relation is similar to our estimate on dimensional grounds in eqn. (7.93), but contains a numerical coefficient.

In the case of this ideal model for aerosol diffusion to a flat plate, we see that the thickness of the diffusion layers varies with the half power of x, the distance from the leading edge of the plate. Furthermore, the flux of particles being deposited on the plate is not uniform with x. Near the leading edge, a large number of particles become absorbed on the surface in contrast to regions farther downstream. We note, of course, that eqns. (7.96), (7.97), and (7.99) do not apply within a length δ_D from the leading edge. This region is a region of discontinuity in which the aerodynamic boundary layer is not well defined.[67]

It is interesting to compare the solution for j_i given by boundary layer theory with the transient diffusion of particles from a stagnant fluid to a wall. To do this, the assumption is made that

$$t \sim x/q_\infty. \tag{7.100}$$

Using eqn. (7.100) and eqn. (7.49), we find the ratio of the flux j_i in the two cases reads

$$\frac{N_\infty (D_i q_\infty/\pi x)^{\frac{1}{2}}}{0.34\, D_i N_\infty (q_\infty/v_G x)^{\frac{1}{2}} (Sc_i)^{\frac{1}{3}}} \approx 1.66\, Sc_i^{\frac{1}{6}}. \tag{7.101}$$

The $Sc_i^{\frac{1}{6}}$ is the order of 0.1–1.0 for practical considerations so that the use of simple arguments like (7.100), which state essentially that the transient solution applies to steady, uniform flow over a plate, gives rise to values of the particle flux to the plate in rather close agreement to estimates taking into account the formation of a diffusion boundary layer over the plate.

Diffusion to a Spinning Disc

If the flat plate discussed above now spins at a speed ω in a stagnant fluid about an axis normal to the plate, somewhat different results are obtained. The aerodynamic problem for the rotating disc has been solved some years ago by von Kármán (e.g. Schlichting[67]). The striking feature of this solution for the fluid motion is that the component of velocity normal to the disc surface depends only on the distance from the disc z and not the distance from the axis of rotation r of the angular position ϕ. For this velocity field, Levich[54] has found

$$N = N(z).$$

That is, the concentration of particles depends only on the distance from the surface of the disc in a manner similar to the velocity field.

Ignoring the region near the edge of a disc of finite radius, the equation for convective diffusion in this case becomes

$$q_i(z)\frac{\mathrm{d}N(z)}{\mathrm{d}z} = D_i\frac{\mathrm{d}^2N(z)}{\mathrm{d}z^2} \tag{7.102}$$

with boundary conditions

$$\left.\begin{aligned} N &= N_\infty \quad \text{at} \quad z \to \infty, \\ N &= 0 \quad \text{at} \quad z = 0. \end{aligned}\right\} \tag{7.103}$$

Levich's solution to eqns. (7.102) and (7.103), with von Karman's relations for the velocity field, yields

$$N(z) = \frac{N_\infty}{1.61\,Sc^{-\frac{1}{3}}(\nu_G/\omega)^{\frac{1}{2}}}\int_0^z \exp\left\{\frac{1}{D_i}\int_0^{z'} q_i(\gamma)\,\mathrm{d}\gamma\right\}\mathrm{d}z' \tag{7.104}$$

for conditions where $Sc_i \gg 1$. Equation (7.104) depends only on the distance z from the disc surface. Furthermore, the diffusional flux to the surface of the disc is uniform radially, as given by

$$j_i = 0.62\,D_i^{\frac{2}{3}}\nu_G^{-\frac{1}{6}}\omega^{\frac{1}{2}}\,N_\infty. \tag{7.105}$$

And the thickness of the diffusion boundary layer has the form

$$\delta_D = \frac{D_i N_\infty}{j_i} \approx 1.61\,Sc_i^{-\frac{1}{3}}(\nu_G/\omega)^{\frac{1}{2}} = 0.5\,Sc_i^{-\frac{1}{3}}\delta_M, \tag{7.106}$$

which is remarkably similar to the relation for the flat plate case, eqn. (7.99). One interesting point here, of course, is that the boundary layer thickness is independent of radial distance from the axis of rotation except near the edge of the disc.

Sparrow and Gregg[168] have checked Levich's results for $Sc_i > 100$ against exact numerical calculations. Equation (7.104) is found to be accurate to 7 per cent with the numerical results in this range of Sc_i. This limit corresponds to particles larger than about 0.004 μ radius at normal atmospheric pressure. The theory is also limited (a) to the regime of laminar flow, corresponding to Reynolds numbers, $\omega R_D^2/\nu_G$ of $\sim 10^4$, where R_D is the disc radius, and (b) to conditions where gravity does not act in the z direction, causing complications resulting from natural convection.

Recently, Friedlander and Pasceri[169] made use of the spinning disc to sample very small aerosols from the atmosphere. When used in still air, the spinning disc sampler has the advantage of obtaining a uniformly distributed sample over the sampler surface. Such a device also could be used to measure accurately the diffusivity of small particles in a gas.

Convective Diffusion in Tubes and Pipes

The diffusional deposition of aerosols in pipes or tubes is a particularly important practical problem since this mechanism effectively underlies the principle of the diffusion battery. When fluid flows into a pipe or a tube a boundary layer develops along the wall of the pipe which grows with distance downstream from the entrance until an equilibrium (laminar) parabolic velocity becomes established (see also Chapter 3). In the region where the aerodynamic boundary layers are still growing, the diffusional deposition can be calculated approximately by the equations for the flat plate case, provided the radius of curvature of the tube is not too great. Downstream in the region of the equilibrium aerodynamic behavior the parabolic profile of gas velocity can be taken into account in the equation for convective diffusion.

The length of pipe required to establish the parabolic profile for velocity is given by Schlichting[67] as

$$h \approx 0.16. \frac{U_\infty \tilde{b}^2}{\nu_G}, \tag{7.107}$$

where U_∞ is the velocity of gas at the centerline of the pipe and $\tilde{b}$ is the radius of the pipe. Beyond this point, the velocity distribution is given by Poiseuille's equation

$$u(r) = U_\infty \left(1 - \frac{r^2}{\tilde{b}^2}\right), \tag{7.108}$$

where $r =$ the radial coordinate in the pipe and

$$U_\infty = \frac{\tilde{b}^2 \Delta p}{4 \rho_G \nu_G l}, \tag{7.109}$$

$\Delta p/l$ being the pressure drop per unit length l in the pipe.

Although the aerodynamic boundary layer development leads to equilibrium Poiseuille flow in the pipe according to eqn. (7.109), the appearance of analogous equilibrium distribution in the diffusional layer may require somewhat longer distances. Thus convective diffusion in tubular flow arbitrarily can be subdivided into three sections.[54] The first, discussed above, covers the development of the aerodynamic boundary layer, and diffusion in this region may be calculated approximately from the flat plate solution. The second zone in which the parabolic velocity profile exists, but the fully developed concentration distribution has not been established. Finally, the third regime consists of fully developed distributions in both velocity and concentration.

For the second regime, the behavior of convective diffusion can be estimated in terms of approximate solutions for the boundary layer equations. Under these circumstances, the layer thickness will be much smaller than the tube radius, and in this region, where $(\tilde{b} - r) \ll \tilde{b}$,

$$u \approx U_\infty \frac{2(\tilde{b} - r)}{\tilde{b}}. \tag{7.110}$$

And, for steady flow, the diffusion equation may be written as

$$\frac{2U_\infty}{\tilde{b}} (\tilde{b} - r) \frac{\partial N}{\partial x} = D_i \frac{\partial^2 N}{\partial r^2} \tag{7.111}$$

with the boundary conditions

$$\left.\begin{aligned} N \to N_\infty \quad &\text{as} \quad (\tilde{b} - r) \to \infty, \\ N = 0 \quad &\text{at} \quad (\tilde{b} - r) = 0. \end{aligned}\right\} \tag{7.112}$$

Levich[54] has given a solution to eqn. (7.111) with (7.112) for the concentration distribution and the diffusive flux to the tube wall, which read

$$\frac{N}{N_\infty} = \frac{\int_0^{\frac{(\tilde{b}-r)}{x^{1/3}}(U_\infty/D_i\tilde{b})^{1/3}} \exp\left[-\left(\tfrac{2}{9}\right)\gamma^3\right] d\gamma}{\int_0^\infty \exp\left[-\left(\tfrac{2}{9}\right)\gamma^3\right] d\gamma} \tag{7.113}$$

and

$$j_i = \frac{D_i N_\infty}{x^{1/3}}\left(\frac{U_\infty}{D_i \tilde{b}}\right)^{1/3}\left\{\int_0^\infty \exp\left[-\left(\tfrac{2}{9}\right)\gamma^3\right] d\gamma\right\}^{-1} = 0.67 N_\infty D_i \left(\frac{U_\infty x^2}{D_i \tilde{b} x}\right)^{1/3}. \tag{7.114}$$

The net rate of particles being deposited on the wall in the region where the diffusional regime is developing can be estimated as

$$\tilde{\Phi}_i = 2\pi\tilde{b}\int j_i \, dx = 2.01\pi D_i N_\infty \left(\frac{U_\infty x^2}{D_i \tilde{b}}\right)^{1/3} \tilde{b}. \tag{7.115}$$

Levich also found the length L over which the equilibrium diffusional regime develops. Noting that this length is determined when $\delta_D \approx \tilde{b}$, then noting that $\delta_D = D_i N_\infty / j_i$, we find that

$$L \approx \frac{U_\infty \tilde{b}^2}{D_i} \approx Re\, Sc_i \tilde{b}. \tag{7.116}$$

This implies that when $Sc_i \gg 1$, $L \gg l$, as assumed in the use of eqn. (7.111).

The third regime where there exists Poiseulle flow and fully developed distributions in concentration of aerosol, the results of the classical Graetz–Nusselt problem[170] in heat transfer apply. For this case, the decay in concentration with distance downstream in a pipe have been estimated by Gormley and Kennedy.[171]

In most practical situations the value of $L/l \gg 1$ so that nearly all of the length of a tube is required for establishing the steady diffusional regime. Therefore, eqns. (7.113) and (7.114) may be applied. For these conditions, the decay in average concentration of particles in the tube may be estimated by the relation:[171]

$$\frac{\bar{N}}{N_\infty} = 1 - 2.56\left(\frac{2D_i x}{\tilde{b}^2 U_\infty}\right)^{2/3} + 2.4\left(\frac{D_i x}{\tilde{b}^2 U_\infty}\right) + 0.177\left(\frac{2D_i x}{\tilde{b}^2 U_\infty}\right)^{4/3}. \tag{7.117}$$

The theoretical results discussed for the pipe do not take into account the effect of settling of particles. Therefore eqns. (7.112)–(7.117) apply to tubes oriented vertically. For horizontal tubes, experimental observations of Davis[172] indicate that gravitational settling has to be included in the analysis for particle deposition. That is, deposition would generally be larger in magnitude for a given value of distance downstream x in a horizontal tube, and the deposit would be heavier on the bottom half of the tube.

Deposition by Natural Convection Near a Vertical Plate

Natural convection frequently occurs inside smoke stacks or in large rooms. It is of considerable practical interest to determine how such slow convective motion induced by gradients in density and gravity affects the diffusional deposition to vertical walls. A simple model for this case consists of a flat plate oriented parallel to the gravitational force vector. The z coordinate is taken positive upwards along the plate, and the x coordinate denotes the normal coordinate directed into the gas–particle suspension. It is assumed in this case that the convection occurring near the wall results from gradients in fluid density caused by differences in temperature. The plate is heated uniformly to a temperature T_w, while the ambient temperature far from the wall remains at T_∞. Assuming that no pressure gradient along the wall exists, the velocity field can be estimated using the boundary layer approximations.[67] Once the velocity field is known, then eqn. (7.90) can be solved for the usual boundary conditions. For $Pe_i \gg 1$, a form of the solution for the concentration distribution similar to that found by Levich[54] is obtained:

$$\frac{N}{N_\infty} = 0.7 c_0 \, Sc_i^{\frac{1}{3}} \frac{x}{z^{\frac{1}{4}}}, \tag{7.118}$$

where

$$c_0 = \left[\frac{g(T_w - T_\infty)}{4 v_G^2 T_\infty} \right]^{\frac{1}{4}}, \tag{7.119}$$

and x is the coordinate measuring distance from the plate. From eqn. (7.118) the flux of particles to the wall read

$$j_i = D_i \left(\frac{\partial N}{\partial x} \right)_{x=0} = 0.7 D_i c_0 Sc_i^{\frac{1}{3}} \frac{N_\infty}{z^{\frac{1}{4}}}. \tag{7.120}$$

Thus the diffusional flux varies only slowly upwards from the leading edge of the vertical plate, and depends primarily on the properties of buoyant convection parameter c_0 and the particle diffusivity. It is found that the boundary layer for natural convection is thicker than encountered for conditions of forced convection, given by eqn. (7.99), and the total diffusional deposition per unit area consequently will be correspondingly smaller in this case.

Convective Diffusion to Curved Surfaces

The discussion of thin diffusion layers at high Péclét number has indicated that problems of this kind can be resolved in principle if the velocity field near the absorbing surface can be calculated. In general, the flow over curved surfaces can be expressed in terms of generalized orthogonal coordinates[173] λ_1, λ_2, and λ_3, where λ_1 and λ_2 lie parallel to the surface, and λ_3 refers to the coordinate normal to the surface. Then the equation of continuity for an incompressible fluid reads

$$\frac{\partial}{\partial \lambda_1} \left(\frac{q_1}{h_2 h_3} \right) + \frac{\partial}{\partial \lambda_2} \left(\frac{q_2}{h_1 h_3} \right) + \frac{\partial}{\partial \lambda_3} \left(\frac{q_3}{h_1 h_3} \right) = 0. \tag{7.121}$$

The velocities q_1, q_2, and q_3 denote the components in the directions of λ_1, λ_2, and λ_3. The parameters h_1, h_2, and h_3 represent geometrical factors relating the coordinate system $\lambda_1, \lambda_2, \lambda_3$ to the common cartesian system x, y, and z. As in previous results for viscous fluids, the velocity components vanish at the surface of the curved body. Provided that h_1, h_2, h_3 and their first derivatives are finite, expansion of eqn. (7.121) indicates that

$$\left(\frac{\partial q_3}{\partial \lambda_3}\right)_{\lambda_3=0} = 0. \tag{7.122}$$

Thus, the velocity very near the surface for fixed values of λ_1 and λ_2 has the form

$$q_3 \approx \left(\frac{\partial^2 q_3}{\partial \lambda_3^2}\right)_{\lambda_3=0} \frac{\lambda_3^2}{2} = \frac{q_3''}{2}\lambda_3^2 \tag{7.123}$$

neglecting terms of third and higher order. The double primed velocity refers to the second derivative of q_3 taken at the surface when $\lambda_3 = 0$. To the same order approximation,

$$\left.\begin{aligned} q_1 &\approx q_1'\lambda_3, \\ q_3 &\approx q_3'\lambda_3 \quad (\lambda_3 \to 0) \end{aligned}\right\} \tag{7.124}$$

Equations (7.123) and (7.124) are readily satisfied for the examples of boundary layers over flat plates discussed previously.

There are two cases of simple geometry that are of interest in idealized models of aerosol deposition. These are the sphere and the cylinder. For cases where the Reynolds number Re_f based on the body radius is unity or less, the stream functions for flow around these bodies are known; the velocity fields are given by Stokes' solution for spheres (3.5), and Oseen's solution for cylinders (3.69).

Upon substitution of the velocity fields of Stokes and Oseen into the diffusion equation written in curvilinear coordinates, approximate solutions may be found for deposition of particles on these bodies. Most of the theoretical efforts in this field have been devoted to seeking solutions within the limits of boundary layer approximations at high Péclét number [e.g. ref. (54)].

In the cases of creeping flow around a sphere, Levich and Friedlander (e.g. Fuchs[(2)]) have found relations for the flux of particles encountering the body. These independent results have the same form, but differ only at a proportionality constant. Levich's[(54)] solution gives

$$j_i = \frac{D_i N_\infty}{1.15}\left(\frac{3U_\infty}{4D_i a^2}\right)^{\frac{1}{3}} \frac{\sin(x'/a)}{\left(\dfrac{x'}{a} - \dfrac{1}{2}\sin\dfrac{2x'}{a}\right)^{\frac{1}{3}}}, \tag{7.125}$$

where the distance x' refers to the distance along the curved surface from the stagnation point, and a is the radius of the sphere (Fig. 3.5). This relation is indeterminate as $x'/a \to 0$, but it can be shown that the function of the angle (x'/a) tends to unity as $x'/a \to 0$. For finite values of x'/a, we note that the deposition of particles is largest at small values of x'/a. The mass flux decreases with increasing x'/a until it reaches zero at $x'/a = \pi$.

Integrating j_i over the spherical surface, the total diffusional deposition is given as

$$\tilde{\Phi}_i \approx 7.98 N_\infty D_i^{\frac{2}{3}} U_\infty^{\frac{1}{3}} a^{\frac{4}{3}}. \tag{7.126}$$

The steady deposition from a stagnant medium onto a sphere by Brownian diffusion alone ($Pe_i \ll 1$) was found in eqn. (7.67) to be $4\pi D_i N_\infty a$. Therefore, the entire range of convective diffusion to a spherical surface for arbitrary Pe_i can be given roughly by

$$\tilde{\Phi}_i = 4\pi D_i N_\infty a(1 + 0.64Pe_i^{\frac{1}{3}}), \tag{7.127}$$

which fits the limits $Pe \ll 1$ and $Pe_i \gg 1$.

The case of slow flow around cylinders has also been worked out by Friedlander and by Natanson [see also ref. (2)]. The rate of deposition of particles per unit length of cylinder in this case reads

$$\tilde{\Phi}_i = \frac{2\pi N_\infty a C_1 Pe_i^{\frac{1}{3}}}{[2(2.002 - \ln 2Re_f)]^{\frac{1}{3}}}, \tag{7.128}$$

where C_1 was determined by Friedlander to be 1.305, but given by Natanson as 1.473.

In the literature dealing with mass transfer, the flux j_i is often given in terms of a mass transfer coefficient $\hat{k}$ defined as

$$j_i = \hat{k}SN_\infty = D_i S\left(\frac{\partial N}{\partial z}\right)_{z=0}, \tag{7.129}$$

where S refers to the surface area of the collector. In terms of the Sherwood number Sh (or Nusselt number for mass transfer), for Pe_i, in this case $U_\infty a/D_i \gg 1$;

spheres

$$Sh \equiv \frac{2\hat{k}a}{D_i} \approx 1.26\, Pe_i^{\frac{1}{3}};$$

cylinders

$$Sh \equiv \frac{2\hat{k}a}{D_i} \approx \frac{1.26Pe_i^{\frac{1}{3}}}{[2(2.002 - \ln 2Re_f)]^{\frac{1}{3}}} \tag{7.130}$$

More generally, Friedlander and Pasceri[174] have indicated on dimensional grounds that convective diffusion to *any* curved surface at low Re_f has an expression for mass transfer coefficients in the form

$$Sh_i = C_2\, \epsilon Pe_i^{\frac{1}{3}}, \tag{7.131}$$

where C_2 is a constant and ϵ denotes a sticking coefficient, the fraction of successful particle encounters with the surface.

Accounting for Finite Particle Size

Strictly speaking the results obtained using boundary layer theory for diffusion of molecules apply only to the limiting situation where the particles are ideally represented as points. For particles of finite size, the boundary conditions like those in (7.91) should read, for example,

$$\left.\begin{aligned} N &= 0, \quad & r &= a + R_i; \\ N &= N_\infty, \quad & r &\to \infty. \end{aligned}\right\} \tag{7.91a}$$

The radius R_i corresponds to that of the depositing particle. The use of these conditions modifies the solutions to the boundary layer equations somewhat.

Using the boundary conditions for particles of finite size, the equations for convective diffusion can be written in dimensionless form in a manner similar to eqn. (7.87). The following dimensionless variables are introduced:

$$N^{\oplus} = \frac{N}{N_{\infty}}, \quad z^{\oplus\prime} = z/R_i, \quad x^{\oplus\prime} = \frac{x}{a}.$$

where z' is the coordinate taken normal to x'. Then from eqn. (7.90) and the thin boundary layer approximation ($Pe_i \gg 1$) for the velocity fields,[175] with the coordinates used in (7.125):

sphere

$$z^{\oplus\prime} \sin x^{\oplus\prime}\left(\frac{\partial N^{\oplus}}{\partial x^{\oplus\prime}}\right) - z^{\oplus\prime} \cos x^{\oplus\prime}\left(\frac{\partial N^{\oplus}}{\partial z^{\oplus\prime}}\right) = \frac{2a^2 D_i}{3R_i^3 U_{\infty}}\left(\frac{\partial^2 N^{\oplus}}{\partial z^{\oplus\prime 2}}\right); \tag{7.132}$$

cylinder

$$4z^{\oplus\prime} \sin x^{\oplus\prime}\left(\frac{\partial N^{\oplus}}{\partial x^{\oplus\prime}}\right) - 2z^{\oplus\prime} \cos x^{\oplus\prime}\left(\frac{\partial N^{\oplus}}{\partial x^{\oplus\prime}}\right) = \frac{D_i a^2}{ZU_{\infty} R_i^3}\left(\frac{\partial^2 N^{\oplus}}{\partial z^{\oplus\prime 2}}\right), \tag{7.133}$$

where

$$Z = [2(2.002 - \ln 2\, Re_f)]^{-1}$$

with the dimensionless boundary conditions

$$\left.\begin{aligned} N^{\oplus} &= 0, \quad z^{\oplus\prime} = 1; \\ N^{\oplus} &= 1, \quad z^{\oplus\prime} \to \infty. \end{aligned}\right\} \tag{7.134}$$

Applying dimensional reasoning to eqns. (7.132)–(7.133) with the boundary conditions (7.134), the dimensionless concentration has to have the functional form:

$$N^{\oplus} = f_1(x^{\oplus\prime}, z^{\oplus\prime}, D_i a^2/ZU_{\infty} R_i^2) = f_2(x^{\oplus\prime}, z^{\oplus\prime}, \hat{K} Pe_i Z), \tag{7.135}\dagger$$

where $\hat{K} = R_i/a$. Thus the dimensionless concentration depends on the geometry of the absorbing surface as well as the Péclét number, as found previously. The flux of particles to the curved surface under consideration, of course will also depend on these parameters.

Analytical solutions to eqns. (7.132)–(7.133) for all values of Pe_i have not been found. However, Friedlander[175] recently published an approximate solution for diffusion near the forward stagnation point of spheres and cylinders. Using an expansion technique for determining $N^{\oplus}$, Friedlander[175] found that the first term in the series for $N^{\oplus}$ reads

sphere

$$N^{\oplus} = \frac{\int_1^{z^{\oplus\prime}} \exp(-\hat{K} Pe_i \gamma^3/2)\, d\gamma}{\int_1^{\infty} \exp(-\hat{K} Pe_i \gamma^3/2)\, d\gamma}; \tag{7.136a}$$

cylinder

$$N^{\oplus} = \frac{\int_1^{z^{\oplus\prime}} \exp(-2Z\hat{K}^3 Pe_i \gamma^3/3)\, d\gamma}{\int_1^{\infty} \exp(-2Z\hat{K}^3 Pe_i \gamma^3/3)\, d\gamma}. \tag{7.136b}$$

† Z = unity for spheres.

These approximate relations correspond closely to solutions obtained for the point particle case, as, for example, the solution for the flat plate given by Levich [eqn. (7.94)]. Friedlander's analysis, in addition, appears to be equivalent to a recent study reported by Stechkina and Fuchs.(176)

Using these expressions, a theoretical approximation for the mass transfer coefficient $\hat{k}_0$ near the forward stagnation point of the bodies can be obtained:(175)

sphere
$$\frac{8\hat{k}_0 R_i}{D_i} = \frac{8(\hat{K}^3 Pe_i/2)^{\frac{1}{3}}) \exp(-\hat{K}^3 Pe_i/2)}{\int_1^\infty \exp(-\hat{K}^3 Pe_i \gamma^3/2)\, d\gamma}; \tag{7.137a}$$

cylinder
$$\frac{2\pi\hat{k}_0 Ri}{D_i} = \frac{2\pi \exp(-2\,.\,Z\hat{K}^3 Pe_i/3)}{\int_1^\infty \exp(-2Z\hat{K}^3 Pe_i \gamma^3/3)\, d\gamma} \tag{7.137b}$$

The curves in Fig. 7.4 give the relations for the dimensionless parameters $8\hat{k}_0 R_i/D_i$ and $2\pi\hat{k}_0 R_i/D_i$ versus $\hat{K}Pe_i^{\frac{1}{3}}$ for spheres and cylinders as estimated by eqns. (7.137a) and (7.137b).

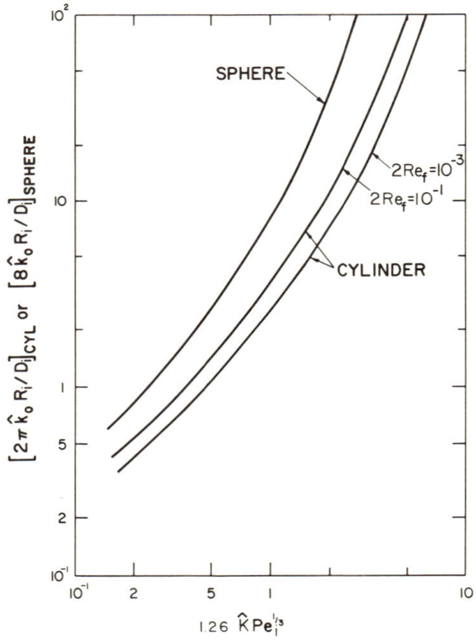

FIG. 7.4. Dimensionless mass transfer coefficients for spheres and cylinders. (From Friedlander.(175))

Convective Diffusion and the Interception of Small Particles

In addition to diffusional deposition, particles following fluid streamlines may become trapped on a surface by virtue of their finite size alone. This mechanism, mentioned in Chapter 3, was named *direct interception* and is accounted for in the boundary conditions (7.91a). As aerosol particles become smaller and smaller, and the flow slows down, diffusion must play an increasingly important role in deposition. With higher velocities and larger particles, interception and impaction control the removal of particles on solid surfaces as discussed in Section 3.2. For fluid velocities lower than a few cm $\sec^{-1}$, however, diffusion is believed to control the deposition of particles less than 0.5 μ in diameter. Between 0.5 μ and 1 μ size direct interception should be important, and above 1 μ impaction should control the removal of particles.

Using the results above for diffusion to curved surfaces, it is possible to get a more quantitative idea of the ranges where convective diffusion and interception have importance to particle trapping.

The removal or collection efficiency of a body η, as given in Section 3.20, has direct relation to the Sherwood number. In the case of cylinders,

$$Sh_i = \frac{2\hat{k}_0 a}{D_i} = \frac{j_i}{2\pi R_i N_\infty} = \frac{2\eta Pe_i}{\pi}. \tag{7.138}$$

Hence the efficiency of removal may be written in terms of the dimensionless variables in eqns. (7.132) and (7.133), or, for cylinders,[175]

$$\eta = \frac{D_i}{aU_\infty \hat{K}} f_3(\hat{K}^3 Pe_i Z). \tag{7.139}$$

In terms of the Sherwood number for cylinders,

$$\frac{2\pi\hat{k}_0 R_i}{D_i} = \hat{K}\pi \,.\, Sh_i = f_4(\hat{K} Pe_i^{\frac{1}{3}} Z^{\frac{1}{3}}), \tag{7.140}$$

This relation corresponds to the similitude rule for the diffusion of particles developed by Friedlander and Pasceri,[174] and later was found by Friedlander[175] for diffusion of particles of finite size much smaller than the size of the collecting cylinder. The theory, applicable in the range of $Pe_i \gg 1$, $Re < 1$, and $\hat{K} \ll 1$, states that at fixed Re_i, $\eta\ \hat{K}\ Pe_i$ should be a single valued function of $\hat{K} Pe_i^{\frac{1}{3}}$).

In the range of diffusion control, $\hat{K} \to 0$, and the solution to eqn. (7.133) may be given in terms of eqn. (7.139) as

$$(\hat{K} \to 0) \quad \eta = C_1 \pi Z^{\frac{1}{3}} Pe_i^{-\frac{2}{3}}. \tag{7.141}$$

where C_1 is a constant.

When $Pe_i \to \infty$, or for control by direct interception, the efficiency of removal depends only on the component of fluid velocity near the normal component of curved surface. Thus, in this limit, η is given in terms of the fluid velocity q_z over the front half of the cylindrical surface:

$$(Pe_i \to \infty) \qquad \eta = -\int_0^{\pi/2} \frac{q_{z', r=a}\, dx}{U_\infty a} = 2Z\hat{K}^2. \tag{7.142}$$

Thus in the limiting regime of diffusion control a plot of $\eta \hat{k}_0 Pe_i$ versus $\hat{K} Pe_i^{\frac{1}{3}}$ should display slope of unity for small values of $\hat{K} Pe_i^{\frac{1}{3}}$. Similarly, the regime of interception control at $\hat{K} Pe_i^{\frac{1}{3}}$ large should display a slope of 3 on such a graph. This limiting behavior is suggested in Fig. 7.4.

As a typical example, consider a spherical collector having a radius of 0.5 cm. For velocities of air $U_\infty \approx 1$ cm $\sec^{-1}$, diffusion controls trapping in particle sizes smaller than ~0.1 μ radius, while interception controls particle deposition for radii greater than ~0.1 μ.

Unfortunately, there has been no check on these results experimentally as yet. It is possible, however, to examine these relations in the light of observations of the performance of aerosol filters. (See Fig. 7.5.)

Diffusion Deposition in the Presence of Electrical Forces

Recently Zebel[80] has extended the analysis of diffusional deposition on spheres to include the influence of electrical forces. Using the method of superposing the solution for the electrostatic forces on the aerodynamic solutions, as indicated for the cylinder in Section 3 (also Zebel[79]), relations for the collection efficiency and the Sherwood number can be obtained for the case where $a \gg R_i$. The example Zebel considered was the case of a sphere falling through a medium with an electrical field oriented parallel to the direction of fall. The mathematical solution for the equation for diffusion within the boundary layer approximations is found in terms of the Péclét number, and the ratios:

$$\tilde{G} = \frac{\mathscr{E}_\infty Q_i B_i}{U_\infty} \equiv \frac{\text{particle velocity in electrical field}}{\text{undisturbed gas velocity}}, \tag{3.96}$$

and

$$\tilde{H} = \frac{Q_i Q' B_i}{U_\infty a^2} \equiv \frac{\text{velocity of a charged particle near a charged sphere}}{\text{undisturbed gas velocity}}. \tag{7.143}$$

The total deposition rate on the sphere is

$$\tilde{\Phi}_i = 2\pi(a + R_i)^2 \frac{U_\infty N_\infty}{Pe_i} \int_{x^{\oplus'}=0}^{x'=x_B'} \left(\frac{\partial N^{\oplus}}{\partial z^{\oplus'}}\right)_{z^{\oplus'}=1} \sin x^{\oplus'}\, dx^{\oplus'}, \tag{7.144}$$

where $x^{\oplus'}$ is the dimensionless intrinsic coordinate, x'/a, and $x^{\oplus'}$ is the location along the sphere from the forward stagnation point where the boundary layer approximations break down. Thus

$$\eta = \frac{2}{Pe_i(1 + \tilde{G})} \int_{x^{\oplus'}=0}^{x^{\oplus'}=x_B^{\oplus'}} \left(\frac{\partial N^{\oplus}}{\partial z^{\oplus'}}\right)_{z^{\oplus'}=1} \sin x^{\oplus'}\, dx^{\oplus'} \tag{7.145}$$

and

$$Sh_i = \frac{\eta}{2} Pe_i(1 + \tilde{G}). \tag{7.146}$$

Listed in Table 7.2 are several results derived by Zebel[80] for different conditions of electrical force strength. As these theoretical results indicate, the deposition rate can vary over quite a wide range depending on the nature of the electrical field, the particle charge and the charge on the sphere. Several interesting features of the particle trajectories are described in Zebel's paper,[80] and in the recent experimental study by Hochrainer, Zebel, and Hidy.[177]

D. Aerosol Filtration

The theory of the filtration of aerosols largely stems from considerations similar to those discussed in the light of boundary layer theory. The early developments in filtration of aerosol particles derive largely from the work of Langmuir and his co-workers[178] during World War II. In this series of studies, it became evident that very porous, fibrous filters function in a fundamentally different way than sieves. Tiny aerosol particles that have radii much smaller than the average distance separating fibers, become imbedded in the filter by diffusion to and direct interception of the fiber surfaces. Langmuir and his colleagues stressed the importance of diffusion and interception in controlling aerosol filtration under conditions of low fluid velocity. But later, Davies extended this theory to particles of sizes larger than about 1 μ diameter by considering deposition by impaction and sedimentation. Increasing particle size and fluid velocity promote removal by impaction, while smaller particles are trapped in filters by the other mechanisms for low

TABLE 7.2. COLLECTION EFFICIENCIES AND SHERWOOD NUMBERS FOR THE DEPOSITION OF CHARGED PARTICLES ON A FALLING SPHERE IN AN ELECTRICAL FIELD. THE FIELD DIRECTION IS ORIENTED PARALLEL TO THE DIRECTION OF MOTION OF THE LARGE SPHERE. BASED ON THEORY OF ZEBEL[80]

$\tilde{G}$	$\tilde{H}/\tilde{G}$	η	Sh_i	Remarks
A. *No Brownian Diffusion*				
>0	$\geqslant 3$	0		Positively charged point particles; repulsive forces on falling sphere
>0	$-3 < \tilde{H}/\tilde{G} < 3$	$\dfrac{(3\tilde{G} - \tilde{H})^2}{3\tilde{G}(\tilde{G} + 1)}$		Positive charge on point particles; repulsive and attractive forces on falling sphere
>0	$\tilde{H}/\tilde{G} \leqslant -3$	$\dfrac{-4\tilde{H}}{\tilde{G} + 1}$		Positive charge on point particles; attractive forces on falling sphere
<0	>0	$-4\tilde{H}/\tilde{G} + 1$		Negatively charged point particles with attractive and repulsive forces on the falling sphere, or attractive forces only on the sphere
<0	$\leqslant 0$	0		Negative charge on point particles with negative charge on the sphere. Dust-free space will appear around falling sphere
B. *With Diffusion*				
0	0	$2{\cdot}52\, Pe_i^{-\frac{2}{3}}$		Uncharged particles[a]
$>0 \ll 1$	$\ll \dfrac{1.5\, Pe_i^{-\frac{2}{3}}}{\tilde{G}}$	$2.52\, Pe_i^{-\frac{2}{3}} + \frac{4}{3}(3\tilde{G} - \tilde{H})$		Positive charged particles. Weak electrical forces compared with aerodynamic forces
$\gg \dfrac{Pe_i^{-\frac{2}{3}}}{2}$	0	$\dfrac{6\tilde{G}}{\tilde{G} + 1}$		Strongly positive charged particles
> 0	$\gg \dfrac{1.5\, Pe_i^{-\frac{2}{3}}}{\tilde{G}}$	$\dfrac{3\tilde{G}}{\tilde{G} + 1}$	$\frac{3}{2} Pe_i\tilde{G}$	$\tilde{H} = 0$, strongly positive charged particles
>0	>3	0	0	$\tilde{H} \neq 0$, strongly positive charged particles
>0	$-3 < \dfrac{\tilde{H}}{G} < 3$	$\dfrac{(3\tilde{G} - \tilde{H})}{3\tilde{G}(\tilde{G} + 1)}$	$\dfrac{(3\tilde{G} - \tilde{H})\, Pe_i}{6\tilde{G}}$	$\tilde{H} \neq 0$, strongly positive charged particles
>0	$\leqslant -3$	$\dfrac{-4\tilde{H}}{\tilde{G} + 1}$	$-2\tilde{H}Pe_i$	$\tilde{H} \neq 0$, strongly positive charged particles; attractive electrical force over entire surface of falling sphere
<0	$\geqslant - [1 + (0.875) \times Pe^{-\frac{2}{3}}]$	$\dfrac{-4\tilde{H}}{\tilde{G} + 1}$	$-2\tilde{H}Pe_i$	$\tilde{H} < 0$, strong negative charging on particles

[a] Zebel's numerical coefficient is somewhat higher than Friedlander's (7.130). However, the coefficient can range from 2.5 to 3.72 depending on the concentration profile approximation assumed.

fluid velocities. The passage from the diffusion range through interception to impaction should be accompanied by a minimum in the curve of fractional removal of particles from a gas stream versus particle size at fixed fluid velocity. Similarly, curves of fractional removal versus fluid velocity for fixed particle size should display a minimum(179) [see also ref. (180)]. These effects have both been observed in many experimental results.

In general, filter devices consist of a wide variety of materials which have different degrees of porosity and different properties of the distribution of surfaces in the trapping bed. Filter beds, of course, can be composed of fibrous materials including paper sheets, or glass mats, membranes, glass beads, sand or lead shot. An interesting example of a membrane type of filter with very uniform pore size is shown in Plate 7.

The requirement, in practice, of filtration theory is the prediction of the mass transfer coefficient $\hat{k}_1$ or the removal efficiency η, in terms of parameters describing properties of the filter, and the fluid flow through the bed. Using dimensional arguments,(174) it may be shown that the ratios,

$$\frac{\hat{k}_1}{\bar{q}} = F(Pe_i, \hat{K}, \hat{I}, Re_d, \hat{\alpha}_1, \hat{c}_1) \tag{7.146}$$

and

$$\frac{\hat{k}_1}{\bar{q}} = c\hat{k}_1^*/\bar{q} \tag{7.147}$$

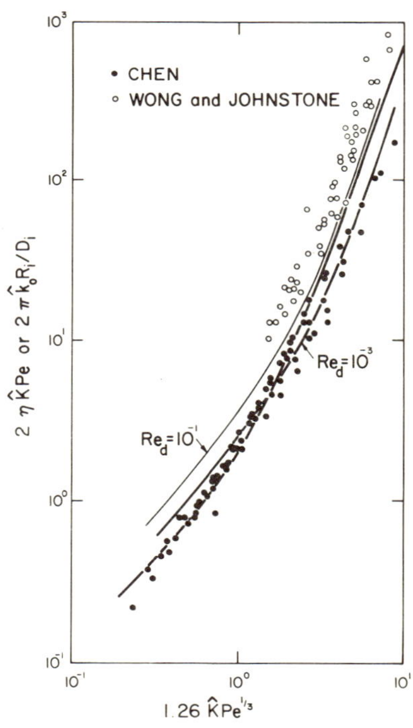

FIG. 7.5. Comparison between experimentally determined deposition rates of dioctylphthalate aerosols on glass fiber mats of Chen(410) and sulfuric acid aerosols (Wong and Johnstone(411, 418) with the theory for transfer to cylindrical surfaces. (From Friedlander.(175))

($\hat{k}_1^*$ = transfer coefficient excluding rebound of particle) if electrostatic effects are unimportant. $\bar{q}$ is the superficial average linear velocity through the filter bed, and $\hat{I}$ is the parameter describing impaction of particles. The ratio $\hat{\alpha}_1$ denotes the fraction of solids in the bed (the porosity), and $\hat{c}_1$ expresses statistically the distribution of fiber sizes in the filter:

$$\hat{c}_1 = \frac{\hat{\sigma}}{\bar{R}_f}, \tag{7.148}$$

where $\hat{\sigma}$ is the standard deviation of the fiber size distribution and $\bar{R}_f$ is the average radius of fibers in the bed.

Historically, an idealized model of a filter bed consisting of banks of smooth cylinders has provided a useful framework for constructing a theory of aerosol filtration. Although this model certainly represents a very inadequate picture of actual filter bed, it does allow rather good correlation of results for $\hat{k}_1$ and η. As an example, Friedlander[175] assumed that for constant α_1, $\hat{c}_1$ and c, the Sherwood numbers estimated for cylinders in the range $10^{-3} < Re_d < 10^{-1}$, where Re_d is based on $\bar{R}_f$, should fit experimental data for fibrous filters where $\hat{\alpha}_1 \ll 1$. Two sets of experimental data for filtration under these conditions are shown in Fig. 7.5. The parameters associated with these data are indicated in Table 7.3.

Friedlander's results for cylinders agree surprisingly well with the cited experimental data. The theory applies to the cylindrical region near the forward stagnation point where the transfer rate is largest. The filtration data, however, are based on an average deposition on the fiber surfaces. Therefore, the data should fall slightly below the theoretical curves. This holds for all of Chen's data, but not for the results of Wong and Johnstone.

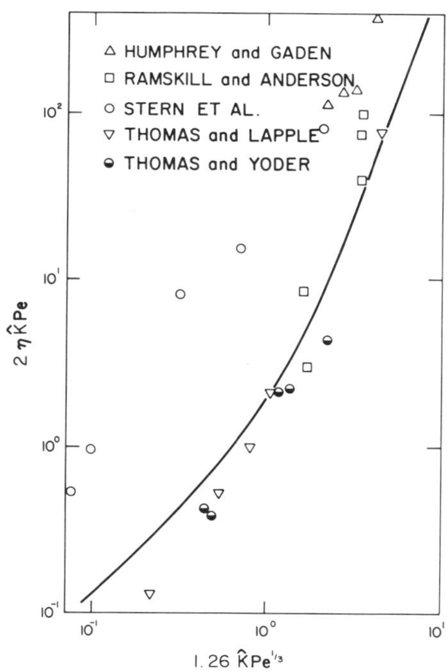

FIG. 7.6. Correlation curve for the efficiency of removal of aerosols on fibrous filters. (From Friedlander and Pasceri.[174]) Table 7.3 gives sources of data for experiments.

TABLE 7.3. PARAMETERS ASSOCIATED WITH DATA FOR AEROSOL FILTRATION BY FIBROUS FILTERS (FROM FRIEDLANDER AND PASCERI[174])

Reference	Filter material[a]	Aerosol[a]
Chen, C. Y.[410]	Glass fiber $\bar{R}_f = 1.3$ $\hat{\alpha}_1 = 0.02$–0.08	bis-2-Ethylhexyl phthalate (DOP) droplets $R_i = 0.08$–0.36 $\rho_i = 0.98$ g cm^{-3}
Humphrey, A. E. and Gaden, E. L., Jr.[412]	Glass, pyrex, wool, cotton, nylon, dacron $\bar{R}_f = 3.8$–8 $\hat{\alpha}_1 = 0.0095$–0.035	B-Subtilis spores $R_i = 0.58$
Ramskill, E. A. and Anderson, W. L.[413]	Glass $\bar{R}_f = 1, 1.5$ $\hat{\alpha}_1 = 0.098$–0.17 (calculated values of Chen.[410]	Dioctyl phthalate droplets H_2SO_4 droplets ($\rho_i = 1.84$ gm cm^{-3}) $R_i = 0.13$–0.5
Sadoff, H. L. and Almlof, J. W.[414]	Glass $\bar{R}_f = 4.23$ $\hat{\alpha}_1 = 0.02$–0.10	T-2 R Bacteriophage $R_i = 0.04$
Stern, S. C., Zeller, H. W., and Schekman, A. I.[415]	Viscose $\bar{R}_f = 8.5$ $\hat{\alpha}_1 = 0.10$	Polystyrene: $R_i = 0.44$–0.58 $\rho_i = 1.02$ g cm^{-3} T-3 E. Coli phage, $R_i = 0.018$ Type 3 Polio virus, $\bar{R}_i = 0.013$
Thomas, D. G. and Lapple, C. E.[416]	Glass $\bar{R}_f = 1.2$–14.7 $\hat{\alpha}_1 = 0.006$–0.06	Dupont oil orange (benzene azo β-naphthol) Supercooled spherical drops $R_i = 0.15$ $\rho_i = 1.26$ g cm^{-3}
Thomas, J. W. and Yoder, R. E.[417]	Glass $\bar{R}_f = 0.75$ $\hat{\alpha}_1 = 0.005$	Dioctyl phthalate $R_i = 0.13$–0.56
Wong, J. B. and Johnstone, H. F.[411] Also Wong, J. B., Ranz, W. E., and Johnstone, H. F.[418]	Glass $\bar{R}_f = 1.7, 3.12, 4.78$ $\hat{\alpha}_1 = 0.045$–0.098	95.5% H_2SO_4 droplets $R_i = 0.22$–0.65 $\rho_i = 1.4$ g cm^{-3}

[a] Fiber and particle radii are in microns.

Re_d	Pe_i	$\hat{K}$	$\hat{I}$	Remarks
1.4×10^{-3}–7.7×10^{-2}	6.2×10^{1}–2.8×10^{4}	0.06–0.29	5.2×10^{-4}–3.7×10^{-1}	
4.2×10^{-2}–5.2	2.7×10^{4}–3.2×10^{6}	0.07–0.15	9×10^{-3}–1.4	
1.1×10^{-2}–5.4×10^{-1}	1.1×10^{3}–6.5×10^{4}	0.10–0.15	7.8×10^{-3}–3.8×10^{-1}	Data for I and J filters only. Several filters were examined: viscose, esparto, wood, cotton. Values of $\hat{\alpha}_1$ taken from Chen[410]
8.2×10^{-4}–2.1×10^{-2}	1.3×10^{1}–3.2×10^{2}	0.0095	10^{-5}–2.6×10^{-4}	
2.3×10^{-2}–4.6	9.1×10^{2}–3.0×10^{6}	0.008–0.069	1.5×10^{-4}–1.05	Data at atmospheric pressure only. Data at reduced pressures are also given in this paper
3×10^{-5}–4×10^{-1}	5×10^{1}–1.7×10^{6}	0.01–0.13	10^{-4}–1.5	
9.4×10^{-5}–9.4×10^{-4}	1.1×10^{1}–5.2×10^{2}	0.16–0.75	2.2×10^{-4}–3.9×10^{-2}	Filter in vertical position with flow upward
0.038–1.39	1.4×10^{4}–6.0×10^{5}	0.049–0.313	3.2×10^{-2}–2.0	

Friedlander and Pasceri[174] on dimensional arguments proposed the semi-empirical curve for aerosol filters, shown also in Fig. 7.6. Their empirical curve can be represented as

$$\hat{K}\pi Sh_i = \eta\hat{K}Pe_i = 1.3\hat{K}Pe_i^{\frac{1}{3}} + 0.7(\hat{K}Pe_i^{\frac{1}{3}})^3. \tag{7.149}$$

For values of $\hat{K}Pe_i^{\frac{1}{3}}$ greater than 3, the slope of this curve is about equal to the theoretical value of 3, for the interception range. For small values of $\hat{K}Pe_i^{\frac{1}{3}}$, the slope approaches unity, as expected from the discussion on p. 197.

Friedlander and Pasceri applied eqn. (7.149) to other filtration data than shown in Fig. 7.6. To do this the results reported in the literature had to be extrapolated to the regime where $\hat{\alpha}_1 \to 0$, and certain other assumptions about the properties of the filter beds had to be made (like $\hat{c}_1$, for example). Although the general shapes of the curves found for these data were similar as indicated in Fig. 7.6, the magnitude of $\eta\hat{K}Pe_i$ varied for nearly an order of magnitude for fixed $K\hat{P}e_i^{\frac{1}{3}}$. The differences are likely to be reflected in variations in electrostatic charge on the particles and the beds, and in the assumptions made about the (unreported) properties of the fibers in the filter.

When the particles become larger and heavier, their trajectories in steady fluid flow will deviate from the streamlines as indicated in Fig. 3.5. Hence, impaction must ultimately play an important part in aerosol filtration. The impaction parameter in eqn. (7.146) is given from the Stokes' settling velocity [eqn. (3.13)] as[174]

$$\hat{I} = \left[1 + \tilde{A}\frac{\lambda_G}{R_i}\right]\frac{\rho_i R_i^2 \bar{q}}{18\mu_G \bar{R}_f}. \tag{7.150}$$

This parameter represents the ratio of the distance traveled by a particle injected into a stagnant gas with velocity $\bar{q}$ to the fiber radius. Friedlander and Pasceri's interpretation of various experimental results suggest that impaction does not become important in filtration for values of $\hat{I}$ smaller than one-half. Some of the data of Wang and Johnstone, for example, fall into the range $\hat{I} > 0.5$. Friedlander and Pasceri[174] have replotted some of the data in Fig. 7.6 for higher values of $\hat{K}Pe_i^{\frac{1}{3}}$ in terms of $\hat{I}$ as shown in Fig. 7.7. The large scatter below $\hat{I} < 0.5$ suggests that impaction does not affect strongly filtration in this range, but direct interception actually controls here. For $\hat{I} > 0.8$, however, the data may be correlated satisfactorily in terms of $\hat{I}$ alone. Over a limited range of $\hat{I}$, the correlation curve can be written as:

$$\eta = 0.075\,\hat{I}^{\frac{6}{5}}\left\{\begin{matrix} 0.8 < \hat{I} < 2 \\ Re_d < 1 \\ \hat{K} < 0.2 \end{matrix}\right\} \tag{7.151}$$

Evidence of different authors has suggested that the filter efficiency changes in different ways with the fraction of solids $\hat{\alpha}_1$. Although the sticking coefficient c can vary with the nature of the filter bed, workers have found that c may be independent of the aerosol being filtered at least for fibrous filters. This is an important result because it implies that both solid and liquid particles can be treated within the same theoretical framework. Experimentation over wider ranges of $\hat{\alpha}_1$, c, and $\hat{c}_1$ would be useful to better determine the applicability of the dimensional arguments of Friedlander and Pasceri.

Pressure Drop Across Fibrous Filter Beds

An important practical consideration for using aerosol filters is the pressure drop across the filter bed. Recently Brinkman's[181] model for flow through unconsolidated porous

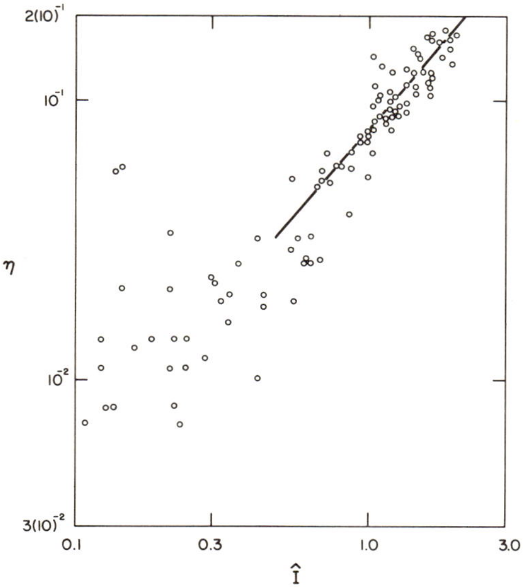

FIG. 7.7. Correlation of filtration data in the range where inertial deposition controls removal. (From Friedlander and Pasceri.[174])

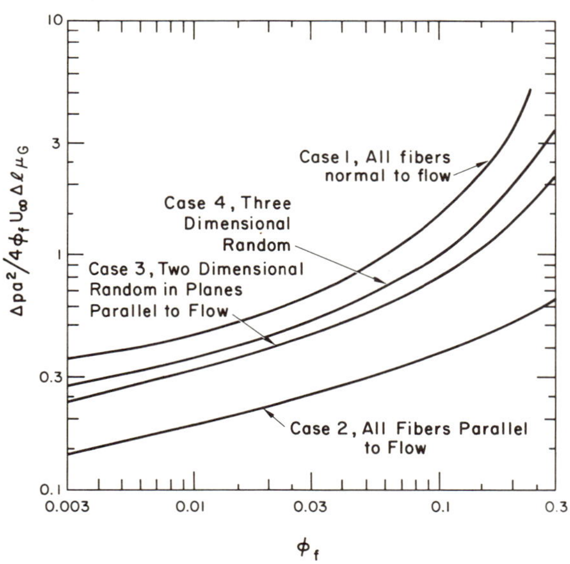

FIG. 7.8. Variation in pressure drop with volume fraction of fibers for fibrous filters. (From Spielman and Goren.[182])

media was applied by Spielman and Goren[182] to predict pressure drop through fibrous media. Brinkman's calculation depends on accounting for the fact that, on the average, gas in the proximity of an obstacle embedded in a porous medium experiences, in addition to the usual aerodynamic forces, a body damping force proportional to the local velocity induced on the flow by neighboring objects.

Spielman and Goren's[182] correlation for pressure drop is shown in Fig. 7.8 for various orientations of fibers. Here the dimensionless pressure drop is plotted as a function of ϕ_f, the volume fraction of fiber in the filter medium. Here Δl is the thickness of the filter. As the volume fraction increases, of course, the dimensionless pressure drop also increases. Spielman and Goren's theoretical results agree satisfactorily with an earlier empirical result of Davies[74] except for $\phi_f \lesssim 0.01$.

Spielman and Goren also made calculations for the effect of decreasing the cylinder radius a or dropping the pressure in terms of the Knudsen number for the filter fibers, $Kn'(=\lambda_G/a)$. Their results for the case of flow past a fibrous mat with all fibers axis aligned perpendicular to the direction of flow are shown in Fig. 7.9. As expected the pressure drop tends to decrease with Kn'. This model is likely to be only qualitative, however, because of the use of a slip flow–continuum approximation for the Knudsen number effect. Until the curves in Fig. 7.9 are verified by experiment, they should be used with reservation in design calculations.

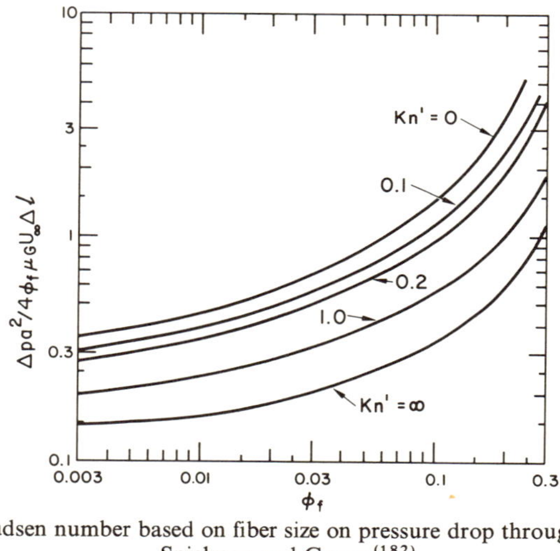

FIG. 7.9. Effect of Knudsen number based on fiber size on pressure drop through fibrous filters. (From Spielman and Goren.[182])

Accumulation of Aerosols on Filter Surfaces

Most models of filtration are limited in the sense that they cannot account realistically for the build up of particles on the filter surface or in the porous areas. In effect, this accumulation will modify the deposition efficiency and pressure drop characteristics of a given filter. Such effects of usage of a given filter for extended periods of time at present have to be evaluated experimentally.

Two interesting examples of the nature of particle deposition on a nucleopore filter are shown in Plate 7. In Plate 7A, the preferential deposition of particles around pores is

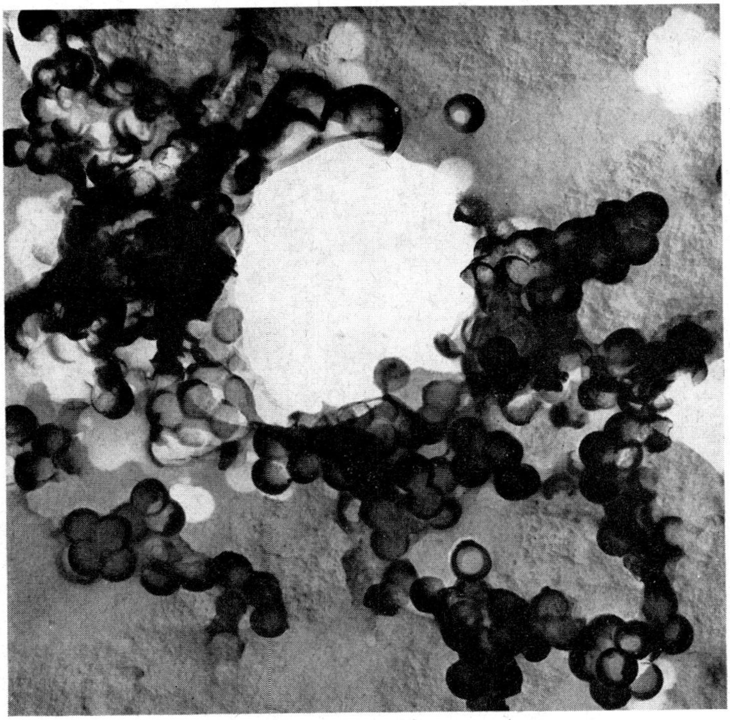

PLATE 7A. Deposits of 2 μ polystyrene spheres on a nucleopore filter with pore size 10 μ diameter. Photomicrograph shows accumulation of spheres around an opening and non-uniform deposition distribution.

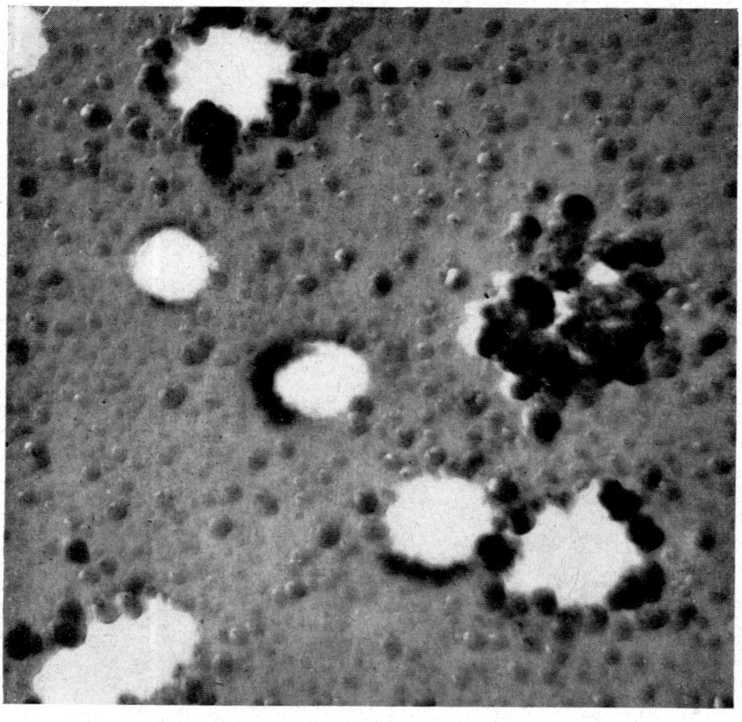

PLATE 7B. Deposits of 2μ polystyrene spheres on a nucleopore filter with pore size 10μ diameter. Accumulation around a pore rim and clogging of a pore. (Photo by E. Frank and K. Spurny.)

shown, with one pore being nearly plugged completely with particles. The photomicrograph taken at higher magnification, Plate 7B, indicates how complicated the structure of the deposit may become near pores. Many agglomerates of polystyrene spheres have built up around one pore as the particles were transported by converging flow towards the hole, and deposited one on top of another at the rim. Such transient modifications in the porous structure of a filter surface make the specification of flow patterns of gas through such systems impossible to estimate except in the basis of rather crude statistical considerations. It is surprising indeed, then, that relatively simple correlations like that of Friedlander and Pasceri's work so well to yield dimensionless values of the deposition efficiency.

7.4. Diffusion of Particles in a Turbulent Medium

The practical importance of turbulent diffusion is widely recognized. There exist many conditions in different natural processes and phases of technology in which particles are transported in a turbulent fluid. Because of the chaotic nature of turbulent flow, the complete specification of particle diffusion in such media becomes exceedingly complicated. At present relatively little is known about the theoretical behavior of aerosols in a turbulent suspension despite the considerable efforts recently exhausted on transport of quantities like heat or gas contaminant in these systems.

Perhaps the major problem in dealing with aerosol diffusion in turbulent flow is that, because of their inertia, one cannot be sure that particles are traveling with the fluid. In general, the behavior of a cloud of particles in a turbulent fluid largely will depend on the particle concentration, and the size of the particles in relation to a characteristic length-scale of the turbulence in the fluid. At high concentrations, the particles interact with one another and with the fluid as a result of particle collisions and influences of the flow near the particles. The motion of particles relative to the fluid can cause additional dissipation of energy through generation of additional turbulence. This effect may contribute to damping of the agitation in the fluid particle systems. At very high concentrations of particles, like the case of fluidized beds, the turbulent flow of fluid may be almost entirely damped out. High concentrations of small particles in a fluid can produce a mixture which behaves in an analogous way to a homogeneous, non-Newtonian fluid.

Even at relatively small concentrations, the particles may influence the viscous dissipation in the turbulent flow. The flow resistance around a particle is normally viscous in nature. Viscous dissipation of turbulence occurs largely in the smallest eddies, which are normally characterized by a large wave number in Fourier space (e.g. Batchelor[183]). As noted by Hinze[55] the additional dissipation results when a lag in turbulent motion exists between the sphere and the fluid. This lag tends to increase with smaller and smaller scale components of the turbulent field; one expects that the presence of aerosols in a fluid will influence the distribution of "eddy" velocities of fluid primarily in the smallest scales or in the larger wave numbers.

Only a few studies have been reported which deal with the motion of finely divided particles in a turbulent medium. Hinze[55] has reviewed much of the important results up to 1959. Relatively little important theoretical work has been reported on this subject since that time.

Before considering the behavior of aerosols in turbulent flow, it is useful to begin by reviewing certain general aspects of the transport of fluid properties (such as heat or contaminant of density equal to the fluid) in a turbulent medium. This is equivalent to the assumption made previously that the particles are points distributed in the fluid, with masses nearly the same as the molecules of suspending gas.

A. Diffusion by Continuous Movements

Perhaps the simplest statistical problem to consider is the specification of the mean square displacement of a marked particle of fluid in a field of randomly varying velocity. This problem, first treated in 1921 by Taylor,[184] is closely related to the treatment of Brownian motion using the Markoff process, and the more general discussion above of Goldstein's "diffusion by dis-continuous movements".[162]

Suppose we track the trajectory of a fluid particle with time. Its (Lagrangian) velocity $v_L(t)$ in one direction x would vary with time in a turbulent medium in a random way. We note that $v_L(t)$ can be divided into a mean value $\overline{v_L}$, and a fluctuating component $v_L'(t)$, whose mean value is zero. In other words,

$$v_L(t) = \overline{v_L} + v_L'(t).$$

The displacement of particle a in time t after t_0 in direction x would be given by

$$x(t_0 + t) = x(t_0) + \int_0^t v_L'(t_0 + t')\,\mathrm{d}t' \tag{7.152}$$

if $\overline{v_L}$ equals zero.†

Since it is not feasible to follow the meandering of each particle in a large group of particles, as in the case of Brownian motion, we look for statistical relations formed from equations of the form (7.152) to express the averaged behavior of many particles. It then is assumed that the field of turbulence is such that one can make use of well known assumed equality between the average behavior of the large number of particles (ensemble average) and the average displacement in time of a single particle.

If we consider dispersion in only one direction x, letting $x(t_0) = 0$, the restriction of homogeneity‡ yields for a marked particle:[55]

$$\overline{x^2(t)} = \frac{1}{T}\int_0^T x^2(t_0 + t)\,\mathrm{d}t = \int_0^t\int_0^t \overline{v_L'(t_0 + t')\,v_L'(t_0 + t'')}\,\mathrm{d}t''\,\mathrm{d}t'$$

$$= 2\int_0^t\int_0^{t'} \overline{v_L'(t_0 + t')\,v_L'(t_0 + t'')}\,\mathrm{d}t''\,\mathrm{d}t', \tag{7.153}$$

† $\bar{v}_L$ is distinguished from the mean thermal speed v_G. The distinction is often made in turbulent flow between the velocity $\mathbf{v}_L(\mathbf{r}_0, t_0; \mathbf{r}, t)$ taken in the Lagrangian frame of reference, and the velocity is an Eulerian frame of reference, $\mathbf{q}(\mathbf{r}, t)$. In the former case, the individual fluid particle is followed for all time t through space from a starting point $\mathbf{r}_0$ at t_0. In the latter, the velocity of all fluid particles passing a point in space is observed. At any instant, of course, the substantial derivatives, or the fluid particle acceleration are equal, $D\mathbf{v}_L/Dt = D\mathbf{q}/Dt$. In a steady flow, $\mathbf{v}_L = \mathbf{q}$, but this is not necessarily true for average velocities in a *statistically* steady flow. Analogous velocities may be defined for the motion of aerosol particles in a turbulent medium.

‡ Homogeneity refers to invariance of statistical properties by translation of coordinate axes. Isotropy means that the system does not change with a rotation or a reflection of axes [see also refs. (183) and (55)].

where the averaging is taken over a large group of particles starting displacements beginning at different times t_0.

For short intervals of time, one expects that a correlation between velocities $v'_L(t_0 + t')$ and $v'_L(t_0 + t'')$ will exist. To express the statistical relation between these velocities, we introduce the (Langrangian) correlation function

$$R_{11}(\tau) = \frac{\overline{v'_L(t)\, v'_L(t+\tau)}}{\overline{v'^2_L}}, \tag{7.154}$$

where $\tau = t'' - t'$ and $\overline{v'^2_L}$ is the mean square velocity of particles. Substituting $R_{11}(\tau)$ into eqn. (7.153),

$$\overline{x^2(t)} = 2\overline{v'^2_L} \int_0^t \int_0^{t'} R_{11}(\tau)\, dt'\, d\tau. \tag{7.155}$$

This result is the classical one deduced by Taylor.[184] An equivalent relation has been given by Kampe de Feriet:[185]

$$\overline{x^2(t)} = 2\overline{v'^2_L} \int_0^t (t - \tau)\, R_{11}(\tau)\, d\tau. \tag{7.156}$$

Both Kampe de Feriet[185] and Batchelor[186] have generalized eqn. (7.156) in tensor notation to cover the three-dimensional case of dispersion in an isotropic field of turbulence.

Using Einstein's relation, eqn. (7.24), we find for the case of Lagrangian dispersion that a "turbulent" diffusivity may be defined as

$$e_G = \frac{1}{2}\frac{d\overline{x^2}}{dt} = \overline{v'^2_L} \int_0^t R_{11}(\tau)\, d\tau. \tag{7.157}$$

Although the value of $\overline{x^2}$ depends in general on the form of $R_{11}(\tau)$, it is possible to find the asymptotic behavior of eqn. (7.157) for very small times and very large times. Taylor noted that since $R_{11}(\tau) \to 1$ as $t \to 0$,

$$\overline{x^2} \to \overline{v'^2_L}\, t^2 \tag{7.158}$$

for very short displacement intervals. In the other limit, for sufficiently large times, $t \to$ large,

$$\overline{x^2} \to \overline{v'^2_L}\, T_{11} t, \tag{7.159}$$

where

$$T_{11} = \int_0^{\mathscr{T}} R_{11}(\tau)\, d\tau; \tag{7.160}$$

$\mathscr{T}$ is the time interval in which $R_{11}(\tau) \approx 0$. T_{11} is called the *Lagrangian integral time scale.* These two limiting relations indicate that the root mean square displacement will display a linear proportionality with time initially, but at large times a parabolic growth in $(\overline{x^2})^{\frac{1}{2}}$ should appear, as in the case of Brownian motion. The correlation time $\mathscr{T}$ is analogous to the correlation time associated with thermal agitation, a lower limit suggested previously for applying the theory of Brownian motion to diffusion.

In analogy to the concepts of kinetic theory given in eqns. (7.36) and (7.35b), one may define an eddy diffusivity e_G as

$$e_G = \tilde{v}_L \ell_L, \tag{7.161}$$

where

$$\ell_L = \tilde{v}_L \int_0^\infty R_{11}(\tau)\, d\tau = \tilde{v}_L T_{11}$$

and

$$\tilde{v}_L = (\overline{v_L'^2})^{\frac{1}{2}}$$

The characteristic length ℓ_L represents a plausible length measure for movement of a tagged fluid particle in one direction, maintaining its identity.

Turbulent Diffusion in the Eulerian Framework

Because of the observed tendency for properties of a turbulent fluid to spread maintaining a nearly Gaussian distribution, and the analogy between the molecular diffusivity and the eddy diffusivity, a statistical equation for turbulent diffusion of the parabolic form has frequently been applied. In the Eulerian frame, the field of dispersing contaminant at any instant is then presumed to be determined by solutions of eqn. (7.86). For an incompressible fluid, where $\nabla . \mathbf{q} = 0$, $\mathbf{q}$ being the Eulerian fluid velocity, the equation for the transport of an ensemble averaged scalar quantity $\bar{\zeta}$ is derived from an ensemble average of eqn. (7.86); or

x coordinate $$\frac{\partial \bar{\zeta}}{\partial t} + \bar{q}_x \frac{\partial \bar{\zeta}}{\partial x} = \hat{\mathscr{D}} \frac{\partial^2 \bar{\zeta}}{\partial x^2} - \frac{\partial}{\partial x} \overline{q_x' \zeta'}; \tag{7.162a}$$

y coordinate $$\frac{\partial \bar{\zeta}}{\partial t} + \bar{q}_y \frac{\partial \bar{\zeta}}{\partial y} = \hat{\mathscr{D}} \frac{\partial^2 \bar{\zeta}}{\partial y^2} - \frac{\partial}{\partial y} \overline{q_y' \zeta'}; \tag{7.162b}$$

z coordinate $$\frac{\partial \bar{\zeta}}{\partial t} + \bar{q}_z \frac{\partial \bar{\zeta}}{\partial z} = \hat{\mathscr{D}} \frac{\partial^2 \bar{\zeta}}{\partial z^2} - \frac{\partial}{\partial z} \overline{q_z' \zeta'}. \tag{7.162c}$$

Here it has been assumed that $\hat{\mathscr{D}}$ is a constant, and the instantaneous Eulerian velocity vector $\mathbf{q}$ can be expressed as a sum of the average $\bar{\mathbf{q}}$ plus a fluctuating portion $\mathbf{q}'$, whose average value $\mathbf{q}'$ is zero; or

$$\mathbf{q} = \bar{\mathbf{q}} + \mathbf{q}'. \tag{7.163a}$$

Similarly, $$\zeta = \bar{\zeta} + \zeta'. \tag{7.163b}$$

At this point, it is customary to assume, in addition, that ensemble averages over many realizations of an experiment equal averages of one configuration taken over a long period of time. That is, there exists spatial and temporal homogeneity in the medium in question. Then eqn. (7.162) can be written in vector form; with the averaged quantities are identified with time averaged processes at a point in space:

$$\frac{\partial \bar{\zeta}}{\partial t} + \bar{\mathbf{q}} . \nabla \bar{\zeta} = \hat{\mathscr{D}} \nabla^2 \bar{\zeta} - \nabla . (\overline{\mathbf{q}' \zeta'}). \tag{7.164}$$

The averages $\overline{\mathbf{q}'\zeta'} = \overline{(q'_x\zeta')}\,\mathbf{i} + \overline{(q'_y\zeta')}\,\mathbf{j} + \overline{(q'_z\zeta')}\,\mathbf{k}$, where $\mathbf{i}, \mathbf{j}$ and $\mathbf{k}$ are unit vectors in the x, y, z directions are then identified with the average gradient in ζ by a classical Boussinesq relation. The x component, for example, takes the form

$$\overline{q'_x\zeta'} = e_G \frac{\partial \bar{\zeta}}{\partial x}. \tag{7.165}$$

Equation (7.165) assumes that the average fluctuation in Lagrangian particle velocity, $(\overline{v_L'^2})^{\frac{1}{2}}$ is essentially equal to the Eulerian r.m.s. turbulent velocity, $(\overline{q'^2})^{\frac{1}{2}}$. The length ℓ_L, then, is identified with the Taylor–Prandtl mixing length for turbulent transfer of momentum[55]. Ordinarily $\hat{\mathscr{D}} \ll e_G$, and therefore the molecular part of diffusion flux may be neglected in eqn. (7.164).

Corrsin[188] has noted that the relation, eqn. (7.165), corresponding to a gradient transport of $\bar{\zeta}$ may be incorrect in principle for diffusion in a turbulent medium. This can be demonstrated by considering the Taylor expansion of an equation equivalent to expansion of eqn. (7.5d). Assuming that $\zeta(\mathbf{r}, t)$ has a Gaussian like distribution, terms of order higher than $(\Delta r)^3$ can be disregarded if the ratio

$$\left(\frac{\ell_L}{2}\right)^2 \ll 6 \left|\frac{(\partial\zeta/\partial x)}{\partial^3\zeta/\partial x^3}\right|_{x=x_0}$$

Identifying ℓ_L with $T_{11}\tilde{v}_L$, Corrsin showed that this criterion becomes

$$T_{11}\tilde{v}_L \ll 3\overline{z^2},$$

where z is the coordinate normal to the downstream direction x. For the case of heat diffusing into a turbulent flow generated in a wind tunnel by a grid, measurements indicated that

$$T_{11}\tilde{v}_L \approx 0.1 \quad \text{and} \quad \overline{z^2} \approx 0.09.$$

Hence the mixing length analogy may not be adequate in general for turbulent diffusion. Yet in practice the parabolic diffusion equation works fairly well for turbulent diffusion though it appears incorrect in principle. For certain problems, however, one may have to seek more elaborate forms for these statistical processes, e.g. in the telegraph equation for diffusion derived earlier in Section 7.1.

Aerosol Particles in Turbulent Flow

To apply the ideas of the diffusion of aerosols in a turbulent medium by Taylor's theory of continuous movements, or by the Eulerian theory, one has to give consideration to the relation between particle acceleration and fluid motion. This question was treated in some detail in Chapter 3. We shall use eqn. (3.55) as the point of departure for treating particle motion in a turbulent medium. Thus, after the assumptions discussed in Section 3.2, the particle trajectory is specified, at least in principle, by a force balance of the form given by eqn. (3.29). Now assuming the analogous relation for $\mathbf{v}_i$ as given in eqn. (7.29),

$$\frac{\mathrm{d}\mathbf{q}_i}{\mathrm{d}t} + \mathscr{A}\mathbf{q}_i = \mathscr{A}\mathbf{v}_L(t) + \hat{F}'(t), \tag{7.166}$$

† For some further details on this complicated question, see, for example, Lumley.[187]

where the "forcing function" of particle motion is

$$\hat{F}'(t) = \mathscr{B}\frac{d\mathbf{q}_i}{dt} + \mathscr{C}\int_0^t \frac{dt'(d\mathbf{v}_L/dt - d\mathbf{q}_i/dt)}{\sqrt{(t-t')}}. \tag{7.167}$$

The terms accounting for external forces and Stefan flow are disregarded in this expression.

As a first approximation we assume that the forcing term $\hat{F}'(t)$ is either negligible,† or at least completely random in the sense that the average value $\hat{F}'(t) = 0$. Thus the remaining portion of the right-hand side is proportional to the fluid velocity $\mathbf{v}_L$ near the particle. Within this approximation, Friedlander[189] considered the problem of the statistics of particle movement in a turbulent medium which is homogeneous in space and time, and is isotropic.‡ If the mean square particle velocity is about the same as the mean square fluid velocity, Friedlander found that the (Lagrangian) relative velocity between the fluid and the particle, $v_r = q_i - v_L$, could be written as

$$v_r = v_0\,e^{-\mathscr{A}_i t} + (v_0\,e^{-\mathscr{A}t} - v_L)\left(\mathscr{A}\,e^{-\mathscr{A}t}\int_0^t e^{\mathscr{A}t}\,v_L(\tau)\,d\tau\right), \tag{7.168}$$

where v_L is the Lagrangian fluid velocity as defined previously, and v_0 is the initial value of v_r. By squaring this relation, and taking averages of each term, one can find an equation for v_r^2 in terms of several correlation functions.

In turbulent media, the correlation functions like $R_{11}(\tau)$ generally reduce to zero over very short time intervals τ. Thus for the restriction $t \gg \mathscr{A}^{-1}$, correlation integrals containing terms of exp $(-\mathscr{A}t)$ and exp $(-2\mathscr{A}t)$ reduce to zero, and a steady state expression for v_r^2 is obtained:

$$\overline{(q_i - v_L)^2} = \overline{v_r^2} = \overline{v_L^2}\left[1 - \mathscr{A}\int_0^\infty e^{-\mathscr{A}\tau}R'_{11}(\tau)\,d\tau\right], \tag{7.169}$$

where $R'_{11}(\tau)$ denotes *the correlation function of fluid surrounding the particle.*§ This relation has certain interesting implications. For example, if the density ratio $\rho_G/\rho_i \to 1$, the relative velocity v_r^2 tends to be zero. For very heavy particles, on the other hand, $\mathscr{A}$ and ρ_G/ρ_i are both small, and $\overline{v_r^2} \approx \overline{v_L^2}$.

When $\mathscr{A} \gg 1$, or for very small particles,

$$\overline{v_r^2} = \mathscr{A}^{-2}\overline{\left(\frac{\partial v_L}{\partial t}\right)^2},$$

or from eqn. (7.168),

$$\frac{\overline{v_r^2}}{\overline{v_L^2}} = \frac{\overline{(q_i - v_L)^2}}{\overline{v_L^2}} \leqslant 1. \tag{7.170}$$

For a more specific example, consider the case where

$$R'_{11}(\tau) = e^{-\gamma_{11}\tau},$$

† This is generally adequate for aerosols because the term is normally much larger than $F'_e(t)$.

‡ In addition the particle radius is assumed much larger than the Eulerian microscale of turbulence (defined, for example, by Hinze[55]).

§ This is distinguished from $R_{11}(\tau)$, the correlation between fluid particles far from the particle.

where γ_{11} is a constant. Then eqn. (7.169) yields with eqn. (7.170)

$$\frac{\overline{v_r^2}}{\overline{v_L^2}} = \frac{\gamma_{11}}{\gamma_{11} + \mathscr{A}}$$

or, for $\mathscr{A} \gg 1$,

$$\frac{\overline{v_r^2}}{\overline{v_L^2}} \to 0. \tag{7.171}$$

Thus for small particles in the size range $R_i \approx 10\ \mu$, the ratio of $\overline{v_r^2}$ to $\overline{v_L^2}$ tends to be very small. This is to be anticipated as the particle inertia becomes smaller and smaller.

Friedlander[189] also obtained a result for the mean square displacement of particles in a medium characterized by homogeneous turbulence. Integrating the square of eqn. (7.166) twice and setting $\overline{\hat{F}'(t)} = 0$, it is found that

$$\overline{x_i^2} = 2\overline{v_L^2}\left[\int_0^t (t - \tau)\, R'_{11}(\tau)\, d\tau\right] - \frac{2\overline{q_i^2}}{\mathscr{A}^2}[1 - R_p(\tau)], \tag{7.172}$$

where the correlation coefficient for the particle velocity is

$$R_p(\tau) = \overline{q_i(\tau)\, q_i(t + \tau)}/\overline{q_i^2}.$$

As $t \to \infty$, $R'_{11}(\tau) \to 0$, and $\overline{x^2} \gg 2\overline{q_i^2}/\mathscr{A}^2$, then

$$\overline{x^2} \approx 2\overline{v_L^2}\, T_r, \tag{7.173}$$

where the correlation for the particle velocity is

$$T_r = \int_0^\infty R'_{11}(\tau)\, d\tau = \text{const.} \tag{7.174}$$

The integral scale T_r equals a constant only as the interval τ becomes very large.

From eqn. (7.173) we note that after the mean square displacement of an aerosol particle is directly proportional to the Lagrangian correlation coefficient for fluid elements *near the particle* and not far from the particle. Thus there can be two extremes of particle behavior. First, if the particle is very light then the particle will very nearly follow the fluid motion, or $\rho_G/\rho_i \to 1$,

$$R'_{11}(\tau) \to R_{11}(\tau).$$

In the second extreme, where the particle is very massive and $\rho_G/\rho_i \to 0$, the particle will tend to remain nearly fixed in space and fluid will flow around it. In this case, $R'_{11}(\tau)$ should approach the Eulerian correlation time for fluid elements. That is, if coordinates are fixed on the particle, the particle will remain fixed in space while the Eulerian fluid velocity will fluctuate around the sphere without causing any perturbations in the acceleration of the sphere.

Defining the eddy diffusivity of the particle e_i as $dx^2/2\,dt$, we then find that

$$\frac{e_i}{e_G} = \frac{T_r}{T_{11}}, \tag{7.175}$$

The eddy diffusivities of the particle and the surrounding fluid in general will be time dependent, and ordinarily will not be the same. From eqn. (7.173), in the limit $\tau \to \infty$, e_i

and e_G can be equal only if the Lagrangian time scales for the fluid around the particle and the fluid particle are the same.

The ratio $e_i/e_G \to 1$, $t \to \infty$ was found by complicated arguments to be identically true by Tchen. However, we note from Friedlander's results that this may not be correct in general because the integral scales T_r and T_{11} may not be the same at any finite interval of time τ.

For the case of particle dispersion over very short traveling times, Friedlander[35] suggested after examining a series expansion of the correlation coefficient $R_p(\tau)$, that

$$\overline{x^2} \approx \frac{\overline{v_L^2}\mathscr{A}^2 t^4}{4}, \quad t \to 0 \tag{7.176}$$

for relatively large particles. However, for particles approaching molecular size, the particles should follow fluid parcels, or

$$\overline{x^2} \approx \overline{v_L^2} t^2, \quad t \to 0. \tag{7.177}$$

To make use of all of the theory outlined above, it is essential to know the relation between the Lagrangian velocity field $\mathbf{v}_L(x, y, z, t)$ of the fluid, and its Eulerian velocity field $\mathbf{q}(x, y, z, t)$.

A key problem in the study of turbulent diffusion lies in interpreting data taken in the Eulerian framework, which is relatively easy to obtain, in the more natural coordinates given by the Lagrangian analysis. The very difficult question of connecting the two analyses has been discussed in some detail for turbulent diffusion of fluids by Hinze,[55] Lumley,[187] and Corrsin.[188].

Experiments in Aerosol Diffusion

Although considerable effort has been expended on turbulent diffusion of gaseous contaminants, very few detailed studies of aerosol diffusion have been reported. Some early qualitative measurements by Longwell and Weiss[190] of the mean square displacement of oil droplets from a jet suggested that larger particles diffused less at a given distance than smaller droplets. Some unpublished experiments of Hidy and Keller[191] for diffusion of 10 μ diameter lycopodia spores, and ~50 μ diameter fluid catalyst particles in turbulent pipe flow indicated similar results. Although the lycopodia spores diffused nearly three times as far as the catalyst particles near the location of injection into the center of the pipe flow, the results failed to confirm quantitatively Friedlander's predictions in eqns. (7.176) and (7.177).

A recent study of Kennedy[192] has become available which represents one of the first attempts to make detailed measurements of aerosol dispersion in turbulence generated behind a grid in a wind tunnel.

An interesting set of experiments of Goldschmidt[193] has shed some light on the question of the magnitude of the ratio e_i/e_G. Goldschmidt utilized a hot wire anemometer to obtain Eulerian distributions of safflower oil droplets, ~6 μ in diameter, in a two-dimensional turbulent jet. He compared his measurements of the lateral diffusion of the aerosol to the diffusion of momentum, as characterized by profiles of average air velocity. Goldschmidt found that the turbulent Schmidt number Sc_T, the ratio of the eddy diffusivity for momentum to the eddy diffusivity for the aerosol, was about 1.1 on the average. Other data for the transport of scalar properties like heat or gas contaminant, has suggested an average value of Sc_T of about 0.8 for turbulent shearing flows. These results imply that

$$\frac{\epsilon_i}{\epsilon_G} \sim 0.7, \tag{7.178}$$

which may be somewhat less than one might expect for particles which are of this size range.

In the absence of better information about the diffusion of aerosols in turbulent gases, it is customary to use the limited results for gaseous contaminants as discussed for example by Corrsin.[188] That is, one assumes that $\epsilon_i \approx \epsilon_G$. One should bear in mind, however, as eqn. (7.178) suggests, that the unrestricted use of such considerations may lead to substantial errors.

With the assumption that the eddy coefficients for particles and gas molecules is the same, combined with the assumptions leading to the use of boundary layer theory for diffusion of particles, a number of calculations have been made for the deposition of particles from turbulent gases. Some of the results of these semi-empirical theories have been reviewed by Davies,[194] but will not be discussed here.

CHAPTER 8

AEROSOL GENERATION BY DISINTEGRATION AND DISPERSAL

To GENERATE an aerosol it is necessary to produce the colloidal suspension in one of four ways: (a) by condensing out small particles from a supersaturated vapor (the supersaturation may be induced either physically or chemically), (b) by direct chemical reaction in flames, (c) by disrupting or breaking up bulk material, or (d) by dispersing fine powders into a gas. A wide variety of devices have been designed to produce aerosols by one of these methods or a combination of them.

In the first case, aerosols are produced by aggregation from a molecular dispersed phase. Perhaps the best known generator using vapor condensation is the LaMer generator (e.g. Sinclair and LaMer,[(195)]). Many other generators using the principle of particle formation by condensation have been designed which involve attainment of supersaturation jet mixing of a moist gas with a cool gas or near adiabatic expansion of gas at high speed through a nozzle. Aerosols have been formed by condensation through chemical means, e.g. via combustion processes, via discharges between electrodes, or by photolysis. Production of aerosol by explosions evidently relies on a combination of breakup and vaporization of material, which subsequently recondenses. One example of such a method utilizes an exploding wire, as described by Kasiosis and Fish.[(196)]

In the bulk of the condensation methods which are uncontrolled, relatively polydisperse aerosols are produced. However, with careful control of nuclei concentration and thermodynamic conditions of the vapor, monodisperse aerosols may be obtained, as in the case of the LaMer generator (see also Section 9.3).

The second case wherein aggregation of single molecules takes place involves the formation of particles by direct chemical reaction akin to polymerization. As an example, Homann[(432)] has reviewed such a mechanism for the formation of carbon particles in a premixed acetylene-oxygen flame. Evidently the particle formation observed in this case does not involve condensation from a supersaturated vapor, but proceeds through pyrolysis of the acetylene, forming in the flame as an intermediate of unstable polyacetylenes. Such mechanisms are mentioned further in Chapter 9.

In the third case where disintegration of coarse bulk material into matter approaching colloidal dimensions is involved, three main types of devices commonly have been used. The first of these is the air-blast or aerodynamic atomizer in which a compressed gas ejects at high velocity into a liquid stream emerging from a nozzle. This type of breakup is found, for example, in paint spray guns, venturi atomizers, and many other sprayers which generate aerosols used for insecticides, for disinfectants, and for therapeutic purposes. Atomizers of this class characteristically yield a cloud with a wide range of droplet size, though this may be controlled by trapping larger particles within the atomizer.

Another class of atomizer depends on centrifugal action wherein a liquid is fed onto the center of a spinning disc, cone, or top and is centrifuged to the outer edge. Provided that the flow rate of liquid into the spinning device is well controlled, sprays produced in this way are rather uniform in size in contrast to results by other methods of atomization.

A third type of atomizer is the hydraulic design in which a liquid is pumped through a nozzle and, upon its exit from the orifice, breaks up into droplets. Disintegration here depends largely on the physical properties of the liquid and the ejection dynamics at the nozzle orifice rather than on the intense mixing between the liquid and the surrounding gas. Perhaps most well known of this class of device is the swirl chamber atomizer which has been used in agricultural equipment, oil-fired furnaces, internal combustion engines, and gas turbines.

In addition to these three main classifications, special methods are available including the electrostatic atomizer and the acoustic atomizer. The former makes use of a liquid breakup by the action of electrostatic forces, while the latter applies high intensity sonic or ultrasonic vibrations to disrupt a liquid.

The fourth group involving the generation of dust clouds by dispersal of fine powders is a straightforward method in principle. All such generators depend on blowing apart a bed of finely divided material by aerodynamic forces, or by a combination of air flow and acoustic or electrostatic vibrations. The size of particulate suspensions produced in this way is limited by the minimum size of material ground up by mechanical or other means, and the nature of cohesive and adhesive forces acting between particles in agglomerates. Generally dust clouds produced by powder dispersal will not fall below a few microns in radius.

Aerosols composed of solid particles, or nonvolatile liquids with sizes much less than those attainable by atomization of pure liquids, or by dispersal of powders, may be produced by atomizing. Breakup of suspensions of a volatile carrier liquid, in which solid particles, an immiscible liquid or a non-volatile solute are suspended, will yield very small particles after evaporation of the volatile liquid.

Several examples, particularly of machines for generating aerosols by disintegration or by dispersal, are reviewed in Green and Lane.[4] Some design methods for liquid atomizors have been described in a review by Fraser *et al.*[226] In the succeeding sections further remarks about more or less general considerations about the mechanics of breakup and dispersal will be given instead of descriptions of particular devices or methods for atomization. In Chapter 9, aerosol formation by spontaneous condensation will be discussed.

Because comparatively little is known about the details of the physics of disintegration and disruption, this chapter is rather brief and is limited largely to qualitative considerations.

8.1. Atomization and the Breakup of Liquids

A. Energetics of Breakup

To subdivide bulk liquids into particles approaching colloidal dimensions, sufficient energy must be added to the bulk material to overcome the intermolecular forces holding the material together. When a liquid is atomized the energy expended in atomization will be used (a) for forming a new surface, (b) for overcoming viscous forces in changing the shape of the liquid, and (c) for balancing losses resulting from inefficient input of energy into the liquid. The energy required to form a new surface upon dividing the bulk phase into droplets of radius R_i is $3\sigma/R_i\rho_L$ per unit mass. For water droplets of $\sim 1\ \mu$ radius this is about 0.215 joules gm^{-1}, or about 0.05 cal gm^{-1}. In addition to this amount there has to be included a small amount of energy used in creating any surface formed during atomization which exceeds that of the final surface. Since the time of droplet formation generally has to be rather short, the rate of deformation of liquid has to be large, and the amount of energy dissipated by the action of viscosity during deformation will be larger.

If a highly idealized model is assumed for droplet formation by atomization, it is possible to estimate the minimum energy required to reshape the liquid. Suppose we adopt Monk's[197] model which assumes that the liquid streams out into a thread or film that collapses under surface tension to yield droplets of average radius half the thread thickness. Assuming that the thread forms after the liquid enters a conical region of transition in which its velocity increases at a constant rate, the minimum energy dissipated per unit mass ϵ_d of liquids is

$$\epsilon_d \approx 8\mu_L d_1^2 \hat{Q}_L / 3\pi d_2^2 L \rho_L, \tag{8.1}$$

where d_1 and d_2 are the diameters of the entry and exit of the transitional cone, L is its length and $\hat{Q}_L$ is the volume flow rate of liquid.

Taking $L = 1$ cm, $d_1 = 1$ cm, $\hat{Q}_L = 1$ cm^{3}sec^{-1}, and $\mu_L = 1$ centipoise, calculated values of $\rho_L \epsilon_d$ have been tabulated in Table 8.1 for the case of water. Monk has suggested that

TABLE 8.1. MINIMUM ENERGY IN CALORIES DISSIPATED IN ATOMIZING 1 CM OF WATER AT A RATE OF 1 cm^3 sec^{-1} (FROM MONK[197])

Thread diameter in microns	1	10	100
1	2×10^6	2×10^5	2×10^4

5	3.2×10^4	320	32

10	200	20	2·0

20	13	1.3	0.13

values for the energy to overcome viscous forces in the liquid below the dotted lines in Table 8.1 may constitute conditions where viscous energy losses will not limit the atomization process. Green and Lane[4] note, however, that Monk's calculations are rather crude and should be regarded only as a qualitative illustration of the importance of viscosity to the generation of fine droplets from bulk liquids.

It is difficult to estimate the losses occurring in applying energy to breaking up a liquid, except in specific configurations. However, it seems clear that energy transfer from a gas to a liquid during atomization by aerodynamic forces is inefficient. However, this inefficiency is offset, at least in part, by the large amounts of energy applied by a compressible fluid, like a gas, as compared with that retained in a liquid under pressure.

B. Mechanics of Droplet Formation

Droplets may be produced by a number of different processes, many of which are quite complicated and poorly understood. Two of the simplest processes for droplet production are (a) the formation of liquid pendants from the tip of a capillary tube, with subsequent detachment, and (b) the development of wave-like instabilities on the surface of a filament or sheet of liquid, with subsequent breakup and droplet evolution. The former mechanism is reviewed by Lane and Green,(198) and the latter represents a classical problem of vibrational instability in jets investigated years ago by Lord Rayleigh.(199) The mechanisms of instability of thin liquid sheets has also been treated more recently by Hagerty and Shea(421) and by Dombrowski and Johns.(422)

The size of droplets falling at a slow rate from circular tips is directly proportional to the external radius of the tip. Furthermore, cylindrical filaments will display bead-like swellings and contractions on their surfaces resulting in an instability of wave number:(199)

$$\tilde{\kappa} = 2\pi/9.016a, \tag{8.2}$$

where a is the mean radius of the filament. Then unstable disturbances will increase in amplitude until the filament breaks up into droplets having principally a diameter of about $9a$. Because it is quite difficult to obtain capillary tubes of radius less than 10's of microns, both of these simple processes have limited direct applicability to the production of swarms of very small droplets in the micron and submicron size range. However, we shall find below that many experimental observations suggest Rayleigh's instability mechanism also plays an important role in more complex atomization processes involving aerodynamic forces.

Aerodynamic Instability of Liquid Sheets

Like the case of breakup of capillary jets, the disintegration of thin sheets of liquid take place as a result of the development of unstable waves on the sheet. To understand better the breakup process involved in several nozzles, Hagerty and Shea(421) studied classes of inviscid instabilities on a plane liquid sheet of uniform thickness. Two classes of wave modes were identified; (a) the sinuous type, where waves on both sides of the sheet are in phase, and (b) the dilational class, in which waves on the two sides are 180° out of phase. Considering only a balance between surface tension and pressure forces, Hagerty and Shea developed a theory for the unstable waves patterned after Rayleigh's results. Assuming that these waves break down to form droplets of the size of the wavelength of the most unstable wave modes, a theoretical mean droplet size could be estimated in terms of physical properties of the fluids. The theory has been used to correlate data for droplet formation during the break up of liquid sheets. Hagerty and Shea(421) found that the lowest stable modes were sinuous waves whose minimum stable frequency was

$$f = \frac{\rho_G \bar{q}_L^3}{2\pi\sigma},$$

where $\bar{q}_L$ is the mean flow of liquid in the sheet. Generally droplets produced by film breakup are a minimum mean radius of 40–60 μ.

More recently Dombrowski and Johns(422) have reported an analysis of the aerodynamic instability and disintegration of viscous sheets with particular application to the performance of fan-spray nozzles. Their conceptual picture of the disintegration of a liquid sheet is shown in Fig. 8.1. Here a sinuous wave is shown initially growing in amplitude, then breaking up first into long filaments with axes roughly parallel to the wave crests. Later the thin filaments develop axial instabilities of dilational waves, possibly related to instabilities

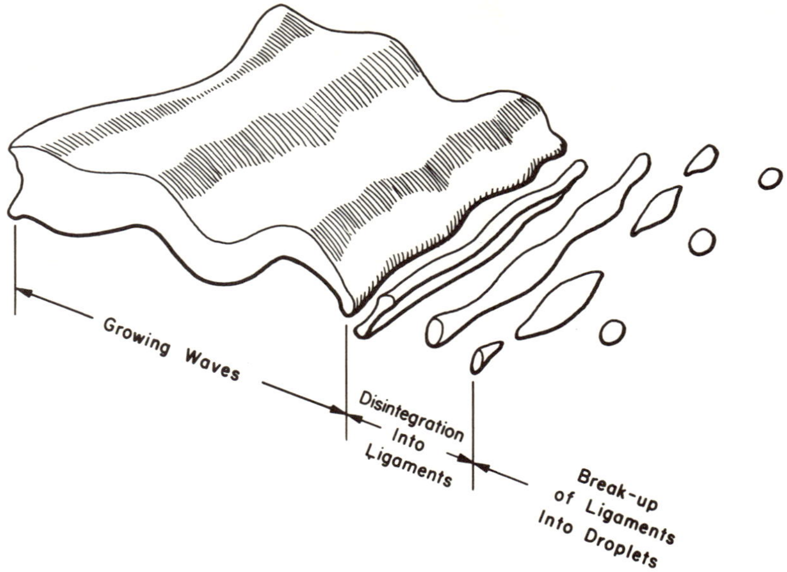

FIG. 8.1. Schematic drawing of the disintegration of a flat liquid sheet traveling through a still gas.

in Rayleigh's capillary jet. These ripples grow and finally break up into droplets. Thus Dombrowski' and Johns'[(422)] model involves two classes of wave formation taking place in sequence to form first ligaments then droplets. The mechanism described by Dombrowski and Johns is readily illustrated in Plate 8. Here a liquid sheet issuing from a nozzle is shown undergoing the wavelike transitional instability before disintegrating into droplets. By taking into account the viscous and inertial forces, as well as pressure and surface tension, these workers found relations for the most probable ligament diameter and the most probable drop diameter. Based on their results, the droplet radius formed from the modes of maximum instability of the film and the ligaments is given by

$$[R_i] = \left[\frac{3\pi}{\sqrt{2}}\right]^{\frac{1}{3}} R_{\text{lig}}\left[1 + \frac{3\mu_L}{\rho_L \sigma 2R_{\text{lig}}}\right]^{\frac{1}{6}}, \tag{8.3a}$$

where the ligament radius is

$$R_{\text{lig}} = 0.9614\left[\frac{(\hbar x)^2\sigma^2}{\rho_G \rho_L \bar{q}_L^4}\right]^{\frac{1}{6}} \times \left[1 + 2.60\left(\frac{(\hbar x)\, \rho_G^4 \bar{q}_L^7}{72\rho_i^2\sigma^5}\right)\right]^{\frac{1}{5}}. \tag{8.3b}$$

Here $\hbar$ is the sheet thickness and x is the distance downstream along the sheet.

Dombrowski and Johns' equations agree reasonably well for observed values of the surface-volume mean particle radius found for wax sprays by Hasson and Mizraki.[(423)] In these data, the mean radius was about 25 μ.

Electrostatic Forces and Breakup

The breakup of droplets issuing from a capillary tip may be modified somewhat by charging the issuing liquid filament and applying an electrostatic field. Droplets falling from a finely drawn out tube become smaller when a high potential is applied to the liquid. If this electrical potential reaches a few thousand volts, the droplets become exceedingly

PLATE 8. The disintegration of a liquid sheet emanating from a spray nozzle. (From Dombrowski and Johns[422].)

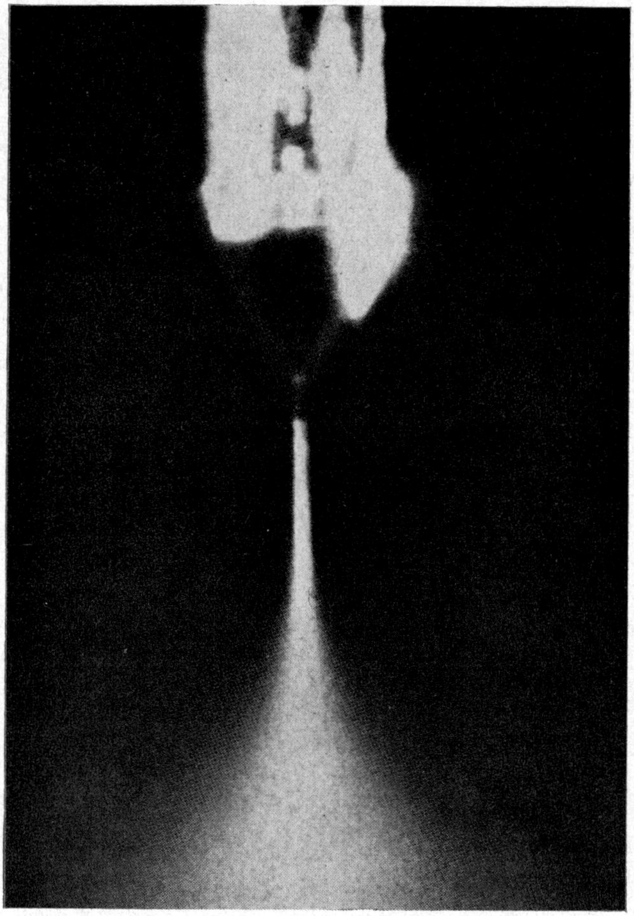

PLATE 9A. Aerosol mist produced by electrical atomization. Continuous exposure photograph. (Courtesy of H. Green and W. Lane).

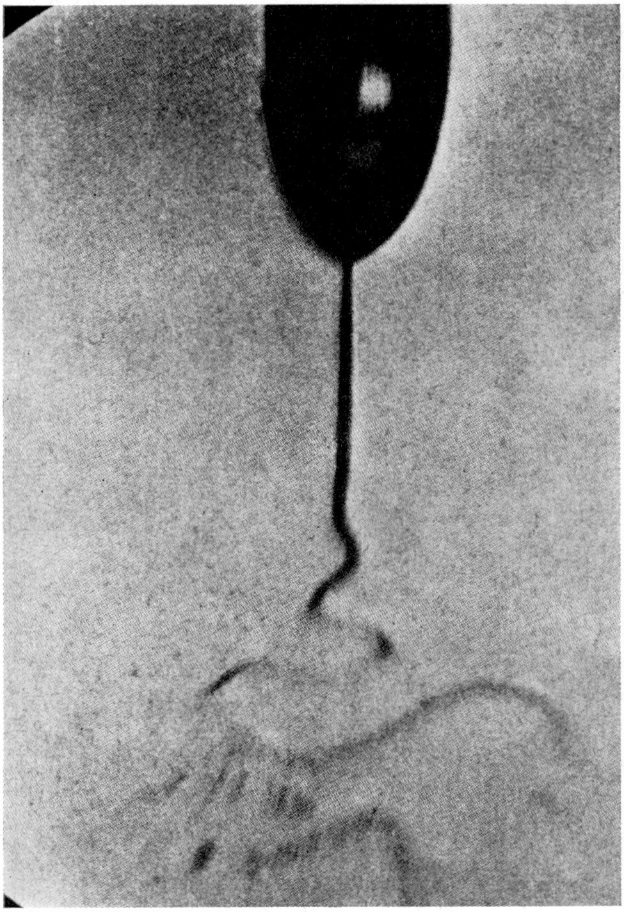

PLATE 9B. Aerosol mist produced by electrical atomization. Disintegration of a thread from a charged water meniscus. (Courtesy of H. Green and W. Lane).

(a)

(b)

(c)

PLATE 10. The formation of droplets from a rotating cup. Rotational direction is clockwise. The cup is two inches in diameter. The liquid being atomized is 165 cS viscosity. (a) Direct droplet formation. (b) Break-down into threads from the cup lip. (c) Disintegration of a liquid sheet emanating from the spinning cup. (From Fraser, Dombrowski and Routley[431].)

small and are expelled from the tip of the tube at a high velocity so that they appear to form a continuous thin stream. The droplets get smaller and fall faster from the tip as the electrical potential is raised.

Under certain conditions of electrical potential and hydrostatic pressure, the liquid flowing from the capillary tip forms an inverted cone. From the apex of this cone streams a fine thread of liquid which breaks up into a brush-like aerosol cloud (Plate 9A). If the electrical potential is increased farther, this thread may disappear leaving only a plume of fine mist. The potential where this occurs depends on the radius, shape, and wall thickness of the capillary tube in the field gradient, and on properties of the liquid.[4] Fine mists of $\sim 10\ \mu$ radius and smaller can be produced with liquids of low conductivity such as distilled water, alcohol, dibutyl phthalate, etc. Experiments have indicated that conducting liquids and non-polar organic liquids of low dielectric constant, like carbon tetrachloride or benzene, cannot be dispersed by this method even at very high electrical potentials.

Because of the interest in electrical atomization for producing fuel oil sprays entering combustion chambers and for ion propulsion of rockets in space, considerable work has been reported on this problem recently, particularly for atomization into a vacuum. Lapple[200] has reviewed this work in detail.

Most investigators believe that electrostatic atomization is achieved as a result of a "Rayleigh-like" instability of the liquid thread issuing from the tube. However, Schultz and Bronson[201] reported some evidence that atomization in this case involves rupturing of molecular bonds during high rates of discharge in their high vacuum experiments. The instability of electrified liquid surfaces, analogous to Rayleigh's hydrodynamic instability, was investigated some time ago by Zeleny.[202] Suppose we consider a spherical surface of radius R_i with charge Q_i. Such a sphere will experience an outward pressure of $Q_i^2/8\pi R_i^4$ which, by opposing surface tension, will result in detachment of a droplet from a tip sooner than an uncharged sphere. If the applied potential is sufficiently high, a spray will be formed whose droplets will be governed at equilibrium by the relation

$$\frac{2\sigma}{R_i} - p - \frac{\mathscr{E}_0}{8\pi R_i^2} = 0, \tag{8.4}$$

where p is (excess) pressure in the droplet and $\mathscr{E}_0$ is the applied potential. Equilibrium becomes unstable when a small decrease in radius gives an outward pressure exceeding that of surface tension, $2\sigma/R_i$. Zeleny found that the criterion for instability of a charged droplet depends on

$$\mathscr{E}_0^2 = f(S, S')\, R_i \sigma, \tag{8.5}$$

where the function f depends on the shape of the droplet and the rate of axial changes, given respectively, for example, by its surface, S, and the rate of change in S with time S'. Thus when the electrical pressure at the liquid surface reaches a critical value, depending on surface tension of the liquid and the radius of curvature of the droplet, the sphere will become unstable. Then slight disturbances will be amplified, and eventually breakup will take place. This condition should be reached first at the lower extreme of the drop where the intensity of electrical field is greatest, and the liquid here is stretched into a thread which breaks apart into droplets forming the brush-like cloud (Plate 9). Zeleny's analysis has been extended by Drozin[203] to incorporate the dielectric constant of the liquid.

Lane and Green[198] remark that the dispersal of droplets into the conical shape shown in Plate 9A appears to be the result of changes in the direction of the electrical force acting on the thread and the mutual repulsion of the charged particles. In Plate 9B a photo taken by spark illumination shows the thread stemming from the conical surface and, less distinctly, its disintegration into droplets.

Experimental data suggest that the greater the conductivity, the easier the atomization. Poor atomization of liquids of very high conductivity is difficult to understand. However, Lapple(200) has noted that it is likely that at least two modes of atomization may exist, the first being Zeleny's instability mechanism and the second being a more intensive disruption for which the mechanism has not been elucidated. The latter may involve corona formation or breaking of molecular bonds as proposed by Schultz and Bronson.(201) While the general guidelines have been established for electrostatic atomization, the details of the physical mechanisms are incomplete, and at present there appear to be no methods for predicting, in detail, atomization performance either by level or charging or in terms of the size distribution of droplets formed. However, if a mechanism like Rayleigh's instability controls the formation of droplets from a jet in an electrical field, the most probable drop size may be estimated from classical stability analysis.

Suppose we assume that a jet of radius a most probably breaks down in the mode of maximum instability to form a droplet of radius R_i. Then in the absence of an electrical field, $R_i = 4.508a$ from eqn. (8.2). If the disturbance amplitude is given by $A_R \sim e^{\pm \tilde{\pi} t}$, then the equation for the mode of maximum instability is(204)

$$\tilde{\pi}^2 = [a\tilde{\kappa} I_0'(a\tilde{\kappa})\, \sigma / I_0(a\tilde{\kappa})\, \rho_L a^3]\, [(a\tilde{\kappa})^2 - 1],$$

where $\tilde{\kappa}$ is the wave number of the disturbance on the jet and I_0 is a zero-order Bessel function of the first kind. The value $\tilde{\kappa}_m$ for maximum A_R is given by

$$\frac{2\pi}{R_i} = \tilde{\kappa}_m = \frac{0.696}{a}. \tag{8.5}$$

This relation is plotted in Fig. 8.2.

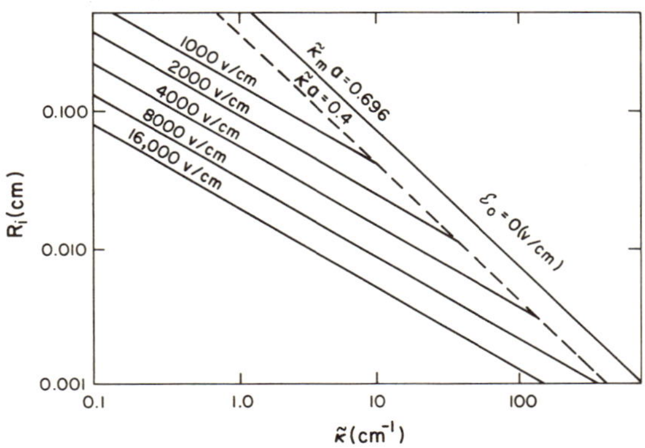

FIG. 8.2. Relation between liquid jet radius and wave number of disturbance for electrical atomization. (From Peskin and Raco.(204))

By a similar analysis, the equation for the disturbance growth on a jet oriented longitudinally in an electrical field $\mathscr{E}_0$ has been calculated by Nayyar and Murty:(205)

$$\tilde{\pi}^2 = \frac{a\tilde{\kappa} I_1(a\tilde{\kappa})}{\pi \rho_L a^2 I_0(a\tilde{\kappa})} \left\{ \frac{\pi\sigma[1 - (a\tilde{\kappa})^2]}{a} - \frac{(\epsilon_G - \epsilon_L)^2\, \mathscr{E}_0^2 a\tilde{\kappa} I_0(a\tilde{\kappa})\, K_0(a\tilde{\kappa})}{4[\epsilon_L I_1(a\tilde{\kappa})\, K_0(a\tilde{\kappa}) + \epsilon_G I_0(a\tilde{\kappa})\, K_1(a\tilde{\kappa})]} \right\},$$

where ϵ_G and ϵ_L are respectively the dielectric constant of the gas and the liquid in the jet, and K_n is the n'th order Bessel function of the second kind for an imaginary argument. The value of $\tilde{\kappa}_m$ in this case may be determined for $\tilde{\kappa}a \ll 1$ from

$$\tilde{\kappa}_m^2 = \frac{2\pi\epsilon_G\sigma}{a^3(\epsilon_G - \epsilon_L)^2 \mathscr{E}_0^2[-0.577 - 0.25 - \ln(\tilde{\kappa}_m/2)]} \tag{8.6}$$

Typical plots of this relation for water in air, $\epsilon_G = 1$, for various values of the electric field are drawn in Fig. 8.2. These curves are terminated at a $\tilde{\kappa} = 0.4$ to indicate that they are valid only for finite $\mathscr{E}_0$ and $a\tilde{\kappa} \ll 1$.

Peskin and Raco[204] have noted from the curves in Fig. 8.2. the ratios

$$\frac{\Delta \log a/\Delta \log \tilde{\kappa}}{\Delta \log a_{\mathscr{E}}/\Delta \log \tilde{\kappa}_{\mathscr{E}}} = \frac{\Delta \log \tilde{\kappa}_{\mathscr{E}}}{\Delta \log \tilde{\kappa}} = \frac{\log \tilde{\kappa}_{f,\mathscr{E}} - \log \tilde{\kappa}_{I,\mathscr{E}}}{\log \tilde{\kappa}_f - \log \tilde{\kappa}_I} = \text{const} = 1.623. \tag{8.7}$$

Here $\tilde{\kappa}_I$ is the initial wave number for an initial jet radius a_I and $\tilde{\kappa}_f$ is the wave number at final radius a_f. An applied electrical field tends to stabilize a liquid jet,[205] so $a_f < a_I$.

Using eqn. (8.7), Peskin and Raco find that the most probable droplet formed from a capillary jet can be calculated from

$$R_i = R_{f,\mathscr{E}} = 2 \quad (\tilde{\kappa}_0^{0.623}/\tilde{\kappa}_{\mathscr{E}}^{1.623}) \tag{8.8}$$

Here $\tilde{\kappa}_0$ is the wave number of a disturbance of maximum instability in the absence of electric field while $\tilde{\kappa}_{\mathscr{E}}$ is the wave number of the disturbance of maximum instability with electric field. From eqn. (8.8) the ratio of droplet size with and without electric field can be estimated from

$$\frac{R_{i0}}{R_{i\mathscr{E}}} = \left[\tilde{\kappa}_{\varepsilon}/\tilde{\kappa}_0\right]^{1.623} \tag{8.9}$$

Experiments reported by Peskin and Raco[204] for the breakup of an amyl alcohol jet verify this relation for longitudinal electric fields greater than 1800 v cm^{-1}.

Sprays by Centrifugation

When a liquid is allowed to flow out over a spinning surface, it will flow to the edge of the surface where it will be flung outward and break up to form droplets. Atomization of liquids by this means has been investigated rather thoroughly over the past 20 years. Some of the first experiments were carried out on flat spinning discs by Walton and Prewett,[206] and on spinning cups by Hinze and Milborn,[424] Boshoff,[425] and Ryley.[426] Later other geometries were tested including vaned and grooved discs.[427] Many of the earlier results for the spray formation performance was reviewed by Friedman *et al.*[428]

The studies of Hinze and Milborn[424] indicated that sprays could be formed from liquid flowing away from the edge of a spinning disc via three mechanisms: (a) by direct drop formation at the edge of the disc, (b) by the formation and breakup of liquid ligaments stretched away from the edge of the disc, and (c) by the formation and disintegration of a liquid sheet spreading radially from the edge of the disc. The latter two situations could be prompted by increasing the liquid flow at the axis of the disc, increasing the angular speed but decreasing the disc radius, or increasing the liquid density, its viscosity, or the surface tension. Experiments generally showed that spray homogeneity could be maintained best by minimizing the liquid flow to aid in direct droplet formation. The sequence of photographs in Plate 10 illustrates qualitatively the three mechanisms for spray formation described by Hinze and Milborn.

Mechanisms of droplet formation by centrifugation have been elucidated further by Emslie *et al.*,[430] and by Fraser *et al.*[431] In the former work, relations were derived to describe the flow of a viscous liquid over the surface of a rotating disc, and conditions for uniform film production were deduced. The studies of Fraser *et al.*[431] involved the exploration of flow characteristics of liquid sheets emanating from spinning cups. They have derived expressions for conditions of sheet formation and for the variation of sheet thickness near the cup to the zone of disintegration. These results were verified by experimental measurements.

As a result of their experiments, Fraser *et al.*[431] found that two principal mechanisms of disintegrations of spinning liquid sheets could be distinguished. When sheets are formed at low peripheral speeds and liquid flow rates, a relatively undisturbed sheet is observed which extends from the cup lip until an equilibrium radius is achieved where the contracting action of surface tension at the free edge equals the centrifugal force. As liquid collected at this outer circular periphery a thick rim develops which is unstable and breaks down into long filaments or threads. These elongated masses of fluid tend to break up into droplets as they spin away from the liquid sheet. In Plates 10A and 10B are shown two stages of droplet generation from the spinning cup. At low liquid flow rates, and low spinning rates, the droplets disengage themselves from protrusions very close to the peripheral layer, forming a nearly monodisperse spray of primary droplets essentially at the cup rim. At somewhat higher spinning rates, indicated in Plate 10B, the liquid filaments develop at the rim layer. The elongated threads then are flung outward away from the cup, eventually breaking up into droplets.

The second mechanism occurs at higher peripheral speeds and liquid flow rates. Under such conditions the shearing of the stagnant gas in the surroundings tends to induce growing waves on the radially propagating sheet of liquid. These disturbances extend from the cup lip and are normal to the tangentially flowing liquid. Sheet disintegration proceeds as waves that are shed in half wavelengths, and these in turn break up into droplets. The wavy disturbances may be sinuous or dilational, but the former appears to be most unstable, as observed by Hagerty and Shea.[421] The break-up of the unstable fluid sheet extending from the spinning cup is shown in Plate 10C. The amplifying wavelike instabilities can be distinguished just inside the zone of disintegration.

Experiments on the behavior of fluids breaking up during centrifugation have been analyzed for simple relations to predict the most probable size of droplets generated. If a liquid is fed at the center of a spinning disc, a spray of nearly uniform droplet size can be produced by centrifugation off the edge of the disc. The size of droplets depends on speed of rotation. For a given disc geometry, Walton and Prewett[206] found that

$$\omega R_i(\tilde{D}\rho_L/\sigma)^{\frac{1}{2}} = \text{const.}, \tag{8.10}$$

where $\tilde{D}$ is the disc diameter, ω is the angular speed of the disc, and R_i is the principal droplet radius. For droplet production in the aerosol range, formed by high speed rotation, experiments have shown that the constant in (8.10) is about 2.2, but for larger droplets ~ 1 mm in radius, the constant is ~ 1.6.

Walton and Prewett found from experiments that in examining the main droplet size produced, the edge profile of the disc was of minor importance over the size range studied. And the principal drop size was independent of liquid viscosity from 0.01 to 15 poise. If the liquid feed rate was too high, the spray became heterogeneous, and a range of particles consisting of the principal size and a variety of satellite droplets merged to form a continuous spectrum of droplets.

Lane and Green[198] have described the formation of droplets by a spinning disc based on observations by Straus,[207] for example. When a nearly monodisperse spray is produced, a liquid torus forms at the outer edge. By the action of surface tension this bulge of fluid gathers into uniformly distributed beadlike protuberances which stream from the edge as indicated in Plate 11. As each protuberance extends outward, a droplet forms which is connected by a filament to the disc. The filament necks down until it is detached from the droplet. The filament then shrinks up into an irregularly shaped droplet which breaks apart into two satellites. Straus[207] has indicated that the process is quite similar to the formation of a droplet on a stationary tip except that, in this case, beads form on the filament which separate and gather into two or more satellite particles.

The droplet-tip analogy may be applied assuming the droplet formation is slow, and that the disc thickness exceeds the equivalent tip diameter necessary to R_i to remain constant for a given value of surface tension. In this limit one finds for droplets falling from a capillary tip:

$$R_i(2g\rho_L/\sigma) = 2.0. \tag{8.11}$$

If g is replaced by $\omega^2\ \tilde{D}/2$, then [8.11] applies to the spinning disc, as given in eqn. (8.10).

Earlier, Walton and Prewett appreciated the implications of the droplet-tip mechanism, but they preferred to interject the droplet formation as a consequence of Rayleigh instability on the surface of the filaments forming during the spinning motion. Application of Rayleigh's theory for the principal modes of instability, yields an expression equivalent to (8.11) except that the constant on the right-hand side is 2.3 instead of 2.0.

The experiments of Straus qualitatively confirmed the prediction of his theoretical model. And his results agreed substantially with Walton and Prewett, except for the fact that he found the edge geometry had a significant influence on the drop size.

Recently, further measurements of the constant in eqn. (8.10) have been reported by Dunskii and Nikitin[208] for two different oils over a wide range of operating parameters. Although no details on the edge geometry of their disc were given, Dunskii and Nikitin found that the atomization constant varied from 1.87 to 2.14 with an arithmetic mean of 2.0 over a range of rotation rates from 4000 to 16,000 rpm. These data apply to oil flow rates from 0.03 to 0.10 $cm^3\ sec^{-1}$, with transformer oil, $\mu_L = 1.95$ poise, and $\sigma = 33.2\ g\ sec^{-2}$, and with diesel fuel, $\mu_L = 0.231$ poise, $\sigma = 30.6\ g\ sec^{-2}$. These results largely confirm the earlier work of Walton and Prewett, and Straus. Furthermore, these data support droplet capillary tip mechanism for controlling the generation of the main droplets by spinning discs.

Under certain operating conditions secondary droplets can be produced in abundance from a spinning disc atomizer. Dunskii and Nikitin[208] have given some interesting suggestions about the mechanism generating these satellite droplets. To follow their arguments, let us apply dimensional reasoning to the performance of the spinning disc atomizer. Suppose it is assumed that droplet radius R_i is governed by the particle velocity q_i (related to the rotational speed ω) the disc diameter $\tilde{D}$, the surface tension σ, the gas and liquid densities ρ_G and ρ_L, and the viscosities μ_L and μ_G. With eight variables and three fundamental units, five independent dimensionless parameters should describe the behavior of these atomizers. Dunskii and Nikitin used the following five parameters:†

Weber number,	$We = \rho_G q_i^2 R_i/\sigma$
Langmuir number,	$La = \mu_L q_i/\sigma$
Density ratio	$= \rho_L/\rho_G$
Viscosity ratio	$= \mu_L/\mu_G$
Time ratio	$= \tau_H/\tau_S$

Here τ_H is the characteristic time of motion of a droplet with high relative velocity, and τ_S characteristic time for disintegration of a droplet.

The experiments of Dunskii and Nikitin[208] indicate that the Weber number is not the controlling parameter for formation of secondary droplets. Instead they determined that the Langmuir number—the ratio of the viscous force to surface tension—is important in this process. Physically one interprets the mechanism of disintegration of droplets as follows: As liquid runs off in filaments from the edge of a spinning disc, larger droplets form with velocities much larger than the surrounding air. The droplets decelerate rapidly by the action of friction, and the dynamic pressure acting on the droplet falls even more quickly. Any deformation of a droplet begun at high pressure will be exhibited as the dynamic pressure drops. If the duration of action of high dynamic pressure on the droplet is less than the time required for disintegration, breakup will not take place. As the deforming forces acting on the droplets decrease, the droplet will tend to take a more or less spherical shape by the restoring force of surface tension. Thus the breakup of primary drops will depend on the ratio of characteristic times τ_H/τ_S.

Using the calculation for the ballistic trajectory for a large droplet, based on equations of the form of eqn, (3.29) but with the viscous drag given an empirical relation for $Re_i > 1$, Dunskii and Nikitin estimate:

$$\tau_H = \frac{2R_i^2 \rho_L}{9\mu_G} \ln \frac{u_0}{u} \left(\frac{1 + 0.771\,(R_i u/\mu_G)^{\frac{2}{3}}}{1 + 0.771\,(R_i u_0/\mu_G)^{\frac{2}{3}}} \right)^{\frac{3}{2}},$$

where u is the horizontal component of the relative velocity q_i, and u_0 is the value of u at the disc edge. For $u_0/u = 2$,

$$T \equiv \tau_H/\tau_S \approx \frac{3.54\,\rho_L/\rho_G}{La\left[1 + \left(\dfrac{La}{We\rho_L\mu_L/\rho_G\mu_G}\right)\right]}. \tag{8.12}$$

Applying Levich's[54] estimate of $\tau_S(\approx 2\mu_L R_i/\sigma)$, Dunskii and Nikitin have evaluated several sets of data for secondary droplet formation in spinning disc atomizers. Their calculations of the dimensionless parameters in the results of several investigators are listed in Table 8.2. They found that in the experiments where subdivision and formation of secondary droplets was observed, the time ratio τ_H/τ_S was very large, hundreds of times greater than cases where no subdivision was found. Clearly, then, the Weber number alone does not control the secondary droplet production, but from eqn. (8.12) such droplet formation depends on both *La* and *We*, as well as the density and viscosity ratios.

† Dunskii and Nikitin also included the Reynolds number for the liquid, but this should not be independent of the other groups listed above.

TABLE 8.2. CRITICAL VALUES OF CHARACTERISTIC PARAMETERS FOR SPINNING DISC ATOMIZATION (AFTER DUNSKII AND NIKITIN[208])

Experiments	We	La	ρ_L/ρ_G	μ_L/μ_G	τ_H/τ_S	Secondary breakup
Volynskii[209]	5.35–7.00	0.185–0.206	670–11,300	0.0076–0.0640	16,400–470,000	Yes
Lane[227]	5.10	0.151–0.480	834	0.0640	8000–19,000	Yes
Bukhman[210]	1·70–5.20	0.115–0.155	834	0.0640	22,000	Yes
Merrington and Richardson[211]	7.50	0.137	834	0.0640	very large*	Yes
Merrington and Richardson[211] (glycerin droplets)	7.50	—	1050	69	very large*	Yes
Basanaev *et al.*[212]	0.05	0.010	1·0	0.85	very large*	Yes
Walton and Prewett[206]	5.20	104	872	1.58	11.5	No
Dunskii and Nikitin[208]	13.0	82.0	743	1.39	16.4	No

* For tests (211) and (212) the quantity τ_H/τ_S cannot be determined from eqn. [8.12], which is only applicable for droplets moving with a large initial velocity relative to the medium.

Droplet Production from Swirl Chambers

Another method for generating sprays using centrifugal forces involves swirl nozzles. In practical devices a hollow chamber is used where a liquid is introduced, tangentially under pressure at the base of the cone, and is spun inside the chamber before issuing through a hole at the apex of the cone. If the liquid pressure is high enough and if the liquid is not too viscous, a vortex with a hollow core of gas is established. The liquid streams out of the orifice as an unbroken tulip-shaped film at low pressures, while at higher pressures the liquid disintegrates into droplets close to the orifice. In the tulip regime droplets break away eventually from the ragged downstream edge of the film. Lane and Green[198] have noted that breakup of the liquid issuing from such nozzles is not the result of turbulent shearing flow in the liquid jet. Evidently the formation of droplets here is linked closely with the thinning of the effluent sheet under the influence of surface tension, viscous forces, and the interaction with surrounding gas.

The eventual production and dispersal of droplets depends on the thickness of the sheet of liquid which in turn is controlled by the flow in the conical chamber. It has been found that theories for the fluid dynamics in a swirl chamber cannot neglect viscosity. However, an approximate analysis based on boundary layer theory, gives results which are at least qualitatively corresponding to observations. Taylor[213] has proposed such a model. The geometry involved in his analysis is shown in Fig. 8.3.

According to Taylor's picture, liquid enters the cone of radius R_3 tangentially and exits through a circular orifice of radius R_2 after passage through the converging cone whose vertex angle is 2α. When the liquid swirls, the bulk of the fluid moves with tangential velocity around the axis ω/r, where r is the distance from the axis of the cone and ω is the rotational speed. At the surface of the cone, a boundary layer will develop in the liquid as it is slowed down by the action of viscosity. In this layer, insufficient centrifugal acceleration persists to hold the liquid in a circular path against the inwardly directed pressure gradient, which results throughout the fluid from a balance of the spinning motion outside the boundary layer. The liquid that flows out of the orifice then streams from this boundary layer as well as from the liquid layer surrounding the air core. Taylor estimated the thickness of the boundary layer in the conical nozzle by a Pohlhausen approximation (e.g.,

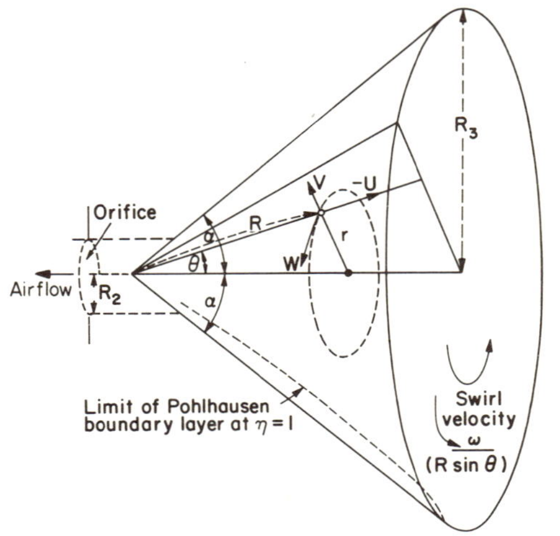

FIG. 8.3. Schematic diagram of an atomizer consisting of a swirl chamber. (After Taylor.[213])

Schlichting[67]). With this analysis the velocity of the liquid toward the vertex was estimated, neglecting the axial component of velocity outside the viscous layer. From his calculations Taylor found that the thickness of the boundary layer δ' was

$$\delta' \approx R_3 \frac{\Omega_1}{\sin \alpha} \left[\frac{\nu_L \sin \alpha}{\omega} \right]^{\frac{1}{2}}, \tag{8.13}$$†

where Ω_1 is a dimensionless constant ranging from 1.14 to 2.66, and ν_L = liquid kinematic viscosity. For this model Green considered the ratio $R_2/R_3 \approx 0.1$, or $\Omega_1 R_3 \approx 20\ R_2$ in practice.

By neglecting the contribution of the cone flow, this relation may be used to obtain an indication of the average drop size produced by the nozzle, R_{noz}, assuming that $R_{noz} \approx \delta'$, the thickness of the film leaving the orifice. Making use of eqn. (8.13) in a different form, Lane and Green[198] have indicated that this relation agrees fairly well with experimental values of (Sauter) mean droplet radius from 125 to 25 μ as determined by Watson.[214]

Atomization by Acoustic Waves

Experiments have indicated that a mist of liquid droplets can be generated by an intense beam of high frequency sound waves. The beam is focused on the liquid surface with a concave reflector or some other type of ultrasonic radiator. When the intensity of the ultrasonic beam is large enough in the focal region, a liquid spout called an ultrasonic fountain rises into the gases overhead, and a dense fog evolves from the base of the spout. Such fogs contain particles which are variable in size down to a few microns radius. McCubbin,[215] for example, has measured the droplet spectrum in water fogs generated by acoustic waves. He found that the droplet radii were grouped mainly around 2–3 μ.

Fog formation by acoustic vibration has been attributed to cavitation, the formation and collapse of cavities induced in the liquid by the intense waves. However, Green and Lane[4] have remarked that fog formation acoustically may proceed by a different mechanism for thin layers of liquid spreading over a violently oscillating surface. If the surface of such a

† For further discussion, see also Green.[216]

layer resting over a vibrating transducer is examined closely, one observes that the surface is covered by tiny ripples. Because of variations in the liquid film and the boundary reflections, the ripple pattern generally may be quite complex. If the amplitude of these capillary ripples becomes sufficiently large, their crests may "break" and shed very tiny droplets. The size of droplets should then be related to the ripple wavelength or the vibration frequency. Some experiments of Bisa, Dirough, and Esche(217) provide some evidence for such a mechanism of atomization. Working at frequencies of 1.2–5.4 mHz, these workers showed that uniform patterns of crossed capillary waves developed on a liquid surface when atomization occurred, and that the median radius of the droplets generated was directly proportional to the capillary wavelength. The latter could be estimated from Kelvin's capillary equation(44) using the existing frequency and properties of the atomized liquid as calculated by Lang.(218)

Peskin and Raco(219) have pointed out that Kelvin's theory assumes no coupling between the surface velocities in the liquid and the forcing function driving the surface. As a result, properties of the atomization system, like the liquid film thickness and the forcing amplitude, do not enter into the expression for the capillary wave number. Peskin and Raco have taken into account such coupling in an analysis using linearized stability equations.

Suppose a film of liquid is considered which lies over a solid surface oscillating at some known frequency, ω_0. Schematically the atomizer is shown in Fig. 8.4. The liquid film has

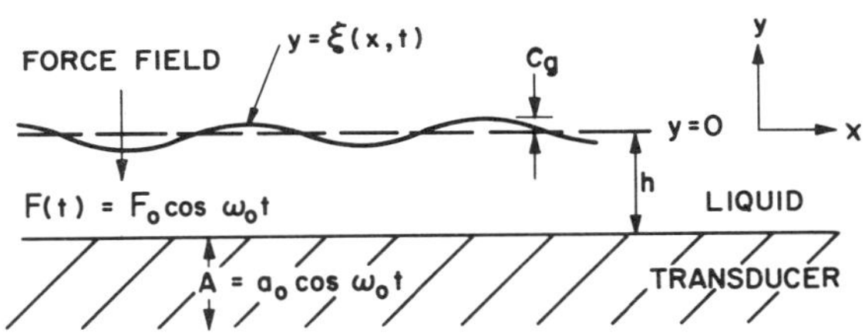

FIG. 8.4. Schematic diagram of ultrasonic atomization.

an average height h, and its displacement from the average value is $\xi(x, t)$. The effect of the oscillation of the liquid film by the solid surface, a driving transducer, is represented by a time dependent inertial force acting in the liquid. This force per unit mass of liquid is $F'(t) = F_0 \cos \omega_0 t$. If the liquid motion is assumed irrotational, viscosity of the liquid may be neglected, and the equations governing the liquid behavior readily may be found (e.g. Lamb,(44) Peskin and Raco.(219)).

The resulting relation for the growth in amplitude of surface displacement $c_g(t)$,† is given by the second order differential equation

$$\frac{d^2 c_g}{dt^2} + \left[\frac{\sigma \tilde{\kappa}^3}{\rho_L \omega_0} \tanh(\tilde{\kappa} h) - \frac{F_0 \tilde{\kappa}}{\omega_0} \tanh(\tilde{\kappa} h) \cos(\omega_0 t)\right] c_g = 0, \tag{8.14}$$

where $\tilde{\kappa}$ is the wave number of the surface wave. Equation (8.14) is Mathieu's equation. Periodic, stable solutions for this relation have been studied extensively, but unstable solutions are less well known.(219)

Peskin and Raco have made the customary assumptions that the most probable drop radius formed is

$$R_{ac} = \frac{2\pi}{\tilde{\kappa}},$$

† The surface displacement is proportional to c_g, as in the function, $\xi = c_g(t) \sinh(\tilde{\kappa} h) \exp(i\tilde{\kappa} x)$.

and have examined in this light unstable solutions to (8.14) of the form

$$c_g(\omega_0 t) = \exp(\tilde{\mu}\omega_0 t)\tilde{p}(\omega_0 t), \tag{8.15}$$

Here $\tilde{\mu}(X, Y)$ and $\tilde{p}$ is a periodic function, $X = (\sigma\tilde{\kappa}^3/\omega_0^2\ \rho_L \tanh(\tilde{\kappa}h)$, and $Y = (F_0\tilde{\kappa}/\omega_0^2) \tanh(\tilde{\kappa}h)$. The growth rate spectrum will be known once (8.14) is solved in the unstable region of the X, Y plane. The droplet size should be related to the growth rate spectrum of the unstable wave numbers.

For interfaces of moderately small surface tension, Peskin and Raco[219] noted that it is sufficient to limit discussion to the first unstable region in the X, Y plane (closest to the X, Y origin). Their numerical solutions of eqn. (8.14) with eqn. (8.15) have yielded curves for maximum instability, which they have expressed as loci of

$$Y = 2.5 - 2X \tag{8.16}$$

within the restriction $0 \leqslant X \leqslant 0.25$. Taking the amplitude of the transducer $a_0 = F_0/\omega_0^2$, and noting that $\tilde{\kappa} \approx \pi/2R_{ac}$, eqn. (8.16) can be rewritten

$$\frac{a_0\pi}{2R_{ac}}\tanh\left(\frac{h\pi}{2R_{ac}}\right) = 2.5 - \frac{\sigma\pi^3}{4\omega_0^2\rho_L R_{ac}^3}\tanh\left(\frac{h\pi}{2R_{ac}}\right). \tag{8.16a}$$

Peskin and Raco also examined the possible influence of viscosity in their model and found that $|\tilde{\mu}| \neq 0$ is not a sufficient criterion for instability and atomization. With viscosity one requires that

$$|\tilde{\mu} - \mathscr{D}\omega_0| \neq 0, \tag{8.17}$$

where $\mathscr{D} = 2\mu_L\tilde{\kappa}^2/\rho_L$.

Figure 8.5 represents essentially performance curves for ultrasonic atomizers where the most probable drop size is given in terms of transducer signal amplitude and frequency,

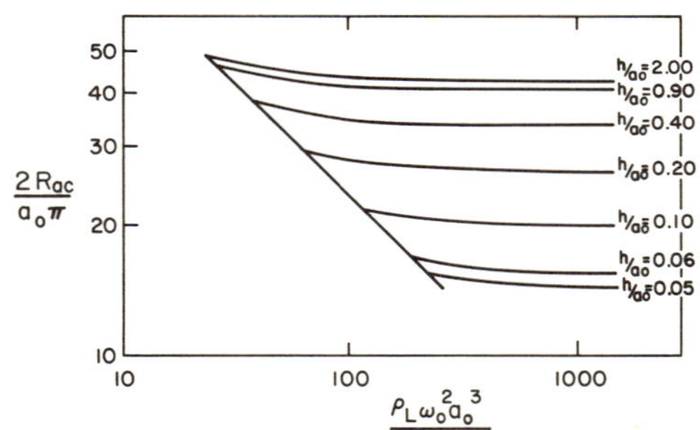

FIG. 8.5. Idealized performance curves for acoustic atomizers. (From Peskin and Raco.[219])

liquid film thickness, surface tension, and fluid density. The bounding curve at the low frequency extreme of each curve of constant h/a_0 denotes the limiting value of X for the linear approximation to the loci of maxima in $\tilde{\mu}$, given in terms of eqn. (8.16a). The equation of this curve is[219]

$$\frac{2R_{ac}}{\pi a_0} = \left[\left(\frac{2\sigma}{\rho_L\omega_0^2 a_0^3}\right) 2\tanh\left(\frac{\pi h}{2R_{ac}}\right)\right]^{\frac{1}{3}} \tag{8.18}$$

For large film thickness, this expression reduces to

$$R_{ac} \approx \tfrac{1}{2}(4\pi^3\sigma/\rho_L\omega_0^2)^{\frac{1}{3}}. \tag{8.19}$$

Equation (8.19) is a correlation that has been reported by several investigators based on experimental results [e.g. refs (218), (220), (221), and (217)]. Equation (8.19) is essentially the result obtained by applying Kelvin's capillary wave theory, and it represents a limiting regime in Peskin and Raco's theory neglecting h and a_0. Unlike the Kelvin theory, however, the results of these latter investigators predict production of a nearly constant drop size at higher frequencies for a fixed value of the ratio of the film thickness to transducer amplitude. An asymptotic upper limit of drop size for larger film thickness is also predicted, given as $R_{ac} \approx 42\, a_0 \pi$ for $\rho_L \omega_0^2 a_0^3/2 > 1000$.

The apparent agreement between Kelvin's theory and various experiments evidently is the result of the use of various film thicknesses and amplitudes in a given drop size-frequency experiment. To resolve the applicability of Peskin and Raco's results, a critical test is needed where experiments are conducted in which the ratio h/a_0 is held constant while the transducer frequency is varied.

Ultrasonic atomization offers advantages over other more conventional methods in that an aerosol of high concentration and reasonably narrow size range is produced which can be controlled easily by varying the frequency of the acoustic waves. In pneumatic atomization discussed below, particle size may be decreased only by diminishing concentration because the air flow has to be increased. In ultrasonic atomization the aerosol concentration is limited only by the ultrasonic power input or the air flow past the vibrating liquid surface. Thus with unrestricted power the limiting concentration of liquid that may be suspended depends ideally only on the fallout rate.

C. Mechanisms of Aerodynamic Atomization

Droplet Breakup in Steady Gas Flow

Aerodynamic or airblast atomization involves the shattering of a coarse liquid stream by interaction with a high speed gas stream. This process is more complex than the ones described above. Consequently, the detailed mechanisms for droplet generation from bulk liquid, in this case, are important in the overall disintegration mechanism. In the first stage small disturbances are initiated on the surface of the liquid in the form of bulges or troughs. The action of aerodynamic forces† on these protuberances continues to deform them until they are drawn out into fine ligaments from the main stream. Like the cases described previously, the ligaments collapse into droplets under the influence of surface tension. As the gas velocity increases the radii of the ligaments formed decrease, their lifetimes become shorter and smaller droplets result. This decrease in size reaches a limit when the ligaments tear apart almost as soon as they form.

Castleman[221] has used Rayleigh's theory[199] to predict the rate of collapse of a ligament which is subjected to a disturbance of the form

$$\eta_d = \eta_{d0} \exp(\mathscr{J} t), \tag{8.20}$$

where η_{d0} is the initial amplitude at time t_0. The rate of shattering is greatest when the exponential term

$$\mathscr{J} = \frac{\sigma^{\frac{1}{2}}}{\rho_L a^3} \mathscr{F}(l/a) \tag{8.21}$$

† For initial instability mechanisms to form bulges or waves that may apply here, see for example, Kelvin[222] or Miles[223, 224]; see also Goren.[22·

is maximum. Here a is again the radius of the unstable filament, and $\mathcal{F}$ is a function of the ratio the length of the filament l to its radius. The maximum value of $\mathcal{F}$ is 0.343 at $l/a = 4.5$.

Using eqns. (8.20) and (8.21), it is possible to estimate the time scale for droplet formation during atomization. If we assume that the filament volume equals the droplet volume, $a \approx 0.53R_i$. Thus for 10 μ radius droplets, $a = 5.3$ μ, or $\mathcal{J} \approx 2.4 \times 10^5$ sec^{-1} for water at 20°C. From eqn. (8.20) the collapse time of a filament of this radius has to be about 10^{-5} sec because η_d cannot exceed about 26 μ, and $\eta_{d0} \geqslant 10^{-7}$ cm. Thus as Green and Lane[4] have noted, spark photographs should reveal the breakup of liquid jets in an air stream of relatively low speed (60 msec^{-1}), while such photographs will not show stretching at higher air velocities, and the small droplets will appear to be bursting directly from the bulk fluid.

Castleman's mechanism provides one means of producing atomization by air blast. However, it is known that droplets of smaller sizes than those produced by ligament collapse are found in sprays as indicated in Fig. 8.6. Droplets much less than ~10 μ radius are likely

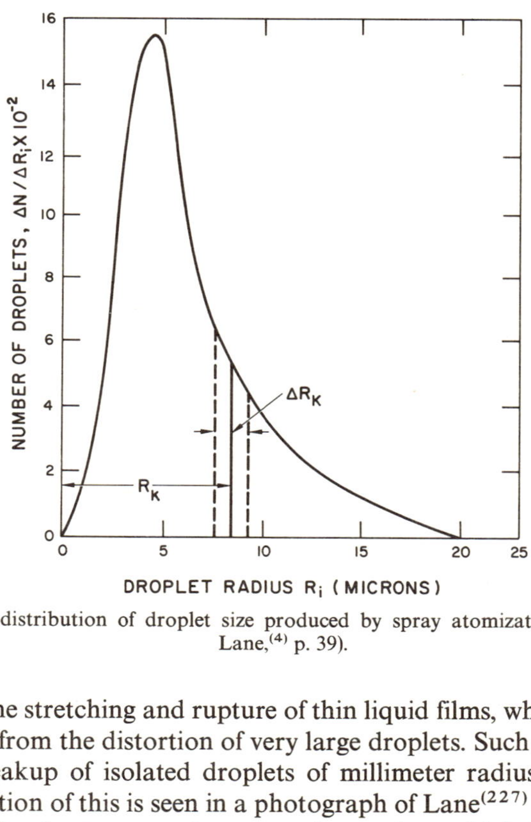

FIG. 8.6. A typical distribution of droplet size produced by spray atomization. (From Green and Lane,[4] p. 39).

to be produced by the stretching and rupture of thin liquid films, which may originate near the liquid nozzle or from the distortion of very large droplets. Such a mechanism has been observed in the breakup of isolated droplets of millimeter radius in an air stream. An excellent demonstration of this is seen in a photograph of Lane[227] which is reproduced in Plate 12. Here the droplet is exposed to a steady gas flow which is just rapid enough to distort its shape. Under the action of the aerodynamic pressure the droplet flattens into a saucer-shaped film and then blows out into a bubble. The bubble in time bursts into a shower of fine droplets leaving temporarily a rim of liquid which disintegrates into a few droplets much larger than formed in the bubble burst.

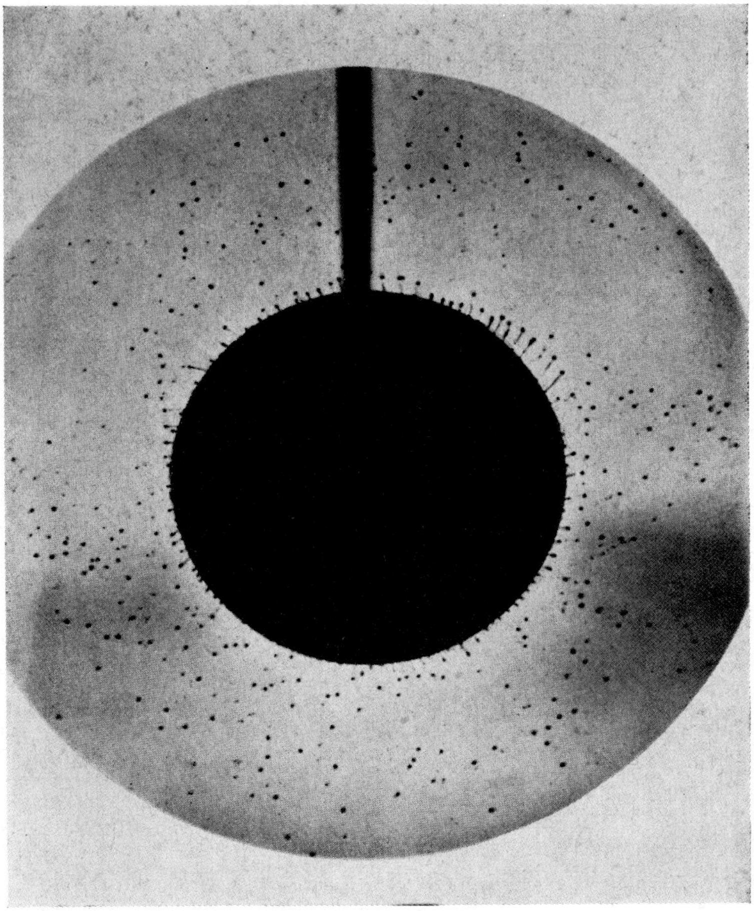

PLATE 11. A flash photograph of liquid filaments and droplets leaving the edge of a spinning disc. (Photo courtesy of H. Green and W. Lane; British Crown copyright, reproduced by permission of the Controller of Her Britannic Majesty's Stationery Office.)

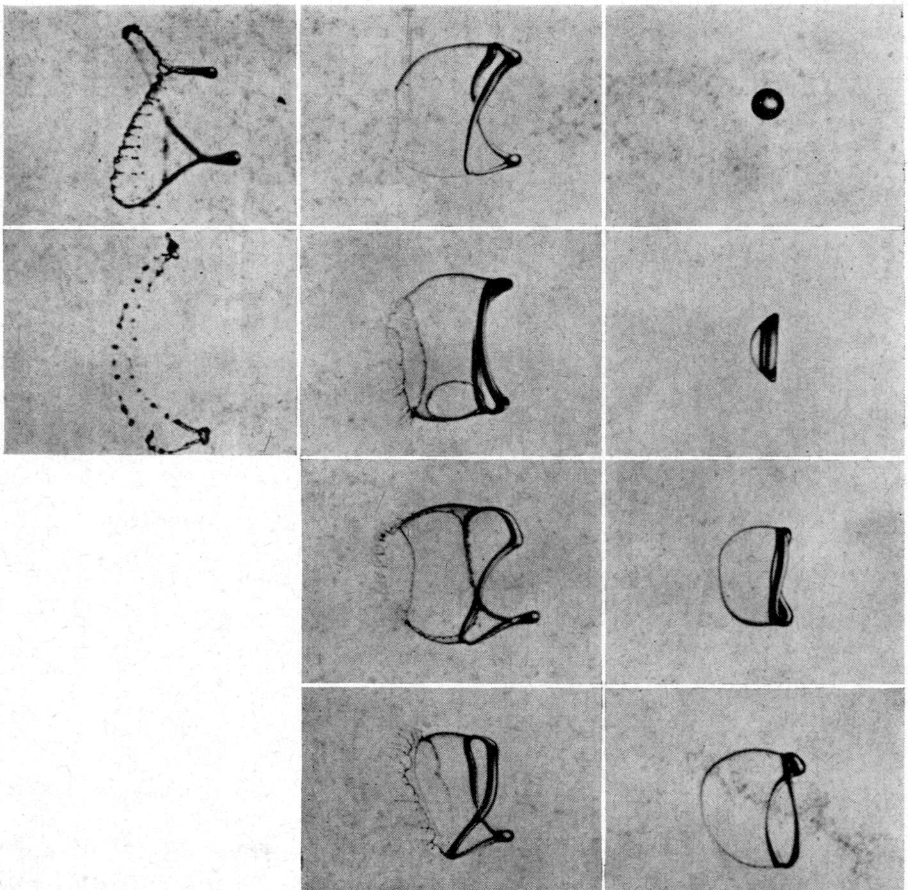

PLATE 12. Disintegration of a water drop in a steady gas flow. A 1.3 mm radius drop falling through a steady upward airflow. At a critical air velocity of 22.5 cm sec^{-1} the drop flattens out and blows into a hollow shell attached to a nearly circular rim. The shell then bursts, producing a shower of droplets, and the rim breaks up later to form layer droplets. (Photo courtesy of H. Green and W. Lane; British Crown copyright, reproduced by permission of the Controller of Her Britannic Majesty's Stationery Office.)

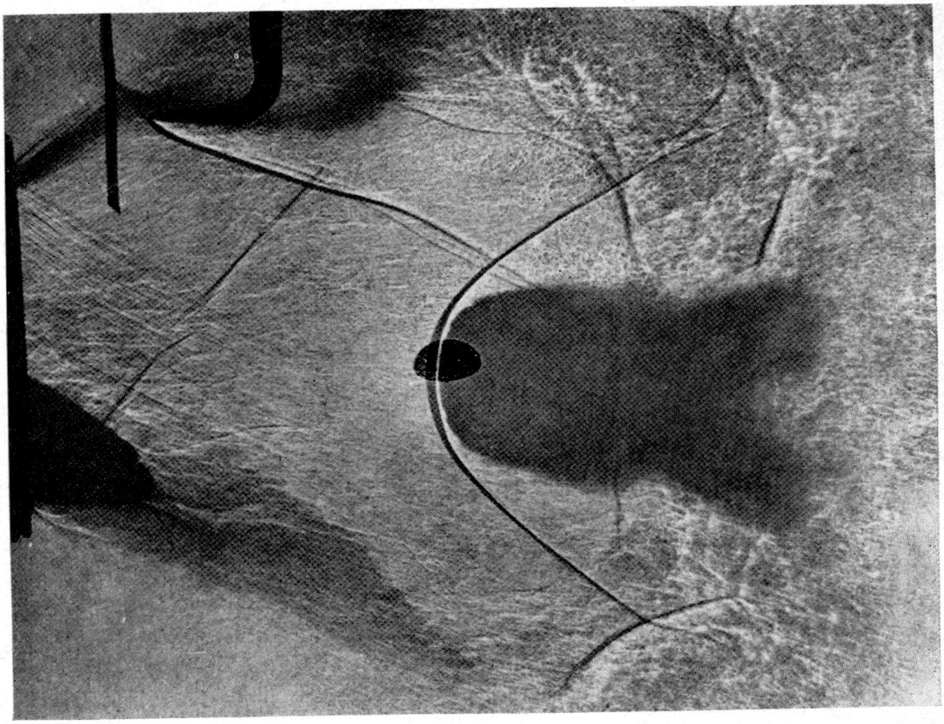

PLATE 13. Break-up of a water drop in a transient air blast. The double exposure shadowgraph shows in the same picture (a) the drop first intact balanced in an upward air flow, then (b) the drop shattered in a blast of air from the orifice of diameter 3.0 cm (left side). The drop deforms into a saucer shape, then atomization proceeds by flow of gas around it, forming a spray of droplets. (Photos courtesy of H. Green and W. Lane; British Crown copyright, reproduced by permission of the Controller of Her Britannic Majesty's Stationery Office.)

The experiments of Lane[227] showed that the critical velocity $(q_G - q_i)$ required to break droplets apart could be estimated from balancing the aerodynamic pressure at the stagnation point of the droplet to the pressure resulting from surface tension

$$\tfrac{1}{2}K_D\rho_G(q_G - q_i)^2 = \frac{2\sigma}{R_i}. \tag{8.22}$$

Here q_G is the critical velocity of the gas stream, q_i is the speed of the liquid droplet at the moment of bursting, and K_D is a constant determined experimentally, based on the drag coefficient of a distorted droplet. For water droplets,

$$(q_G - q_i)^2 R_i = 6.12 \times 10^3. \tag{8.22a}$$

A relation similar to eqn. (8.22) has been suggested for the critical velocity, for example, by Littaye[228] and later by Hinze,[229] who indicated that the Weber number

$$We = \rho_G(q_G - q_i)^2 R_i/\sigma$$

should reach a critical value for disintegrating droplets. Hinze estimated the critical Weber number We_{crit} to be about 6.5 for liquids of low viscosity. This compares favorably with 5.3 deduced from eqn. (8.22a) and $We_{crit} = 5.3$–7.0 from data of Volynskii.[209]

Lane and Green[198] also have pointed out that the critical Weber number depends somewhat on the variation of relative velocity with time. Higher values of $We_{crit} \approx 10$ have been observed for freely falling droplets. High viscosity will also tend to increase the critical *We*.

In designing air blast atomizers it was found many years ago that the polydispersivity of a mist from such devices could be reduced by adding a baffle [e.g. refs (230, 231)]. Installation of a baffle will remove larger droplets from the spray formed by ordinary air jets. Green and Lane[4] have noted that it is sometimes claimed that the fineness of aerosol produced by baffled atomizers results from secondary shatter by impingement of the primary spray on the walls of the baffle. However, it appears now that the baffle only serves as a barrier to collect coarser material, while the finer material travels on out of the device. It is noted, however, the fineness of spray from baffled atomizers may be controlled to a degree by varying the distance between the jet and the baffle surface.

An interesting combination of air blast atomization and centrifugal atomization has been reported recently by Fraser *et al.*[433] Using their spinning cup, combined with an air jet directed normally onto the free liquid sheet, breakup could be induced in a controlled way. They found that the sheet of spinning liquid originating from the cup does not disintegrate immediately on impact with the airstream. It is deflected away from the gas flow. Waves are initiated at the point of impact and the sheet breaks down into droplets through unstable ligaments. The resulting droplet distribution depends primarily on the thickness of the liquid sheet. Thus the thinner the sheet the finer the atomization for this device.

The method of disruption of liquids by air blast atomization also can be used to produce aerosols with size ranges considerably smaller than shown in Fig. 8.6. By atomizing solutions of soluble salts or mixtures of miscible liquids it is possible to generate clouds of submicron-sized particles after evaporation of larger droplets produced by the atomization process. Typical of such distributions is shown in Fig. 8.7. Here a 2 per cent solution of sodium chloride, when atomized, yields a mean radius of about 0.15 μ compared with droplets ten times as large before evaporation. The total number of particles generated by this method has been found to vary considerably with solution concentration, but ranges from $N \approx 10^{-4}$ to 10^5 cm^{-3}, as found typically by Preining, Sheesley, and Djordjeric.[232]

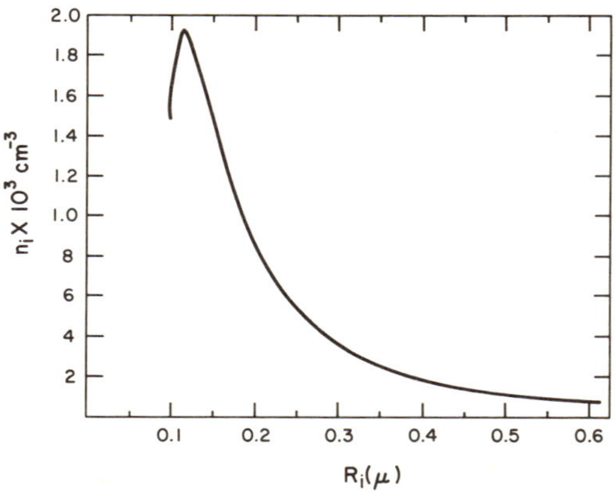

FIG. 8.7. Typical size distribution of solid salts produced from evaporation of spray droplet from an atomizer. (From Preining *et al.*[232])

Transient Air Blasts and Shattering

When subjected to transient pulses of high speed gas streams, drops appear to break up in a manner distinctly different from the steady flow case described above. Instead of the drop blowing out into a thin hollow bag suspended to a thicker rim, it is deformed in the opposite direction as a surface convex to a gas flow. According to Lane and Green[198] the edges of the convex saucer are drawn out into a thin sheet that will disintegrate into a spray of droplets if the gas speed is fast enough. A shadowgraph of droplet shattering is shown in Plate 13. This picture was taken by Lane and colleagues by spark photography. In the first part of the double exposure, the drop appears intact, balanced in an upward-directed stream. Superimposed on this image is the droplet being shattered into a mist of fine droplets as indicated by the darkened cloud to the right of the picture. A detached shock wave situated just upstream of the mist cloud shows that part of the droplet remains coherent enough to obstruct the air flow.

Taylor[233] has made some calculations for the shape which a drop might develop when accelerated in a gas stream before disintegration. These calculations indicate that the drop should be flattened into a plano-convex lenticular body of a radius about two times the radius of the original sphere. This prediction qualitatively represents the observations taken in the laboratory.

Measured gas velocities for disintegration by transient blasts are lower than in steady flows. The difference in critical velocity increases as the drops get smaller. Taylor[233] has suggested that this may be expected if the drop behaves as a vibrating system. If, for example, we make an analogy to a loaded spring, the force F_T applied suddenly is equal to a force F_s applied steadily to produce the same extension. For a maximum extension which would be analogous to the maximum distortion, a drop could allow, without breaking up, $F_T = \frac{1}{2}F_s$. Since the force tending to distort the drop is proportional to the square of the relative velocity,

$$\frac{U_T}{U_s} = \left(\frac{F_T}{F_s}\right)^{\frac{1}{2}} \approx 0.71. \tag{8.23}$$

In Table 8.3 several cases for breakup of water drops are listed as given by Lane and Green.[198] The ratio 0.71 is approached for drops of radius 1 mm or less, but is exceeded for larger drops having oscillation periods much larger than the duration of air blast (4 msec).

TABLE 8.3. CRITICAL VELOCITIES FOR SHATTER OF WATER DROPS IN TRANSIENT BLASTS AND STEADY STREAMS OF AIR (FROM LANE AND GREEN[198])

Diameter of drop (mm)	Transient (U_T) (m/sec)	Steady (U_s) (m/sec)	$U_s^2 d$	U_T/U_s	Period of oscillation of drop (sec)
4.0	12.0	12.5	625	0.96	0.023
3.0	12.1	14.1	622	0.83	0.015
2.0	12.6	17.5	612	0.72	0.008
1.0	16.0	24.7	610	0.65	0.003
0.5	24.0	35.0	612	0.69	0.001

Taylor's[233] theoretical calculations have demonstrated that the thickness of the liquid sheet shed by a gas flow of given velocity from a drop of a given density and viscosity may be estimated by applying boundary layer theory. At comparatively small gas speeds the predicted thickness of the sheet is in satisfactory agreement with estimates from photographs, but at high gas speeds the calculated value is larger than deduced from the droplet size observed in the mists produced. If the surface sheet of the drop is removed by tangential stresses, while at the same time the drop is accelerated by the pressure distribution in the gas, the life history of the drop may be deduced, in principle, by incorporating the rate of removal of liquid and the drag coefficient for the lens-shaped drop.

Taylor and Hodgkinson[198] found that the ratio of speed of the drop to that of the gas stream would be governed by:

$$\frac{q_i}{q_G} = 1 - \left[1 + 0.6C_D\left(\frac{\rho_G}{\rho_L}\right)^{\frac{2}{3}}\left(\frac{\nu_L}{\nu_G}\right)^{\frac{1}{6}}\left(\frac{q_G R_0}{\nu_L}\right)^{\frac{1}{2}}\frac{R_i}{R_0}\right], \tag{8.24}$$

where R_i is the radius of the lenticular drop at speed q_i and R_0 is the drop radius at rest. Lane and Green[198] have estimated this ratio for 0.5 cm drops of dibutyl phthalate disintegrating in an air blast of 3.4×10^4 cm sec^{-1}. In this case, $q_i/q_G = 0.9$ when R_i/R_0 is negligibly small. Thus, in general, the drops do not attain the air speed before disintegration.

Disruption of droplets by aerodynamic forces also may be linked to the formation of unstable waves forming on their surface by Kelvin's capillary wave[44] mechanism. If the droplet radii produced by the shearing flow of air over the droplet are the order of the wavelengths of the most unstable waves, a theory analogous to Peskin and Raco's[219] would apply. Some years ago Taylor (Lane and Green[198]) examined this question and found that the most probable drop size formed by such an instability mechanism would be

$$R_i \sim \frac{\pi\sigma}{\rho_G q_G^2},$$

where the proportionality factor is a function of ρ_L, ρ_G, and q_G^2. Taylor's estimate gives much smaller droplet radii than encountered in practice.[198] The discrepancy, however, may be linked to the fact that q_G is much smaller near the droplet surface than the free stream value normally taken for estimating R_i on theoretical grounds.

Simultaneously with mechanisms of breakup, some droplets will collide and recoalesce. The evidence of such agglomeration in atomizers, based on calculations of Dunskii,[234] for example, suggest that this may be an important consideration in the performance of these devices, but further studies would be needed to clarify this point.

Aerosol Sprays by Liquified Gas Propellants

Aerosol clouds generated by the transient action of liquified gas expanding under pressure through a nozzle are used in a variety of applications such as for dispersal of insecticides, disinfectants, and insect repellants. The effectiveness of the aerosol in accomplishing its intended purposes depends largely on the size distribution of droplets. Until recently, relatively little has been reported about the nature of the size distribution of sub-micron particles from pressurized aerosol cans. However, Liu[235] has reported some interesting data which may be used to characterize these atomization devices.

Liu's experiments were made on several samples of cans containing 2 per cent dioctyl phthalate and a liquid Freon propellant. The pressure containers were standard units produced by the Johnson Company for an insecticide spray called "Raid". In Table 8.4

TABLE 8.4. CHARACTERISTIC PARTICLE RADII IN CLOUDS PRODUCED FROM PRESSURIZED AEROSOL CANS WITH NOZZLE ORIFICE DIAMETERS ~ 0.3 mm. RADII ARE NUMBER MEDIAN VALUES (FROM LIU[235])

Sample	Container pressure (psig)	Minutes after spray	$R_i(\mu)$	Minutes after spray	$R_i(\mu)$
1	19	5	0.02	15	0.02
2	29	2	0.033	15	0.022
3	31	2	0.275	15	0.032
6	55.5	2	0.023	15	0.035

are the properties of the various samples. The measured sizes, given in median radii, are listed in this table. These spray devices yield aerosols of very small median size compared with other disintegration methods discussed above. There seems to be no systematic effect of the pressure in the container.

Aerosol clouds generated in exhaust from pressurized containers are relatively polydispersed. Remarkably enough, Liu[235] found that the size distributions of the aerosols from his sample cans could be correlated well in the form for "self-preserving" distributions discussed in Section 10.4B. Use is made of the dimensionless variable $\tilde{\psi}_R$ proportional to the number concentration in the range of radius $\tilde{R}$, $\tilde{R} + \mathrm{d}\tilde{R}$, and the dimensionless particle radius proportional to $\tilde{R}$. These variables are defined on p. 357. Such dimensionless size distributions for samples of aerosols generated from the pressurized containers are shown in Fig. 8.8. These distributions are quite stable over storage times up to 15 min, and they follow a form

$$\tilde{\psi}_R \sim \tilde{\eta}_R^{-4}$$

over a wide range of $\tilde{\eta}_R$.

As discussed in section 10.4B, this form of dimensionless size spectrum has been found for many cases of aged aerosols, especially those in the atmosphere. Although the details for a physical mechanism causing this type of behavior are still not fully understood, it appears that coagulation of the aerosol particles is certainly involved (see also Section 10.4B).

Size Distribution and Physical Variables

Because of the complicated mechanism for droplet formation by aerodynamic forces, one can expect a wide range of droplet sizes to be present in mists formed by aerodynamic

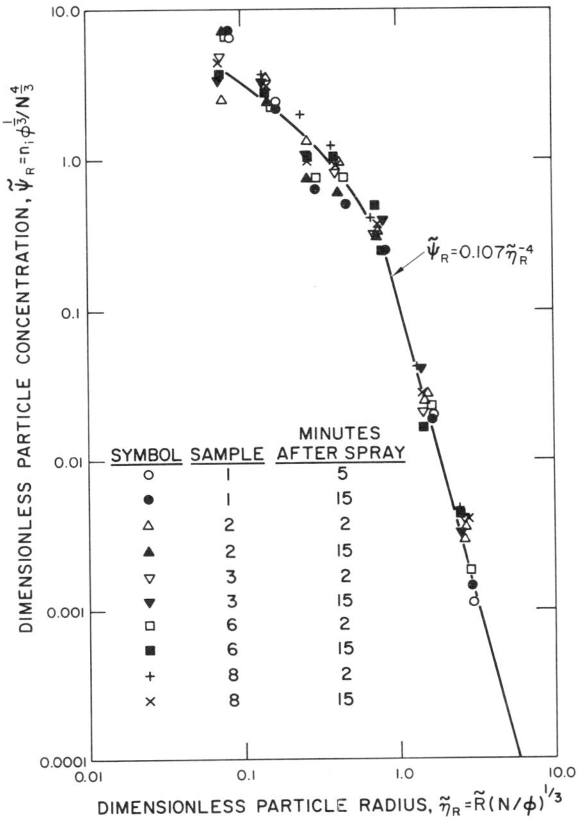

FIG. 8.8. Dimensionless size distributions of dioctyl phthalate aerosols produced from pressurized spray cans. (From Liu.[235])

forces. This is borne out by many experimental studies. Typically, droplet spectra produced by air blast atomization, such as the one in Figs. 8.6 or 8.7, can be placed in the form[4]

$$\frac{\Delta N(R_i)}{\Delta R_i} = aR_i^m \exp(-bR_i^n), \tag{8.25}$$

where ΔN represents the number of droplets in the size range R_i, $R_i + \Delta R_i$, and a, b, m, and n are constants for a particular atomizer. These, of course, have to be determined experimentally. Nukiyama and Tanasawa[236] have reported, for example, that $m = 2$ for conditions where the volume flow of gas per unit volume of liquid sprayed, $V'_G/V'_L > 5000$, and the gas velocity, $U_G > 180$ m sec^{-1}. Though the Japanese authors indicate n in eqn. (8.25) is unity under these conditions, Green and Lane[4] have recommended that this parameter be checked by experiment.

Nukiyama and Tanasawa[236] investigated the relationship between the droplet size distribution generated from air blast atomizers and the operating variables and physical properties of the fluids. Their findings, based on several hundred tests under different conditions, yielded an empirical equation for the average droplet radius in such mists:

$$[R_i] = \frac{292}{U_G - U_L}\left(\frac{\sigma}{\rho_L}\right)^{0.5} + 298\left(\frac{\mu_L}{\sqrt{(\sigma\rho_L)}}\right)^{0.45}\left(\frac{1000\,V'_L}{V'_G}\right)^{1.5}. \tag{8.26}$$

Here

$$[R_i] = \sum_{i=1}^{S} R_i^3 \sum_{i=1} R_i^2,$$

$[R_i]$ refers to the mean volume to surface radius in microns. U_L refers to the liquid velocity in m sec^{-1}.

The Nukiyama and Tanasawa relation is dimensionally inconsistent, but Green and Lane[4] have indicated that it is a very useful equation for the performance of air jet atomizers when the liquid properties are in the range $19 \leqslant \sigma \leqslant 73$ dyne cm^{-1}, $0.7 \leqslant \rho_L \leqslant 1.2$ g cm^{-3}, and $0.005 \leqslant \mu_L \leqslant 0.5$ poise, and where the air speed is subsonic. The equation implies that when the ratio V'_G/V'_L is very large, the mean droplet size produced is governed by the ratio $(\sigma/\rho_L)^{\frac{1}{2}}/(U_L)$, the viscosity being of little importance in this regime. However, when V'_G/V'_L is small, the mean droplet size will be governed by the second term on the right-hand side of eqn. (8.26), and both surface tension and viscosity of the liquid enter in. Figure 8.9 shows operating curves for a typical air blast atomizer based on eqn. (8.26). These curves are calculated for dibutyl phthalate ($\rho_L = 1.05$ g cm^{-3}, $\sigma = 34$ dyne cm^{-1}, and $\mu_L = 0.20$ poise).

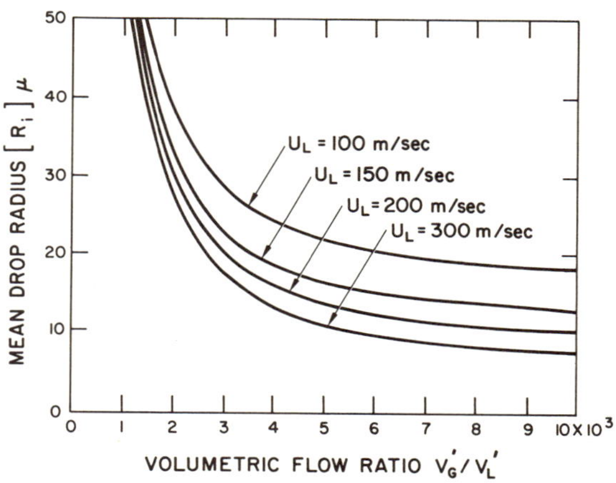

FIG. 8.9. The Nukiyama and Tanasawa equation for atomization of dibutyl phthalate. (From Green and Lane,[4] p. 41.)

Nukiyama and Tanasawa's experiments were all performed for conditions of subsonic gas velocity. More recently Bitron[237] undertook further work on air blast atomizers in which the gas velocity could exceed supersonic speed. Bitron's results indicate that eqn. (8.26) may be applied to a Mach No. of 2, implying that the mechanism for droplet formation in atomizers is essentially the same for subsonic and supersonic flows. It is of interest, however, that geometrical configurations may be important to maximizing the production of small droplets. Hrubecky,[238] for example, found that other conditions being the same, atomization was finest if the liquid was injected parallel to the gas flow in the zone of maximum gas speed. For conditions where $U_G > 140$ m sec^{-1}, and $V'_G/V'_L >$

10^4, Hrubecky found somewhat lower values of the mean volume/surface radius, $[R_i]$, than (8.26) predicts for water atomization.

In the case of the combined spinning cup–air blast atomizer of Fraser *et al.*,[433] their results indicate that pre-filming devices can produce smaller drop sizes than jet atomizers where the Nakiyama and Tanasawa relation applies. Thus, controlled production of thin liquid films can aid in finer atomization. Fraser *et al.* found that their device could produce droplets with a surface-volume mean radius of about 50 μ with a size distribution given by a Rosin–Rammler curve.[434]

8.2. Aerosols by Dispersal of Powders

A. Formation of Powders by Attrition or Impulsive Forces

Dust clouds may be produced by the injection into a gas of particulate material originating either from the disruption of bulk solids or from the dispersion of finely divided powders. Examples of the first group are the dust clouds that evolve when rocks are disintegrated by processes of grinding or blasting. Layers of natural dust clouds in the Earth's atmosphere rising over arid regions and artificial dust clouds of insecticides, for example, fall into the latter category.

Mechanical destruction of a solid substance may be carried out by applying shearing or tensile stresses to the material. This may be done slowly in milling processes, or by crushing material by means of rollers or ball grinders. More rapid disruption may be done by impact crushing, or by blasting with explosives. In both cases the applied forces disintegrate the solid by splitting or cracking along zones of weakness in the material, like cleavage planes in polycrystalline substances. Green,[239] for example, has distinguished between two kinds of mineral dusts represented by small fragments, and by finely divided material released from freshly formed surfaces by microscale cracking as material is broken down.

Apparently there is a limit to the extent to which solids may be fragmented by *comminution processes*. Green and Lane[4] have pointed out that if a material is ground indefinitely, no change in the particle size distribution will be achieved after a certain time no matter how much longer the grinding is prolonged. This means that even though particles of all sizes continue to be broken up, some particles, probably the smallest ones, reagglomerate with each other or with larger particles as quickly as they are formed so that a kind of dynamic equilibrium evolves. The studies of Bowden and Tabor[240] suggest a possible mechanism in which smaller particles unite with larger ones by welding resulting from frictional heating. The continual rubbing together of particle surfaces probably produces sufficiently high temperatures to soften small particles and to allow welding together. Since particles increase in size as well as decrease in size during grinding, the particle size distribution arising from continued disruption results from a dynamic equilibrium rather than a static one.

At first thought one might expect that the shock of a high explosive charge in a solid material would be especially effective in disintegration. However, it has been found that in fact the size distribution resulting from explosive shattering of minerals is about the same as less violent processes of disruption. Green and Lane[4] note that large numbers of tiny particles are present in dust raised from mining blasts, but these are spherical in shape and are believed to be released from the explosive itself rather than the broken stone. Explosion involves break up at very high velocities, but the mechanism of disintegration remains the same; i.e., fragmentation with release of finely divided material along fresh surfaces. Only the rate of disintegration is vastly different in the explosive case. The fragmentation by explosion is more complete, of course, but the sizes of finer particles are quite similar to industrial dusts.[4]

Particulate clouds also may be formed by explosive breakdown caused by gases generated within a solid material. The copious quantities of dust emitted from volcanoes are believed to originate in this way. The pressures in volcanic magma saturated with steam and hot gas at high pressure is released suddenly to the atmosphere, and the molten or partly solidified material explodes violently.

B. Cohesive and Adhesive Forces

Particles in dusts or powders tend to stick together remarkably well. It is difficult to break aggregates of particles up to produce clouds of non-agglomerated material. The ability of particles to stick together indiscriminately is the result of relatively weak attractive forces between molecules as well as attractive electrostatic forces. These forces have been named cohesive and adhesive depending on the heterogeneity of material at the boundary between particles. The distinction between cohesion and adhesion in the literature on fine powders is somewhat fuzzey, but we shall adopt the following conventions which are consistent with classical definitions in physics. *Cohesion* is the tendency for parts of a body of *like composition* to hold together. This implies that cohesive forces arise between like molecules in a solid or between small particles of the same composition. *Adhesion*, on the other hand, refers to attraction across the boundary or interface between two *dissimilar* materials. Thus adhesive forces are likely to be the most common attractive forces in all but artificially generated aerosols. The mechanisms for adhesion of particles to surfaces has been reviewed in detail recently by Corn,[244] so only a brief summary of this topic is included here.

Molecular Attractive Forces

The combined effect of the London–Van der Waals interaction between two particles containing several million molecules may be appreciable despite the fact that these forces are very weak between individual molecules and fall off with the 7th to the 8th power of the distance between the molecules. An estimate of the order of magnitude of the attractive forces between two microscopic spheres has been carried out by integrating the London–Van der Waals attraction over all pairs of interacting molecules. The calculations of Bradley[241] and Hamaker[242] indicate the net attractive force between two spheres of radii R_1 and R_2 is, in a vacuum,

$$F_{at} = \frac{\pi^2 n_A^2 c_{at} R_1 R_2}{6 r_m^2 (R_1 + R_2)} \text{ dynes,} \tag{8.27}$$

where n_A is the number concentration of atoms, C_{at} is a constant, and r_m is the minimum distance between the particles. Similar estimates have been made for non-symmetrical particles. From eqn. (8.27) one readily calculates that the net force between two 0.25 μ spheres, for example, separated by ~5 Å, is ~10^{-4} dyne.

Calculations also have been made for the influence of non-sphericity of particles on the magnitude of attractive force. For example, Vold[(243)] has found estimates of F_{at} for flat plates, rectangular rods, and cylinders. Particle orientation is important in these cases because these particles tend to seek an orientation of smallest potential energy, which is normally that in the largest plane areas opposed.

It has been pointed out some time ago (e.g. Casimir and Polder[(245)]) that the London theory has certain severe limitations in that its predictions will err at very short interatomic distances where the atomic wave functions overlap, and at large distances where electromagnetic retardation develops. For the attraction force per unit area between two flat plates close together, the retardation effect is:

$$\left.\begin{array}{lll}\text{non-retarded} & F_{at} = & A_F/6\pi r^3; \\ \text{retarded} & F_{at} = & -B_F/r^4,\end{array}\right\} \tag{8.28}$$

where r is the distance between the plates and A_F is a constant $\sim 10^{-12}$ erg. Casimir and Polder[(245)] have estimated the retardation coefficient to be

$$B_F = \frac{23\hbar c}{8\pi}\left(\sum_{1,2} n_{AS}^{\alpha}\right)^2, \tag{8.29}$$

where $\hbar$ is Planck's constant, c is the velocity of light, n_{AS} are numbers of atoms per unit area in Plates 1 and 2, and α is the polarizability of atoms in 1 and 2. Black *et al.*[(246)] have indicated that B_F can be related to the dielectric constants of the plate material, or

$$B_F = 0.162\frac{\hbar c}{\pi^3}\left(\frac{\epsilon - 1}{\epsilon + 2}\right)^2. \tag{8.30}$$

This relation yields a value of $B_F \approx 10^{-19}$ which is several orders of magnitude smaller than the parameter A_F. A remarkable macroscopic theory of Lifschitz[(249)] also gives a form for B_F

$$B_F = \frac{\pi\hbar c}{480}\left(\frac{\epsilon - 1}{\epsilon + 1}\right)^2 \phi(\epsilon) \tag{8.31}$$

for large distances r. Here $\phi(\epsilon)$ is given as a tabulated function of the dielectric constant ϵ. Lifschitz's result is nearly equivalent to Casimir and Polder's above.

Several investigators have looked at the retarded attraction between two flat quartz plates, and a micron-sized sphere.[(246, 247, 248)] They found satisfactory agreement between the experimental results and the theory of Casimir and Polder[(245)] and Lifschitz.[(249)]

Corn[(244)] has summarized the results of various workers for the constant $\pi^2 n_A^2 C_{at}$ in eqn. (8.27). These results are listed in Table 8.5. The large variation and high values obtained by Overbeek and Spaarnay evidently have been attributed to electrostatic enhancement of intermolecular forces, and lower figures have been obtained subsequently[(246)] agreeing with lower values in Table 8.5. Thus on the basis of these results, the constant $\pi^2 n_A^2 C_{at}$ is the order of 10^{-13} erg for clean, pure surfaces in a dry gas free of electrostatic charge.

In a recent study not included in Corn's review, Tabor and Winterton[(420)] have reported direct measurements of Van der Waals' forces between two mica cylinders. Their results indicate that the normal Van der Waals interaction predominates only at separations of

TABLE 8.5. SUMMARY OF INVESTIGATIONS OF THE VALUE OF $(\pi n_A C_{at})$ IN EQN. (8.27) (AFTER CORN[244])

Author	Gaseous media	Materials	Minimum separation (Å)	$\pi^2 n_A C_{at} \times 10^{12}$ (ergs)
Bradley[250]	Vacuum	Borate and	3	4.7
		quartz spheres	3	2.2
Overbeek and Sparnaay[251]	Air	Glass plates 1	200	0.01–22
		Glass plates 2	200	0.015–15
		Glass plates 1	2500	1.1–30
		Glass plates 2	5000	60–230
		Quartz plates	3000	11–30
Derjaguin *et al.*[247]	Air	Quartz sphere and plate	1000	0.05
Bailey and Courtney-Pratt[252]	Air	Cleaved mica	5–25	0.1–10 (estimated)
Prosser and Kitchener[253]	Air	Quartz sphere and plate	—	0.05
Howe *et al.*[254]		Pyrex and glass plate	Close approach	0.1–10 (estimated)
Theoretical (Casimir and Polder)[245]	Vacuum			1.0

less than 100 Å. The retarded forces tend to dominate at separations greater than 200 Å. Tabor and Winterton's measurements for mica–mica interaction indicate that $A_F \approx 10^{-12}$ erg in agreement with Hamaker's constant, while $B_F \approx 10^{-19}$, as calculated for the retarded forces by Lifschitz.[249]

The theory discussed above can only provide an order of magnitude estimate for the attractive forces between irregular particles in a bed. However, it is reasonable, based on eqn. (8.27), to expect that both the adhesive and cohesive forces are proportional to R_i, the particle radius, if r_m is constant. This is verified by an extensive review of Fuchs.[2] Other than this piece of information, scarcely anything is known about the nature of particle contact and attractive forces in powders or dusts. It is often assumed that the equilibrium distance is a few Ångstrom units, but the lack of any information about sub-micron structure and surface irregularity of particles prevents making quantitative estimates of the effective distance of separation. Molecular attraction, however, will become increasingly important as particles become smaller. If r_m remains constant, for example, the attractive forces vary with R_i. But forces tending to separate loose agglomerate vary with R_i^3 (for vibration), or as R_i^2 (aerodynamic forces). Thus, small particles will be more difficult to pull apart than larger ones, as we shall see in the discussion in the next sections.

Electrical Attractive Forces

The calculations of Bradley, Hamaker, Casimir and Polder, and Lifschitz elucidate the role of interatomic forces between bodies, but other attractive forces may also come into play. Generally, particles can carry substantial electrical charge. For particles charged with opposite sign, attraction forces will exist. Thus in some cases electrical effects may contribute to the mutual adhesion of particles forming aggregates. The larger the particle, the larger the charge it may carry. The importance of electrical charge is readily appreciated in the remarks of Black *et al.*[246] dealing with experimental difficulties in obtaining measures of the attractive forces between plates. Furthermore, we may note Beisher's work[255] which showed the obvious importance of electrical forces when the force between

0.5 μ particles carrying only one electron unit of charge will consist mainly of electrostatic interaction and not molecular interaction. Other examples of qualitative electrostatic influences on adhesion have been cited by Corn.[244]

In his treatment on electrification, Loeb[10] has noted that there is a well defined mechanism for charge accumulation on small particles called spray electrification. This mechanism probably stems from the orientation of electrical dipoles in liquid substances at a gas–liquid interface with negative polarity outwards. Yet, the question of electrification during solid–solid contact is poorly understood. The electrification phenomenon in solid particles is complicated by the influence of moisture, electrolytic effects, and the formation of diffusive double layers, or by films of impure materials. In addition, the nature of contact of solid surfaces such as sliding, rolling, or static contact may affect the charging process. The nature and state of the surface along with mechanical and thermal stressing can alter charging. Because of the complexity of electrification between solids, little of a quantitative nature is known at present about electrical forces and adhesion.

By making a crude estimate on the basis of electrostatic theory, however, one can obtain an idea of the relative magnitude of the electrostatic forces and the intermolecular forces. Suppose a particle carrying a charge Q_i is separated a distance h from an adhering particle in a dielectric medium. The electrical force will be

$$F_{\text{el}} = \frac{\epsilon Q_i^2}{h^2}.$$

With an "equilibrium" charge of 15 e.s.u. for a particle of 0.5 μ radius and $h \approx 10$ Å, $F_{\text{el}} \approx 5 \times 10^{-3}$ dynes. With eqn. (8.27) and $\pi^2 n_A C_{\text{at}} \approx 10^{-13}$, $F_{\text{at}} \approx 10^{-4}$ dynes for this case. Thus the electrostatic forces may easily exceed the intermolecular forces as an adhesive attraction. Such a conclusion may also be reached from Derjaguin and Smilgas'[256] calculations of the adhesive force between metal semi-conductor crystals based on electronic theory. Their calculations suggest that the electronic force of adhesion for such materials is 10^8–10^9 dynes cm^{-2}. For a zone of contact of 100 Å, the net force of adhesion would be $\sim 10^{-4}$ dyne.

Derjaguin and Smilgas' theory indicates the importance of surface conditions in controlling high adhesion. They have suggested ways of conditioning adhesion by chemical reactions on the surface of contacting bodies. Indeed, modification of the adhesive force has been accomplished recently by irradiation of contacting surfaces with a beam of electrons (Dukhoskoi, Kragelskii, and Silin[257]).

From the discussion above, it is clear that future work to study the fundamental nature of adhesion must incorporate studies of electrical interactions between solids, as well as intermolecular interactions.

Effect of Surface Films

If a surface is wet or dirty, an additional attractive force may develop as a result of capillary effects of surface films. It has been found, for example,[2] that increase in humidity always increases adhesion of particles to a surface. A film of water apparently forms by capillary condensation (Fig. 8.10) between the surface and the particles which adds to the attractive interaction between the two bodies.

The magnitude of the capillary force or surface tension can be estimated considering the diagram in Fig. 8.10. Suppose a sphere lies on a flat surface and a liquid meniscus is formed

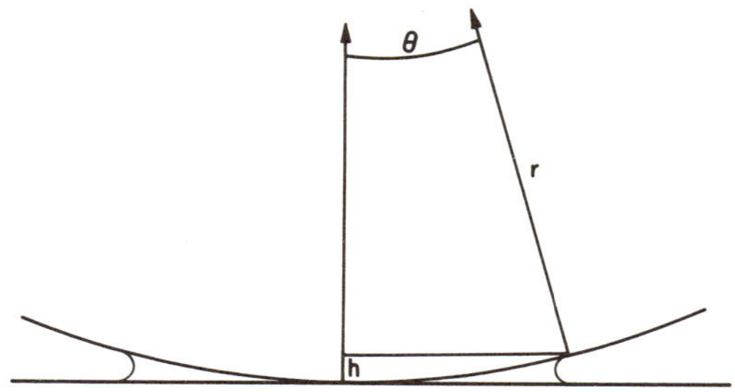

FIG. 8.10. Liquid meniscus between a sphere and a flat surface.

around the zone of contact. When the radius of the film is small relative to the radius of curvature of the sphere ($r = R_i$), the distance h (Fig. 8.10) is[2]

$$h = R_i(1 - \cos\theta) \approx 0.5\, R_i\theta^2,$$

where θ is the angle between the vertical and the radial line through the top of the meniscus is $\sim h/2$. Then the negative pressure with the liquid film is $2\sigma/h$ (σ is the surface tension of the liquid film). And, according to Laplace's formula,[2] the adhesive force is estimated as:

$$\text{(sphere–plane)} \quad F_{at} = (2\sigma/h)\, R_f^2 \approx 4\pi R_i\sigma, \tag{8.32}$$

where R_f = radius of the film at the point of contact. For two equal-sized spheres in contact, h is twice as large as the value in Fig. 8.10 so that

$$\text{(sphere–sphere)} \quad F_{at} \approx 2\pi R_i\sigma. \tag{8.33}$$

MacFarlane and Tabor[258] verified this relation by direct experimental measurements of the attractive force between wetted spheres. Equations (8.32) and (8.33) represent the maximum force between two surfaces exerted by a moisture film. It is not realized until the relative humidity of the environment reaches 100 per cent. From these results we find that the adhesive force associated with wetted surfaces will be as high as the order of σR_i, which can exceed 10^{-3} dynes between two smooth spheres, for example. Thus the effect of surface films may be larger than either the attraction of intermolecular forces or electrical forces.

Other Factors Influencing Adhesion

In addition to the influences of intermolecular forces, electrical forces, and surface tension, Corn[244] has pointed out that other known mechanisms also alter the adhesive properties, for example, polarization of charge on particles may be modified by an external electrical field. The microscopic surface structure of bodies are important, as well as the nature and extent of surface contamination. Particle surface contact plays an important role in adhesion. This may be demonstrated very well by data of Bowden and Tabor[240] listed in Table 8.6. These workers found a wide range of net adhesive force could exist when glass spheres are placed on a flat surface at 100 per cent relative humidity. The larger the height of the surface irregularities, the smaller the adhesion because of reduction in the surface of contact.

TABLE 8.6. ADHESION OF GLASS SPHERES TO GLASS PLATE IN ATMOSPHERE OF 100 PER CENT RELATIVE HUMIDITY BASED ON DATA OF BOWDEN AND TABOR[240] (FROM CORN[244])

	Mean approximate height of surface irregularities (Å)	Adhesion as per cent of value observed with highly polished surfaces
Highly polished	150	100
500 carborundum paper	1,000	79
320 carborundum paper	4,000	51
150 carborundum paper	100,000	0

C. Entrainment of Particles from Surfaces

In practice the production of natural and artificial dust clouds involves the disruption of a bed of particles by various means like blowing a gas over the top of the bed or by passing gas up through the bed. Before taking up the cases, it is useful to consider the mechanisms of entrainment of particles from a smooth flat surface.

To dislodge a single particle from a surface several possible processes may be invoked. In all cases, however, the principal requirement for removal is that the dispersing force acting on the particle must exceed the adhesive force of attraction. We have seen that strength of adhesion (as well as cohesion) depends on several factors including the nature of the particle material, its size and shape, surface roughness, and relative humidity of the surrounding gas. In addition, the presence of electrostatic charge must be taken into account.

Particles adhering to a surface can be entrained into surrounding air by several means including (a) mechanical concussion or vibration, (b) explosive dislodgement, or (c) aerodynamic dispersement.

Mechanical Concussion or Vibration

Dislodgement by mechanical means such as vibration requires that the normal component of the adhesive attraction be overcome. As we have seen above, adhesion depends on many factors. However, Corn and Stein[259] have cited typical values of the attractive force for glass spheres on a glass slide from which may be estimated the typical mechanical force for entraining particles into air. For example, for an 18.5 μ radius glass sphere sticking to a clean, smooth flat glass surface, the component of cohesive force normal to the slide is 0.60 dynes to remove 50 per cent of the particles in air at 35 per cent relative humidity, 20°C, and 1 atm pressure. For a particle density of 2 g cm^{-3}, the acceleration equivalent to this force is 1.25×10^7 cm sec^2. If the surface on which the particle is lodged moves at 1 m sec^{-1}, the deceleration to a standstill of the particle in removal from the surface would have to take place in $\sim 10^{-4}$ cm; at a surface speed of 10 m sec^{-1}, the deceleration distance is $\sim 10^{-2}$ cm. Corn and Stein[259] note that under these circumstances, removal of particles is not easily carried out by deceleration. However, particles can easily be dispersed by transfer of momentum from a moving body. Dust may be swept away by a brush or a hand. Transfer of the 0.66 dynes and dispersal of the 18 μ particles may be accomplished as in the action of a shoe scraping across a floor.

Vibration of a substrate may also induce entrainment. For constant acceleration a,

$$a = 4\pi\omega_0^2 a_0,$$

where ω_0 = frequency of vibration and a_0 = vibration amplitude. To remove an 18 μ glass sphere from the glass surface the frequency is 565 Hz for a 1 cm amplitude vibration.(259) Such rapid oscillations are not likely to occur commonly.

Explosive Dislodgement

Several years ago Billings and co-workers(260) found experimentally that the explosive release of energy could dislodge dust particles from a slag wool filter. A blast wave of 127 mmHg over-pressure with a wind speed of 42 msec^{-1} and a dynamic pressure of 10 cm of water was enough to remove adhering particles from the filter fibers. The exact mechanism of dislodgement is still uncertain. Particle removal may be due in part to aerodynamic effects, as discussed below, and possibly to vibration of the substrate. However, the work of Gerrard(261) dictated that energy transfer from the blast wave also may be associated with reflection of the shock wave from the substrate and subsequent raising of the particles with the reflected wave. This appears to be analogous to vibrating the substrate.

Aerodynamic Dispersement

Perhaps most common of all mechanisms for dispersing powders is the blowing away of particles by a gas flow. Prediction of the onset of particle pickup from a powder bed is difficult because of uncertainties in specifying the velocity distribution near the particles as well as the cohesive and adhesive force. In the somewhat simpler case of aerodynamic removal of single particles from surfaces, more is known based on the studies of Bagnold(262) of sand grain migration by wind. Above the range in particle size where cohesive bonding is important, particles on a smooth surface go through successive stages of movement induced by the air flow overhead. Under the action of the wind the particles roll over other grains. The continuing rollover, with a bouncing action, was referred to by Bagnold as *saltation.*

Particles evidently leave horizontal surfaces by wind action in the following way. The grains first begin to slide or roll along until their speed is sufficient for small bumps on the surface, or on the particles themselves, to cause them to jump. The process of making grains airborne is sketched with the distribution of air flow near the surface in Fig. 8.11. Since the air flow near the surface generally will be turbulent, particles may jump out of the viscous sublayer very close to the surface. Then the particle may either be caught up by turbulent fluctuations associated with eddying motion or, if it is spinning, may be affected by the so-called Magnus force,† and may be carried clear of the surface. Fine particles held mainly by molecular or electrostatic attraction probably break loose in a similar way provided that the attractive force between the particle and the surface is overcome by the aerodynamic lifting.‡ Since the aerodynamic forces oriented tangentially to a smooth surface will be greater than those normal to the surface, one expects that the

† The Magnus force is an apparent force acting on a spinning body. The spin may be induced on a smooth body by a velocity gradient (see also Fuchs(2)).

‡ Calculations of the lifting force on a body in a viscous medium is difficult, but estimates can be made for inviscid potential flow, say by means of a Kutta–Joukowsky theorem. These calculations will not apply exactly, of course, to particles of $Re_i \approx 1$.

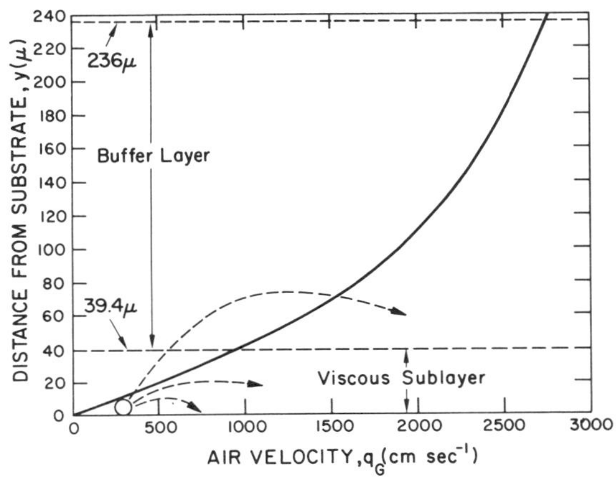

FIG. 8.11. Rolling and lifting of small particles in a turbulent boundary layer.

particle will begin its dislodgement by sliding or rolling. This will inevitably give rise to the reduction in the number of points of contact between the particle and the surface and, as a consequence, will cause a reduction in the net adhesive force. Thus the probability of a particle leaving the surface after beginning to roll is considerably enhanced.

The calculation of the force required to carry away particles from a smooth surface is uncertain. It is difficult to estimate quantitatively the adhesive and electrostatic forces, and the values of static friction are not well known. Though the static friction of large bodies is easy to measure, there are considerable difficulties in performing similar measurements for bodies smaller than about 0.5 mm in radius.

The net force of static friction, which is the tangential component of the adhesive force, is proportional, like the normal component, to the area of contact. The extent of this area depends largely on the plastic deformation of microprojections at the points of contact. Bowden and Tabor[(240)] have shown that for larger bodies, the contact area is approximately proportional to the normal force on which the frictional theory is based. However, when the size of the body decreases, the normal force $m_i g$ (its weight) decreases rapidly as well, and the influence of the adhesive force becomes more important. The actual contact area, and consequently the frictional force, is proportional to the sum of the normal forces. Since the adhesive force roughly is proportional to R_i, the coefficient of friction will increase with decrease in R_i.

Recent experiments of Löffler,[(263)] using two independent methods for measuring the force of adhesion, indicated that F_{at} depends on particle size as well as deposition velocity on a surface and relative humidity. For 5–15 μ dry particles filtered onto monofile polyamide fibers F_{at}, as determined by centrifugation or blow off ranged from 10^{-4} to 5×10^{-2} dynes, with both experimental methods agreeing with each other.

Differences of roughness in zones of contact of large bodies having the same normal force will be averaged out, but for fine particles the normal component of the adhesive force and the contact area will depend largely on the microgeometry of the surfaces and should vary greatly from particle to particle, from one orientation to another, and from one point on the surface to another. Experiments on the removal of dusts from flat surfaces have shown considerable differences in results, which tend to confirm the inherent variability in adhesive forces. The above remarks also explain why, in a given set of measurements of dust removal, relations between flow rate vs. fraction of particles blown off evolve in which complete removal requires several times greater gas flow than the one where removal begins.

Because of the lack of knowledge of surface geometry, as well as in the adhesive and cohesive forces, estimation of the force required to blow off particles from powder beds is even more uncertain. Based on the consensus of many measurements, however, a rough rule of thumb has been given for the adhesive or cohesive force[2] between particles in beds:

$$F_{ad} \approx 2R_i \text{ to } 5R_i \text{ dynes.} \tag{8.34}$$

R_i is in cm. This magnitude is the same order as the frictional component ($m_i g$), but appears to be substantially greater than predicted by the Bradley–Hamaker theory. One would expect that this intermolecular force, given by eqn. (8.27), would be the minimum force required to lift a particle out of a powder bed.

An alternative approach to estimating or measuring the adhesive force for various configurations has been to try and establish a purely empirical correlation between aerodynamic parameters and a susceptibility for removal, say in terms of the ratio of the drag force action on the particle, has recently been presented by Zenz[264] for particle removal from powder beds with Bagnold's results for pick up of sand grains, and Jordan's[265] data for blow off of individual 5–10 μ radius quartz and coal particles from glass surfaces. Jordan's curve is based on an extrapolation to the flow velocity when only 1 per cent of the adhering particles are removed. Zenz's correlation is shown in Fig. 8.12 with some of

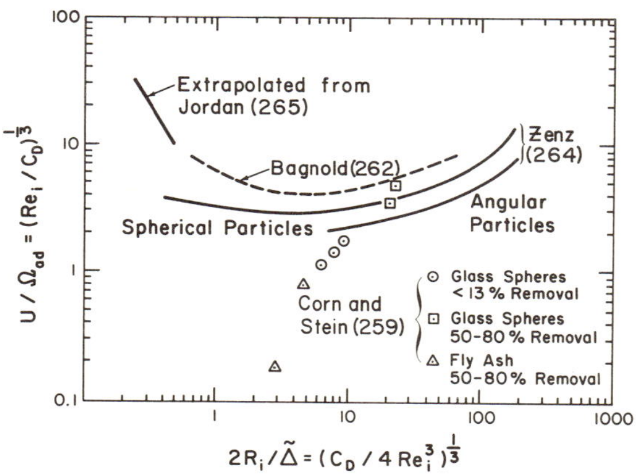

FIG. 8.12. Correlation of Zenz[264] for the velocity to lift particles from a surface. (From Corn and Stein.[259])

Corn and Stein's results[259] for pick up of glass spheres and fly ash from glass surfaces (<50 μ radius particles). Here is the superficial stream velocity for initiation of removal without saltation,

$$\Omega_{ad} = [4g\mu_G(\rho_i - \rho_G)/3\rho_G^2]^{\frac{1}{3}}$$

and

$$\Delta = [3\mu_G^2/4g\rho_G(\rho_i - \rho_G)]^{\frac{1}{3}} \tag{8.35}$$

Although Zenz's results tend to indicate qualitatively an agreement in correlation between his results and those of Jordan and Bagnold, Corn and Stein's data do not fit such a correlation. The latter authors have speculated that such a correlation will not be adequate because of differences between the local nature of the air flow over individual particles or near the surfaces as contrasted to particles lying in beds. However, this cannot be the whole story of the differences between individual experimental results. Certainly the differences in the adhesive and cohesive forces, as well as electrostatic influences, must be taken into account in any improved correlation of results.

Because of the inherent difficulties in making quantitative calculations for the aerodynamic lifting of particles from surfaces or powder beds, advances in understanding this aspect of aerosol production undoubtedly will come from extensive experiments on the gas flow over rough surfaces and the nature of adhesive and cohesive bonding. Improved empirical correlations of the type proposed by Zenz should certainly play a role in presenting the results of new experimental investigations.

D. Dispersal of Powders by Air Jets

When gas flows at relatively low speeds over layers of powder, particles become detached from the surface and are carried away by the turbulent motion. As particles lift away from the powder bed, the bed is compacted by the gas flow and begins to assume a rounded, streamlined shape until erosion no longer occurs. To renew the removal of particles from the powder layer, much greater gas speeds are required. After the onset of the higher gas flow, similar erosion and streamlining occur. However, at sufficiently high gas speeds not only are individual particles removed, but whole aggregates may be shed away from the bed and immediately disintegrated by the action of the airstream. Under these conditions bulges in the surface of the bed are produced, the surface becomes more and more uneven, aiding, in turn, further particle removal. Fuchs[(2)] has pointed out this constitutes the transition from surface to bulk atomization of the powder. The latter occurs, for example, in nozzles used for injecting coal dust with furnaces in the pneumatic separation of powders and in the dispersal of insecticide dusts.

Fluidization of Powders

Bulk atomization of powder beds by a gas jet is closely akin to the fluidization of powders enclosed in a vertical pipe. Fluidization has been studied extensively in recent years because of the interest in chemical technology in using such systems for carrying catalysts for chemical reaction. One comprehensive survey of understanding of fluidization mechanisms has been given by Davidson and Harrison.[(266)]

Qualitatively the process of fluidization of a powder bed in a tube is described as follows: gas is passed through the bed via a porous bottom plate until critical velocity is reached when the bed begins to expand in depth. Up to critical gas velocity the pressure drop increases steadily; after reaching critical flow the pressure drop becomes nearly constant. At low gas velocities, where particles remain stationary, the pressure drops follow the Kozeny relation

$$\Delta p = K_p \frac{C_p L \mathscr{U} \mu_G \phi^2}{\bar{R}_i^2 (1 - \phi)^3}, \tag{8.36}$$

where K_p is the factor depending on particle shape, L is the depth of the bed, ϕ is the volume fraction of particles, and $\mathscr{U}$ is the gas velocity above the bed. The critical gas flow

$\mathscr{U}_{cr}$ is given by the situation when the pressure gradient of the medium equals the gradient in hydrostatic pressure of the powder, $g\phi(\rho_i - \rho_G)$, or

$$\mathscr{U}_{cr} = \frac{\Delta p R_i^2 (1-\phi)^3}{K_p L \mu_G \phi^2}. \tag{8.37}$$

At this condition the particle bed begins to expand and the flow resistance (given by Δp) becomes essentially constant because the product $L\phi$ remains constant.

In finer powders, where cohesion between particles plays an important role, the Kozeny formula is only qualitatively correct. The ratio $\mathscr{U}/\Delta p$ increases more rapidly than eqn. (8.36) predicts. This is related to the observation that fluidized beds do not expand uniformly, but the bed breaks down into separate cells with slitlike open passages between them. Much of the gas goes through the passages, so $\mathscr{U}$ is greater than predicted by eqn. (8.36).

After a certain amount of expansion of 5–20 per cent of the original volume, particles begin to move and the resistance of the bed decreases a little, possibly as a result of reduced friction between the particles. Beyond this stage fluidization takes place in which particles are at least partially suspended in the turbulent gas. In this regime the gas bubbles through the expanded bed, and the layer depth continues to expand but fluctuates wildly. The bubbling causes rather vigorous agitation of the powder promoting a circulation rising at the center of the tube and settling along the walls. This state has been aptly made analogous to a boiling liquid. As the gas flow is further raised, particles whose settling rate is less than the flow velocity begin to leave the "fluid phase" and are drawn out into a "gaseous" or "aerosol" phase.[2] The particle concentration in the aerosol phase keeps increasing until the boundary between phases fades away, and the powder is completely dispersed in the gas.

Dispersability of Powders

The successive stages of aerosol formation in a powder bed being fluidized by gas flow give a picture of the interaction between the fluid and the particles in dispersal. Similar interaction no doubt can be associated with any dispersal by high speed jet air blowing horizontally along a layer of powder. The main uncertainties in fluidization processes lie in the influences of particle shape and nature of the material just as in the previous discussion of aerodynamic entrainment at relatively low gas speeds. In the "boiling" state dispersability again depends on the cohesive and adhesive properties of the powder as well as on the size and shape of the particles.

Apart from particle size, shape, and mechanical properties, plasticity and the density of packing of particles, or their number of contacts with a neighbor are all important. Fuchs[2] has noted, for example, that it is more difficult to disperse powders of soft plastic materials than hard ones. Isodisperse powders atomize easier than polydisperse powders because in the latter more intergranular space is filled, so more points of contact exist between individual particles. It is especially difficult to disperse powders whose size range is large since their packing density is so much greater than isodisperse powders. Decrease in packing density or in the bulk density of a powder, on exposure to sonic vibrations, for example, apparently substantially facilitates dispersion.

Apart from studies of fluidization the investigation of mechanisms for powder dispersal is virtually untouched in the literature. Even empirical information on the relative dispersability of different powders is very scarce. One classical study of this kind is available by Anderson.[267]

In Anderson's experiments 2 cm^3 of powder were poured through a slit into a vertical tube of height 250 cm and diameter 4.5 cm. The particles separated partially during fall to the bottom of the tube. The percentage of powder not settled on the tube bottom in 6 sec was measured. Anderson estimated that individual particles would not fall through the tube to the bottom in this time so that his numbers would represent a percentage of dispersed powder, or the *dispersibility*. Fuchs[2] has remarked that almost certainly some of the unsettled fraction also contained some aggregates so that Anderson's experiments were only of qualitative significance. Nevertheless, one can readily see from his data in Table 8.7 that typically there exists quite a wide range in dispersibility for various classes of particulate material of roughly the same size.

TABLE 8.7. DISPERSIBILITY OF SOME POWDERS (FROM ANDERSON[267])

	Particle radius limits (μ)	Dispersibility (%)
Lycopodium	12	100
Wood charcoal dust	0–25	85
Wood charcoal dust	0–7	23
Aluminum powder	0–15	66
Talc	0–20	57
Carbon black	0–15 (?)	47
Potato starch	0–35	27
Graphite dust	0–25	17
Pulverized slate	0–25	13
Cement	0–45	5.5
Prepared chalk	0–6	1.5
Polydisperse silica dust (coarse)	—	21
Polydisperse silica dust (fine)	—	8
	11.5	68
Isodisperse silica dust	8	83
	5.6	45
	7	50
Porcelain dust with fine fractions	2.7	52
	1.1	21
	0.45	12
Porcelain dust without fine fractions removed	—	5

Of all the powders listed in Table 8.7, lycopodia spores apparently have the greatest fluidity. This may be attributed to the uniformity of size, lightness and spherical shape of the particles as well as by the presence on their surfaces of tiny ridges which will favor a loose structure in the powder.

CHAPTER 9

FORMATION OF AEROSOLS BY NUCLEATION OF SUPERSATURATED VAPOR

IN THE introductory portion of the last chapter, two fundamental mechanisms for forming aerial dispersions were reviewed. The first involves the agglomeration of molecules, either by physical change from vapor to condensed phase, or by chemical reaction leading to polymerized species; the second involves the breakup of material to form particles of colloidal size. Having discussed further the process of disintegration by various means, let us now turn to the molecular agglomeration process.

Best understood, and most widely used in the formation of aerosols of the molecular agglomeration processes, is the nucleation of a supersaturated vapor. Nuclei may consist either of groups of molecules of condensable vapor (embryos), or "foreign" material such as ions or dust ranging in size from a few Ångstroms to several microns. Supersaturation in a pure gas, or a mixture of gases may come about by physical processes of adiabatic expansion or mixing a cool gas with a warm gas, for example. Supersaturation also may be achieved by means of chemical reaction. That is, a gas mixture may consist of chemically reactive components that yield polymerized material of sufficiently low volatility for saturation to be exceeded at a given temperature or pressure in the mixture.

Perhaps the most common of the natural physical activity leading to cloud formation are the developments of collections of water droplets in the earth's atmosphere; the supersaturation in this case largely stems from adiabatic expansion of masses of air rising by buoyant convection from zones near the earth's surface to higher altitudes. Aerosol generation in the atmosphere may also occur by chemical means, as for example, by photochemical oxidation of certain hydrocarbons or by combustion of material. Such reactions yield layers of smog which have become well known to the urban dweller.

There is some recognition of an additional mode of particle production by a process which we may term "chemical" nucleation. Perhaps the most familiar example of this process is found in the formation of carbon particles in the combustion of hydrocarbons. The feature which distinguishes "chemical" nucleation from "physical" nucleation is principally the existence of chemical rather than the usual physical adsorptive energies which come into play during particle growth processes. It would appear likely that in chemical nucleation certain free radical species act as nuclei upon which various molecular species may undergo collision and reaction. Growth to a large particle by chemical processes will depend on both propagation of the free radical reaction and coagulation of the initial particles.

Particle formation in premixed hydrocarbon flames has been investigated extensively by Wagner and his colleagues.[(432, 441)] They have found in some excellent studies that acetylene is formed at an early stage in the combustion. The acetylene is believed to undergo a series of reactions with free radicals to give a distribution of polyacetylenes and their free radicals. These polyacetylene free radicals then react with acetylene and polyacetylenes and yield particles through a complicated series of steps involving breaking of linear polyacetylene chains, ring closure, and additional accretion of free radicals and other molecular species.

The subject of "chemical" nucleation is only beginning to receive attention. Many important advances in our knowledge of the mechanisms of particle growth and production in atmospheric and other processes will no doubt occur as "chemical" nucleation begins to be better understood.

In the laboratory, aerosol clouds have been prepared in a variety of devices by making use of sudden expansions or mixing processes. Among the former cases, perhaps best known is the Wilson[(268)] cloud chamber, a closed vessel in which a sudden expansion is used to induce relatively uniform supersaturations in a volume of gas. Such units have had extensive application to the study of nucleation processes as well as to the investigation of certain kinds of elementary particles having high kinetic energies, including cosmic rays. Perhaps less often cited are the clouds of condensate that are formed by expansion of gas in high speed flow through a converging–diverging nozzle. The formation of particulate suspensions in such situations was investigated extensively in connection with problems arising from condensation in supersonic wind tunnels (e.g. Wegener and Mack[(269)]). Later, fundamental studies were undertaken in supersonic flows to elucidate features of homogeneous nucleation (e.g. Wegener and Pouring[(270)]). Supersonic nozzles are particularly useful in that they provide a device for developing very high supersaturations continuously. Like the tubular chemical reactor using steady fluid flow, the kinetic process of phase change can be "stretched out" over the nozzle in a time scale proportional to the ratio of the nozzle length and the mass velocity of the fluid.

The idea of making use of the mixing of a cool gas with a warmer, moist gas has provided a second method for generating aerosols in the laboratory. Polydispersed clouds may be formed very simply, for example, by passing a gentle current of gas over an open dish of heated material, or by blowing air over liquid or solid heated on a small boat. Aerosols of reproducible size distributions have been produced in this way as described by Whytlaw-Gray and Patterson.[(271)] More monodisperse clouds have been formed under carefully controlled conditions of mixing and nuclei concentration using the well known La Mer generator (Sinclair and La Mer[(195)]).

Workers also have produced smokes by jetting a hot moist gas into cool surroundings. This technique was used extensively to produce screening clouds for military purposes during World War II. The mixing principle also has been applied for generating insecticide fogs, and it is involved in the main mechanism for producing aircraft condensation trails in the atmosphere. Mixing by molecular diffusion has been used to produce supersaturation in diffusion cloud chambers. An example of this class of devices has been described by Powell and Oswald[(272)] (See also, Franck and Hertz[(309)]).

Turbulent transport processes involved in condensation by mixing in jets have been discussed in detail by Amelin[(273)] and by Hidy and Friedlander.[(274)] These methods and their practical application have been reviewed more extensively in a recent monograph by Amelin.[(275)]

Aerosols are produced in the laboratory as well as in the atmosphere by a variety of chemical reactions. Some of the more common methods make use of the reaction of hydrogen chloride and ammonia to form ammonium chloride smoke, or the reaction of sulfur trioxide and water vapor to form sulfuric acid mists. Other chemically formed aerosols may develop from combustion processes, e.g., burning magnesium strips from metallic oxides produced by electric arcs struck between electrodes of a metal, by photolysis of iron carbonyl, etc. The practical aspects of these methods are outlined in somewhat more detail by Green and Lane.[4]

To discuss the process spontaneous nucleation of a supersaturated vapor, it is customary to begin with the homogeneous case even though most nucleation phenomena are heterogeneous, or chemical in nature. The treatment of nucleation idealized as a homogeneous process, nevertheless, provides a useful fundamental framework for understanding at least qualitatively the more complicated heterogeneous effects. With this point of view, we begin by discussing in some detail the thermodynamics and kinetics of homogeneous nucleation.

9.1. Nucleation of Vapour in the Absence of Foreign Nuclei

When a pure vapor or a component of a gas mixture exceeds the saturation condition in an unconfined space, spontaneous condensation—called *nucleation*—may be observed depending on the degree of supersaturation achieved, and the nature and concentration of nuclei present in the gaseous medium. If very few foreign nuclei are present, or if the rate of increase in supersaturation is very rapid, conditions in the gas may far exceed the state of saturation before stable particles begin to appear. In a pure vapor, free of dust or ions, for example, the vapor phase will become thermodynamically unstable when supersaturated. But collapse of this instability to a condition of lower energy will not take place until the supersaturation is sufficiently high to insure that tiny particles of condensed phase will grow in time rather than re-evaporate if formed initially by a molecular scale fluctuation in the vapor. Because the formation of particles is a rate process, a very rapid expansion, or very rapid cooling by mixing may yield very high apparent supersaturations in a gas even though many nuclei are present. If changes in temperature and pressure occur in a dusty medium on scales of less than 10^{-3} sec, aerosol cloud formation at very large supersaturation have been observed, e.g., in nozzle expansions by Stodola[276] and in turbulent jets by Hidy and Friedlander.[274] If the cooling rate is less strong, however, relief of supersaturation may occur by diffusion of vapor on larger particles of dust or on ions if they are present (see, for example, Plates 14 and 15) or onto walls of a confining vessel. Growth of single small particles by absorption of diffusing vapor was discussed in detail in Chapter 4.

A. Supersaturation and Thermodynamic Equilibrium

Let us consider first the case of a pure vapor or a gas mixture that is free of all foreign nuclei. Without confining walls, we expect that nucleation will proceed by means of the diffusional or effusional growth of vapor on embryos formed by fluctuations in the gas. As vapor saturation is exceeded, clusters or embryos will form in the supersaturated medium. Because of their small size, their surface-free energy is critical to their stability. These submicroscopic agglomerates thermodynamically have greater probability of splitting apart than continuing to grow in the weakly supersaturated vapor. The vapor then may remain in a metastable gaseous state (macroscopically) until higher supersaturations are attained. With continued increase in supersaturation, the molecular clusters of con-

PLATE 14. A photograph of nucleation of methyl alcohol vapor in a large diffusion chamber. The large ellipsoidal cloud on the right center was triggered by an α particle. The long thin tracks are the result of nucleation by electrons passing through the chamber. (Photo by P. Luger.)

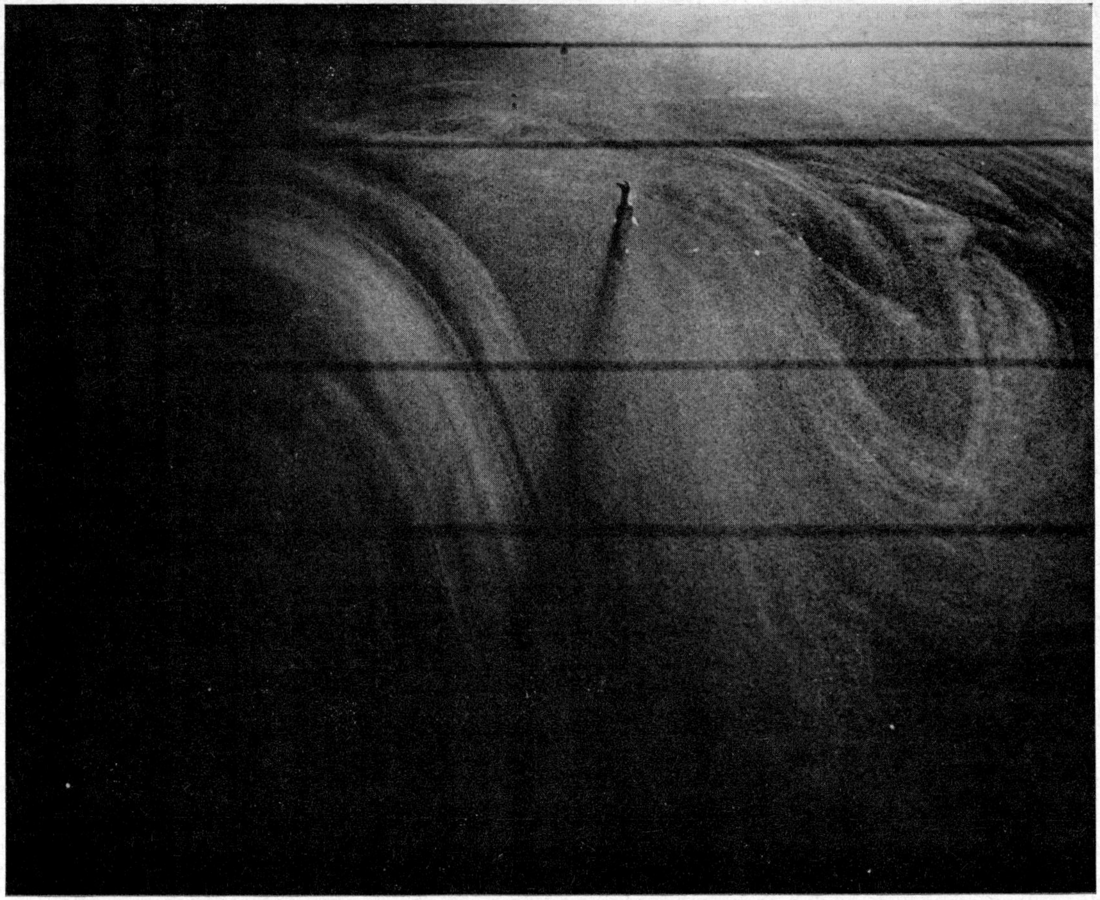

PLATE 15. A picturesque pattern of nucleated droplets of methyl alcohol in a large diffusion chamber. Patterns are associated with heterogeneities in temperature and vapor distributions associated with vapor motion in the chamber. (Photo by P. Luger.)

densed phase are created on a microscale until the rate of formation of particles surpasses their rate of disintegration, giving rise to appreciable condensation to form droplets as an aerosol cloud. The conditions of stability of the supersaturated medium can be described by thermodynamic arguments (see for example, Prigogine and Defay[(84)]).

Before outlining the requirements for stability of phases, we shall list briefly the definitions of thermodynamic variables which are normally used to describe phase equilibria. The first of these is the Helmholtz free energy, or the work function

$$\hat{F} = \hat{U} - T\hat{S}, \tag{9.1}$$

where $\hat{U}$ is the internal energy per mole and $\hat{S}$ is the entropy per mole of material in the system considered. Similarly, the Gibbs free energy is

$$\hat{G} = \hat{U} - T\hat{S} + pV, \tag{9.2}$$

where T is the absolute temperature, V is the volume per mole, and p is pressure. With the first law of thermodynamics, differential changes of the two free energy functions are given by

$$\mathrm{d}\hat{F} = -\hat{S}\,\mathrm{d}T - p\,\mathrm{d}V \tag{9.3}$$

and

$$\mathrm{d}\hat{G} = -\hat{S}\,\mathrm{d}T + V\,\mathrm{d}p. \tag{9.4}$$

From these relations it follows that a system will be in equilibrium at constant temperature and volume if $\mathrm{d}\hat{F} = 0$ or if $\hat{F}$ is a minimum. Similarly, a system in equilibrium at constant temperature and pressure requires that $\hat{G}$ be a minimum or $\mathrm{d}\hat{G} = 0$. In any spontaneous infinitesimal change, the free energy functions will vary in a way which will depend on the path taken by the system as well as constraints placed on it. In moving at constant volume to move stable equilibrium, for example, $\hat{F}$ must decrease. On the other hand, approach to stable equilibrium at constant pressure requires that $\hat{G}$ decrease.† Usually pressure is held constant in an "open" system like the atmosphere undergoing nucleation so that our concern will be limited arbitrarily to the Gibbs function $\hat{G}$.

If a multiphase system is being considered which contains n molecules of phase i, then the partial molar quantity for a single component

$$\hat{\mu}_i = \left(\frac{\partial \hat{G}}{\partial n_i}\right)_{T,\,p,\,n_j} \tag{9.5}$$

is useful. This term, $\hat{\mu}_i$, is often called the chemical potential of phase i.
When the system contains only two phases i and j that are in equilibrium, the requirement $\mathrm{d}\hat{G} = 0$, means that

$$\hat{\mu}_i = \hat{\mu}_j. \tag{9.6}$$

Stability of Phases

Let us consider briefly some implications of thermodynamic stability and phase change in an unrestricted volume of pure vapor. We examine the three possibilities:‡

† For a clarification of the role of the function $\hat{F}$ and $\hat{G}$ within arbitrary thermodynamic constraints, see Reiss.[(277)]

‡ Reiss[(277)] has implied in his discussion on stability that there is nothing absolute about the kind of stability of a given equilibrium configuration. Like the behavior of the chemical potentials, on whose curvatures the stability conditions depend, the question of stability itself is bound to the constraints applied to the system and the kind of work it can perform.

(a) *Stable equilibrium* in which the initial phase is stable with respect to all phases different from the initial phase.

(b) *Metastable equilibrium* wherein the initial phase is stable with respect to all phases infinitesimally different from it, but there exists at least one other phase with respect to which it is not stable.

(c) *Unstable equilibrium* in which the initial phase is unstable in relation to all phases infinitesimally different from it.

If stable equilibrium exists in a vapor, no spontaneous change of phase can ever develop in macroscopic amounts in the gas. However, an unstable phase will immediately disappear and give rise to one or more neighboring phases. Because of molecular fluctuations the adjacent phases always form continuously so that unstable phases transform spontaneously to a more stable condition, and the unstable conditions are not realizable physically. In the case of metastable phases, the system may remain indefinitely at equilibrium without appearance of a new phase. This corresponds to the condition of relatively low supersaturation without the presence of nuclei of a more stable phase. If nuclei are introduced, a change over to the more stable conditions will occur.

To illustrate the criteria for thermodynamic stability, consider the pure vapor under conditions of constant internal energy, and constant volume. This system will be in stable equilibrium if

$$\left(\frac{\partial^2 \hat{S}}{\partial \hat{U}^2}\right) < 0$$

and

$$\left(\frac{\partial^2 S}{\partial V^2}\right)\left(\frac{\partial^2 \hat{S}}{\partial \hat{U}^2}\right) > \left(\frac{\partial^2 \hat{S}}{\partial V \partial \hat{U}}\right)^2. \tag{9.7}$$

This implies that[277] the condition of thermal stability is

$$C_v > 0, \tag{9.8}$$

where C_v is the heat capacity at constant volume, whereas the condition of mechanical stability is

$$-\left(\frac{\partial p}{\partial V}\right)_T > 0. \tag{9.9}$$

Higher order conditions of mechanical stability, as for example, near the critical point of the material are treated by Prigogine and Defay.[84]

When one deals with conditions of thermodynamic stability in a multicomponent mixture, the criterion of diffusional stability must be taken into account in addition to (9.8) and (9.9). This condition requires that

$$\sum_{\alpha} \sum_{\beta} \left(\frac{\partial \hat{\mu}_\alpha}{\partial n_\beta}\right) \delta\xi_\alpha \delta\xi_\beta > 0, \tag{9.10}$$

where $\hat{\mu}_\alpha$ is the partial molal free energy (the chemical potential) of (gaseous) species α, and n_β is the number of moles of species β in the system. ξ_α is the extent of change occurring in the system, given by the law of definite proportions as the ratio of the change in number of moles of α, n_α, in a time dt to the stoichiometric coefficient of α, ν_α ($\nu_\alpha = 1$ unless chemical reaction occurs in the gas). Relation (9.10) implies that, for binary systems,

$$\frac{\partial \hat{\mu}_B}{\partial x_B} > 0 \quad \text{or} \quad \frac{\partial \hat{\mu}_B}{\partial x_A} < 0, \tag{9.10a}$$

where x_B is the mole fraction of component B.

To see how these conditions of stability apply to changes in a supersaturated vapor, suppose we examine a typical p–v diagram of a pure substance, as sketched in Fig. 9.1. Consider isothermal changes only. Since real systems are usually constrained such that their heat capacities are positive, we need only regard the condition of mechanical stability, i.e. the coefficient of compressibility is greater than zero.

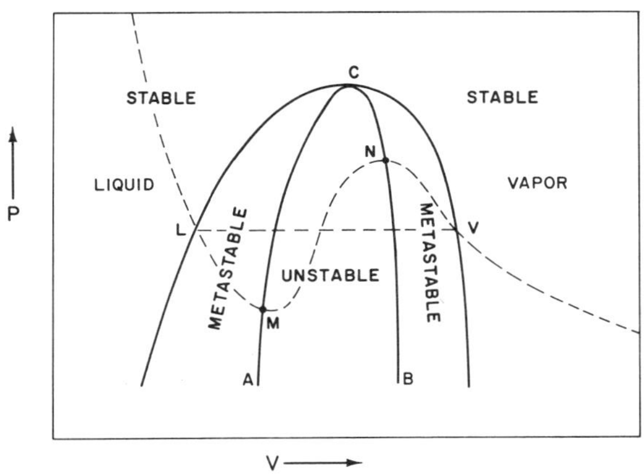

FIG. 9.1. Stability of phases as pictured on a pressure–volume diagram. (From Prigogine and Defay,[84] p. 231.)

In Fig. 9.1, the region in which two phases may be in equilibrium is given by the solid curve LCV, where C is the critical point. A dotted curve corresponds to a typical isotherm passing through the two-phase region. We see that where the isotherm passes through zones of single phase there exists stable equilibrium. Regions where the slope of the isotherm is negative in the two phase zone are metastable. The portion MN, where $(\partial p/\partial V)_T > 0$ refers to the unstable regime of the two-phase zone. Thus the condition

$$\left(\frac{\partial p}{dV}\right)_T = 0$$

marks the transition between metastabie and unstable thermodynamic states. When a vapor becomes supersaturated isothermally, the path of states attained would follow upward from V to N into the metastable zone. The extent to which the vapor would proceed towards

N once V is passed would depend on the rate of change of p and V along the isotherm, and the nuclei present in the system. It has never been possible to realize conditions of equilibrium macroscopically in the unstable region. Thus in the absence of nuclei within an infinitesimal distance of N along the isothermal, thermodynamic reasoning would indicate that the vapor state would collapse to liquid–vapor equilibrium corresponding to the location L.

It is possible to use equations of state like that of Van der Waals[84] to construct isotherms displaying zones of instability and metastability, as drawn in Fig. 9.1. However, it is important to realize that the question of phase changes does not depend upon the existence of an equation of state common to the two phases. Indeed, classical thermodynamics alone cannot indicate whether localized regions of two phases exist in a system, or how many of these zones develop. A more complete analysis requires a statistical study of the properties of the free energy functions, and this sort of investigation remains incomplete for all substances.

B. The Kinetics of New Phase Formation

Despite the inadequacies of the general theory of two-phase systems, considerable effort has been devoted to constructing a dynamic model of particle formation from limited knowledge of the thermodynamic properties of droplets or crystals in equilibrium with a supersaturated suspending medium.

In analogy to the theory of chemical reactions the formation of condensed phase in a supersaturated vapor is considered to be a chain reaction composed of two-body collisions, and thus, in the absence of spurious nuclei, the agglomeration process in a pure vapor may be written formally as

$$\begin{aligned}
e_1 + e_1 &\rightleftharpoons e_2 \\
e_1 + e_2 &\rightleftharpoons e_3 \\
e_2 + e_2 &\rightleftharpoons e_4 \\
e_3 + e_1 &\rightleftharpoons e_4 \\
&\cdot \\
&\cdot \\
e_j + e_i &\rightleftharpoons e_k.
\end{aligned} \tag{9.11}$$

Here e_i refers to agglomerates of additive mass, e_1 being a molecule of vapor.

Like the chemical analogy, the rate of formation of molecular clusters has an energy barrier associated with it. The estimation of this energy barrier, the transition probabilities between species i and the approximate form of the rate equation are the central results required of the theory.

The Activation Barrier and Equilibrium of Droplets in a Vapor

The significance and origin of the activation barrier to particle formation is most easily understood conceptually through the thermodynamic properties of a small droplet suspended in its own vapor.

Following Gibbs,† the work required to form a spherical embryo of k molecules, or the free energy of formation of a spherical embryo in a system containing n_G molecules of vapor and n_L in the liquid phase is[278]

$$\Delta \hat{G}_k = 4\pi R_k^2 \sigma + k(\hat{\mu}_L - \hat{\mu}_G), \tag{9.12}$$

where σ = the surface free energy per unit area of the embryo, $\hat{\mu}_L$ = the chemical potential of a molecule in the condensed phase, and $\hat{\mu}_G$ = the chemical potential of the vapor molecule.

For an isothermal process at an arbitrary vapor pressure p,

$$d\hat{\mu}_L = V_L \, dp \quad \text{and} \quad d\hat{\mu}_G = V_G \, dp$$

or

$$d(\hat{\mu}_L - \hat{\mu}_G) = (V_L - V_G)\, dp, \tag{9.13}$$

where V_L is the volume of a vapor molecule in the liquid phase and V_G is the volume of a vapor molecule in the gas phase. If the two phases are in equilibrium, the chemical potential

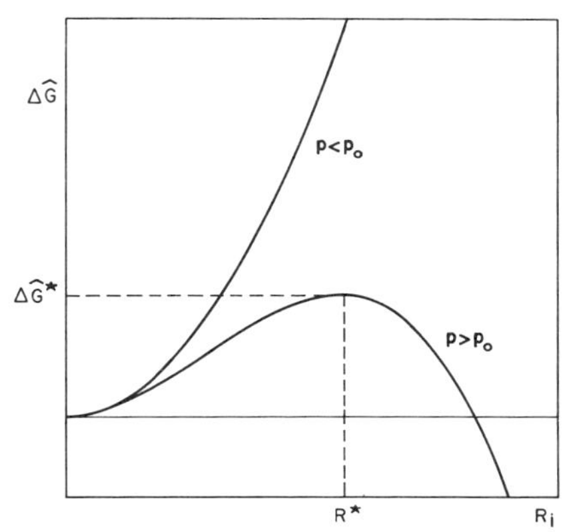

FIG. 9.2. The change in free energy between liquid and vapor phases as a function of droplet radius.

is the same for molecules in each phase. Thus in our example, the liquid will exist in equilibrium with a saturated vapor with pressure over a plane liquid surface. The vapor under consideration is at an arbitrary pressure so that the change in chemical potential $(\hat{\mu}_L - \hat{\mu}_G)$ can be obtained by integration of eqn. (9.13).

In the case of the vapor being a perfect gas, integration of (9.13) yields

$$\hat{\mu}_L - \hat{\mu}_G = -kT \ln (p/p_0), \tag{9.14}$$

noting that $V_G(= kT/p) \gg V_L$. The ratio p/p_0 is often called the *supersaturation ratio*, S_p.

† Abraham[279] recently has given arguments for the use of the Helmholtz free energy instead of the Gibbs function $\hat{G}$ in specifying the thermodynamics of small droplets. However, we adopt here the classical arguments since the final conclusions of both methods are essentially the same. Abraham's conclusion is not a new one since this distinction has been recognized by a number of workers (see, for example, Reiss[437]).

From eqn. (9.14) we may deduce that $(\hat{\mu}_L - \hat{\mu}_G)$ is positive when the vapor is unsaturated but negative for supersaturated vapor. This sort of behavior is illustrated for spherical droplets in Fig. 9.2. When the vapor is unsaturated, the free energy change increases rapidly with increasing size of the embryo. However, when the vapor becomes supersaturated the free energy of formation of the embryo displays a maximum value $\Delta\hat{G}^\star$. at a radius $R^\star$. The conceptual nature of an activation barrier now is clarified. We identify this barrier $\Delta\hat{G}^\star/kT$ with the maximum value of the free energy of formation of an embryo or an agglomerate of molecules in its own vapor. Under conditions of supersaturation where $p > p_0$, embryos of radii less than $R^\star$ are unstable and tend to disappear, whereas droplet embryos of radii larger than $R^\star$ tend to grow without limit and become macroscopic droplets. Agglomerates having critical radii $R^\star$ exist in an unstable equilibrium with the supersaturated vapor.

In terms of the chain mechanism of (9.11), the activation barrier implies that for a given supersaturation ratio the probability of the forward reaction is smaller than the backward reaction for embryos less than critical size. For an agglomerate of radius larger than $R^\star$, however, the probability of a condensation reaction is greater than that of evaporation, and the small nuclei tend to grow larger and larger.

The value of the critical radius may be determined from eqn. (9.12) by maximizing the free energy change, i.e. setting $\partial\Delta\hat{G}k/\partial R_k = 0$. For spherical embryos, this results in

$$R^\star = \frac{2\sigma V_L}{kT \ln S_p}. \tag{9.15}$$

Since droplets of radius $R^\star$ are in equilibrium with vapor at pressure p, we can rewrite eqn. (9.15) as

$$\ln S_p = 2\sigma V_L/kTR^\star. \tag{9.16}$$

This is the well known *Kelvin–Gibbs relation* for the vapor pressure ratio at equilibrium over a curved surface.

Using eqn. (9.15) in eqn. (9.12) we find after maximizing eqn. (9.12) that the free energy of formation of the critical sized embryo reads

$$\Delta\hat{G}^\star = \tfrac{4}{3}\pi R^{\star 2}\sigma = \frac{16\pi\sigma^3 V_L^2}{3(kT \ln S_p)^2} \tag{9.17}$$

Formulation of the Rate Expressions

The sequence of chain reactions in eqn. (9.11) express the fact that, in general, the process of spontaneous condensation, involving collision between molecules and between molecules and clusters of molecules, is accompanied by collisions and sticking of clusters (*coagulation*). The process of aerosol formation then is viewed as the result of many binary collisions occurring simultaneously, with the added possibility for splitting of single particles into two. In other words, nuclei may enter the kth class (k is the number of molecules in an embryo cluster) by two mechanisms: (a) by collisions among particles smaller than the

kth species, and (b) by the splitting of larger clusters such that one of the products is of the kth class. The former case includes the special case of $k-1$ particle with a single molecule while the latter mechanism incorporates evaporation, wherein one of the products split off is a molecule.

Agglomerates may leave the kth class (a) via collisions with any other class including molecules (class 1), or (b) via splitting into two new species such that the total number of species of 1 is k.

The rate of generation of particles in the kth class in particle per unit time per unit volume takes the form

$$G_k = \tfrac{1}{2} \sum_{i+j=k} \mathcal{N}_{ij} + \sum_{i=1} \mathcal{S}_{ik} - \sum_{i=1} \mathcal{N}_{ik} - \tfrac{1}{2} \sum_{i+j=1} \mathcal{S}_{ij}, \qquad (9.18)$$

where $\mathcal{N}_{ij}$ is the number of effective collisions in a unit of time per unit volume between two classes of clusters composed of i and j molecules, and $\mathcal{S}_{ij}$ denotes the number of clusters split into the ith and jth components per unit time per unit volume.

Friedlander[280] has noted that both the classical theory of nucleation and Smoluchowski's theory of coagulation[281,282] represent special limiting cases of eqn. (9.18). In the theory of coagulation, for example, the terms for evaporation and splitting are disregarded and the collisional process leading to growth is considered irreversible. As discussed later in Chapter 10 the term $\mathcal{N}_{ij}$ is derived as

$$\mathcal{N}_{ij} = b_{ij} n_i n_j,$$

where b_{ij} is the collisional parameter, given for Brownian motion, in a continuum, for example, by eqn. (10.10), and n_i is the concentration of species i in number per unit volume. Thus the rate equation becomes:

coagulation:

$$\frac{dn_k}{dt} = G_k = \tfrac{1}{2} \sum_{i+j=k} b_{ij} n_i n_j - n_k \sum_{i=1}^{\infty} b_{ik} n_i. \qquad (9.19)$$

In the classical theory of particle generation by nucleation (e.g. refs. (278), (280)] all of the terms in the summations of eqn. (9.18) are disregarded except for those involving the single molecule species and clusters of molecules. In other words:

nucleation:

$$G_k = \mathcal{N}_{1,k-1} + \mathcal{S}_{1,k+1} - \mathcal{N}_{1,k} - \mathcal{S}_{1,k}. \qquad (9.20)$$

This assumption may be justified by the fact that the concentration of single molecules present in a supersaturated medium will overwhelm that of any other species. Collisions between single molecules and embryos will occur much more frequently than encounters between embryos. Furthermore, evaporation of single molecules is presumed to be very much more probable than the splitting of clusters of molecules. Thus the analytical model describing nucleation in a supersaturated medium requires that the effective time of collision and sticking of vapor molecules is much shorter than the time of collision between agglomerated species. We note, of course, that unless relatively large supersaturations are maintained by external means, there will be an advanced stage of the process in which coagulation may become important.

With the restrictions of eqn. (9.20), we then confine our attention only to the sequence

$$e_1 + e_{k-1} \rightleftharpoons e_k$$

in eqn. (9.11). We further assume for the moment that nucleation occurs in a pure vapor.

Equation (9.18) may be rewritten in the following form:

Nucleation:

$$\frac{dn_k}{dt} = G_k = j_{k-1} S_{k-1} n_{k-1} + j'_{k+1} S_{k+1} n_{k+1} - j_k S_k n_k - j'_k S_k n_k. \tag{9.21}$$

In this modification, j_k denotes the flux of molecules of the condensing species onto kth sized embryos of surface S_k and j'_k refers to the flux of molecules evaporating from the kth species of embryos. Then $j_k S_k n_k$ and $j'_k S_k n_k$ are interpreted as the overall rate at which embryos are removed from the kth class by condensation or evaporation respectively. And $j_k S_k$ equals, for example, the probability per unit time and concentration of e_k of a transition from e_k to e_{k+1}.

The flux of condensing species may be calculated in several ways. For example, the term $j_k S_k n_k$ has been determined in Smoluchowski's coagulation theory by a diffusion mechanism, as discussed in Section 10.1. For molecular transfer, however, the kinetic theory of gases is applied, and we find with rearrangement of the collision rate given in (10.24) that

$$j_k = j = \left[\frac{(1 + k^{\frac{1}{3}})}{k^{\frac{1}{3}}}\right]^2 \left(\frac{1 + k}{k}\right)^{\frac{1}{2}} p/(2\pi m_G kT)^{\frac{1}{2}}. \tag{9.22}$$

Here m_G is the molecular mass of condensable vapor.

Account is usually taken for the non-equilibrium effect of accommodation on surfaces of species by multiplying eqn. (9.22) by α_c, the accommodation coefficient for condensation on the bulk phase. Care has to be taken in such semi-empirical corrections, however, to remember that identification of an α_c with an individual molecule involves an assumption of an average temperature of a molecule. Thus, it is necessary here to qualify the use of one value of α_c as an average taken over embryos having various kinetic energies given by a statistical distribution. Furthermore, equations like (9.22) cannot be correct, in general, if there is non-thermal accommodation. Hence, to keep the classical theory relatively simple, we shall assume for the subsequent argument that $\alpha_c = 1$, and there is thermal accommodation.

Most theories of nucleation have used for j_k the effusion rate to a flat surface[278]

$$j_k = p/(2\pi m_G kT)^{\frac{1}{2}}. \tag{9.23}$$

We note, of course, that this reflects the implicit assumption that the embryos are quite large compared with the size of condensing molecules. That is, (9.22) reduces to (9.23) when $k \gg 1$.

The probabilities for backward reaction are determined by applying the principle of microscopic reversibility, or

$$j_{k-1} S_{k-1} = j'_k S_k n'_k / n'_{k-1}. \tag{9.24}$$

To estimate j'_k then the equilibrium concentrations of the various cluster species n'_k have to be known. These may be estimated with the aid of the expressions for a series of equilibrium constants K_k.

The Evaporation Rate and the Distribution of Small Clusters at Equilibrium

To formulate the kinetic equations for embryo formation one uses the usual assumptions stemming from chemical kinetics that the rate of formation of nuclei proceeds under conditions applicable to the restriction of microscopic reversibility (near equilibrium). Hence a value of the concentration of various species k suspended at equilibrium in the vapor has to be estimated. To carry out such a calculation one has to elucidate further the equilibrium properties of various agglomerated species. If we assume that the clusters are non-interacting, the chemical potential of the kth species is

$$\hat{\mu}_k = -kT \ln \frac{\tilde{Q}_k}{V_{\text{syst}}} + kT \ln n_k, \tag{9.25}$$

where $\tilde{Q}_k$ is the partition function for species e_k, V_{syst} is the volume of the system, and n_k is the concentration of e_k in the system. At equilibrium the change in free energy is zero, hence the equilibrium concentration of e_k, n'_k, is given by

$$n'_k = K_k n_G^k, \tag{9.26}$$

where K_k is the equilibrium constant

$$K_k = \frac{\tilde{Q}_k/V_{\text{syst}}}{(\tilde{Q}_G/V_{\text{syst}})^k}. \tag{9.27}$$

Such a treatment of the vapor as an ideal mixture of clusters has a formal resemblance to the virial expansion for an imperfect gas using mathematically defined non-interacting clusters. This approach has led to an extension of Van der Waals' work to higher order approximations of imperfect gases. The formal identification of the physical embryos with mathematical clusters cannot be taken too far, however. For instance, an exact expansion of the equation of state of a real gas yields terms resulting from the interactions between various species (see, for example, Hill[(6)]). Similarly, the mathematical clusters of Mayer, for example,[(6)] do not display properties of physical embryos in that they may have negative partition functions.†

The embryos are then considered to have a stable lifetime as a molecular configuration which is long compared with collision times of gas molecules. If the gas molecules contain no internal degrees of freedom, the equilibrium constant may be evaluated from classical statistical mechanics.

$$K_k = (k!)^{-1} \int_{\Sigma_k} \ldots \int \exp(-\Phi_k/kT)\, dx_{21}\, dy_{21} \ldots dz_{i1}, \tag{9.28}$$

after integrating over the momenta of all molecules.‡ Here Φ_k denotes the potential energy of interaction of the k molecules in e_k, x_{i1}, y_{i1}, z_{i1} equal the spatial coordinates of the kth molecule in relation to the first molecule, and Σ_k refers to the configurations for which the gas molecules are bounded in the cluster.

† Nevertheless, it is possible with assumptions to derive a kinetic theory of nucleation from Mayer's approach (e.g. Parlange[(284)]) which yields a result equivalent to the classical thermodynamic treatment.

‡ At equilibrium, the contributions to the momentum integrals in (9.28) by $\tilde{Q}_k$ just equal those of $\tilde{Q}_1^k$. Integration then proceeds over all coordinates of single molecules and one of the molecules in the cluster. Because the clusters do not interact, this gives a contribution of V_{syst} for each molecule and each cluster. These factors exactly cancel the V_{syst} present already to yield eqn. (9.28).[(283)]

To evaluate the integrals in the expression for the equilibrium constant K_k, in general, would require many tedious computations, and the criteria for determination of the binding molecules becomes difficult if k exceeds 2. So far only the values of K_k for dimerization reaction, $e_1 + e_1 \rightleftharpoons e_2$ have been estimated from eqn. (9.28) by Stogryn and Hirschfelder.[285] However, an approximate technique for obtaining K_k up to $k = 8$ has been reported by Reed.[286] This method disregards all but the nearest neighbor interactions by treating the embryonic clusters as polymers of a configuration of lowest energy and greatest symmetry, and by confining the molecular vibrations to harmonic oscillations about their equilibrium positions.

Andres[283] has indicated that Reed's approximation agrees with the more exact estimates of Stogryn and Hirschfelder for $k = 2$ only at very low temperatures. The higher the temperature, the more the deviation between the two calculations. This seems reasonable since the high temperatures should give rise to stronger perturbations from equilibrium configurations, and higher probabilities of cluster states that exceed minimal energy.

Because of the difficulties in evaluating the configuration integrals for K_k beyond $k = 2$, most workers have reverted to a substantial simplification by identifying the thermodynamic properties of the bulk (liquid) phase with parameters in $\tilde{Q}_k$ to evaluate n'_k.

Several methods have been proposed for calculating the equilibrium distribution of embryos using the thermodynamics of the bulk phase.[278, 287] However, they all result in a form for n'_k equivalent to a Van't Hoff isotherm. Making use of the fact that the standard chemical potential $\hat{\mu}_k^\circ$ is related to the partition function as

$$\frac{\hat{\mu}_k^\circ}{kT} = -\ln \frac{kT\tilde{Q}_k}{V_{\text{syst}}}$$

and considering the embryos in the supersaturated vapor as a dilute, ideal solution, we find that

$$n'_k = N_{\text{mol}} \exp\left(-\Delta\hat{G}_k^\circ/kT\right). \tag{9.29}$$

N_{mol} here is the total number of molecules in the system per unit volume where the standard free energy change $\Delta\hat{G}_k^\circ$, is identified with $\Delta\hat{G}_k$ in eqn. (9.12). As a first approximation, N_{mol} is often set equal to n_G or p_0/kT, the concentration of vapor molecules in the system under equilibrium condition. Since $n_G \gg n_k$ prior to the rest of nucleation, there is little error introduced by this assumption.

Experimental observations of the onset of spontaneous nucleation in supersaturated vapor have suggested that embryos become stabilized in the supersaturated vapor for further growth (i.e. reach critical size) after agglomerating to 100 molecules or less. It is conjectured[288, 289] that the application of the thermodynamic properties of the bulk phase to clusters of this microscopic size range breaks down. Therefore, dissatisfaction in the early theories of cluster formation has led back to the further search for successive approximations of the configurational integrals like eqn. (9.28). Some theoretical extensions have made use of corrections for changes in surface tension with size of droplets.[287] These models follow the reasoning using the macroscopic surface energy but apply statistical mechanical corrections for size of Tolman or Kirkwood and Buff, as described, for example, by Hill.[6]

Reiss,[288] on the other hand, has attempted to construct the equilibrium distributions of clusters with an idealized model of a vapor consisting of arbitrary spatial divisions each containing a set of molecules equal to the number necessary to form a critical embryo.

This simplified statistical model failed to permit exchange of vapor between cells so that it could not be considered a useful calculation. Though Reiss' model predicted a distribution of the same form as eqn. (9.29), the mathematical complexity of removing the cell restriction precluded efforts to bring the theory to completion.

Recently, further exploration of certain subtleties in specifying the partition functions of the embryos has been reported. For instance Kuhrt[290] and Dunning[291] have re-examined the relations for $\tilde{Q}_k$, but intuitively split the partition function into a part corresponding to the translation and rotation of the embryo in its entirety, and a remainder approximated by a contribution of $(k - 2)$ molecules in the condensed phase multiplied by the usual macroscopic free energy change. Lothe and Pound[292, 293] have also argued that a correction for $\tilde{Q}_k$ should be included to estimate the equilibrium concentration of embryos. However, their model differs from that of Kuhrt and Dunning in that the constraints placed on the partition functions stems from the degrees of freedom unavailable to molecules in the bulk liquid phases as contrasted with molecules in the vapor phase. Lothe and Pound[293] have indicated that, instead of eliminating the complete bulk contribution of two molecules, account should be taken only of the six vibrational modes characteristic of the bulk liquid phase. The difference between Kuhrt and Dunning's model for the embryo and that of Lothe and Pound leads to a small correction in the equilibrium distribution in the former case, but a very large one up to 10^{17} for water vapor in the latter case.

The intuitive arguments of Lothe and Pound have been criticized by Reiss and Katz,[294] who attempted to derive a statistical mechanical correction factor for n'_k from considerations of the evaluation of $\tilde{Q}_k$ in a stationary system of coordinates, but allowing for variation within a droplet of its center of mass. More recent calculations of both Abraham and Pound[295] suggest that the correction or replacement factor multiplying the right side of eqn. (9.29) should be of the order of 10^8, but is strongly dependent on the model assumed for the molecular cluster. This factor appears to be much closer in agreement with the newer calculation of Reiss *et al.*,[296] who suggest for water vapor at 300°K, that the correction factor should lie between 10^3 and 10^6, again depending on the cluster model. Even though there can be a difference of several orders of magnitude in nucleation rate predicted by the theories, such corrections produce a relatively small adjustment in the estimate of supersaturation ratio for the onset of observable condensation. Until recently, it was not possible to distinguish experimentally between the rates forecast by the different models. However, Jaeger, *et al*[444], Dawson, *et al*[446], and Katz[443] have reported measurements showing distinctions between the claculated nucleation rates of several materials based on the models.

From this brief discussion, we can readily see that the estimation of equilibrium properties of molecular clusters, equivalent to the smallest of aerosol particles, is in a somewhat unsatisfactory state. Despite the fact that these properties are essential to formulation of the kinetic theory of nucleation, as we shall see below, only those properties of the largest and the smallest extremes in size can be estimated. The application of the Van't Hoff relation, or an equivalent form of eqn. (9.29) may be applied in the former case, while the tedious method of Stogryn and Hirschfelder with (9.26) may be used. Between these limiting regimes, we must rely on rather crude approximations for evaluating the properties of clusters. These approximations are largely an outgrowth of the classical theory of Volmer and Weber, Becker and Döring, Zeldovich and Frenkel, and others.[278, 287]

Andres[283] has illustrated the variety of results obtained in the different methods for calculating the equilibrium constant K_k. Several computations have been made for Argon at 85°K, and are tabulated in Table 9.1.

TABLE 9.1. EQUILIBRIUM CONSTANTS FOR ARGON AT 85°K (FROM ANDRES[283])

	$\ln K_2$	$\ln K_4$	$\ln K_6$	$\ln K_8$
Andres and Reed[283]	48.7	148.8	250.4	352.1
Courtney[289]	56.0	153.3	249.6	346.5
Kuhrt and Dunning[290, 291]	58.7	153.2	248.0	343.7
Lothe and Pound[292]	39.7	134.2	229.0	324.7

Andres' estimates come from an extension of Reed's model by evaluation of the set of integrals in eqn. (9.26) serially. Using all of Reed's assumptions and neglecting the off diagonal terms in the matrix for the interaction force constant,

$$K_k = (8\pi^2 r_e)^2 (2^{\frac{1}{2}})^{k-3} \left[\frac{2\pi kT}{(\partial \Phi_k/\partial r)_{r=r_e}} \right]^{3(k-2)/2} \exp(\gamma_k d_\phi/kT), \tag{9.30}$$

where r_e = separation of molecules in the embryo at equilibrium, γ_k = number of bonds of (assumed) length r_e in the embryo, d_ϕ = depth of the minimum in the potential energy curve $\Phi_k(r)$.

Assuming the approximations buried in (9.29) give a reasonable picture of cluster behavior at equilibrium, we find that the extensions of the classical theory by Courtney,[289] Kuhrt,[290] and Dunning[291] are remarkably good evaluations of K_k for this monatomic gas. This is of special significance because the results of these three investigators are nearly the same as the older "classical" treatment of Volmer and Weber.[297]. Lothe's and Pound's[292] equations appear to give estimates several orders of magnitude too low as compared with the other estimates. If Abraham and Pound's[295] newer results were used, however, better consistency with Reed's method would be obtained.

Encouraged by evidence such as that in Table 9.1, if we adopt the Van't Hoff isotherm for the equilibrium concentration of species k, we then find that

$$j_k' = \frac{\alpha_c p}{(2\pi m_G kT)^{\frac{1}{2}}} \left[\frac{S_{k-1}}{S_k} \exp\{[-\sigma(S_k - S_{k-1}) + \ln S_p]/kT\} \right] \tag{9.31}$$

assuming the accommodation coefficient for evaporation is the same as that for condensation.

C. Solutions to the Kinetic Equations for Nucleation

Having a set of differential equations for the rate of change of cluster or embryo species suspended in a supersaturated vapor, all that remains is to determine the rate of production of stable embryos.

No general analytical solution to eqns. (9.21) with (9.31) has been obtained as yet. However, in the case of a steady state distribution of k species, eqn. (9.21) may be integrated once to give

$$I = jS_{k-1}n_{k-1} - j_k' S_k n_k, \tag{9.32}$$

where I is the (constant) rate of formation of clusters of size k. This equation is a first order difference equation so that its solution will contain an arbitrary constant determined by a boundary condition. One natural assumption would fix the number of gas molecules N_1. Friedlander[280] has noted under these circumstances that the solution to eqn. (9.32) cannot be sufficient to determine the rate I. For given values of temperature and supersaturation, the spectrum will assume a shape in steady state corresponding to any value of I.

The existing solutions for the steady state case[(278)] assume a certain form of the embryo spectrum, which is given by the equilibrium distribution, eqn. (9.29). The attention is centered on the embryos of critical size $R^\star$. It is assumed that the rate of production of stable embryos, or clusters of molecules that will continue to grow in the supersaturated medium, depends on the rate of formation of embryos of initial size $R^\star$. In other words, all clusters attaining a size $R^\star$ are sufficiently stable to evaporation that they are eliminated from the system and are replaced by an equivalent number of single molecules.

In connection with this question, the arguments about removal of stable species from the system after their appearance are not absolute requirements of the steady state theory. Suppose we write eqn. (9.32) in continuous form and carry out the integration over all species (see also Frenkel,[(278)] pp. 394–395) with $\partial n(\tilde{R})/\partial t = 0$:

$$\frac{1}{j}\int_1^\infty \frac{I(\tilde{R}, t)\,\mathrm{d}\tilde{R}}{S(\tilde{R})\,n(\tilde{R})} = \frac{n(1)}{n'(1)} - \frac{n(\infty)}{n'(\infty)}. \tag{9.32a}$$

Here $\tilde{R}$ is the continuous variable for species in the radius range $\tilde{R}$, $\tilde{R} + \mathrm{d}\tilde{R}$, equivalent to R_k, I is the flux of embryos through size $\tilde{R}$ and time t. Equation (9.32a) is subject to the boundary conditions $n(1)/n'(1) \equiv 1$, and $n(\infty)/n'(\infty) = 0$, with n' equal to the distribution at equilibrium [Eqn. (9.32a) with these boundary conditions is essentially Zeldovich's[(278)] approximation.]. The substitution of the two boundary conditions into the right-hand side of eqn. (9.32a) indicates that the integral on the left must be essentially constant within the steady state assumption. Suppose that I is non-zero at a value corresponding to the smallest stable embryo size $R^\star$. Since $1/n$ must be nearly a delta function with its peak at $R^\star$, the flux I of embryos through $\tilde{R} = R^\star$ can be taken out of the integral. The integral is no longer a function of time so that $I(R^\star)$ must tend to be constant. Reiss has pointed out that the use of this particular argument does not require that $I(\tilde{R})$ be constant, and that a universal or global steady state prevail. Yet $I(R^\star) \approx$ const. does amount to the requirement that an incubation period or time lag is involved in the nucleation mechanism (see below). After this time has elapsed, the constriction at stable embryo size tends to force the system to a quasi-steady state. By accounting for this effect, one can avoid the requirements for removal of $\tilde{R} > R^\star$ from the nucleating system.

For a given temperature and concentration of single molecules, the steady state rate of nucleation may be determined from the difference equation (9.32) after considerable algebraic manipulation only in terms of the flux for condensation:[(298)]

$$I = \left(\sum_{k=1}^{J} \frac{1}{j_{k+1} S_{k+1} n'_k}\right)^{-1}. \tag{9.33}$$

This expression has been attributed to Becker and Döring.[(299)] Here the reciprocal of the net rate is set equal to the sum of the reciprocals of all the equilibrium rates for the various transitions. Frenkel[(278)] has described a simpler approximate analysis of Zeldovich which yields essentially the same expression for the rate of cluster production.

Since no sharp change in the free energy of formation is found corresponding to a transition of vapor molecules to an agglomerate, the summation in (9.33) has to be extended past the nucleus of critical size until additions of terms in the sequence produce a negligible modification in I. As a first approximation, however, the principal term in the summation, equivalent to the equilibrium rate of the reaction.

$$e_N + e_I \rightarrow e_{N+1}$$

is assigned as the rate of nucleation where e_N is an embryo of size $R^\star$. This transition, of course, simply contributes one of the largest terms in the sum of eqn. (9.33) since the nucleus has the lowest equilibrium concentration. Thus the crudest approximation for the solution to (9.33) assumes that

$$I^\circ \equiv I_{eq} \approx j_{N+1} S_{N+1} n'_N \quad \text{and} \quad n'_N \approx n_G \exp(-\Delta\hat{G}^\star/kT) \quad \text{from eqn. (9.29).}$$

And

$$I^\circ \approx \frac{\alpha_c p}{(2\pi m_G kT)^{\frac{1}{2}}} 4\pi R^{\star 2}\, n_G \exp[-\Delta\hat{G}^\star/kT]. \tag{9.34}$$

This result is the one obtained from the classical analysis of Volmer and Weber.[297]

By taking additional terms around the zone of critical size N in the summation of (9.33), the result of Becker and Döring[299] and Zeldovich[278] gives the steady state solution

$$I^1 = \frac{\alpha_c p}{(2\pi m_G kT)^{\frac{1}{2}}} Z(4\pi R^{\star 2})\, n_G \exp[-\Delta\hat{G}^\star/kT], \tag{9.35}$$

where the non-equilibrium correction is

$$Z = \left(\frac{\Delta\hat{G}^\star}{3\pi kTn_G^2}\right)^{\frac{1}{2}} = \frac{(kT)^2 \ln^2 S_p}{8\pi\sigma(\sigma kT)^{\frac{1}{2}}}. \tag{9.36}$$

The steady state solution of Becker and Döring, and Zeldovich accounts for the net rate of nuclei formation while the earlier model of Volmer and Weber considers only the "forward" rate of nuclei production. Hence the correction in eqn. (9.36) indicates that the equilibrium rate of nuclei formation I° is always larger than the steady state value I^1. We also observe[283] that the contributions to the various reaction steps to the overall rate are independent of their position relative to the nucleus. This does not correspond to the usual assumption that passage through the "critical" state is equivalent to reaction.

Non-steady State Approximations

Some time ago, workers recognized that the steady state assumption might be inadequate for determining the rate of particle formation in a supersaturated vapor, especially during rapid cooling. Although complete analytical solutions of the rate equations have not been found, several estimates of the time to reach steady state in the nucleation process have been determined. Several early theories for the time of induction are treated in the review of Hirth and Pound.[287] More recently, Andres and Boudart[301] examined this question and found that indeed steady state solutions to equations of the form of eqn. (9.33) can exist as long as $\Delta\hat{G}_k/kT \gg 1$. They determined that the time lag for establishment of steady nucleation is given by

$$\tau_k = \sum_{k=1}^{J} \frac{n_k^{ss} - n_k(0)}{I_{k-1}}, \tag{9.37}$$

where n_k^{ss} is the concentration of k species at steady state, and $n_k(0)$ is the initial value of the kth concentration.

Courtney[289] has solved numerically the set of equations (9.21) for several cases of water vapor condensing at different temperatures. His measure of the induction time corresponded to the time required for the nucleus concentration to reach 95 per cent of its steady state value. Using Courtney's transition probabilities, Andres[283] has calculated values of τ_k from (9.37) for water vapor. Comparison between the two results is shown in Table 9.2. These computations display only order of magnitude agreement, but they illustrate, as in the case of earlier calculations, that the time to reach steady state is the order of microseconds.

TABLE 9.2. TIME LAG τ_k AND 95 PER CENT TIME FOR SUPERSATURATED WATER VAPOR[a] (FROM ANDRES[283])

T(°C)	S_p	$n^\star$	τ_k (μsec)	95% time, (μsec)
−10	5	69	0.9	1.4
−40	10	43	3.8	19
−60	20	30	11.2	27

[a] T is the temperature of the system, S_p is the ratio of the actual pressure of water vapor to the equilibrium vapor pressure, $n^\star$ is the number of molecules in the nucleus.

The time lag of microseconds is negligible for most experiments involving cloud chambers or for the formation of aerosols in nature. However, this lag must become important under conditions where particulate clouds are generated during combustion, by rapid expansion, or by rapid mixing in supersonic nozzles or turbulent jets.

D. The Formation of Solid Particles by Nucleation

Since only spherical embryos were considered in the theory, this approach applies primarily to the generation of liquid droplets. The nucleation theory can be modified, however, to account for the formation of solids of arbitrary shape, at least approximately.

In the theory of nucleation of liquids, it was assumed that the chemical potential could be defined completely by a single continuous parameter, either the radius of curvature, or the number of (spherical) molecules in the embryo. Furthermore, because of the high mobility of molecules in a liquid, the only molecular process considered was the absorption of vapor at the embryo surface.

Unlike the case of liquid droplet formation, crystallite nucleation must involve a surface free energy that is inherently anisotropic. And the relatively low mobility of molecules requires consideration of the addition of molecules to a faceted surface by adsorption, with surface diffusion and accretional steps.

The question of the dependence on shape of the surface free energy has been examined in light of the well known Gibbs–Wulff theorem.[(287)] This approximation should apply best to larger faceted bodies of metastable equilibrium shape. The surface energy of small clusters having kinks, steps, edges, and corners whose dimensions are the order of the particle radius may not be amenable to description by the macroscopic theory.

Analysis for the nucleation process incorporating various mechanisms of surface accretion or adsorption of molecules requires a detailed theory covering the growth processes of solid lattice structures. Approximate theories of nucleation involving rather drastic assumptions have been summarized by Hirth and Pound.[(287)] The key problem lies in evaluation of the non-equilibrium factor. They note that these treatments all yield a rate expression for nucleation that is analogous to the expression for liquids, eqn. (9.35).†

Depending on the temperature and pressure of the vapor, nucleation to form droplets or crystallites may develop. It sometimes is of interest to know whether the sublimation or condensation mechanism may control the clustering process. Fletcher[(302)] has given a rough analysis of this question. Following his argument, we may rewrite the approximate rate relation for liquids, eqn. (9.34), as

$$I^\circ = \mathscr{a}_L \exp(-\Delta\hat{G}_L^\star/kT), \tag{9.35a}$$

where $\mathscr{a}_L$ is the pre-exponential factor for liquids. The nucleation of solid particles takes the form

$$I^\circ = \mathscr{a}_S \exp(-\Delta\hat{G}_S^\star/kT), \tag{9.38}$$

where $\mathscr{a}_S = \left(\dfrac{\alpha_c p}{2\pi m_G kT}\right)^{\frac{1}{2}} 4\pi R^{\star 2} \Lambda n_G$,

Λ = correction factor for the surface area of the solid.

$\Delta\hat{G}_S^\star = \dfrac{16\pi\sigma_{SG}^3\,\xi V_S}{3(kT\ln S_p)^2}$,

σ_{SG} = free energy per unit area of the solid-vapor surface.

V_s = molecular volume in the solid phase.

$\xi = \Lambda^3/\Lambda'^2$, the correction Λ' accounts for the modification in surface volume of the non-spherical crystallite.‡

If the clusters are relatively large, $\mathscr{a}_S \approx \mathscr{a}_L$ within an order of magnitude. Comparison between the rates of formation of crystals and droplets then suggests that crystals will predominate if

$$\Delta\hat{G}_S^\star < \Delta\hat{G}_L^\star. \tag{9.39}$$

† It is of interest to note here that the treatment of Reiss *et al.*[(296)] is restricted implicitly to liquids because of the way the phase integral is calculated. For solid embryos, the large rotational contribution has to be estimated separately, giving a factor the same order as found by Lothe and Pound.[(293)] However, the latter authors appear to have applied their arguments largely to cases where the embryos are likely to be liquid.

‡ We note that, in general Λ and Λ' are larger than unity, but the approach one as a polyhedral shape becomes spherical.

Making use of the expression for $\Delta\hat{G}_S^\star$, the inequality of (9.39) can be written as

$$\frac{\ln S_p\,(\text{solid})}{\ln S_p\,(\text{liquid})} > \xi^{\frac{1}{2}} \frac{V_S}{V_L}\left(\frac{\sigma_{SG}}{\sigma_{LG}}\right)^{\frac{3}{2}}, \tag{9.40}$$

where the subscripts to σ are added to distinguish this surface free energy for the liquid–vapor interface from σ_{SG}.

For appreciable rates of nucleation, finite values of I° must exist, and this fixes the pressure p as some function of temperature. Then eqns. (9.38) and (9.40) can be solved simultaneously to find T and S_p for which sublimation becomes important.

In the case of water vapor, for example, the right-hand side of eqn. (9.40) has approximately a constant value larger than one. The left-hand side is greater than unity at temperatures below 0°C, but its value depends on the vapor pressure of the water. If it is assumed that $\sigma_{SG} = \sigma_{SL}$, one finds that sublimation occurs mainly below −65°C for water vapor.

E. Nucleation from Multicomponent Gases

The analysis of homogeneous nucleation of a supersaturated vapor in a multicomponent gas has not appeared as yet.† However, theoretical work along the lines of the classical theory above has been reported for binary systems.

Ignoring the calculation of the non-equilibrium correction, a rough approximation for the rate of formation of nuclei may be derived that is analogous to eqn. (9.34). Flood[303] demonstrated that with the assumptions used to obtain eqn. (9.34), the Gibbs free energy of formation of a critical nucleus in a binary vapor has the form of (9.12) wherein the surface energy $\bar{\sigma}$ and $\Delta\hat{G}^\star$ are the values for the bulk liquid solution assumed equal to cluster solution. Denoting x_A as the mole fraction of the first component in the critical sized embryo and $\bar{V}_L$ as the average molecular volume of the solution,

$$\Delta\hat{G}_k = -\frac{kT}{\bar{V}_L}\left[x_A \ln (p_A/p_{oA}) + (1 - x_A)\ln (p_B/p_{oB})\right], \tag{9.41}$$

where p_{oA} is the equilibrium vapor pressure of component A over a bulk solution of the same composition as the nucleus.

The composition of the critical cluster may be calculated from the thermodynamic condition that the chemical potential of each component must be equal to that in the supersaturated vapor. Hence, making use of the conditions that the critical radius is determined by $\partial\Delta\hat{G}_k/\partial R_k = 0$ and $\partial\hat{G}_k/\partial x_A = 0$, a rate expression is obtained that is the same form as eqn. (9.34).

One of the uncertainties in applying the equilibrium model for calculations of nucleation rates in binary systems is the form and magnitude of the pre-exponential factor in the steady state rate equation. Some time ago Reiss[438] devised a kinetic theory for binary vapors along the lines suggested by the analyses of Becker and Döring, and Zeldovich. Reiss constructed a rate equation of the same form as eqn. (9.35) for the binary system, giving an explicit relation for Z in terms of the properties of the two components of the system. For a vapor (A)–gas(B) combination where the concentration of A is dilute, for example, Reiss finds that[437]

$$Z \approx \frac{p_A}{(2\pi m_A kT)^{\frac{1}{2}}} 4\pi R_{AB}^{\star 2} n_G \Omega_{AB},$$

† The effect of association of molecules on the nucleation of a pure vapor has been worked out by Katz *et al.*[300] This extension of the classical theory may bear on the behavior of multicomponent systems. They found that the influence of association suggests that the nucleation rate is decreased over the unassociated vapor for a given supersaturation.

where $R^{\star}_{AB}$ is the embryo of critical size, composed of A and B molecules, $n_G = n_A + n_B$, and Ω_{AB} is a complicated function of the surface energy of the binary liquid embryo.

Reiss' model has been applied by Doyle[439] to the self-nucleation of sulfuric acid–water vapors. He found, typically, that rapid nucleation could occur in such systems with 50 per cent relative humidity at partial pressures in the range of 10^{-8}–10^{-10} mmHg. Considerable uncertainty in Doyle's calculations exist, however, because inadequate data on the partial pressure of sulfuric acid over its aqueous solutions, and on the dominant sulfuric acid bearing species in a humid gas.

F. Experimental Observations of Homogeneous Nucleation

To investigate the nucleation of condensed phases in a supersaturated vapor, supersaturation has to be produced without introducing cold surface into the vapor. The presence of such surfaces will allow for relief of the supersaturated regime by vapor diffusion to the surfaces. Perhaps the most common method employed for generating highly supersaturated vapor is the expansion cloud chamber. This device, first introduced by Wilson[268] at the turn of the century, has been used by many investigators in attempts to obtain measurements for verifying the theory of nucleation. These kinds of chambers chill a vapor by approximately adiabatic and reversible expansion, permitting the estimation of the temperature and vapor pressure change and consequently the supersaturation during the stages of expansion.

In addition to devices using expansion cooling, supersaturation may be attained in chambers by diffusion of vapors from a warm saturated surface to a cooler saturated surface in a stagnant medium. The degree of supersaturation realized in such diffusion cloud chambers depends on the relative transport rates of heat and vapor as well as the curvature of the saturation vapor pressure versus temperature curve.[304] Although these devices generally have been used to obtain supersaturations of only a few per cent, some recent results of Katz and Ostermier,[305] for example, indicate that very high supersaturations may be achieved in diffusion chambers.

It has not been possible to measure the rate of nucleation directly in cloud chambers. However, a property of the equation for the nucleation rate makes it possible to link a gross observation of the onset of fog formation to a supersaturation ratio which is critical to the appearance of the fog. The suddenness of observation of droplets in a supersaturated vapor as predicted on theoretical grounds readily may be illustrated by calculating the rate of particle formation from eqn. (9.35). If it is assumed that α_c is unity, the curve for water

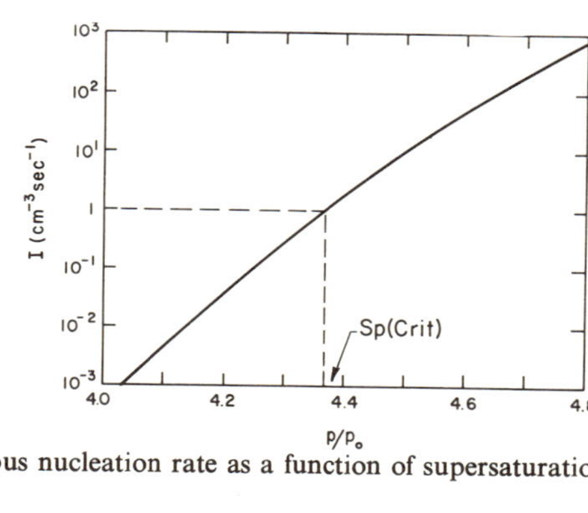

FIG. 9.3. Homogeneous nucleation rate as a function of supersaturation for water vapor at 0°C.

vapor at 0°C is given in Fig. 9.3. The nucleation rate changes exponentially with the supersaturation ratio. Thus one expects that the collapse of the vapor phase to form a cloud is a rather sudden one in which the rate of formation of particles is negligible up to a critical supersaturation, and very large at supersaturations above the critical value. The different corrections in the pre-exponential factor in eqn. (9.35) may change the overall rate of nucleation by an order of magnitude. But the sudden onset of condensation for the "critical" supersaturation remains nearly the same because the overall rate depends so strongly on the parameter $\ln S_p$. This prediction on the basis of the theory is almost universally confirmed by experiments.† Thus, nearly all of the more meaningful experimental observations in expansion cloud chambers have been reported in terms of S_p(crit), defined where $\ln I = 0$.

Ideally, the determination of the temperature (and hence the supersaturation) for which nucleation begins to occur is a straightforward matter in an expansion chamber. If the rate which heat penetrates the core of the chamber from the walls is sufficiently slow, the expansion will be essentially adiabatic.‡ Then the final temperature is given simply by the thermodynamic equation

$$T = T_o \left(\frac{p}{p_o} \right)^{\gamma - 1/\gamma}$$

where the subscript 0 here refers to the initial conditions, and $\gamma = C_p/C_v$.

The calculation of critical supersaturation in a diffusion chamber also is straightforward, in principle. In such a device the bottom and top are maintained at vapor saturation for two different temperatures. By vapor and heat diffusion between the walls, critical supersaturation can be attained at some unique height in between. A critical chamber condition corresponding to the temperatures of the bottom and top plates (the evaporator and condensor) at a given total pressure then refers to the onset of droplets by nucleation at a particular distance between the plates. To obtain a distribution of critical supersaturation for various temperatures, for example, Katz and Ostermier(305) have shown that data for several critical chamber states are required. For each critical state, an iterative computer solution of the heat and vapor diffusion equations is needed to specify the variation of temperature and supersaturation in the chamber. Since the critical state in the chamber is that which droplets are first seen, then at only one thin plane in the chamber will the supersaturation be sufficiently large to approach the value for homogeneous nucleation. The curves of supersaturation vs. temperature should touch to the critical supersaturation curve at only this temperature. Hence Katz and Ostermier's data show the experimental variation with temperature of the critical supersaturation as the envelope curves for critical chamber states.§

Results of Cloud Chamber Experiments

The early experimental technique of Wilson(268) is typical of procedures used by many subsequent investigators. In a typical experiment a volume of inert gas might be saturated with water vapor and then expanded. A sequence of preliminary expansions at a nucleation temperature of 12°C caused condensation on any dust or other foreign nuclei present, and these droplets were removed by falling out of the chamber. By steadily increasing the

† See also the discussion of Carstens *et al.*(429) on this subject.

‡ The principal problem at present in evaluation of all experimental data relating to homogeneous nucleation is to determine whether or not the onset of observable condensation really corresponds to diffusional growth on embryos rather than on a few "foreign" nuclei that always appear to be present.

§ Katz(443) has recently clarified this procedure in a new paper.

expansions, conditions were reached where no droplets appeared until a supersaturation ratio exceeding 4 was attained. From this point until a supersaturation ratio of approximately 8 was reached, relatively few droplets were observed, but after exceeding this supersaturation ratio, fog formed by vapor condensation as successively smaller droplets. The results were interpreted as heterogeneous nucleation on ions up to a supersaturation ratio of 7.9, after which homogeneous nucleation controlled the droplet generation. This general procedure, or variations of it, has generally been used to determine the values of the critical supersaturation ratio at the "gross" temperature of the mixture after expansion.

The experimental procedure for the diffusion chamber is somewhat similar to that of the expansion devices. In any given experiment, the temperature gradient in the chamber is set by fixing the upper and lower plates such that homogeneous nucleation just takes place, as observed visually. After the chamber comes to a steady state condition, the zone of supersaturated vapor has been cleared of dust and other foreign nuclei. It is kept clear of weakly energetic ions with a vertically oriented electrical field. Homogeneous nucleation can be detected as the range of operation wherein two to three droplets per cubic centimeter are randomly distributed in the lighted region as they fall out. When critical supersaturation has not been achieved, only occasional clusters of droplets are observed resulting from nucleation on spurious ions penetrating the chamber. When the critical supersaturation is greatly exceeded, heavy rainout of droplets is detected. Nucleation by energetic particles is distinguished easily by a path of droplets forming suddenly across the chamber (Plate 14). The tracks are formed by nucleation on ions left in the wake of the particle.

In a given run, the critical supersaturation will be achieved in a diffusion chamber at essentially one height. Repeating the observations of the same droplet formation rate many times for different bottom and top temperatures, and different pressures, generates operating curves from which the S_p *(crit)* vs. temperature envelope curve is derived. This gives the experimental result outlined previously. The envelope curves can then be compared with the

TABLE 9.3. COMPARISON OF MEASURED AND CALCULATED VALUES OF CRITICAL SUPERSATURATION RATIO. DATA OF VOLMER AND FLOOD[306] TAKEN IN AN EXPANSION CLOUD CHAMBER

Vapor	Number of molecules in stable nucleus	Nucleus radius R_k in Angstroms	S_p(crit) calculated	S_p(crit) measured
Water, 275.2°K	80.0	8.9	4.2	4.2 ± 0.1
Water, 261.0°K	72.0	8.0	5.0	5.0
Methanol, 270.0°K	32.0	7.9	1.8	3.0(1.9)[309]
Ethanol, 273.0°K	128.0	14.2	2.3	2.3
n-propanol, 270.0°K	115.0	15.0	3.2	3.0
Isopropyl alcohol, 265.0°K	119.0	15.2	2.9	2.8
n-butyl alcohol, 270.9°K	72.0	13.6	4.5	4.6
Nitromethane, 252.0°K	66.0	11.0	6.2	6.0
Ethyl acetate, 242.0°K	40.0	11.4	10.4	8.6 to 12.3

theory using a numerical solution for the chamber conditions, with accounting for the variation in physical properties of the nucleating vapor as a function of pressure and temperature.

Typical of the early results for S_p(crit) found in expansion chambers are those of Volmer and Flood[306] listed in Table 9.3. These measurements indicate that the observed appearance

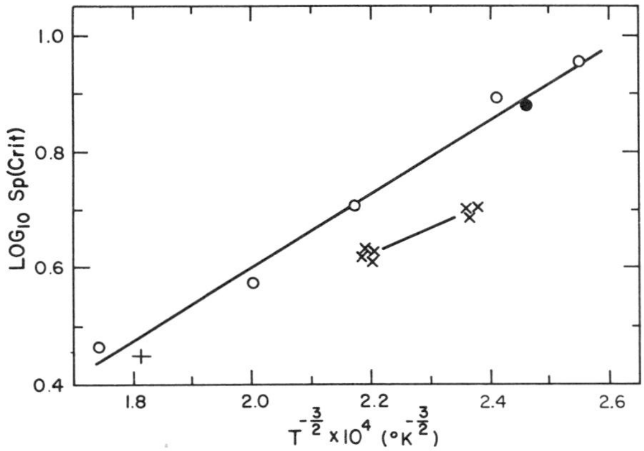

FIG. 9.4. Temperature dependence of the critical supersaturation ratio for water vapor in the ranges −47° to −26°C. ○ Powell,(307) ● Wilson,(268) × Volmer and Flood,(306) + Amelin.(273) (From Dunning.(308))

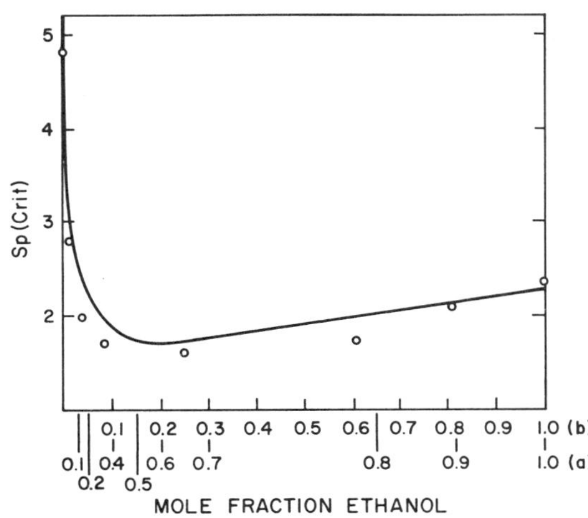

FIG. 9.5. Critical values of the supersaturation ratio for nucleation from a binary vapor in a cloud chamber. (From data of Flood.(303))

of droplets in substantial numbers is at least qualitatively consistent with the theoretical predictions. The recent data taken by Katz and Ostermier(305) in a diffusion chamber substantially verify Volmer and Flood's earlier experiments. In particular, Katz and Ostermier found agreement within 5 per cent between measured critical supersaturation and the Becker–Döring theory for ethanol and methanol over a 60°K range. Katz' more recent experiments(443) indicate that homogeneous nucleation of n-hexane, n-heptane, n-octane, and n-nonane takes place as predicted by the Becker–Döring theory, but does not agree with estimates from either the Lothe–Pound model, or the Reiss *et al.*(296) model.

The temperature dependence of the exponential factor $\Delta\hat{G}^{\star}/kT$ has been tested to some extent using data from expansion chambers. If it is assumed that for a given apparatus the critical supersaturation where cloud appears (a fixed concentration of embryos) equals the critical rate of nucleation, then the factor $T^{\frac{3}{2}}$ ln (S_p) should be nearly constant. This will be the case provided that one can disregard variations in the pre-exponential factors when the critical supersaturation is measured at various temperatures. A plot of ln S_p vs. $T^{-\frac{3}{2}}$, shown in Fig. 9.4, using Powell's(307) (see also Dunning(308) or Andres(283)) results indicates that the theoretical prediction for the temperature dependence is again consistent with experiment. Remarkably enough, experimental points of Wilson(268) and Amelin,(273) taken in different equipment, also fall on approximately the same curve. However, Volmer's and Flood's(306) results fall somewhat lower than the others, but they appear to be self-consistent.

Experiments to deduce the critical supersaturation ratio in a binary vapor have also been reported by Flood.(303) Using the method of Volmer and Flood,(306) values of S_p(crit) for the ethyl alcohol–water system were determined, and are indicated in Fig. 9.5. The results are in adequate agreement with the theoretical curve estimated from eqn. (9.41).

The experimental results on nucleation in expansion chambers have been criticized on the grounds that the observations cannot be quantitative because of uncertainties in the detailed fluid dynamics of the expansion device and the inability to measure embryos until they have grown to macroscopic size (e.g. Mason(310)). The transient heating of gas because of warm walls in an expansion chamber(440), the possible introduction of turbulence, and the pressure oscillation induced by bouncing of the piston, all may cast doubt on the application of simple thermodynamic estimation of local conditions in the chamber as they change with time. In addition, the impossibility of measuring embryos at their inception requires calculation of their growth to macroscopic size. This is a difficult problem in itself, and would rigorously necessitate calculations bridging the Knudsen range of growth by condensation (Chapter 5). Furthermore, the growth process involves the need to account for vapor depletion during condensation (e.g. Carstons *et al.*(429); see also Section 9.3A). Recently Kassner and his colleagues(311, 312) have made some progress towards more realistic interpretation of experiments in expansion chambers. By defining a uniform criterion for determining the vapor temperature after expansion, Kassner and Schmitt(312) found satisfactory agreement between the Becker–Döring theory and measurements of the nucleation of water vapor in helium gas. Schuster(103) has reported that it is possible to trace the nucleation and growth of water droplets in an expansion chamber using a laser beam technique. His nucleation rates agree satisfactorily with the Becker–Döring theory and rates predicted by eqn. (9.35) provided an empirical relation for the variation in surface tension with embryo size is used, and it is assumed that the evaporation coefficients $\alpha_c \approx 1.0$ for the water droplets.

The discrepancy between Wilson's early observations and the results of Volmer and Flood has been discussed in some detail by Hirth and Pound.(287) In their view, the fact that Volmer and Flood took their data in the presence of an electrical field suggests that these data really represent critical supersaturation ratios for condensation in ions. However, Katz and Ostermier's(305) recent work by a different experimental method, and Kassner's and Schmitt's(312) data would tend to confirm the validity of the earlier observations as applying, at least qualitatively, to homogeneous nucleation.

Methods Using Steady Flow

The ambiguities in the results from cloud chambers has led some investigators to consider methods in which supersaturation develops under conditions of steady flow. Essentially three different techniques have been examined. These are: (a) supersaturation by adiabatic expansion in a converging–diverging nozzle, (b) supersaturation by mixing of a cool dry gas with moist warm gas in a free, turbulent jet, and (c) supersaturation in a low density nozzle beam (a molecular beam). All three of these methods offer certain advantages over the chamber experiments. Since the nucleation process is spread out spatially instead of in time, detailed measurements can be made of the "time" history of condensation. This information combined with lengthy computations based on the theoretical behavior of fluid in such devices provides greater insight into the rate processes taking place than the cloud chamber experiment. Furthermore, supersaturations may be achieved sufficiently rapidly that homogeneous nucleation can occur in the first two cases despite the presence of significant numbers of foreign nuclei.[269, 274]

The onset of nucleation in a supersonic nozzle is generally detected by a sharp change in light scattering as the gas expands in the diverging section. The condition of nucleation may also be detected in measurements of static pressure along the longitudinal axis of the nozzle. Typically, the static pressure in the supersonic flow tends to decrease continuously with expansion of the gas. At the location where droplets are observed to first form a jump in static pressure can be detected if the vapor concentration in the gas mixture is high enough.

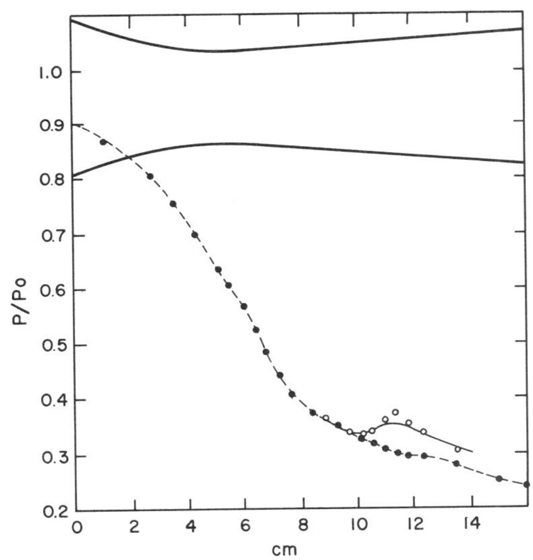

FIG. 9.6. Profiles of the static pressure with distance downstream in a supersonic nozzle flow based on data of Binnie and Woods[315] compared with calculations of Oswatitsch.[314] (From Dunning.[308])

The location where the pressure increase is observed is the same as the zone where droplets are detected optically. Some typical experimental results for the distribution of the ratio of static pressure to stagnation pressure p_0 in a converging–diverging nozzle are shown in Fig. 9.6. The continuous decrease in pressure for flow of dry air is indicated by the dotted line. The onset of condensation of nitrogen in the air is marked well by the pressure "bump" 10 cm from the inlet. Because the zone of the onset of nucleation is a sharp front accompanied by the pressure increase, this region has been called a *condensation shock*.

It is possible to calculate theoretically the static pressure distribution for the supersonic flow of dry gas in the converging–diverging nozzle. Using a heat balance, combined with the rate equation for formation of nuclei, one can estimate the location of the zone of first appreciable nucleation theoretically; formation was detected by light scattering.

Oswatitsch[314] applied the theory of compressible flow through the nozzle to estimate the change in thermodynamic properties of the gas mixture. He then made calculations of the nucleation rate based on the Becker–Döring theory and found the solid curve shown in Fig. 9.6. The experiments of Binnie and Woods[315] shown in the figure are believed to be quite accurate, and Oswatitsch considered that the deviations shown between his theoretical calculations exceeded the experimental errors. However, perhaps the values of the surface tension used in the calculations were at fault. Despite this possibility, the agreement between the theory and experiment for the converging–diverging nozzle seemed encouraging because of the intricate calculations required for integrating the rate theory with the fluid dynamics.

Further efforts to make use of adiabatic expansion in nozzles have been reported by Wegener and Pouring.[270] They studied the formation of ice crystals by condensation in a nozzle flow. The measurements of static pressure change in the nozzle combined with observations of the onset of cloud formation failed to agree quantitatively with the known rate theories based on the steady state model. However, the results could be correlated qualitatively with the theory. Like Oswatitsch, these workers believed that the lack of quantitative agreement may be the lack of knowledge about the material properties of small ice crystals at temperatures of 200–220°K.

Recent experiments on the nucleation of liquids in supersonic nozzle flows reported by Jaefer, *et al.*[444] and Dawson, *et al.*[446] add further contradictory evidence about the applicability of various theoretical models. These workers, using essentially Oswatitsch's methods, found that homogeneous nucleation of water vapor in a nozzle flow followed the classical theory, but ammonia nucleation agreed better with the Lothe–Pound[293] approach. Similar experiments on the homogeneous nucleation of benzene, chloroform and Freon 11 agreed with the latter theory, yet the nucleation of ethanol was more consistent with the Reiss, *et al.*[296] model. Dawson, *et al.* concluded that polar or rod-like molecules nucleate as predicted by the Becker–Döring model, while fluids of compact molecules without a substantial permanent dipole tend to follow the Lothe–Pound relation. These conclusions do not agree, of course, with Katz'[443] results, or a variety of other experimental evidence by different methods.

The use of turbulent mixing in free jets as a system to study nucleation rates apparently was first suggested by Amelin[273] when a stream of warm moist gas is mixed with drier, cooler gas, say around the edges of a jet, nucleation may occur if supersaturation accompanies the mixing. As the warm stream is stirred into the cool environment, the temperature and vapor concentration are reduced in a direction normal to the flow of the jet. If the initial vapor concentration is sufficiently large, a zone of supersaturated vapor can form roughly in an ellipsoidal shape around the jet nozzle. Like the condensation shock observed in supersonic nozzle flow, a similar nucleation front is observed at the zone of high apparent supersaturation in a turbulent free jet. If one knows the temperature and vapor distribution in the jet, one can, of course, calculate where the zone of "critical" supersaturation is located based on the theory for homogeneous nucleation. Amelin first attempted this, and found that the critical supersaturation calculated for a free jet based on the Becker–Döring theory agreed qualitatively, at least, with the region of first appreciable nucleation determined optically in a free jet. Higuchi and O'Konski[316] made use of such a jet to test the Becker–Döring theory of nucleation of materials of rather low volatility compared with those in Table 9.3. In an analogous manner to the method of Oswatitsch, Higuchi and O'Konski applied the theory of jet mixing to estimate the average distributions of vapor pressure and temperature in the jet. Using an integral technique for estimating the overall rate of nucleation by the theory, the experimental measurements by light scattering of the rate of droplet formation could be compared. Since the surface free energies of small particles is of key interest in the theory, Higuchi and O'Konski chose to compare the values of the bulk values of the surface tension with those estimated from the apparent values of $\hat{G}_k$ from the experiments. Their results are shown in Table 9.4. Accepting the validity of the

TABLE 9.4. COMPARISON OF RESULTS OF HIGUCHI AND O'KONSKI[316] WITH THE BECKER–DÖRING THEORY

Aerosol	T (°K)	S_p (avg)	Literature (dyne cm^{-1})	est. from B.D. theory	(no. of molecules in cluster of critical size)
Dibutyl phtalate	332	70	29.4	28.8	21
Triethylene glycol	324	25	42.8	36.8	29
n-Octadecane	319	63	26.1	24	25
Sulphur	325	420	67.6	54	14

jet technique, the agreement between theory and experiment is qualitatively satisfactory except for the case of triethylene glycol. The difference in this case was interpreted as the result of possible effects of hydrogen bonding on the behavior of a very small cluster containing about 30 molecules.

A detailed study of nucleation in a turbulent jet by Hidy and Friedlander[274] suggested that the method used by Higuchi and O'Konski may not be as quantitative as might be hoped. The influence of intense fluctuations in vapor properties induced in a turbulent medium are difficult to interpret by mean values alone. At any instant spatial and temporal variations in temperature and vapor pressure may cause supersaturation ratios to deviate substantially from those predicted by a mixing theory like the one used by Higuchi and O'Konski. Therefore more extensive experiments should be undertaken on condensation in turbulent media before accepting this method as a satisfactory technique for studying nucleation quantitatively.

Like the cloud chamber experiments, the nucleation in a converging–diverging nozzle and in turbulent jets suffer the same disadvantage that it is not possible to trace the growth of individual clusters. The ideal experiment for the study of the nucleation process, as Andres[283] has pointed out, would be one in which the rate process could be frozen at some stage and the concentration of all clusters determined. Such a possibility is offered by the beam of low density gas issuing at supersonic speeds through a nozzle. This device consists of a converging nozzle through which a gas flows into a vacuum. The supersonic stream of vapor that develops expands radially in a known fashion up to a skimmer orifice placed at the axis, but normal to the stream. At this point, only the center of the stream is allowed to pass on into a second vacuum chamber to form a nearly collisionless flow field. The resulting stream then is collimated into a beam whose properties may be estimated by conventional methods (see, for example, Ramsey[317]).

Because of the very low mass densities of vapor achieved in the beam, the populations of clusters cannot retain a steady state concentration, and it is likely then that they freeze with respect to the nucleation process at a location downstream of the skimmer.

In 1956 Becker and his colleagues[318] apparently first noticed nucleation in a nozzle beam. When they cooled their nozzle, they found that for a particular source pressure a critical source temperature could be determined for which there was an abrupt increase in apparent beam intensity which they associated with the appearance of embryos. Intensity increases would be anticipated because the molecular clusters, having a high molecular weight, would tend to remain near the centerline of the beam more than single species.

Later developments in observing clusters in supersonic beams have provided direct evidence for multimolecular species. Andres[(283)] has indicated that mass spectroscopy used with the beam should be a powerful technique for studying the kinetics of nucleation. However, there still remains the problem of unscrambling the uncertainties in the true concentration of clusters from ion currents. Such uncertainties arise from possible fragmentation during ionization and from mass separation effects in the process of forming the beam.

No quantitative comparisons between the observations in a beam and the theory of nucleation have been reported as yet. However, Andres[(283)] has discussed qualitative agreement between experiment on argon beams, and a freezing theory for dimer formation using equilibrium constants calculated by the method of Stogryn and Hirschfelder[(285)]. These preliminary results give hope for tracing the rate constants for individual transitions in the chain process of cluster formation. This kind of information may be helpful in interpreting the curious results of Leckenby, Robbins and Trevalion,[(319)] who found that a distinct difference appeared between the stability of even and odd number clusters of condensed water vapor. Such observations cannot be reconciled at all with the classical theory which uses bulk thermodynamic properties of the bulk phase.

9.2. Heterogeneous Nucleation

When a supersaturated vapor contains a number of small particles, these "foreign" nuclei can act as centers of spontaneous condensation. Depending upon the cooling rate and the number of foreign nuclei, the resulting transition of phase of the metastable supersaturated vapor may take place at considerably lower supersaturation ratios than expected on the basis of the assumption of homogeneous nucleation. In a very "dirty" gas medium like the Earth's atmosphere, for example, nucleation to form clouds of water droplets proceeds under supersaturation conditions rarely exceeding a few per cent.

Because nucleation on foreign matter involves at least one additional degree of freedom, the analysis of nucleation under these conditions is much more complicated than the homogeneous case. The nucleation rate will depend on the nature of the nucleus surface, including its geometry and chemical properties, its chemical compatibility with the vapor, and the condition of the surface in addition to the supersaturation. The detailed treatment of nucleation on foreign matter requires considerable discussion of surface chemistry, which is beyond the scope of this chapter. Much of the research on this subject has been summarized, however, by Hirth and Pound.[(287)]

In this section we shall take up only the simplest theories of foreign nucleation. It will be seen readily that this is substantially consistent with our initial assumption that we are dealing with idealized smooth, spherical particles, which are chemically inert.

A. Condensation of Liquids on Ions

The early experiments of Wilson in his expansion chamber suggested that the presence of ions in the supersaturated system can reduce considerably the supersaturation required to produce an aerosol cloud. Wilson's later experiments in which he separated positive and negative ions by an electric field indicated that these two species had different nucleating properties for water vapor. For example, at $-6°C$ he found that nucleation occurred at $S_p \approx 4$ for negative ions, while positive ions became active with $S_p \approx 6$ at a slightly reduced temperature. Later repetition of these experiments by several other investigators has largely confirmed these results at least qualitatively.

Though no satisfactory theory for nucleation in the presence of ions has yet been developed, the work of Volmer and colleagues(320) appears to provide a basis for more elaborate work.

Like the theory for homogeneous nucleation, the analysis for nucleation on ions centers around equilibrium considerations of the chemical potential of an embryo $\hat{G}_k$. For non-conductive spherical aggregates condensing on an ion of radius R_1 and charge Q_1, the Gibbs free energy of formation has the form

$$\Delta\hat{G}_k = -\tfrac{4}{3}\pi R_k^3 kT \ln S_p + 4\pi R_k^2 \sigma + \frac{Q_1^2}{2}\left(1 - \frac{1}{\epsilon}\right)\left(\frac{1}{R_k} - \frac{1}{R_1}\right).$$

The last term on the right represents the change in electrostatic energy of the ion assuming that the dielectric constant of the suspending medium is one.

The radius of critical size is determined as before by requiring that $\partial\hat{G}_k/\partial R_k = 0$, which yields

$$\ln S_p(\text{crit}) = \frac{2\sigma V_L}{kTR^\star} - \frac{Q_1 V_L}{8\pi R^{\star 4} kT}\left(1 - \frac{1}{\epsilon}\right). \tag{9.42}$$

This relation, quoted by Volmer,(320) reduces to the result found in 1888 by J. J. Thomson for conducting spheres, where $\epsilon \to \infty$.

Comparison between eqn. (9.42) and the Kelvin–Gibbs relation (9.16), shows that the influence of charging tends to decrease the supersaturation ratio needed to maintain a droplet of critical size. In addition, the dependence of vapor-pressure on radius behaves quite differently for the charged droplet, as drawn, for example, for a water droplet in Fig. 9.7. The equilibrium saturation ratio for the charged droplets branches away from the curve for uncharged species at $R_k \approx 16$ Å. After displaying a maximum value at $R_k \approx 6$ Å, the charged droplet curve rapidly reduces to zero near $R_k \approx 4$ Å.

There has been a temptation to suggest the radius corresponding to a maximum value in (p/p_0) (Fig. 9.7) as the threshold for the onset of nucleation. However, Fletcher(302) has noted that the theory of fluctuations in the vapor–ion system again has to be applied to find the nucleation rate. Consider a cut through the curve in Fig. 9.7 referring to a constant value of (p/p_0). The point A refers to a condition of metastable equilibrium so that all ions at this state will tend to gain molecules of vapor and form droplets of this radius. The rate process of growth then is composed of some droplets making a transition by statistical fluctuations in the medium from point A to the location of unstable equilibrium at point C. This process will take place, of course, when the line AC lies below the maximum in the equilibrium vapor pressure curve at B.

In contrast, the interpretation of phase transition by homogeneous nucleation would correspond to the change from point D, the radius of a single molecule of vapor, to point E, the critical radius of the uncharged droplet, as given by the Gibbs–Kelvin relation. Thus the ions must assist the nucleation process.

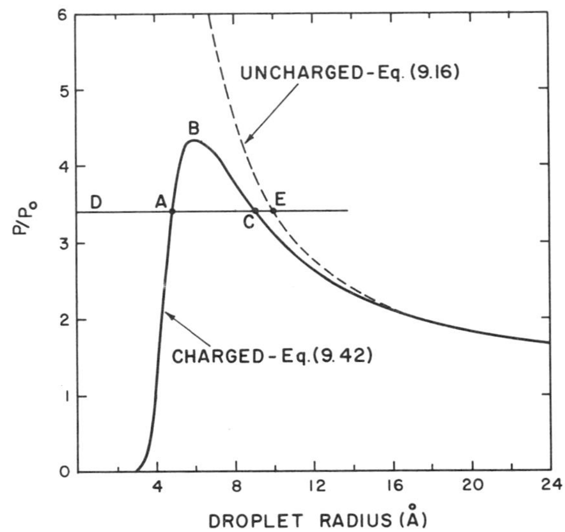

FIG. 9.7. Vapor pressure variation with particle radius for uncharged and charged particles. (From Fletcher,[302] p. 50.)

To express the phase transition more quantitatively, we could write a material balance for the kth species again as in eqn. (9.20), and proceed to find steady state solutions analogous to those of Volmer and Weber, or Becker and Döring. The result yields an equation for the nucleation rate in a form similar to eqn. (9.34):

$$I' = \frac{\alpha_c p}{(2\pi m_G kT)^{\frac{1}{2}}} Z_c(4\pi R^{\star 2})\, N_c \exp\left[-\Delta\hat{G}^{\star}/kT\right], \tag{9.43}$$

where N_c is the ion concentration, $\Delta\hat{G}^{\star}$ is the free energy of formation for a droplet of critical radius, as calculated from (9.42), and the non-equilibrium factor is[287]

$$Z_c = \left\{\frac{[4\pi\sigma R^{\star 2} - (1 - 1/\epsilon)\, Q_1^2/R^{\star}]}{9\pi k^{\star 2} kT}\right\}^{\frac{1}{2}}.$$

Like the value of Z for homogeneous nucleation, Z_c has a value the order of 10^{-2}.

At 265°K, estimation of the values of the droplet radii at A and B in Fig. 9.7 from (9.43) with $\epsilon \approx 80$ yields 4.8 Å and 10.2 Å respectively.[302] This gives S_p (crit) ≈ 3.2, which is the right order of magnitude, but considerably smaller than Wilson's experimental estimate of about 4.1 for water vapor condensing negative ions.

Wilson's first experimental results for nucleation on ions have been confirmed by later investigations. A list of observations for nucleation of water vapor on ions, taken in expansion chambers from several experiments, is given with Wilson's value in Table 9.5. There is some scatter in the results, but they appear to indicate that conditions where nucleation occurs on ions are more sharply defined experimentally than for homogeneous nucleation. Yet agreement with the theoretical model is rather poor.

To explain the discrepancy between theory and experiment, Volmer[320] has argued that it may be expected since the bulk dielectric constant of water was used. Using a value of $\epsilon \approx 2$, better agreement between the theory and the experiment may be expected. The substantial reduction of the dielectric constant ϵ with droplet size has theoretical basis from

TABLE 9.5. CRITICAL SATURATION RATIOS FOR SPONTANEOUS CONDENSATION ON IONS BASED ON DATA FROM EXPANSION CLOUD CHAMBERS (FROM MASON[310])

Investigator	Sign of change	Final temperature (°K)	Ratio of final to initial chamber volume	S_p (crit)
Wilson[268]	−	265	1.25	4.1
	+		1.31	6.0
Przibam[323]	−		1.24	3.7
	+		1.31	6.0
Laby[324]	−	268	1.26	4.2
Andren[325]	−	268	1.25	4.1
Powell[307]	−	267	1.25	3.98
Flood[303]	−	265	1.25	4.1
Scharrer[326]	−		1.25	4.14
	+		1.28	4.87
Sander and Damköhler[327]	−	265		3.9

the work of Debye.[321] However, Fletcher[302] has remarked that ϵ should not decrease below the optical value of 3, and this reduction is still inadequate to yield good agreement between theory and experiment.

Perhaps a more important shortcoming in the existing theoretical analysis is that there is no account taken of the difference between positive and negative ions. Substances like water are quite polar, and there exists evidence that such materials may have preferential orientation in surface layers (e.g. Henniker[322]). For water the preferred dipole direction requires negative ions nearest the surface.[302] Thus when condensation occurs on a negative ion, the free energy of the surface dipoles is lowered and condensation is enhanced. The reverse would be true of water nucleation on positive ions. Because the theory fails to give quantitatively correct values of S_p(crit), and does not distinguish between the role of different kinds of ions, it must be considered quite incomplete at this time.

B. Nucleation on Foreign Particles

Most natural aerosols, and those often formed in the laboratory by condensation, are not generated by homogeneous nucleation or nucleation on ions. There are sufficient quantities of natural nuclei such as dust in the atmosphere to provide centers for spontaneous condensation if a vapor becomes slightly supersaturated. The behavior of natural contaminants, called *condensation nuclei*, remains somewhat obscure because very little information about the physical and chemical properties is available. However, it is possible to carry on with our idealized theoretical calculations to some extent without considering the characteristics of individual nuclei.

As first approximation, one expects that an insoluble particle might act as a nucleation embryo having the same radius as one generated by heterophase fluctuations in a vapor. This sort of idea led Volmer[320] to modify the original Volmer–Weber theory to include nucleation on a smooth plane surface. This path largely has served as the basis of subsequent work in this field.

If we consider first the equilibrium configuration of a droplet on a flat surface, we find that the droplet will take a shape of a spherical cap provided that it does not wet the surface (Fig. 9.8). The two key geometrical parameters for the droplet are the radius of curvature r and the angle of contact between the liquid and the substrate ϕ. For equilibrium on the

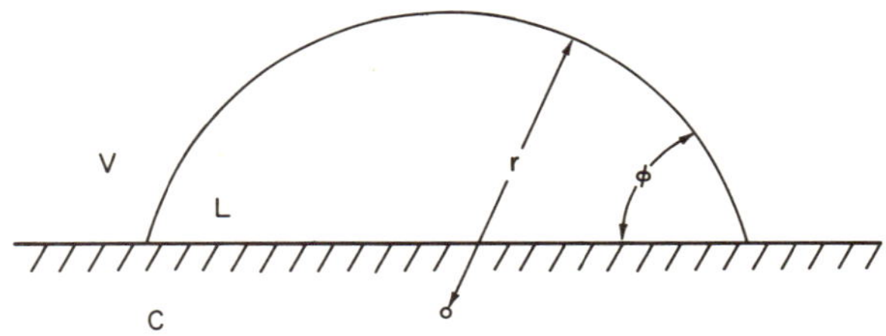

FIG. 9.8. An embryo in the form of a spherical cap developing on a flat surface. (From Fletcher,[302] p. 52.)

surface, the line common to the three phases—substrate C, liquid L, and vapor $V(G)$—has to be restricted by

$$m \equiv \cos\phi = (\sigma_{CG} - \sigma_{CL})/\sigma_{LG}. \tag{9.44}$$

The free energy of formation for an embryo as a spherical cap then is given by

$$\Delta\hat{\mu}_c = \frac{\pi r^3}{3}(2 + m)(1 - m)^2\,[\hat{\mu}_L - \hat{\mu}_G] + 2\pi r^2(1 - m) + (\sigma_{CL} - \sigma_{CG})\,2\pi r^2(1 - m). \tag{9.45}$$

Applying the criterion that the free energy of formation of the critical sized cap should be maximum, we find that the critical radius of curvature, as determined from eqn. (9.45) will be the same as that given by the Kelvin–Gibbs relation. We then find that

$$\Delta\hat{G}_C^{\star} = \frac{16\pi\sigma_{LG}^3 V_L}{3[kT\ln S_p]^2}\,f(m), \tag{9.46}$$

where

$$f(m) = (2 + m)(1 - m)^2/4. \tag{9.47}$$

Thus the coefficient $f(m)$ represents a correction to the case of homogeneous nucleation. We note that $0 \leqslant f(m) \leqslant 1$ so that any foreign material acting as a nucleus will reduce the free energy necessary to form an embryo of critical size.

To calculate the rate of nucleation, one assumes that the distribution of critical sized embryos is given by an equilibrium distribution like eqn. (9.29). As a rough approximation the surface area of the critical embryo is $r^{\star 2}$, and the rate of impact of vapor molecules per unit area will be given by eqn. (9.23). Hence, the "equilibrium" rate of nucleation per unit area of substrate surface is

$$I^0 \approx \frac{p}{(2\pi m_G kT)^{\frac{1}{2}}}\,\pi r^{\star 2} n_{\text{ads}} \exp\left(-\frac{\Delta\hat{G}_C^{\star}}{kT}\right), \tag{9.48}$$

where n_{ads} is the number of (single) vapor molecules adsorbed per unit surface area of substrate.

The calculation of the non-equilibrium correction for this expression requires consideration of the adsorption and desorption rate on the surface combined with the rate of migration of the vapor molecules on the substrate surface. If it is assumed that the net adsorption on the surface represents an appreciable part of a monolayer at the supersaturation attained, then the pre-exponential coefficient can become 10^{24} to 10^{27} cm^2 sec^{-1} assuming a correction similar to Becker and Döring's in eqn. (9.35).

Taking into account the expression for the rate of deposition of vapor on the surface of the substrate in eqn. (9.48), Fletcher[302] has given an illustration of the variation in super-

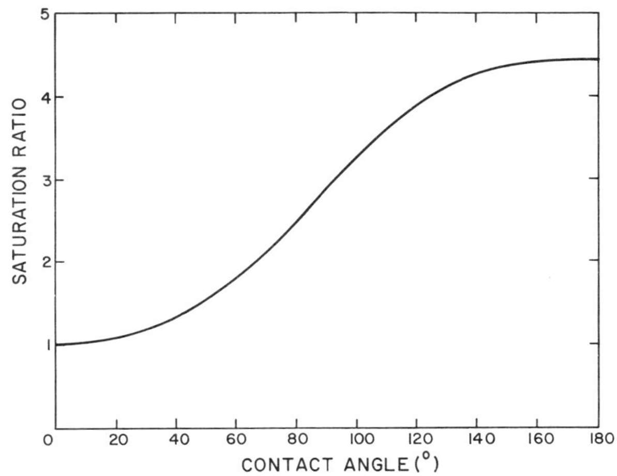

FIG. 9.9. Critical supersaturation ratio for water on a plane substrate of given contact angle at 0°C. (From Fletcher,(302) p. 55.)

saturation ratio with contact angle for water at 0°C (Fig. 9.9). We see for conditions of complete wetting $\phi = 0$, nucleation recurs at supersaturations of ~ 100 per cent. However, if the foreign nuclei are not wetted, $\phi = 180°$, the nucleation becomes effectively homogeneous.

Twomey(328) has substantially verified the curve in Fig. 9.9 by using a glass slide covered with films of more or less hydrophobic substances. The slides were placed in a cloud chamber and the onset of nucleation was observed for various values of p/p_0.

Under normal conditions, the surfaces of foreign nuclei will not be flat but will be irregular or curved. Fletcher(302) has summarized the geometrical corrections that have to be made to account for variations in size of the nucleus. Instead of the correction $f(m)$ being a function of m only, in this case, the curvature ratio $a/r^\star$ has to be included. Thus for curved surfaces, $f(m)$, the correction to $\Delta \hat{G}_C^\star$ becomes

$$\left. \begin{aligned} f'(m, a/r^\star) = \frac{1}{2}\Bigg\{1 + \left[\frac{1 - m(a/r^\star)}{\tilde{g}}\right]^3 + \left(\frac{a}{r^\star}\right)^3\left[2 - 3\left(\frac{a/r^\star - m}{\tilde{g}}\right)\right. \\ \left. + \frac{(a/r^\star - m)}{\tilde{g}}\right]^3 + 3m\,(a/r^\star)^2\left[\frac{(a/r^\star) - m}{\tilde{g}} - 1\right]\Bigg\}, \end{aligned} \right\} \quad (9.49)$$

with $\tilde{g} = [1 + (a/r^\star)^2 - 2m(a/r^\star)]^{\frac{1}{2}}$, and a = radius of the nucleating particle. The nucleation rate per particle may then be given as a rough approximation by(302)

$$I^0 \approx \frac{p}{(2\pi m_G kT)^{\frac{1}{2}}} 4\pi^2 a^2 r^{\star 2} n_{\text{ads}} \exp\left[-\frac{\Delta \hat{G}_C^\star}{kT}\right]. \quad (9.50)$$

The pre-exponential factor here has a value the order of $10^{25}(4\pi a^2)$. Fletcher has noted that eqn. (9.50) breaks down if $a < r^\star$ wherein it is better to set $a = r^\star$

C. Nucleation of Liquids on Soluble Particles

When vapors nucleate on particles that are soluble in the liquid phase of the condensing material, the effect of the particles is analogous to ions. The main difference lies in the fact that the nucleating particles are usually large enough to neglect the influences of heterophase fluctuations.

The treatment of the growth of droplets upon a dissolving substrate was developed originally by Kohler[329] as early as 1921. His ideas were further extended by Wright.[330] The key to the theoretical treatment involves the fact that the vapor pressure changes over a solution as contrasted to a pure liquid. For example, the vapor pressure of water is less over aqueous solutions than over pure water. Water vapor then may condense on the surface of solution at lower values of (p/p_0) than expected for pure water. This solution effect then tends to offset the increase in vapor pressure associated with surface curvature of small droplets.

If we assume that Raoult's law governs the equilibrium vapor pressure over the solution, then[302] in the case of aqueous solution surrounded by an ideal gas,

$$\frac{p_A}{p} = x_A = \left[1 + \frac{\Omega_H \tilde{m}_B M_A}{M_B(\frac{4}{3}\pi R_k^3 \rho_s - \tilde{m}_B)}\right], \tag{9.51}$$

where x_A is the mole fraction of water (component A). Here p_A and ρ_s refer to the partial pressure of the vapor over the solution, and the density of the solution. The term $\tilde{m}_B$ is the mass of substance of molecular weight M_B dissolved in a mole of water to form a droplet of solution of radius R_k. The factor Ω_H has been named for Van't Hoff and accounts for the degree of dissociation of salts. Equation (9.51) thus indicates that the vapor pressure over the aqueous solution has been decreased by x_A as given in (9.51).

Taking into account the vapor pressure reduction by (9.51), the Kelvin–Gibbs relation for a droplet of radius R_k then has to be modified as

$$\frac{p_{AR}}{p_0} = \left[\exp\left(\frac{2\sigma V_L}{kTR_k}\right)\right]\left[1 + \frac{\Omega_H\, \tilde{m} M_A}{M_B(\frac{4}{3}\pi R_k^3 \rho_s - \tilde{m}_B)}\right], \tag{9.52}$$

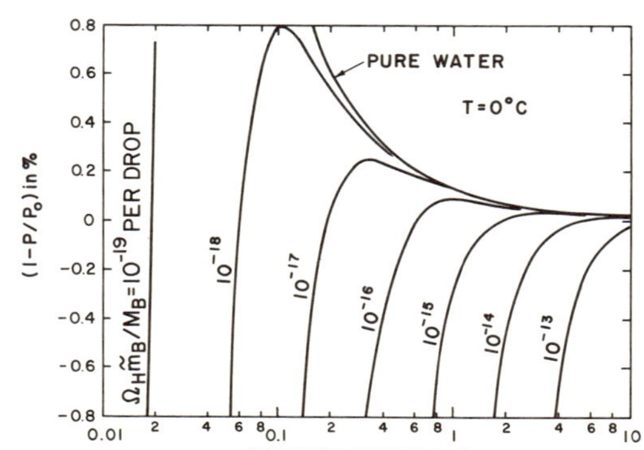

FIG. 9.10. Equilibrium supersaturation as a function of water droplet size for different amounts of dissolved material in the droplet. (From Fletcher,[302] p. 60.)

where p_{AR} refers to the vapor pressure over the droplet solution, and is σ the surface energy of the droplet of solution. Normally this relation is approximated by

$$\frac{p_{AR}}{p_0} \approx 1 + \frac{2\sigma V_L}{kT}\frac{1}{R_k} - \left(\frac{\Omega_H \tilde{m}_B M_A}{\frac{4}{3}\rho_s M_B}\right)\frac{1}{R_k^3}. \tag{9.53}$$

To illustrate how the ratio p_{AR}/p_0 varies with particle radius, eqn. (9.53) has been used to calculate several curves for water at 0°C, shown in Fig. 9.10. From this set of curves, we can readily see that as the supersaturation ratio is increased the droplet of solution increases with a definite radius, but is metastable. However, after the peak is reached, the size will increase without limit for a given supersaturation ratio greater than 1.

The critical radius and supersaturation ratio may be determined from eqn. (9.53) as

$$R^\star = \frac{3}{8}\left(\frac{\Omega_H\, \tilde{m}\, M_A kT}{\pi\rho_s M_B \sigma V_L}\right), \tag{9.54a}$$

$$\left(\frac{p_{AR}}{p_0}\right)_{\text{crit}} = 1 + \frac{2}{3}\left[\frac{2(\sigma V_L/kT)^3}{\Omega_H \tilde{m}_B M_A/\pi\rho_s M_B}\right]^{\frac{1}{2}}. \tag{9.54b}$$

The curves in Fig. 9.10 are similar to the one calculated for ions in Fig. 9.7, but the radius where maxima occur is shifted to much higher radii. For practical cases, the maxima lie in a range of $R_k \approx 0.1\ \mu$ so that nucleation of such particles largely may be treated as a process of diffusional growth as discussed in Chapters 4–6, and the need of accounting for molecular statistical fluctuations in the vapor may be avoided.

Fletcher[302] has noted that one important effect has been disregarded in calculating the curves in Fig. 9.10. It has been found that it actually is not possible to increase the concentration of solutions to arbitrarily large values. When a certain value of concentration is exceeded salt crystals begin to separate out, and at this point complete desiccation of the droplet may occur almost immediately if the droplet is in equilibrium with aqueous vapor. It has been observed by Junge,[331] for example, that there can be a hysteresis effect involved in this phase change because the production of crystalline phase may require an amount of supersaturation in the solution. This effect gives some uncertainty as to the evaluation of the growth rates based only on a model of vapor diffusion to a liquid solution.

D. Heterogeneous Nucleation to Form Crystalline Materials

In a manner similar to the enhancing of nucleation of liquids by foreign nuclei in a supersaturated vapor, sublimation of crystalline material may be aided by the presence of foreign surfaces. To construct a simple theory for the sublimation process, let us consider first the formation of a solid embryo on a flat, smooth surface. In analogy to the liquid case shown in Fig. 9.8, we can suppose the solid embryo forms a spherical cap shown in Fig. 9.11. Again the geometry of the cap depends on the radius of curvature r and the cosine of the contact angle ϕ.

Without attaching too much macroscopic significance† to surface energy ratio m, it can be defined as before as

$$m = \sigma_{CG} - \sigma_{CS}/\sigma_{SG}, \tag{9.55}$$

where the subscripts now refer to the solid material sublimated onto the substrate C.

There can be two distinct ways that the embryos can grow on the substrate. In the first case, the sublimating solid can retain its normal lattice dimensions right down to the interface. Under these circumstances, there will be dislocations in the sheets of condensed

† In dealing with the process of nucleation on a molecular scale on solid surfaces it is very likely that any macroscopic theory will be only a very crude first approximation. Therefore, the theory presented here is included primarily to provide a simple, qualitative approach for understanding the nature of heterogeneous nucleation of this type within the framework of the theory of homogeneous nucleation.

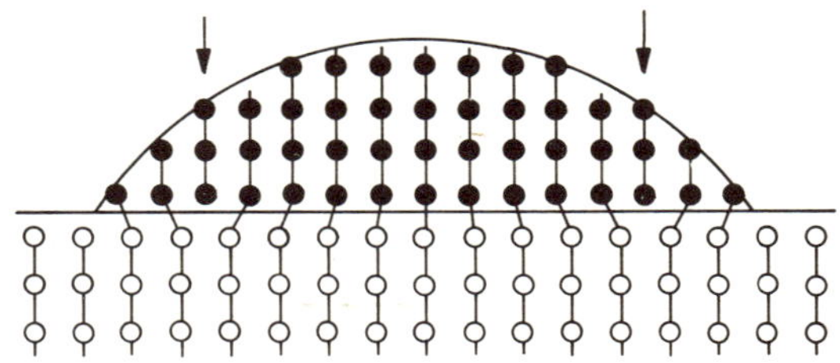

FIG. 9.11. Schematic diagram of an ice embryo growing on a crystalline substrate with a misfit of 10 per cent. The interface dislocations are indicated with arrows. For coherent epitaxy the ice embryo would be deformed elastically to eliminate the dislocations. (From Fletcher,[302] p. 211.)

atoms. The second case will develop when the crystal lattice of the condensing material joins coherently to the lattice of the substrate. There will always be a dislocated boundary when the substrate differs markedly from the condensed solid in its symmetry. However, a coherent attachment will be favored on substrates which have similar crystalline properties to the condensing substance. In general, one might expect a combination of the two modes of growth.

Fletcher[302] has pointed out that the dislocation at the interface will raise the surface energy σ_{CS}, and hence will lower m. Elastic strain within the embryo, in contrast, will raise the bulk free energy of the condensing material. Both of these influences will act to reduce the nucleating efficiency of the substrate as the number of dislocations and the amount of elastic strain increases.

Another requirement for the nucleation agent develops on further consideration of the binding mechanism of molecules to the substrate.[302] Because of the polar nature of some molecules like water, the binding may be largely electrostatic. Thus nucleation of ice from supersaturated vapor, for example, may be preferred on surfaces with relatively intense ionic fields.

Considering only the misfit geometry sketched in Fig. 9.11, we can develop a qualitative theory for crystal nucleation. Suppose that the average elastic strain e_s produced within the ice embryo depends only on the concentration of dislocations at the interface and the parameter m. Turnbull and Vonnegut[332] then demonstrated that the free energy of formation of the embryo would have the form

$$\Delta\hat{G}_s = -n_s kT \ln S_p + \tilde{C}_s e_s^2, \tag{9.56}$$

where n_s is the concentration of molecules of condensing material in the solid phase, and $\tilde{C}_s$ is the constant of proportionality depending on the elastic properties of the condensing material. For the case of ice nucleating on a basal plane at temperatures just below 0°C, $\tilde{C}_s \approx 1.7 \times 10^{11}$ dyne cm^{-2}.[302] Using eqn. (9·55) for $\Delta\hat{G}_s$, the formal theory for nucleation then can be carried out as before. For a nucleating particle of radius a, causing an average elastic strain e_s, with a contact angle of $\cos^{-1} m$, the critical free energy is

$$\Delta\hat{G}_s^{\star} = \frac{16\pi\sigma_{SG}^3 f(m, a/r^{\star})}{3(-n_s kT \ln S_p + \tilde{C}_s e_s^2)^2} \tag{9.57}$$

Here, $f(m, a/r^{\star})$ is defined by eqn. (9.49).

As in the case of the treatment for nucleation of liquid droplets, one can write the equilibrium rate of nucleation with $\Delta\hat{G}_s^{\star}$ in the form of eqn. (9.50).

Surface Effects and Mixed Nuclei

In the analyses above we have dealt with idealized smooth surfaces, which is consistent with the initial assumptions described in Chapters 1 and 2. However, the completion of the theory of nucleation of liquids has to include physical imperfections like pits, holes, or steps in nuclei. If one adds these macroscopic changes in the cavity or step geometry of nuclei surfaces, one finds that the nucleation threshold depends on the details of the surface geometry, but appears to be lower than the supersaturation necessary for nucleation on a plane or convex surface.

Other imperfections that have characteristic lengths that are small compared with the radius of the critical embryo also have to be considered. The incorporation of microscopic growth steps like screw dislocations on crystal faces, for example, have been examined. Although these sites may provide favored nucleation sites, they appear to have relatively little influence on the critical supersaturation ratio, at least for the nucleation to form liquids. Their alteration of the simple theory can be much more important for the formation of crystalline materials by sublimation, however.

In any conditions where one is concerned with the nucleation of water vapor in a natural condition one must deal with the possible importance of mixtures of insoluble and soluble nuclei. The nature of condensation nuclei in the Earth's atmosphere has occupied investigators for many years. Because of the wide variety of sources of nuclei in the atmosphere, it has not been possible as yet to obtain a clear quantitative picture of the extent to which natural variations in atmospheric aerosols affect the formation of clouds of water droplets. For more detailed discussion of this subject and related areas, we refer the reader to books on cloud physics such as Fletcher[(302)] and Mason.[(333)]

9.3. Some Features of Aerosols Created by Spontaneous Condensation

All of the complete theories for aerosol formation by spontaneous condensation are threshold theories. That is, the analyses discussed above only give the rate of formation of very small particles which pass the critical size, and which then should grow to larger and larger sizes. What is of interest in practice, however, is the nature of the size distribution of aerosols generated by condensation. Experimental observations (e.g. Green and Lane[(4)]) have suggested that the distributions produced by nucleation devices will be rather polydisperse unless care is taken to control carefully the nuclei concentration and the rate of cooling of the vapor near saturation conditions. Although the available information is rather scanty, there are indications based on Schuster's[(103)] observations, for example, that the fogs obtained by homogeneous nucleation are relatively monodisperse. Similarly, the observed higher order Tyndall spectra in an aerosol formed by mixing near the nozzle of a turbulent jet also indicated that initially monodisperse aerosols form.[(274)] However, the more extensive experiments of controlled mixing in aerosol generators of the LaMer type have indicated that the presence of foreign nuclei are necessary for good control of the monodispersivity of aerosols formed in these devices.

To calculate the size distributions of aerosols produced by spontaneous condensation, solutions of the sets of equations like (9.20) are required, and these are not available. With the exception of Courtney's[(298)] numerical integrations, very little work has been done on the behavior of these equations beyond the critical regime of supersaturation. Schuster's[(103)] calculations suggest that a simple model for the diffusional growth of single particles gives satisfactory results for the observed aerosol, provided the particle concentration is very small. (See also Chapter 4.) However, it is useful to have a better statistical picture of the growth of the entire collection of particles in a depleting supersaturated vapor.

A. Diffusional Growth of a Collection of Particles

When the growth of an assembly of particles by diffusion of vapor is considered, we can readily see that the relative growth of the entire collection will depend on the initial supersaturation as well as the concentration of particles present in the suspension. If the number density of droplets is sufficiently small, one may argue that the entire assembly may be treated roughly in terms of the behavior of a single particle. In other words, for the first approximation, one need not take into account the competition between particles for depletion of vapor surrounding a particle. Under these circumstances, it is possible to apply an expression such as eqn. (4.43) provided that we assume that the initial size is larger than about 1 μ radius. Then, for a single growing particle,

$$\frac{\mathrm{d}R_k^2}{\mathrm{d}t} = \text{const.}$$

or

$$R_k^2 - R_0^2 = (\text{const.})\, t, \tag{4.45}$$

where R_0 denotes the initial radius of the particle. The part (const.) $\times\, t$ is normally much larger than R_0^2, and is essentially the same for all particles. Suppose during the growth process, the droplets are of initial radii 1 μ and 2 μ. During a certain time interval the first nucleus grows to 10 μ say, giving (const.) $\times\, t = 99\ \mu^2$. In the same time interval, the 2 μ nucleus will grow to about 10.2 μ. Thus a 100 per cent difference in initial size reduces to about a 2 per cent difference after growth in the supersaturated medium. Qualitatively, then, we expect on the basis of single particle behavior that growth by vapor diffusion will always be accompanied by a reduction in polydispersity when the particle concentration is sufficiently small. Such an argument has been generalized for a cloud of particles by Reiss[(334)]† and Frisch and Collins.[(335)] In Reiss' approach, competition between growing particles is accounted for even in very dilute clouds by considering the vapor density far from any one particle a function of time.

To illustrate how the interaction between growing droplets tends to be self-regulating such that larger particles are retarded in size increase, let us reconsider the above situation wherein the diffusional growth is governed by the quasi-stationary theory in the continuum approximation. Suppose we introduce Reiss'[(334)] approximation that the concentration of vapor in the medium far from any particle is a function of time

$$n_{A\infty} = n_{A\infty}(t),$$

but the rate of vapor transport through a spherical shell of radius r_C and around a reference particle is given by the steady state form

$$\tilde{\Phi}_A^c(t) = -4\pi r^2 D_{AB}\left(\frac{\partial n_A(t)}{\partial r}\right). \tag{9.58}$$

† Zung[(336)] recently has extended Reiss' cell model to estimate cloud life times.

Implicit in assuming the transport rate in the form of eqn. (9.58) are the requirements that the concentration gradients of vapor are not severe except in the immediate vicinity of the particles (vapor sinks). The neighborhood of influence of a sink then does not extend beyond the order of 10 particle diameters from the sink. This is a rather important assumption in that it allows for a simple treatment of competition between particles; it makes it possible for the vapor concentration at any instant to be the same at any two points which are outside the sphere of influence of two particles, provided that the two particles are located outside the vicinity of other particles. This conceptual model, as Reiss notes, allows for a random spatial distribution of growing droplets.

Growth of a Cloud of Particles in an Initially Monodisperse Aerosol

Equation (9.58) holds in the continuum regime for any particle, with the above assumptions. Integration of eqn. (9.58) at any distance from the reference point yields the distribution of vapor concentration, or

$$n_A(r, t) = [\tilde{\Phi}_A^c(t)/4\pi D_{AB} r] + \hat{\phi}(t), \tag{9.59}$$

where $\hat{\phi}(t)$ is an arbitrary function of time. The expression for the flux at the droplet surface, and at a distance r_c‡ from the droplet, then is[334]

$$\tilde{\Phi}_A^c(t) = -4\pi D_{AB} R_1 [n_A(r_C, t) - n_A(R_1, t)] \tag{9.60}$$

with

$$\hat{\phi}(t) = n_A(r_c, t) \tag{9.61}$$

provided that

$$1/r_c \ll 1/R_1.$$

Equation (9.60) is an analogy to Maxwell's solutions for the diffusive flux to a single spherical particle in the quasi-stationary approximation.

The condition for assuming that the concentration is approximately constant at large distances from any arbitrary particle ($1/r_c \ll 1/R_1$) should be satisfied when the total number of particles $N \leqslant 10^6$ cm^{-3}, and $R_1 \leqslant 5$ μ. Of course, when $N \ll 10^6$ cm^{-3}, R_1 can exceed 5 μ and still retain this assumption.

The rate of change in mass of the particle is related to the condensation rate by the relation

$$\frac{\rho_L}{m_A}\frac{d}{dt}(\tfrac{4}{3}\pi R_1^3) = \frac{\rho_L}{m_A} 4\pi R_1^2 \frac{dR_1}{dt} = -\tilde{\Phi}_A^c \tag{9.62}$$

Hence if the supersaturation is maintained by cooling the vapor–particle mixture such that

$$n_A(R_1, t) = n_{AC} \exp(-\gamma_c B t),$$

where γ_c is the cooling rate, B is an experimentally determined constant, and n_{AC} is the critical vapor density,

‡ In the choice of r_c, the condition that $n_A(r_c, t) = n_A(t)$, the ensemble average vapor concentration is used instead of requiring that the net gradient in n_A a large distance from an arbitrary particle is zero.

$$\frac{\mathrm{d}R_1^2}{\mathrm{d}t} = (2m_A D_{AB} n_{AC}/\rho_L)\,[1 - \exp(-\gamma_c Bt)] - \tfrac{8}{3}\pi D_{AB} N R_1^3. \tag{9.63}$$

Here the increase in surface area of any droplet in the assembly does not grow linearly with time but depends on the cooling rate (the rate of supersaturation) as well as the total number of particles in the cloud, N.

Growth of an Initially Polydispersed Aerosol

When the initial distribution of particles does not have the same size, eqn. (9.62) will be rewritten as

$$\frac{\partial[R_i^2(R_{i0}, t)]}{\partial t} = \frac{2m_A D_{AB}}{\rho_L}\,[n_A(r_c, t) - n_A(R_i, t)]. \tag{9.64}$$

However, from the previous remarks, the vapor density difference in the expression has to be a function of time alone. Reiss[334] then points out that we must reach the essential conclusion that, within the approximation $1/r_c \ll 1/R_i$, the rate of increase in R_i^2 is the same for all particles regardless of their initial radius R_{i0}. Analytically,

$$R_i^2(R_{i0}, t) - R_{i0}^2 = \Upsilon(t), \tag{9.65}$$

which is directly analogous to eqn. (4.45) for the case of single particles. Equation (9.65) is the regulation mechanism that always tends to sharpen the distribution of particles growing by vapor diffusion.

$$\left.\begin{aligned} \frac{\partial[R_i^2(R_{i0}, t)]}{\partial t} &= \frac{\mathrm{d}\,\Upsilon(t)}{\mathrm{d}t} \\ &= \frac{2m_A D_{AB} n_{A0}}{\rho_L}\,[1 - \exp(-\gamma_c Bt)] - \tfrac{8}{3}\pi D_{AB} \sum_{i=1}^{n} [(R_{i0}^2 + \Upsilon(t))^{\frac{3}{2}} - R_{i0}^3]\, n_i(R_{i0}), \end{aligned}\right\} \tag{9.66}$$

where $n_i(R_{i0})$ denotes the concentration of particles of the ith particles initially present. If $n_i(R_{i0})$ is known, eqn. (9.66) may be integrated with the initial condition $\Upsilon(0) = 0$ to yield an expression for $\Upsilon(t)$.

Narrowing of the Particle Spectrum

The tendency for a size distribution of a cloud of particles growing by diffusion of vapor to become more narrow can readily be illustrated as follows: Suppose we consider any m moment of the size distribution. The breadth of the initial distribution will be given in terms of particles of limiting radii R_{10} and R_{20}, where $R_{20} > R_{10}$,

$$(R_{20}^m - R_{10}^m)/\overline{R_k^m}, \tag{9.67}$$

where the overbar denotes the mean of the mth power of R_k. After a certain amount of growth has taken place, the breadth will be measured by

$$(R_2^m - R_1^m)/\overline{R_k^m}. \tag{9.68}$$

If the distribution tends to become narrow relative to the initial spectrum, the ratio of (9.68) to (9.67) must be less than one, or

$$(R_0^m/R_k^m)\,[(R_2^m - R_1^m)/(R_{20}^m - R_{10}^m)] < 1. \tag{9.69}$$

This ratio will always be less than

$$(R_{20}^m/R_1^m)\,[(R_2^m - R_1^m)/(R_{20}^m - R_{10}^m)]. \tag{9.70}$$

But eqn. (9.65) insures that

$$\left.\begin{aligned} R_1^m &= (R_{10}^2 + \Upsilon(t))^{m/2}, \\ R_2^m &= (R_{20}^2 + \Upsilon(t))^{m/2}. \end{aligned}\right\} \tag{9.71}$$

When the growth is significant $\Upsilon(t) \gg R_2$, so that substitution of expanded forms of these relations into (9.70) gives

$$\frac{1}{[2\,\Upsilon(t)/NR_2^2 + (R_{10}/R_{20})^2]}\,\frac{[1 - (R_{10}/R_{20})^2]}{[1 - (R_{10}/R_{20})^m]} < 1 \tag{9.72}$$

for

$$\frac{2\,\Upsilon(t)}{N} > (R_{20}^2 - R_{10}^2). \tag{9.73}$$

Since (9.72) is valid within the restriction of (9.73), the inequality (9.69) must also be true.

Reiss[334] has remarked that the restriction of (9.73) does not demand that the growth be very large before the narrowing in the spectrum is observed. However, this cannot be independent of the assumption that $\Upsilon(t) \gg R_2$, so that in fact considerable vapor diffusion to particles may take place before a significant change in the distribution will be seen.

It is of interest to observe that natural clouds of water droplets in the Earth's atmosphere grow initially by vapor diffusion. The size spectra of these clouds are usually rather polydisperse rather than narrow in breadth. Indeed, the broadness of their size distribution is the apparent requirement for formation of precipitation. The polydispersivity in natural clouds then must come about by mechanisms other than vapor diffusion on particles in an isolated, rising air parcel. The currently accepted mechanism involves the formation of many droplets of large size by collision and coalescence of smaller ones. The dynamical theory of collisional processes between aerosol particles is discussed in the next chapter.

B. Modeling the Size Distribution of Nucleating Aerosols

In the absence of detailed theoretical analyses for the formation and growth of aerosols by condensation, some useful information may be obtained by applying dimensional arguments to correlate the behavior of a condensing vapor. To illustrate this technique, let us consider the formation of an aerosol by mixing a hot, moist gas with a cooler, dry gas. After saturation is attained in the mixture, the particle creation will depend on the temperature time history of an element of the gaseous suspension, and the concentration of nuclei. Friedlander[280] has discussed this problem when the mixing is rapid enough, and the foreign nuclei concentration is low enough for homogeneous nucleation to control.

Friedlander assumes the following conditions:

(a) In analogy to the macroscopic equations for gases, the equation for the conservation of species in the system may be written in the form equivalent to (4.2):

$$\rho_G \frac{\mathrm{d}n_k/\rho_G}{\mathrm{d}t} = \nabla \cdot \rho_G D_k \nabla n_k/\rho_G + G_k, \tag{9.74}$$

where D_k is the particle diffusivity, and G_k is given by eqn. (9.21).

(b) The concentration of condensable vapor is sufficiently small so that the release of latent heat may be disregarded. This implies that the temperature–time history of any element is controlled by the mixing, and is effectively an externally imposed parameter.

(c) The vapor pressure of condensable material follows the law of corresponding states, i.e. the vapor pressure made dimensionless by the saturation vapor pressure p_{0c} at critical temperature T_c is given by

$$p_{0k} = p_0/p_{0c} = f_1(T/T_c) = f_1(T_R). \tag{9.75}$$

(d) The surface free energy may be assumed to be associated with the property of the bulk fluid, and has a form

$$\sigma S_1/kT = f_2(T/T_c). \tag{9.76}†$$

(e) The mixing and the associated cooling take place at constant pressure, and the gas mixtures conform to ideal gases under the conditions of interest.

(f) Initially the embryo spectrum is given by an equilibrium distribution equivalent to eqn. (9.29).

(g) The diffusion coefficient of embryos of the kth class may be given by the expression for a rigid sphere from the kinetic theory of gases.‡

$$D_k = \frac{3}{8n_G(R_k + R_1)^2}\left[\frac{kT(m_k + m_G)}{2\pi m_k m_G}\right]^{\frac{1}{2}}.$$

(h) The ambient temperature of the cooler environment is sufficiently low for the system to reach a frozen equilibrium before the size spectrum has been measured.

(i) The evaporation coefficient α_c is a function of reduced temperatures T_R only.

Using the assumptions (a) to (i), one can rewrite eqn. (9.74) in dimensionless form as

$$\left.\begin{aligned} \frac{\mathrm{d}n_k^{\oplus}}{\mathrm{d}t^{\oplus}} &= \frac{1}{Pe_1}\left(\frac{T}{T_0}\right)\nabla^{\oplus}\left(\frac{T}{T_0}\right)^{\frac{1}{2}}.\nabla^{\oplus} n_k^{\oplus} \\ &+ \frac{n_1}{n_{10}}[2{\cdot}84\alpha_c \mathcal{N}_t^{-1}]\left(\frac{T}{T_0}\right)^{\frac{1}{2}}\left\{\Delta\left[(k+1)^{\frac{2}{3}} n_k^{\oplus} \exp[(k+1)^{\frac{2}{3}} f_2\right.\right. \\ &\left.\left. + 2\ln S_{p0}\frac{f_{10}}{f_1}\cdot\frac{T}{T_0}\right] - \Delta(k^{\frac{2}{3}} n_k^{\oplus})\right\}, \end{aligned}\right\} \tag{9.77}$$

† Eötvos' law is a special form of this relation.

‡ If the molecules of condensable vapor are much larger than the molecules of the remaining gas molecules, $D_k \approx D_1/k^{\frac{2}{3}}$.

where $n_k^{\oplus} = n_k \rho_{G0}/n_{10}\rho_G$, the subscript 0 referring to the initial conditions. $t^{\oplus} = Ut/L$, and $\nabla^{\oplus} = L\nabla$. L is a characteristic length of the fluid dynamical system; U is a characteristic fluid velocity; Pe_1 is the Péclét number LU/D_{AB}; $\mathcal{N}_t = Ut/L$, where $t = (n_{G0}R_G^2)^{-1} \times (m_G/\pi k T_0)^{\frac{1}{2}}$, a mean time between molecular collisions. S_{p0} is the supersaturation ratio P_{A0}/P_{A00} based on the initial conditions (usually less than unity), and the operator Δ is defined by $\Delta f_k = f_k - f_{k-1}$. We note that the term in brackets containing Δ is a function only of index k and T_R as a result of our restrictions, including condition (d) above.

$$n_{k0}^{\oplus} = n_{10}^{\oplus} \exp\left[(k^{\frac{2}{3}} - 1) f_2(T_{0R}) + \ln S_{p0}\right] \tag{9.78a}$$

with the ambient condition

$$n_{k\infty}^{\oplus} = n_{1\infty}\rho_{G0}/n_{10}\rho_G \approx 0. \tag{9.78b}$$

Mixing in a fluid system is often dominated by convective disturbances. With the mixing controlled by the fluid mechanical regime as determined by the Reynolds number, $Re\,(= UL/\nu_{G0})$, the spectrum function $n_k^{\oplus}$ in the state of frozen equilibrium will be a function only of the following groups:[280]

$$n_k^{\oplus} = \psi_1(k, T_0/T_c, T_\infty/T_c, Re, \mathcal{N}_t, S_{p0}, Sc_1). \tag{9.79}$$†

The Schmidt number Sc has been introduced here as a combination of the Péclét number and the Reynolds number, $Sc_1 = Pe_1/Re = \nu_{G0}/D_{10}$.

The result given in eqn. (9.79) has some practical importance. Its utility may be demonstrated by considering the jet mixing problem. For a particular condensing substance flowing out from a nozzle into a cold gas, the average particle size of the aerosol formed by spontaneous condensation should be given in the form

$$\overline{R}_k/R_G = \psi_2(T_0/T_c, T_\infty/T_c, \mathrm{Re}, \mathcal{N}_\pi, S_{p0}, Sc_1). \tag{9.80}$$

where $\mathcal{N}_\pi = Re/\mathcal{N}_t S_{p0}$, a number that is independent of the jet velocity and the vapor concentration. Thus, Friedlander has noted that the influence of the Reynolds number alone (the aerodynamic variation) may be evaluated by varying the average fluid velocity at the U while keeping other conditions constant. Similarly, the influence of the supersaturation ratio should be found by keeping the jet velocity constant while changing the vapor concentration at constant nozzle temperature T_0. Making use of well-known methods for correlating observational data, one should find functional curves of similar form for substances abiding by the assumptions made in deriving eqns. (9.79) and (9.80).

Other properties that depend on the size distribution of the aerosol, such as the optical density, also should be amenable to modeling in the same areas as the average particle size.

The modeling theory of aerosol formation by condensation has not been tested in any detail. However, some experimental observations taken by Hidy[33] in a cloudy jet suggested that such a theoretical framework would be applicable. In these experiments, the height above the nozzle where the cloud of droplets first appeared was very sharply defined under given light conditions. This height, used as the characteristic length L, could be correlated relatively well for aerosols of glycerine, dibutyl phthalate and propylene glycol.

† Fixing the Schmidt number in a given gas will yield a unique relationship through the Lewis number between n_1/n_{10} and T/T_0 so that specification of one of these distributions will be sufficient for both.[304]

CHAPTER 10

INTERACTION BETWEEN AEROSOL PARTICLES—GROWTH BY COAGULATION OR COALESCENCE

IN CHAPTER 9 a rather general, formal equation for the rate of change in concentration of the kth sized particle n_k was presented, from which both the special cases of nucleation and coagulation could be derived. The limiting model of the coagulation mechanism followed if collisions between particles are limited to binary cases, and they are irreversible. Thus the net rate of change in n_k is

$$\frac{\mathrm{d}n_k}{\mathrm{d}t} = G_k = \tfrac{1}{2}\Big[\sum_{\substack{j=1\\ i=k-j}}^{i=k-1} b_{ij}n_in_j\Big] - n_k\Big[\sum_{i=1}^{\infty} b_{ik}n_i\Big], \tag{10.1}$$

where the first term on the right-hand side represents the gain to species k by collisions between i and j particles, while the second term denotes the loss by collisions of k species and all other particles. If the particles stick together with a known probability which is independent of surface properties and size, the determination of G_k constitutes a purely mechanical problem centered on the evaluation of the collision parameter b_{ij}.

Relatively little is known about the efficiency of sticking once particles collide. Generally one expects that coagulation or coalescence of particles hitting one another will depend on the nature of the surface of the particles, as well as their size. In addition, the influences of changes in shape during collision of liquid particles, and the effects of electrostatic forces or other induced forces like Van der Waals' interaction between the particles are not too well understood. In the absence of better information, one assumes that the probability of sticking after collision is unity, or at least some constant factor less than 1 to be taken into account empirically in matching experimental results with the purely mechanical theory. We shall adopt this procedure, and consider the collision of inert spheres whose sticking probability is unity.

10.1. Brownian Motion of Electrostatically Charged and Uncharged Particles

As a beginning, let us consider the classical, idealized problem of coagulation of aerosols undergoing Brownian motion in a stagnant or uniformly flowing medium. The particles are assumed to move randomly in the inert, un-ionized fluid.

A. Smoluchowski's Theory

Perhaps the first theory for the coagulation of sols in Brownian motion originated with Smoluchowski[282] in the early twentieth century. Smoluchowski applied the theory of Brownian diffusion to the statistical problem of the collision of hard non-interacting spheres thermally agitated in a continuum. An expression for the rate of collision between particles per unit volume, the *collision rate density*, was inferred from the solution of a simplified diffusion problem analogous to the one in Section 7.2(A4) (see also Chandrasekhar.[5])

Smoluchowski's early theory has been generalized by different workers (e.g. Fuchs,[338] Debye,[339] and Kramers[340] to account for the effect of a drift velocity along the line of centers between the colliding particles. To determine the collision rate density in this case one considers first the problem of determining the flux of particles to a central, perfectly absorbing particle in a monodisperse cloud of uniform spatial concentration N_∞ cm^{-3} (Fig. 10.1). From the arguments leading to eqn. (7.39), the total rate of transport to the fixed body is

$$\tilde{\Phi}_i^c = 4\pi r^2 \left(D_i \frac{\partial N}{\partial r} + q_i N \right), \tag{10.2}$$

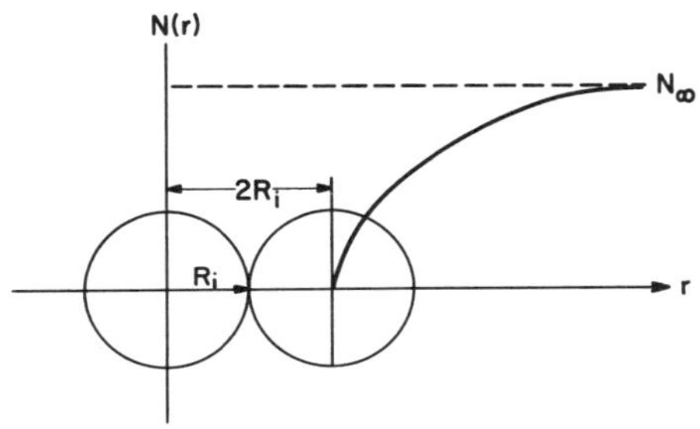

FIG. 10.1. Geometry of collision model as diffusion to a central sphere.

where D_i is the Stokes–Einstein diffusivity, eqn. (7.31), and r is the radial coordinate.

The drift velocity q_i can be defined in terms of a potential field Φ:

$$q_i = -B_i \frac{\partial \Phi}{\partial r}; \quad \Phi \to 0 \quad \text{as} \quad r \to \infty. \tag{10.3}$$

Here B_i denotes the mobility, given in Section 3.2 for particles moving in a continuum as the Stokes' coefficient, $(6\pi\mu_G R_i)^{-1}$.

To find the rate of transport $\tilde{\Phi}_i^c$, $N(r, t)$ must be determined first. This may be done by taking the divergence of eqn. (10.2), and solving the resulting equation:

$$\frac{\partial N}{\partial t} = \frac{\partial}{\partial r}\left[D_i \frac{\partial N}{\partial r} + q_i N \right] \tag{10.4}$$

for the boundary conditions:

$$\left.\begin{array}{lll} N = N_\infty = \text{const,} & \text{at} \quad t = 0 & r > R_i; \\ N = N_\infty & r \to \infty \\ N = 0 & r = 2R_i \end{array}\right\} \begin{array}{l} \\ t > 0. \\ \end{array} \tag{10.4a}$$

The complete solution to eqn. (10.4) with conditions (10.4a) is complicated, but is similar to the solution given in Section 7·2(A.4). Usually, however, we restrict ourselves to cases where a characteristic time $R_i^2/D_i \gg 1$, and a steady flux $\tilde{\Phi}_i^c$ = const., is assumed. For the conditions (10.4a), the steady flux of particles to the fixed body may be determined from (10.2), or

$$\tilde{\Phi}_i^c = 4\pi D_i N_\infty \Big/ \int_{2R_i}^{\infty} \frac{1}{r^2} \exp\left(\frac{\Phi}{kT}\right) dr \tag{10.5}$$

This expression is generalized to account for different sized particles by first assuming that all particles are j multiples of singlet or the unit sized particles. Then the rate of collision per unit volume between i and j species L_{ij}^c is estimated from eqn. (10.5) after taking into account the relative motion between the particles.

Suppose that i and j move in space respectively a distance Δx_i and Δx_j in a time interval Δt. Then their relative square displacement, on the average, is

$$\overline{(\Delta x_i - \Delta x_j)^2} = \overline{\Delta x_i^2} + \overline{\Delta x_j^2} - \overline{2\Delta x_i \Delta x_j}.$$

When there is no interaction between the particles, and their motion is independent, $\overline{\Delta x_i \Delta x_j} = 0$. Hence, with use of Einstein's expression for the diffusivity, the average relative displacement becomes

$$\overline{(\Delta x_i - \Delta x_j)^2} = 2(D_i + D_j)\,\Delta t = 2D_{ij}\Delta t. \tag{10.6}$$

However, when a force field is induced between particles and $q_i \neq 0$, $\overline{\Delta x_i \Delta x_j}$ cannot necessarily vanish. Nevertheless, it generally has been presumed that this term is very small compared to $\overline{\Delta x_i^2}$ and $\overline{\Delta x_j^2}$, and the collision rate between two interacting particles can be expressed by a relation containing only a form similar to (10.6).

The value of R_i for colliding particles has been called "the sphere of influence" of the particles. This implies that R_i may have some vague definition analogous to a "soft shelled" molecule to account empirically for interaction between the spheres. However, for hard spheres, the collisional radius R_{ij} of interest is that which corresponds to the contact of particles:

$$R_{ij} = (R_i + R_j), \tag{10.7}$$

where R_i denotes the radius of a sphere of the ith size, where $R_i = i^{\frac{1}{3}} R_1$. Hence, with eqns. (10.5) and (10.6) the collision rate density for the ith and jth species reads

$$L_{ij}^c = 4\pi M_{ij} D_{ij} R_{ij} \Omega_{ij}^c n_i n_j, \tag{10.8}$$

where

$$\Omega_{ij}^c = \left[R_{ij} \int_{R_{ij}}^{\infty} \frac{1}{r^2} \exp\left[\frac{B_{ij}\Phi_{ij}}{D_{ij}}\right] dr \right]^{-1}, \tag{10.9}$$

and n_i is the concentration of species i, taken at a distance r far from any species (the subscript ∞ is dropped here). Then the collision parameter b_{ij} for this model is

$$b_{ij}^c = 4\pi M_{ij} D_{ij} R_{ij} \Omega_{ij}^c. \tag{10.10}$$

The term M_{ij} is the symmetry number

$$\left.\begin{aligned} M_{ij} &= \tfrac{1}{2}, \quad i = j, \\ M_{ij} &= 1, \quad i \neq j. \end{aligned}\right\} \tag{10.11}$$

This factor is included so that collisions per unit volume per unit time between particles of the same size are not counted twice.

If equal sized particles do not interact, $\Omega_{ii}^c = 1$, and eqn. (10.8) reduces to Smoluchowski's original expression for the collision rate of a monodispersed sol,

$$L_{ii}^c = 8\pi D_i R_i N^2 = \frac{4}{3}\frac{kT}{\mu_G} N^2. \tag{10.12}$$

Although the collision rate density given by Smoluchowski has not been verified by direct experiment in gases, there is considerable evidence that eqn. (10.12) satisfactorily describes the agglomeration of non-interacting particles in liquids. Typical of the more recent evidence is that of Swift and Friedlander.[341] They determined the value of b_{ii}, assuming a nearly monodispersed sol, from the slopes of curves of the decay in the inverse of the total number of particles $1/N$ with time [see also eqn. (10.97) and related discussion]. Similar experiments have been carried out on coagulating aerosol clouds. The collision rate observed from measurements of the decrease in N suggest that the collision rate of aerosols by Brownian motion is generally larger than Smoluchowski's theory would predict (e.g. Green and Lane[4]). There are several possible explanations for the difference as we shall see in the subsequent sections.

Effect of Electrostatic Charge

To further illustrate the effect of an induced field of potential energy between colliding particles, consider as an example simple coulombic interaction, where

$$\Phi_{ij}^{(1)} = \frac{z_i z_j \tilde{e}^2}{r}, \tag{10.13}$$

where z_i is the number of unit electron charges on the ith particle, and $\tilde{e}$ is the unit of charge. The factor Ω_{ij}^c then becomes

$$\Omega_{ij}^{c(1)} = \frac{z_i z_j \tilde{e}^2}{R_{ij} kT[\exp(z_i z_j \tilde{e}^2/R_{ij} kT) - 1]}. \tag{10.14}$$

As expected, eqn. (10.14) indicates qualitatively that repulsion of like charged species should decrease the rate of coagulation, whereas the attraction of oppositely charged particles should increase the agglomeration rate.

Since the Coulombic potential involves two body interaction only, eqn. (10.14) should have certain restrictions of applicability. It is not really clear what criteria should be used to determine quantitatively the limits of applicability of this relation. Qualitatively, however, eqn. (10.14) cannot be expected to give reliable results for any cases other than for very dilute or highly dispersed aerosols, where the concentration of particles is much less than the concentration of gas molecules. There must also be a limitation involving the intensity of charging of the aerosol particles. One criterion for the strength of charging involves the ratio of the collision radius R_{ij} and a characteristic length of electrostatic influence, $z_i z_j \tilde{e}^2/kT$. Zebel[342] has indicated that strong charging of aerosols may be defined as

$$|z_i z_j \tilde{e}^2/R_{ij} kT| \approx 1. \tag{10.15}$$

Then the Coulombic interaction factor $\Omega_{ij}^{c(1)}$ probably becomes increasingly inaccurate as the left-hand side of eqn. (10.15), named the *Coulomb number*, *Co*, becomes larger than unity.

The predictions of the factor $\Omega_{ij}^{c(1)}$ can be checked using more sophisticated models for the influence of charging in clouds. A well known theory which accounts in an approximate way for the effects of two-body interaction, combined with interaction of all other charged particles in the cloud, is the Debye–Hückel model (see, for example, Hill[(6)]). Since this model stems from a linearized theory, it can be expected to apply rigorously only to dilute clouds of unpolarized particles near "electrochemical" equilibrium. Nevertheless, the Debye–Hückel treatment extends the simple Coulomb model by one order of approximation.

In contrast to the qualitative feeling for the applicability of eqn. (10.14), definite conditions for the applicability can be placed on the Debye–Hückel model. For unpolarized spheres, the interaction potential from this theory reads

$$\Phi_{ij}^{(2)} = \frac{z_i z_j \tilde{e}^2 \, \mathrm{e}^{-\kappa r}}{r}, \qquad (10.16)\dagger$$

where the Debye reciprocal length κ is defined as

$$\kappa = \left[\frac{4\pi\tilde{e}^2}{kT} \sum_{i=1}^{s} n_i z_i^2\right]^{\frac{1}{2}}. \qquad (10.17)$$

The exponential term in eqn. (10.16) represents a "shielding" effect of all the particles in the cloud, which tends to reduce the effectiveness of the Coulombic interaction at large distances. This potential is a good approximation to an "ionized" system of aerosols when

$$\left|\frac{z_i z_j \tilde{e}^2 \kappa}{kT}\right| \ll 1. \qquad (10.18)$$

For the purposes of evaluation of electrostatic interaction, the important values of r are those in the neighborhood of the length κ. Therefore, a criterion for the applicability of the Debye–Hückel model should utilize eqn. (10.18).

The left-hand side of eqn. (10.18) defines another characteristic number for a system of charged particles. This parameter will be called the *Debye number*, *De*. It should be much less than unity for the Debye–Hückel model to be an accurate picture of interaction in a cloud of charged particles. The criterion for application of the Debye–Hückel treatment, eqn. (10.18), then implies that, if $z_i = z_j = 1$, N cannot exceed 10^{10} cm^{-3}. However, when $z_i = z_j = 10$, the cloud must be much more dilute, i.e. $N \leqslant 10^4$ cm^{-3}.

To obtain a relation for the interaction factor based on the Debye–Hückel model, eqn. (10.16) is substituted in eqn. (10.9), the exponential term in $(B_{ij}\Phi_{ij}^{(2)}/D_{ij})$ is expanded, and the integration is carried out term by term. κR_{ij} is usually less than 1 under ordinary circumstances. And as a first approximation, terms in ln R_{ij} and of the order κR_{ij} or higher may be dropped. Hence,

$$\Omega_{ij}^{c(2)} \approx \frac{z_i z_j \tilde{e}^2 \mathrm{e}^{-\kappa R_{ij}}}{R_{ij} kT\{\exp\left[(z_i z_j \tilde{e}^2/R_{ij} kT)\,\mathrm{e}^{-\kappa R_{ij}}\right] - 1\}}. \qquad (10.19)$$

† Rewriting eqn. (10.16)

$$\Phi_{ij}^{(2)} = \Phi_{ij}^{(1)} \mathrm{e}^{-\kappa r},$$

we see that the Debye potential may be considered the second approximation to an n-body interaction problem which contains the first approximation, the Coulombic potential.

Thus the correction for Debye shielding can be taken into account approximately in eqn. (10.14) by multiplying the electrical terms in the Coulombic expression by the exponential factor, $e^{-\kappa R_{ij}}$. To see how severe this correction is, the ratio $\Omega_{ij}^{c(1)}/\Omega_{ij}^{c(2)}$ has been calculated and is tabulated in Table 10.1 for different values of charging of 10 μ and 0.1 μ radius particles. These results indicate that the Debye correction to the Coulombic interaction factor $\Omega_{ij}^{c(1)}$ generally is quite small and may be disregarded.

From the estimates of the correction factor Ω_{ij}^{c} in Table 10.1 we can see that the effect of electrostatic charging is very small for a weakly charged aerosol cloud of 1 μ particles, where $C_0 < 1$. However, when the charging is increased to 10 electron units per particle (strong charge), a severe reduction is estimated for charging of the same sign (unipolar charging), and a strong increase in coagulation is predicted for charging of the opposite sign (bipolar charging). The estimates for strongly charged clouds in particular should be

TABLE 10.1. ELECTROSTATIC EFFECTS ON COAGULATION

$T = 20°\text{C}, N = 10^4\ \text{cm}^{-3}$

R_i(μ)	R_j(μ)	z_i	z_j	$\Omega_{ij}^{c(1)}$	$\Omega_{ij}^{c(1)}/\Omega_{ij}^{c(2)}$	z_i	z_j	$\Omega_{ij}^{c(1)}$	$\Omega_{ij}^{c(1)}/\Omega_{ij}^{c(2)}$
1.0	1.0	+1	+1	0.990	1.00	+1	−1	1.01	1.00
1.0	0.1	+1	+1	0.983	1.00	+1	−1	1.03	1.00
0.1	0.1	+1	+1	0.869	1.00	+1	−1	10.2	1.00
1.0	1.0	10	10	0.173	1.00	+10	−10	1.15	1.00
1.0	0.1	10	10	0.028	1.00	+10	−10	1.28	1.00
0.1	0.1	10	10	0.93×10^{-11}	1.00	+10	−10	28.7	1.00

viewed with caution since polarization of the electric field should be taken into account under these circumstances. Polarization will tend to reduce the effect of charging in coagulation, as indicated by Fuchs.[(2)]

Experimental studies of coagulation in charged clouds have yielded ambiguous results. Early experiments on bipolar charged aerosols indicated that, for particles in the micron and submicron range of size, charging up to 4–6 units per particle had no detectable effect on coagulation. However, a study of Gillespie[(343)] has indicated that bipolar charging of $z_i \approx 6$–14 units on particles of ~1 μ radius can increase coagulation rates by an order of magnitude. Fuchs[(2)] has remarked that these results may not be reliable because of complications resulting from electrostatic deposition of particles on the walls of Gillespie's chamber. Gillespie, of course, did attempt to account for losses of particles on the walls of his chamber in analyzing his data. Careful experiments recently reported by Devir (Weinstock)[(344)] of more monodisperse dioctylphthalate aerosols ($R_i \approx 0.7$ μ, $z_i = 1 - 3$) have indicated no observable effect of charging. This is certainly in agreement with early results, and indicates that the simple Coulomb interaction model is an adequate theory for the effect of weak charging.

Measurement of the coagulation of unipolar charged aerosols is complicated by the effect of electrostatic dispersion, discussed in Section 3.2E. As a result of the concentration of like charge, there is a tendency for the entire aerosol cloud to expand because of the mutual repulsion induced locally between each particle. Because of this effect, virtually no reliable experimental studies concerning unipolar charging and aerosol coagulation are available. However, it still may be possible to test for decreased coagulation of unipolar charged aerosols by a method similar to Gillespie's[343] or Devir's,[344] taking into account losses on the walls of the apparatus.

B. Free Molecule Theory

The classical continuum theory provides a satisfactory mechanical theory of coagulation of non-interacting particles in liquids. However, the continuum theory must have severe limitations when applied to aerosols whose Knudsen number exceeds a few hundredths. For example, Smoluchowski's diffusion theory should predict coagulation rates that are too low if a correction for the particle mobility is made where $Kn_i > 0.01$. Indeed, the experimental observations suggest that this is the case.

In examining the problems of collision rates of aerosols, it is clear that, as Kn_i gets large, the limiting value of b_{ij} based on a mechanical theory should approach the rate given by the kinetic theory of (binary) collisions for gas molecules (e.g. Hirschfelder, Curtiss, and Bird[17])

In cases where $Kn_i \geqslant 10$, the kinetic theory of gases may be applied to aerosol coagulation even though the particle radii are greater than molecular radius of the suspending gas. This implies, of course, that the aerosols are "dissolved" in the gas in such a way that the aerosol particles constitute just another component of a multi-component gaseous system.

Thus the behavior of uncharged free molecule aerosols, regarded as large, spherical elastic molecules, should be correctly described by the Boltzmann equation. The collision rate density L_{ij}^* of such an aerosol is then given by the expression developed in Section 2.2.

$$L_{ij}^* = \int_{\ell}\int_{\varepsilon}\int_{\mathbf{v}_i}\int_{\mathbf{v}_j} M_{ij} f_i^1 f_j^1 g\, \ell \, d\ell \, d\varepsilon \, d\mathbf{v}_i \, d\mathbf{v}_j, \tag{10.20}$$

where f_i^1 and $\mathbf{v}_i$ are respectively the distribution function and the velocity vector of the ith species of particle. As in eqn. (2.70), g is the magnitude of the difference in particle velocities $\mathbf{v}_i - \mathbf{v}_j$, and $d\ell \, d\varepsilon$ denotes the parameters associated with the collisional geometry (Fig. 2.5).

Within the restriction where the total concentration of particles N is much less than the concentration of gas molecules n_G, the evolution of the distribution functions for uncharged particles is determined principally by particle–gas molecule encounters. Collisions between aerosol particles take place relatively less frequently. Hence, as an approximation, the evolution of f_i^1 involves a relation of the Boltzmann type analogous to eqn. (2.70). Thus for uncharged free molecule aerosol–gas systems with the influence of external forces, the equations for the distribution functions are

$$\frac{\partial f_G^1}{\partial t} + \mathbf{v}_G \cdot \nabla f_G^1 = J_{GG}\{f_G^1[\mathbf{v}(1)], f_G^1[\mathbf{v}(2)]\}, \tag{10.21}$$

$$\frac{\partial f_i^1}{\partial t} + \mathbf{v}_i \cdot \nabla f_i^1 = J_{iG}\{f_G^1[\mathbf{v}(1)], f_i^1[\mathbf{v}(1)]\}. \tag{10.22}$$

The solutions to eqns. (10.21) and (10.22) have been studied as cases of a quasi-Lorentzian gas.[345, 346, 347] These investigations have indicated the f_i^1 have the form

$$f_i^1(\mathbf{r}, \mathbf{v}_i, t) = n_i(\mathbf{r}, t) \left(\frac{m_i}{2\pi kT}\right)^{\frac{3}{2}} \exp\left[-\frac{m_i(\mathbf{v}_i)^2}{2kT}\right], \tag{10.23}$$

where m_i and $\mathbf{v}_i$ are the mass and the velocity of the ith particle, and T is the local temperature.

For uncharged aerosols, the interparticle forces extend over short distances. Therefore, the motion of these species can be assumed independent of each other up to the instant of collision. Thus, with the collisional cross-section of interest R_{ij} with eqn. (10.23) and with the assumption that binary collisions predominate, eqn. (10.20) may be integrated, giving[348]

$$L_{ij}^{*} = M_{ij}\left[\frac{8\pi kT(m_i + m_j)}{m_i m_j}\right]^{\frac{1}{2}} R_{ij}^2 n_i n_j, \qquad (10.24)$$

and the collision parameter corresponding to this case is

$$b_{ij}^{*} = M_{ij}\left[\frac{8\pi kT(m_i + m_j)}{m_i m_j}\right]^{\frac{1}{2}} R_{ij}^2. \qquad (10.25)$$

As anticipated, this expression for the collision rate density of uncharged, free molecule aerosols in Brownian motion is directly analogous to the collision frequency estimated for different *molecules* of gases colliding, as given by Hirschfelder *et al.*[17]

Equation (10.24) represents the same parameter as the one given by eqn. (10.10), even though their forms are very different. In particular, these two relations for limiting regimes of Kn_i vary quite dissimilarly with respect to the particle size. For example, in the case of coagulation of uncharged spherical particles of the same density, and initially of the same size, the collision rate in the continuum limit is independent of particle size. However, as $Kn_i \geqslant 10$, the collision rate for $i = j$ as found in eqn. (10.24) is proportional to $R_i^{\frac{1}{2}}$. We note also that the often used modification of Smoluchowski's theory using the diffusivity based on kinetic theory rather than the Stokes–Einstein relation to force this model to apply to the free molecule regime leads to completely erroneous results. Therefore, knowledge of the particle diffusivity alone cannot lead to prediction of correct theoretical collision rates.

The first attempt to verify the theory for coagulation of highly dispersed aerosols where $Kn_i > 10$ was reported recently by Fuchs and Sutugin.[349] Their data, taken in a continuous flow reactor, for particles of average radius 25 Å and 45 Å yielded values for the collision rate of 13.4×10^{-10} cm^3 sec^{-1} and 15.0×10^{-10} cm^3 sec^{-1} with standard deviations of 11.5 per cent in the former and 12.5 per cent in the latter case. These values are about two to three times too high based on estimates from eqn. (10.25), but they reflect qualitatively the right dependence on particle size. Fuchs and Sutugin have attributed the deviation between theory and experiment to the influence of intermolecular forces. These are known to be negligible for much larger aerosols in the submicron range of size because of the retardation effect discussed for example by Overbeek.[350] The retardation effect is responsible for a much stronger decrease in force potential $\Phi(r)$ acting between two particles, at distance between their line of centers r, than given by Hamaker's[242] relation. Hamaker's equation [Eq. (8.27)] assumes that the Van der Waals forces induced between two particles is additive. The retardation effects between two interacting particles becomes large only beyond an interaction distance of 200 Å. Fuchs and Sutugin assumed that for particles of radius ~ 25 Å, this influence on b_{ij}^{*} could be neglected. Thus, Hamaker's theory could be applied for estimating the influence of Van der Waals' interaction on the coagulation of their highly dispersed aerosols.

The increase in coagulation rate by attractive molecular forces can be estimated[350] by considering geometry of collision of two spheres of the same radius (Fig. 10.2). Fixing

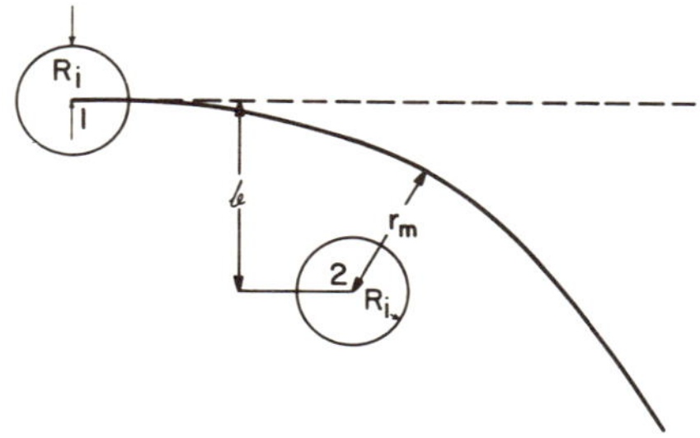

FIG. 10.2. Collision between two spheres.

the coordinate system on the particle whose center is 2, the attractive potential based on relative motion of the particles is $2\Phi_H(r)$. We assume that $\Phi_H(r) = 0$ at the center of the other particle 1 and the distance b denotes the shortest path of a tangent to the trajectory of particle 1 from the center 2 (the collision cross section). The distance r_m refers to the minimum distance from 2 to the curving trajectory of 1. By conservation of momentum, Fuchs and Sutugin deduced that the distance b may be found from

$$b^2 = r_m^2 \left[1 + \frac{2\Phi_H(r_m)}{E_1}\right], \tag{10.26}$$

where E_1 is the total energy of the moving particle 1. For particles in relative motion, E_1 is the kinetic energy where $\Phi_H = 0$. The average value of E_1, then, is $3kT$, k denoting Boltzmann's constant. When $r_m \to 2R_1$, and the particles actually touch, $\Phi_H \to \infty$, and $b \to \infty$. When $r_m \to \infty$, $\Phi_H \to 0$, but b still approaches infinity. Since $\Phi_H(r_m)$ is positive for $2R_1 < r_m < \infty$, b will yield a minimum value b_m in this range. If $b \to b_m$, r_m is real, and the particles never collide. When $b < b_m$, r_m becomes imaginary and no minimum value of r_m exists. In other words, there can be no minimum distance from 0 to the trajectory.

The analysis of Fuchs and Sutugin indicates that the minimum of the right-hand side of (10.26) for reasonable values of Hamaker's interaction constant $(\pi n_A^2 C_{at})$ shows minimum values of $r_m > 2R_i$; i.e. the particles will not collide if $b > b_m$.

Without molecular forces the collision cross section $b = 2R_i$. Thus the probability of collision between particles with inclusion of molecular forces must increase by a factor of $b_m^2/4R_1^2$. Assuming a value of 8.98×10^{-13} erg for $(\pi n_A^2 C_{at})$, based on London's coefficient estimated from ultraviolet spectrum of sodium chloride, Fuchs and Sutugin found b_m graphically using eqn. (10.26). They determined the ratio $b_m^2/4R_1^2 \approx 2.19$; the theoretical coagulation rates for 25 Å and 45 Å (average) radius particles respectively were 10.5×10^{-10} and 14.1×10^{-10} cm^{3}sec^{-1} taking into account the Van der Waals forces. Their theoretical values are still about 15 per cent lower than the experimental results. However, there is certainly some error in their estimate of the constant C_{at} and no account was taken of the enhancement in coagulation rate because of polydispersivity.

Electrostatically Charged Free Molecule Aerosols

When free molecule aerosols become charged, the long range nature of the Coulombic interaction must be taken into account. In the general case, multiple interactions and delocalization of charge then must be considered in estimating the evolution of the particle and gas distribution functions, and in evaluating the collision rate density.

When charged particles exist in equilibrium within a suspending gas, Balescu's[7] results, based on Liouville's theorem, have indicated that the Debye–Hückel potential should be an exact expression for the electric field developing in the cloud. Hence, using the Debye potential in eqn. (10.20) with eqns. (10.21) and (10.22), and assuming the particles possess a fixed charge during the effective time of collision, the collision parameter for charged free molecule aerosols in Brownian motion is

$$b_{ij}^{*} = M_{ij}\left[\frac{8\pi kT(m_i + m_j)}{m_i m_j}\right]^{\frac{1}{2}} \Omega_{ij}^{*} R_{ij}^{2}, \tag{10.27}$$

where the interaction factor reads

$$\Omega_{ij}^{*} = \exp\left[\frac{-\, z_i z_j \tilde{e}^2\, \mathrm{e}^{-\kappa R_{ij}}}{kTR_{ij}}\right]. \tag{10.28}$$

It is interesting to note that Ω_{ij}^{*} has a form differing from the continuum result $\Omega_{ij}^{c(2)}$.

At one atmosphere pressure of gases, aerosol particles will have radii of about 0.01 μ or less in the free molecule regime. This implies from the criterion of the Coulomb number that these particles are strongly charged even with only one unit of charge per particle. Furthermore, in this size range Ω_{ij}^{*} becomes very small for clouds containing unipolar charged particles at 295°K with $N = 10^4$ cc^{-1}. However, one must bear in mind that increases in thermal agitation can reduce the effectiveness of the potential barrier between particles, and the force induced by particle polarization must always be attractive. Therefore, increases in temperature of an aerosol cloud combined with the influence of polarization would reduce the "suppressing power" of the Coulombic field in the cloud. As in the case of the continuum results, the net effect of bipolar charging must take into account the separate probabilities between charged particles of like sign and opposite sign. Again assuming equal probability for both classes of encounter, one finds that the strengthening effect of bipolar charging on coagulation is less than the inhibiting influence of interaction between like charged particles.

So far, there are no experimental results available which deal with electrostatic effects in free molecule aerosols. Therefore, at present, we have no criteria for the validity of these theories other than some relative security in the extension of the well understood kinetic model of dilute gases.

C. The Region of Transition in Knudsen Number

To formulate a complete theory even for the simplest cases of coagulation of uncharged particles in Brownian motion, the two theories for the limiting regimes of Knudsen number must be spliced together. In the transition regime where $0.2 \lesssim Kn_i \leqslant 10$, the physical processes of collision become rather difficult to specify because neither the continuum model nor the simple kinetic theory model apply. That is, on one extreme of the transition regime, disturbances in the collisional orbits cannot be accurately computed from the continuum equations, whereas on the other extreme, the mechanics of a simple two body collision are disturbed by interactions with the surrounding gas. The physics of the transition regime for aerosols involves many of the inherent difficulties found in the statistical mechanical theories for the dynamics of dense gases. As indicated in previous chapters, this no man's land remains one of the fields containing important unsolved problems in modern fluid dynamics. Without completely understanding the fundamental problems of interaction between gas molecules and bodies in this regime, relatively little can be done in formulating an "exact" statistical mechanical theory as applied to transition aerosols.

The inherent difficulties involving the effect on collisions of Knudsen number were recognized some years ago by aerosol workers. Experimental observations of the coagulation rate of smokes of submicron size indicated that the coagulation was somewhat faster than predicted by Smoluchowski's original theory. Part of the increased coagulation rate could be accounted for by electrical effects. But certainly the influence of the increased mobility of sub-micron particles resulting from slippage of gas flow around these bodies had to influence the rate of coagulation.

The first attempts to extend the results of diffusion theory to the regime of transition centered attention on the Stokes–Cunningham correction for particle mobility. Whytlaw-Gray and Patterson[271] extended Smoluchowski's results using the Stokes–Einstein diffusivity by empirically introducing the Stokes–Cunningham correction into D_i of eqn. (7.35), giving

$$D_i = \frac{kT}{6\pi\mu_G R_i}\left(1 + \frac{\tilde{A}\lambda_G}{R_i}\right). \tag{10.29}$$

where $\tilde{A}$ was given previously in eqn. (3.9b). Incorporation of the Stokes–Cunningham correction into eqn. (10.8) gives rise to a continuous increase in collision rate with increasing Knudsen number instead of a relation independent of Kn_i. This can be illustrated easily by considering the case of a monodisperse aerosol, containing equal sized particles. We define a ratio β as

$$\beta = \frac{L}{L^c} \tag{10.30}$$

where L is the actual collision rate and L^c is defined in eqn. (10.12). Then for the case of the monodisperse aerosol, β increases as indicated in Fig. 10.3 for $Kn_i > 0.01$.

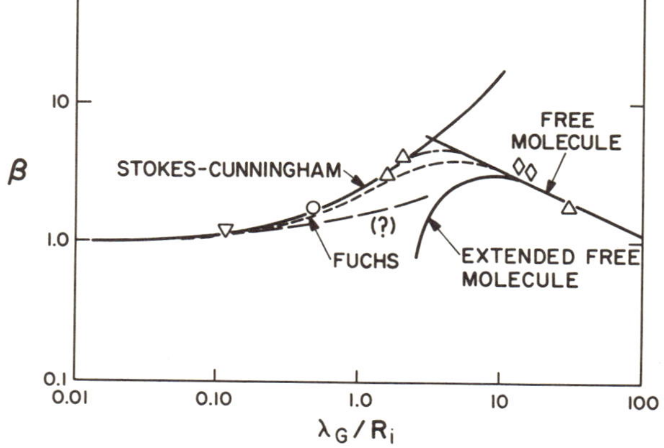

FIG. 10.3. Collision rates for equal sized spheres normalized by Smoluchowski's result for the continuum. Experimental data: ○ Patterson and Cawood,[357] △ Cawood and Whytlaw-Gray,[358] ▽Devir (Weinstock)[344] ◇Fuchs and Sutugin.[349]

With the realization that the Stokes–Cunningham correction could not be extended to aerosol coagulation at large Kn_i, came other attempts to modify the diffusion theory. For example, Junge[351] and more recently Friedlander[352] reasoned that particles should diffuse like gas molecules in the regime $Kn_i > 10$. Therefore, as a rough empirical approximation, the Stokes–Einstein diffusivity should be replaced in eqn. (10.29) by the form for the molecular diffusivity of gases. This model leads to a collision rate which is inversely proportional to the particle radius, and not the correct $R_i^{\frac{1}{2}}$ proportionality indicated in eqn. (10.24) for the free molecule regime.

Fuchs' Concentration Jump Theory

A semi-empirical extension of the continuum theory for the collision rate of uncharged aerosols has come from Fuchs'[2] calculations. He realized that for $Kn_i \geqslant 0.1$, Smoluchowski's results incorporating the Cunningham correction could be seriously in error after incorporating boundary conditions differing from eqn. (10.5). Fuchs suggested that transfer of particles to an absorbing sphere at low pressures must involve a discontinuity concentration near the boundary of the sphere in analogy to the concentration jump near an evaporating droplet,[353] or the temperature jump condition for heat transfer from a hot sphere to a low pressure gas (see also Section 5).

The region where the discontinuity in concentration exists has a thickness the order of the mean free path λ_i of the approaching aerosol particles. Thus, Fuchs has argued that at low pressures, the continuum diffusion equation applies over distances large compared to λ_i, or only in the regions around the sphere in Fig. 10.4 greater than another surface of

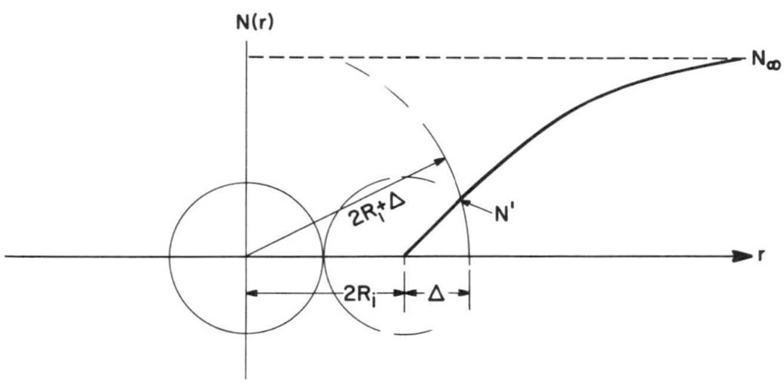

FIG. 10.4. Geometry of Fuchs' diffusion model for coagulation using the concentration jump hypothesis.

radial distances $\sim\lambda_i$ from the absorbing particle. Inside the thickness $\sim\lambda_i$, particles will not move as if they are in a continuum, but, instead as if they are in a vacuum. Thus, if N' is the concentration of particles at arbitrary thickness Δ, the boundary conditions of (10.4a) should read

$$N = N_\infty, \quad r \to \infty,$$

$$N = N_\infty - (N_\infty - N')\frac{(R_i + \Delta)}{r}, \quad r > 2R_i + \Delta,$$

for the steady state case. This model using a concentration jump also was proposed independently by Zebel.[342] However, Fuchs has criticized Zebel's slightly different results on the basis that Zebel retained conditions (10.4a) at the particle surface rather than at $r = 2R_i + \Delta$.

The number of (non-interacting) particles then diffusing by Brownian motion across the outer surface at $2R_i + \Delta$ towards the absorbing sphere is

$$\tilde{\Phi}_i^{\Delta} = 4\pi(2R_i + \Delta)^2 D_i \left(\frac{\partial N}{\partial r}\right)_{r=2R_i+\Delta} \approx 4\pi D_i(2R_i + \Delta)^2 \frac{(N_\infty - N')}{(2R_i + \Delta)}. \tag{10.31}$$

For the case of steady state, the particles diffusing across the sphere at $2R_i$ must also equal that given by the free molecule theory

$$\tilde{\Phi}_i^{\Delta} = \tilde{\Phi}_i^* = 4\pi(2R_i + \Delta)\, D_i(N_\infty - N') = 4\pi(2R_i)\, N'\bar{v}_i/4. \tag{10.32}$$

This relation allows the concentration N' to be solved for in terms of N_∞. By subsequent rearrangement, Fuchs[2] determined that

$$\Phi_i \approx L_{ii} = 8\pi R_i D_i N_\infty \frac{1 + \tilde{A}Kn_i}{R_i\Big/\left[R_i + \frac{\sqrt{(2)}}{2}\lambda_i\right] + (\pi\lambda_i)/[2\sqrt{(2)}\,R_i]} \tag{10.33}$$

provided that it is assumed that

$$\Delta \approx \frac{\sqrt{2}}{2}\lambda_i, \quad \frac{8D_i}{\pi} = \lambda_i\bar{v}_i, \tag{10.34}$$

and $\sqrt{(2)}\,\bar{v}_i$ is used in place of $\bar{v}_i$ to account for the relative motion between the particles. The term in brackets on the right-hand side of eqn. (10.33) is equivalent to the ratio β as previously defined.

We note that when $\lambda_i/R_i \ll 1$, β reduces to

$$\beta \approx (1 + \tilde{A}Kn_i)/(1 + 0.76\lambda_i/R_i), \tag{10.35}$$

which is similar to the empirical correction introduced by Whytlaw-Gray and Patterson,[271] but gives a smaller correction for Kn_i. When $Kn_i \gg 1$,

$$D_i(1 + \tilde{A}Kn_i) \approx \frac{\pi}{8}\left(\frac{8kT}{\pi m_i}\right)^{\frac{1}{2}}\lambda_i,$$

and eqn. (10.33) in this limit yields the form

$$L_{ii}^* \approx 2\left[\frac{\pi kT}{m_i^2}\right]^{\frac{1}{2}} R_i^2 N^2$$

in agreement with the free molecule theory, eqn. (10.24).

By using a form analogous to (10.33), Fuchs[2] has suggested that the generalized β_{ij} for collisions between species of different size may be written as:

$$\beta_{ij} \approx \frac{R_{ij}}{R_{ij} + (\lambda_i^2 + \lambda_j^2)} + \frac{4D_{ij}(1 + \tilde{A}Kn_i)(1 + \tilde{A}Kn_j)}{(\bar{v}_i^2 + \bar{v}_j^2)^{\frac{1}{2}} R_{ij}} \tag{10.36}$$

To see how the coefficient β looks for Fuchs' theory as compared to the empirical correction using the Stokes–Cunningham correction values of β have been calculated for a cloud of equal sized particles. These have been tabulated with Kn_i in Table 10.2, and are shown as a function of Kn_i in Fig. 10.3 assuming for air at 1 atm pressure, $T = 293°\text{K}$; $\lambda_G \approx 6 \times 10^{-6}$ cm.

TABLE 10.2. COAGULATION OF AEROSOLS IN TRANSITION REGIME OF KNUDSEN NUMBER

$\lambda_G = 6 \times 10^{-6}$ cm[a]

Kn_i	β (Fuchs)	β (Astakhov)	β (Brock)
0.01	1.00	1.00	—
0.05	1.06	1.06	—
0.10	1.13	1.14	—
0.50	1.79	1.76	—
1.0	2.42	2.45	—
5.0	4.04	3.95	4.04
10.0	3.03	2.14	3.03
50.0	1.51	—	1.51
100.0	—	—	—

[a] Values of λ_i from Table 7.1.

Although Fuchs regarded his theory only as a qualitative one, some more recent evidence of one of his colleagues has suggested that simple calculations may be better than originally anticipated. To construct a more accurate theory for coagulation in the transition regime where $Kn_i \approx 1$, Astakhov(354) has examined Smoluchowski's diffusion problem in the light of the solutions to the steady state Fokker–Planck equations. Assuming that a spatially homogeneous cloud of equal sized aerosols in Brownian motion could be treated as points distributed in space, Astakhov found a solution for the diffusion of "points" to a spherical absorber by means of the Fokker–Planck equations. His results indicate that under conditions where Kn_i is small,† for equal sized particles, β should have the form

$$\beta = \frac{(1 + \tilde{A}Kn_i)(1 + 0.08\lambda_i/R_i)}{1 + 0.56\lambda_i/R_i + 0.21(\lambda_i/R_i)^2 + 0.015(\lambda_i/R_i)^3} \tag{10.37}$$

And for $\lambda_i/R_i \ll 1$,

$$\beta \approx (1 + \tilde{A}Kn_i)[1 - 0.48(\lambda_i/R_i)]. \tag{10.38}$$

Values of β calculated from Astakhov's results are shown with those of Fuchs in Table 10.2. There is remarkably good agreement between these calculations up to values of $Kn_i \approx 5$.

Extending the Free Molecule Theory

The free molecule results may be revised to encompass part of the regime for $Kn_i < 10$ by applying a first collision correction developed by Brock (see Hidy and Brock(355)). In this model, the collisional process for two uncharged particles is pictured as if the bodies interact only through streams of gas molecules being reflected or "emitted" from each

† Fuchs and Astakhov have presented corrections in terms of the ratio λ_i/R_i rather than Kn_i. Astakhov notes that his results are rigorous only for cases where $\lambda_i/R_i < 1$.

particle. The gas molecules near the aerosols are then perturbed from an assumed Maxwellian distribution of velocity by the motion of the particles. In effect, the emission and associated perturbation of gas molecules sets up a mutual repulsive force between the particles, acting for practical purposes along the line of centers between the bodies. The formal expression for the collision rate density is given again by eqn. (10.20). And the distribution functions for particles and gas are represented formally by eqns. (10.21) and (10.22). However, the apparent repulsive force resulting from opposing streams of gas molecules emitted from the aerosols must be taken into account in the collisional cross-section.

The assumptions inherent in this procedure require that the motion of the two particles be "dynamical" in character. That is, the collision time t_c for two particles must be less than a characteristic relaxation time for the particles t_r. The collision time in the instance of equal sized particles is

$$t_c \sim R_i/\bar{v}_i. \tag{10.39}$$

The characteristic relaxation time for attainment of a steady state has the form given in eqn. (7.35) as

$$t_r \sim m_i D_i/kT = \mathscr{A}^{-1} \tag{10.40}$$

of the particle mass times its mobility. The criterion that $t_c < t_r$ effectively means that the ratio $R_i/\lambda_i < 1$. Hence, the parameters tabulated in Table 7.1 suggests that the so-called "dynamical" behavior of the particles as defined here will occur generally in the upper and transition region, where $Kn_i > 7$. The range of validity of this restriction, of course, will depend strongly on the particle size, the gas density, and less importantly, on parameters such as the particle density and the weight of gas molecules.

An approximate analysis of this problem with the assumptions previously noted gives for the magnitude of the net repulsive force F_r between two colliding particles of different size (i and j), resulting from the opposing molecular streams:

$$F_r = \tfrac{8}{3}(1 + \pi/8)\, N_G(2\pi m_G kT)^{\frac{1}{2}}\, g\, \frac{e^{-\alpha r/\lambda_G}}{2} R_i^2 R_j^2, \tag{10.41}$$

where $\boldsymbol{r}$ is the distance between centers of the two approaching particles, g is the magnitude of the relative velocity of the two colliding particles, α is a constant whose value may vary from about 0.6 to about 1.1 depending on the model chosen for molecular scattering. Using this relation, the collision rate density can be calculated by performing the integrations in eqn. (10.20). The result reads

$$\left.\begin{aligned} L_{ij} = M_{ij} n_i n_j R_{ij}^2 \left[\frac{8kT(m_i + m_j)}{m_i m_j}\right]^{\frac{1}{2}} \Bigg\langle 1 - \frac{\pi}{6}\left(1 + \frac{\pi}{8}\right)\left[\frac{(m_i + m_j)\, m_G}{m_i m_j}\right]^{\frac{1}{2}} \frac{16 N_G (R_i R_j)^2}{R_{ij}} \\ \times \left\{ e^{-\alpha R_{ij}/\lambda_G} + \frac{\alpha R_{ij}}{\lambda_G}\left[0.577 + \ln\left(\frac{\alpha R_{ij}}{\lambda_G}\right) + \sum_{\gamma=1}^{\infty} \frac{(-1)^\gamma (\alpha R_{ij}/\lambda_G)^\gamma}{}\right]\right\}\Bigg\rangle \end{aligned}\right\} \tag{10.42}$$

This expression approaches the form of eqn. (10.24) for uncharged particles when $Kn_i > 10$. Equation (10.42) thus represents a first order correction to the free molecule theory for the effect of molecular collisions on collisions between aerosol particles in a "slightly dense medium". In the same way, of course, Astakhov's results described above constitute a first order correction for a "slightly non-continuum" collision process.

For comparison, values of β from Brock's estimates by eqn. (10.42) have been added to Table 10.2 in the case where $\alpha = \pi/4.8$. These results are also indicated in Fig. 10.3. The calculations of Brock show a sharp drop of $Kn_i \approx 6$. Evidently, the effects of interaction between particles and the subsequent perturbation in aerodynamic forces become increasingly important in this range of Kn_i, making eqn. (10.42) increasingly inapplicable below $Kn_i \approx 7$. Such a limitation fits the restrictions of the arguments requiring that $t_c < t_r$. The extended free molecule calculations, of course, agree with Fuchs' theory above $Kn_i = 10$.

In the region $Kn_i \leqslant 1$, Astakhov's calculations confirm an increase in coagulation rate as estimated earlier by Fuchs. For situations where $1 \leqslant Kn_i \leqslant 10$, there presumably must be a maximum in β, occurring at approximately $Kn_i = 5$. We may speculate that the correct relation above $Kn_i = 1$ must match Astakhov's results for $Kn_i \approx 1$ and Brock's estimates, $Kn_i \approx 10$ so that the maximum value of β may be somewhat less than predicted by Fuchs' theory and the estimates of Astakhov. Thus, in the transition regime for $Kn_i \approx 5$, collision rates may be more than three times as large as in either limit of Kn_i.

The maximum in β for transition aerosols may be significant in considering the observations of the size distributions of aerosols in the Earth's troposphere. In air, $\lambda_G \approx 6 \times 10^{-6}$ cm. at atmospheric pressure. Therefore, the region where $Kn_i \approx 8$ exists for tropospheric particles of radii of approximately 0.005 μ. This region is precisely the zone where the observed size distributions of tropospheric aerosols decrease rapidly.[356] It has been found that below $R_i \approx 0.01$ μ virtually no "continuous" distribution of particles exists. Only occasional "spikes" of particulates are observed. The fact that highly dispersed aerosols are not continuously spread through this regime of size, and their relative scarcity (small concentration) in this range has been attributed to the relatively rapid growth of material by thermal coagulation as it enters the troposphere. The theoretical maximum in the rate of coagulation shown in Fig. 10.3 may add further credence to this explanation.

Experiments of Thermal Coagulation in Transition Aerosols

There are very few experimental data with which to test the behavior suggested in Fig. 10.3. Most of the experiments have utilized estimates of the collision rate from the decay of the total concentration of polydispersed aerosols in a vessel. The problems associated with measurements of this kind will be discussed later in Section 10.4. However, it is worth noting some values of the coagulation rate for use of a qualitative test at least of the curves in Fig. 10.3. In the extreme low ranges of Kn_i, the recent experiments of Devir[344] show satisfactory agreement with the theory as indicated in Fig. 10.3. In the free molecule regime, Fuchs' and Sutugin's[349] data, reduced by the correction for the influence of intermolecular forces, fit the theoretical curve closely. Hidy and Brock[355] have used some early data of Patterson and Cawood,[357] and Cawood and Whytlaw-Gray[358] estimate experimental values of β for $Kn_i \lesssim 0.5$. These results qualitatively suggest that the maximum around $Kn_i \approx 5$ exists. These data are indicated in Fig. 10.3. Some other data in the range $0.9 \lesssim Kn_i \lesssim 5.0$ has been reported by Quon.[359] His results show an increase with Kn_i, but give values of β nearly an order of magnitude too high. It appears, however, that Quon's results may have contained strong electrostatic effects which would severely increase values of β calculated from his data.

Bearing in mind that both polydispersivity and electrostatic effects increase values of collision rate density, and consequently give values of β greater than expected from the theory of non-interacting particles, we can nevertheless get an indication that the bulk of the available experiments tend to be at least qualitatively consistent with the theoretical predictions in the transition regime. Further more careful, extensive experiments on coagulation rates of aerosols for ranges of $Kn_i > 0.1$ seem to be essential, of course, to verify the existing theory in any detail.

10.2. Influences of Non-uniformities in the Suspending Gas and External Forces on Coagulation

A. Free Molecule Aerosols

The coagulation by the relative thermal motion of aerosols represents a simple explanation for the agglomeration of particles in a volume of gas based on equilibrium (Maxwellian) considerations. Observations have indicated, however, that very large particles of size greater than 10 μ agglomerate rapidly in a gravitational field despite the fact that they undergo very small displacements by Brownian motion. Furthermore, we have noted that several forces resulting from non-uniformities act on particles. Because the forces are proportional to particle size, they can induce components of relative motion between aerosols which are not accounted for in the simple theory of Brownian motion.

The exact influence of gaseous non-uniformities and external forces is difficult to calculate theoretically in a completely general way. However, it is possible to incorporate these effects in an approximate way by generalizing the theory of Brownian motion, or by using specific, simplified mechanisms for collision. The first approach has been used in the system of Boltzmann equations to give a more complete theory for the collisions of free molecule aerosols.(348) The second, more classical technique has been applied to some particular cases for collision of aerosols in a continuum. To obtain a feeling for the complexities of a generalized treatment, let us first consider the case of coagulation of hard sphere, free molecule aerosols suspended in a non-uniform gas, and subjected to an external field of force.

Extended Free Molecule Theory

The framework of kinetic theory discussed in Section 2.2 can again be used to provide a unified approximation to an exact collision theory for aerosols in the free molecule regime. Equation (10.20) remains valid, provided that an estimate can be made of the distribution functions f_i^1, f_j^1 for aerosols suspended in the non-equilibrium system.

If $N \ll n_G$, the evolution of the distribution functions for charged particles in determined by particle–gas molecule encounters as long as charging of particles is low enough to neglect multiple interactions and the suspending gas remains un-ionized. Thus, the distribution function of the *ith* species can be given approximately in terms of the Boltzmann equation

$$\frac{\partial f_i^1}{\partial t} + \mathbf{v}_i \cdot \nabla f_i^1 + \mathscr{E}_i \cdot \nabla_{\mathbf{v}_i} f_i^1 = J_{iG}\{f_G^1[\mathbf{v}(1)], f_i^1[\mathbf{v}(1)]\}, \tag{10.43}$$

where f_G^1 is given by eqn. (10.21), and $\mathscr{E}_i$ denotes the generalized external force acting on the ith particle. Solutions to eqn. (10.43) with eqn. (10.21) as before have been given by

Mason,[345] Kihara,[346] and Mason and Chapman.[347] And f_i^1 has been shown to have a form similar to the equilibrium case, eqn. (10.23):

$$f_i^1(\mathbf{r}, \mathbf{v}_i, t) = n_i(\mathbf{r}, t)\left(\frac{m_i}{2\pi kT}\right)^{\frac{3}{2}} \exp\left\{-\frac{m_i}{2kT}(\mathbf{v}_i - \mathbf{q} - \mathbf{q}_i)^2\right\}, \tag{10.44}$$

where $\mathbf{q}_i$ is the local diffusion or drift velocity of species i relative to a coordinate system moving with a mean velocity of the suspending gas $\mathbf{q}$.

The diffusion velocity may arise either from non-uniform states of the suspending gas, or from the action of external forces. The essential feature of considering the non-uniformities in the suspending gas is that the effect of these states on the motion of aerosol particles is equivalent to the effect of an external force (see also Mason and Chapman[347]).

Using the assumption of quasi-stationary states, the drift, velocity may be related to forces acting on particles by the mobility relation

$$\mathbf{q}_i = B_i \mathbf{F}_i, \tag{10.45}$$

where, for free molecule aerosols, the mobility relation has the form from eqn. (5.41),

$$B_i^* = \left[\tfrac{8}{3} R_i^2\, n_G (2\pi m_G kT)^{\frac{1}{2}}\left(1 + \frac{\pi \alpha_m}{8}\right)\right]^{-1}. \tag{10.46}$$

It follows from the estimates of Kihara,[346] which deal with heavy positive ions moving in a light gas in an external electric field, that eqns. (10.45) and (10.46) apply to free molecule aerosols moving under external forces as long as $q_i^2 \ll kT/m_G$. Otherwise, the mobility itself becomes a function of the external force.

The mathematical forms of the external forces acting on aerosol particles has been discussed elsewhere. They are summarized as follows:

(a) *Electric fields*

$$\mathbf{F}_{\mathscr{E}} = z_i \tilde{e}\, \mathscr{E}_0, \tag{10.47}$$

where $\mathscr{E}_0$ is the self-consistent intensity of the electric field.

(b) *Magnetic fields*

$$\mathbf{F}_M = z_i \tilde{e}(\mathbf{q}_i' + \mathbf{q}) \times \mathbf{H}, \tag{10.48}$$

where $\mathbf{q}_i'$ is the diffusional velocity of an aerosol particle exclusive of that arising from the magnetic field, and $\mathbf{H}$ is the self-consistent intensity of the magnetic field.

(c) *Gravitational fields*

$$\mathbf{F}_g = -\, m_i \nabla \Phi_g, \tag{10.49}$$

where Φ_g is the gravitational potential.

(d) *Centripetal fields*

$$\mathbf{F}_c = (4/5) R_i^2\, m_i \omega \times (\omega \times \mathbf{r}), \tag{10.50}$$

where $\mathbf{r}$ is the distance from the center of rotation of the system at a particular point, and ω is the angular velocity of the system.

The forces arising from non-uniform states in the suspending gas which may act on particles have been given in previous sections. They are summarized as follows:

(a) *Temperature gradients*

$$F_t = (4/5)\, R_i^2(2\pi k/m_G T)^{\frac{1}{2}}(k_G/C_v)\, \nabla T, \tag{10.51}$$

where k_G is the thermal conductivity of the gas and C_v is the gas heat capacity at constant volume.

(b) *Concentration gradients*

For a binary gas system around aerosol particles,

$$\mathbf{F}_d = -(8/3)\, R_i^2 n_G(2\pi kT)^{\frac{1}{2}}\hat{\mathscr{F}}(m_G)^{\frac{1}{2}} D_{AB}\nabla(n_A/n_G), \tag{10.52}$$

where D_{AB} is the binary diffusivity of gaseous species A in gaseous species B.

$$\hat{\mathscr{F}}(m_G)^{\frac{1}{2}} = (m_A)^{\frac{1}{2}}[1 + (\pi/8)\,\alpha_{mA}] - (m_B)^{\frac{1}{2}}\,[1 + (\pi/8)\,\alpha_{mB})] \tag{10.53}$$

(c) *Pressure gradients*

$$\mathbf{F}_p = kT(m_A/m_G)\,\nabla \ln p, \tag{10.54}$$

where p is the pressure of the gas.

(d) *Shear fields*

$$\mathbf{F}_s = (8/5)\, R_i^2(2\pi m_G/kT)^{\frac{1}{2}}\, \mu_G[1 - (4/3)\,\alpha_m]\,\mathbf{q}.\nabla\mathbf{q}, \tag{10.55†}$$

where μ_G is the gas viscosity.

Assuming again that the spatial correlation between charged particles is approximated by the Debye-shielded potential function, and adding a term to account approximately for the effects of polarization,[348] the collision rate density is found to have the form of eqn (10.20), but the collisional parameter is given by

$$b_{ij}^* = M_{ij}R_{ij}^2\left(\frac{2\pi kTm_{ij}}{m_i m_j}\right)^{\frac{1}{2}}\left\{\exp(-\xi_{ij}^2) + \left[\frac{1}{2\xi_{ij}} + \xi_{ij}\right][\pi^{\frac{1}{2}}\,\mathrm{erf}(\xi_{ij})] - \left[\frac{\pi^{\frac{1}{2}}\,\mathrm{erf}(\xi_{ij})}{\xi_{ii}}\right] \times \left[\frac{z_i z_j \tilde{e}^2 \exp(-\kappa R_{ij})}{kTR_{ij}} - \frac{(z_i^2\tilde{\alpha}_i + z_j^2\tilde{\alpha}_j)}{2kTR_{ij}^4}\tilde{e}^2\right]\right\}, \tag{10.56}$$

$$\frac{z_i z_j \tilde{e}^2 \exp(-\kappa R_{ij})}{kTR_{ij}} \ll 1,$$

where

$$\xi_{ij} = \left(\frac{m_i m_j}{2kTm_{ij}}\right)^{\frac{1}{2}} |\mathbf{q}_i - \mathbf{q}_j|,$$

and $\tilde{\alpha}_i$ is the polarizability of species i, and κ is the Debye reciprocal length.

It is also possible to estimate L_{ij}^* for other types of molecular models. Calculations for other attractive fields such as Van der Waals differ little from the results in eqn (10.56). Therefore, other idealized models will not be considered further.

The form of b_{ij} given by eqn. (10.56) reduces to the expression for collisions of charged particles in a Maxwellian gas, eqn. (10.25), when ξ_{ij} is very small. However, we note that when the term ξ_{ij} becomes large, the directing effects of gaseous non-uniformities and external fields of force become more important, and eqn. (10.56) ultimately becomes:

† The contribution of gradients of gas velocity is expected to be small, except when the velocity change is large over a distance comparable with the particle diameter. However, for completeness, this expression has been included here.

$$L_{ij}^* \approx n_i n_j R_{ij}^2 M_{ij} \left[\frac{2kTm_{ij}}{m_i m_j}\right]^{\frac{1}{2}} \pi^{\frac{1}{2}} \xi_{ij} \tag{10.57}$$

$\xi_{ij} \to$ large.

As an example, for spherical particles of unit density suspended in a gas at atmospheric pressure, with $R_i = 0.05\ \mu$, and with $|\mathbf{q}_i - \mathbf{q}_j| \approx 100$ cm sec^{-1}, the collision rate is ten times the value estimated for $|\mathbf{q}_i - \mathbf{q}_j| = 0$. Thus, directed collisions can play a primary role in coagulation, even for aerosols of very small size.

Another interesting feature is implied in the results, eqns. (10.56) and (10.57). The directing effects evidently do not influence the beginning of coagulation in an initially monodisperse, uniformly charged free-molecule aerosol. However, after some thermal coagulation in the cloud, the size and charge distribution tends to become non-uniform. Then, the directing effects can begin to affect the coagulation process. If ξ_{ij} is moderately large, the directing effects will contribute to inducing collisions between larger and smaller particles at a rate which increases with the difference in particle size. This means that once a few aggregates of $i > 1$ are formed by thermal coagulation, these multiplets can grow more and more rapidly with time by the action of non-uniformities in the suspending gas, and by the influences of external fields.

Equation (10.56) provides a general, unified basis for computing collision rates between aerosol particles for large values of Kn_i, provided no surfaces or boundaries of discontinuity exist in the medium. Some examples of discontinuities generated in gases include: (a) a gas jet directed into a stagnant gas, (b) two or more jets of gas directed toward each other, (c) shock waves standing in or travelling through a gas, and (d) movement of a gas near a surface where mass transfer is occurring. Equation (10.56) then may be applied in these situations as long as the aerosol distribution is locally Maxwellian. If this condition is not satisfied, additional analysis is necessary.

The use of the free molecule theory in cases where the suspending medium contains surfaces of discontinuity requires a redefinition of $\mathbf{q}_i$.

$$\mathbf{q}_i = \mathbf{q}_i' + \mathbf{q}_{Gi}. \tag{10.58}$$

The term $\mathbf{q}_i'$ represents the previous drift velocity associated with external forces and gaseous non-uniformities, while $\mathbf{q}_{Gi}$ denotes the average velocity of a particular element of suspending gas. Practical calculations that account for the discontinuities require knowledge of both terms in eqn. (10.58).

In view of these comments it is seen that collisions resulting from strongly directed motion, given in eqn. (10.57), can develop in two different ways. Large values of the relative velocity may occur when external fields are present or gaseous non-uniformities are extreme. For example, $|\mathbf{q}_i - \mathbf{q}_j|$ may be quite large for oppositely charged particles in magnetic or electric fields. On the other hand, large values of $|\mathbf{q}_i - \mathbf{q}_j|$ may also be achieved easily in systems where surfaces of discontinuity are generated.

The use of the kinetic theory of dilute gases as developed here permits a relatively accurate and unified formal account of a number of phenomena which affect collisional processes of aerosols in the free molecule regime. The effect of electrical charging, external fields of force, non-uniform states in the suspending medium, including surfaces of discontinuity, are described by this theory.

It is relatively straightforward to incorporate many influences on aerosol coagulation in a unified way into the free molecule theory. However, it has not been possible to do similar calculations for collision processes for non-equilibrium media modelled as a continuum. Nevertheless, several special models have been worked out, and some important examples are discussed in the following sections.

B. Examples of Continuum Theory for Coagulation of Suspensions in Non-uniform Gases and in External Fields of Force

Effect of Velocity Gradients

One of the earliest attempts to account for the influence of fluid non-uniformities on coagulation is found with Smoluchowski's calculations for Brownian motion.[282] Smoluchowski considered the idealized case shown in Fig. 10.5 for laminar flow of a sol

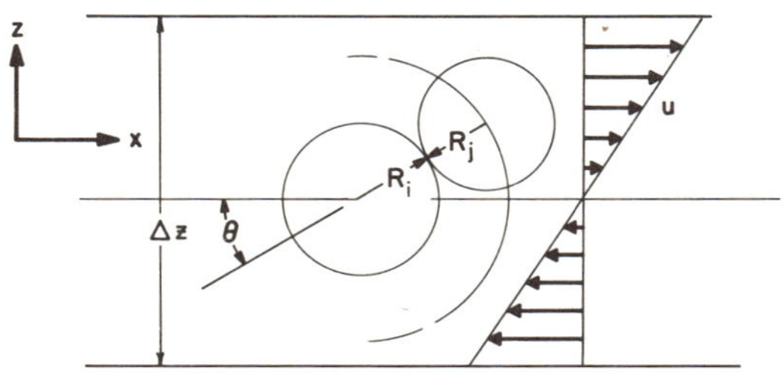

FIG. 10.5. Collision of two spheres in a shearing flow.

between two parallel plates moving at different speeds. A particle in such a flow situated in a region farther from the slower moving plate will move with a greater speed than a particle closer to the slow wall. If the distance Δz, measured normal to the plates, does not exceed the sum of the particle radii, the particles will collide. The simplest theory for this mechanism assumes that Brownian diffusion may be disregarded, and there is no aerodynamic interaction between particles. For the two particles i and j, fixing the coordinates on i, the relative velocity between them is approximately

$$u(\Delta z) = g_x(\Delta z) = (\partial g_x/\partial z)\,\Delta z$$

on geometrical considerations, the collision between particles occurs if the distance between them measured along the z axis is

$$\Delta z \leqslant R_{ij} \sin \theta.$$

As before, we estimate the number of collisions per unit time by calculating the number of particles reaching the spherical surface of radius R_{ij}, which equals

$$\tilde{\Phi}_i = N \int 2uR_{ij} \cos\theta \,\mathrm{d}\,\Delta z = 4N\left(\frac{\partial q_x}{\partial z}\right)R_{ij}^3 \int_0^{\pi/2} \cos^2\theta \sin\theta \,\mathrm{d}\theta = \tfrac{4}{3}N(R_{ij}^3)\frac{\partial q_x}{\partial z}. \quad (10.59)$$

For a monodisperse sol, the collision rate is given by

$$L_{ii}^{(s)} = \frac{32}{3} N^2 R_i^3 \frac{\partial q_x}{\partial z}, \quad (10.60)$$

which suggests that coagulation should increase markedly in a shearing flow as the particle size increases.

The ratio between the collision by Brownian motion and collisions in a shearing flow is for equal sized particles:

$$\frac{L^{(s)}}{L^{(c)}} = \frac{32N^2(\partial q_x/\partial z)\,R_i^3}{24\pi D_i R_i N^2} = \frac{4(\partial q_x/\partial z)\,R_i^2}{3\pi D_i}. \tag{10.61}$$

Thus the effect of shear on coagulation can be large only in regions of velocity gradients

$$\frac{\partial q_x}{\partial z} > \frac{\pi D_i}{R_i^2}$$

For an aerosol of radius 10 μ, the velocity gradient must then be $\sim 4 \times 10^{-2}$ sec^{-1}, which is relatively small. However, if $R_i = 1\,\mu$, $\partial q_x/\partial z \approx 12$ sec^{-1} so that relatively small decreases in particle size change substantially the role of velocity gradients in influencing coagulation. Although the free molecule theory leads to a different form for Eq. (10.61), we see nevertheless that velocity gradients qualitatively must be very large indeed to affect the coagulation of aerosols when the pressure is high.

Gravitational Coagulation. Under the influence of external fields, aerosol particles can experience velocities relative to one another, provided the force exerted on them by the external field is proportional to particle size. In the case of free molecule theory, a general expression for the influence of external forces was developed which included the part played by Brownian motion. Although no counterpart of this theory exists for the continuum regime, the limiting regime in which the effect of Brownian motion may be disregarded can be estimated approximately by the following sequence of arguments.

Suppose we consider two particles moving steadily in a field of external force ($Re_i < 1$). The particles travel at different speeds that are proportional to the exerted force. In the absence of any electrical or aerodynamic interaction between the spheres, the bodies will collide if their separation distance equals R_{ij} or less. The path of a smaller particle moving near a larger particle is illustrated schematically for this situation in Fig. 3.7. The coordinates are fixed on the larger particle, and the (uniform) external force field is oriented along the x coordinate. On geometrical grounds, the larger particle j, will sweep out all smaller particles i in the cross-sectional area πR_{ij}^2. The rate of sweep out will depend on their components of relative velocity in the direction of the force field

$$v_{sr} = v_{sj} - v_{si}, \tag{10.62}$$

or the collision parameter will be given as

$$b_{ij}^{(g)} = \pi R_{ij}^2 (v_{sj} - v_{si}). \tag{10.63}$$

In general the steady particle velocities v_{sj} and v_{si} may be given by a mobility force relation like eqn. (10.45).

When the particles fall by the action of gravity alone, v_{sj} and v_{si} are given by the Stokes settling velocity and eqn. (10.63) reads

$$b_{ij}^{(g)} = \frac{2\pi\rho_i(\mathscr{g})}{9\mu_G} R_j^4\,[(\hat{K}+1)^2 - \hat{K}^2(\hat{K}+1)^2], \tag{10.64}$$

where $\hat{K} = R_i/R_j$.

Equations (10.63) and (10.64) do not account for the curvature of streamlines about the larger particle accurately. A more careful examination of the collision process requires a better estimate of the part corresponding to direct interception of the small particles on the larger one. We assume that the particle Reynolds number for both i and j species is less than unity. Then the streamline of fluid passing the larger sphere will be given by the Stokes stream function, with the velocity, at an infinite distance away equal to the terminal velocity induced by the force field. This corresponds to the Stokes settling velocity (3.13) if the external field is gravity.

Fixing the coordinate system on the larger particle, the collision parameter then is obtained by integrating the velocity component of the smaller particle in the direction of the force field over the surface of a hemisphere of radius R_{ij},

$$b_{ij}^{(g)} = \int_S v_{sr}\, ds = \int_S v_{sj}\, ds - \int_S v_{si}\, ds. \tag{10.65}$$

Accounting for the Stokes flow around j, the first term on the right of this expression has been evaluated by Fuchs[360] (see also Table 3.2) and independently by Friedlander[98] for $\hat{K} \ll 1$:

$$\int_S v_{sj}\, ds = \pi v_{sj}(R_{ij})^2\left[1 - \frac{3}{2(1+\hat{K})} + \frac{1}{2(1+\hat{K})^3}\right]. \tag{10.66}$$

The second integral in the above expression is

$$\int_S v_{si}\, ds = \pi R_{ij}^2 v_{si}. \tag{10.67}$$

When the external force is gravity, and v_{sj} and v_{si} are the Stokes settling velocities,

$$b_{ij}^{(g)} = \frac{2\pi\rho_i(g)}{9} R_j^4\left\{(1+\hat{K})^2\left[1 - \frac{3}{2(1+\hat{K})} + \frac{1}{2(1+\hat{K})^3}\right] - \hat{K}^2(1+\hat{K})^2\right\},\quad \hat{K} < 1. \tag{10.68}$$

The early theory of removal of small particles by a larger falling body neglected the finite size of the smaller particle, but included the effect of relative velocity. Thus, the simplest expression for $b_{ij}^{(g)}$ would be $\pi R_j^2 \eta_c (v_{sj} - v_{si})$, where η_c is the collision efficiency.† Considering eqns. (10.63) and (10.64), we note that, for gravitational settling, the crudest evaluation of the effect of finite sized smaller particles (interception) is a geometrical one giving

$$\eta^{(0)} = (1 + \hat{K})^2. \tag{10.70}$$

In the limit where the smaller particles are assumed points, $\eta^{(0)} \to 1$; when the particles are of equal size, $\eta^{(0)} = 4$.

A better approximation accounting for viscous flow around the larger sphere is

$$\eta^{(1)} = \frac{\hat{K}^2}{2}\cdot\frac{(2\hat{K}+3)}{(1+\hat{K})} \qquad \left.\begin{matrix}\hat{K} < 1\\ Re_i < 1.\end{matrix}\right\} \tag{10.71}$$

These two approximations for η are independent of particle size, but depend on the size ratio $\hat{K}$.

† The efficiency of collision is sometimes distinguished from the collection efficiency η in that

$$\eta = \Omega_c \eta_c, \tag{10.69}$$

where Ω_c equals the probability of sticking upon collision. We assume that Ω_c is always unity, or $\eta = \eta_c$. In Chaps. 3 and 7, the treatment assumed implicitly that $\Omega_c = 1$.

We observe an interesting feature of (10.71), in that for very small values of the ratio $\hat{K}$,

$$\eta^{(1)} \approx 1.5\,\hat{K}^2 \qquad (\hat{K} \to 0) \tag{10.72}$$

for gravitational fall. Thus the rate coefficient will reduce to zero as the radius of the smaller particle i vanishes instead of the geometrical removal predicted by eqn. (10.70). In reality, this cannot be exactly true because thermal motion will increase with decreasing R_i, and diffusional deposition on j will become important. However, if we neglect the effect of diffusion, the collision rate resulting from gravity alone should decrease with decreasing R_i for a small range, but it must also approach zero when R_i approaches $R_j (v_{si} \to v_{sj})$. In other words, the collision rate for sweep out by gravity $L_{ij}^{(g)}$ must have a maximum in the range $R_j < R_i < 0$. Friedlander[361] took these limits into account in suggesting a simplified form of η that is identical to eqn. (10.71), with eqn. (10.63).

The theory for capture of particles of differing size during settling in an external field as considered above is effectively a kinematic one. These calculations exclude the inertial displacement of the smaller particles, and the influence of aerodynamic or electrical forces induced between the bodies as they pass one another. If the particles in the cloud have radii below 15 μ, the inertial displacement will cause only a small increase in collision rate. However, the influence of aerodynamic forces can have an appreciable effect on particle collisions even if the particles are uncharged. Experiments[2] suggest that the collision efficiency approaches zero as $\hat{K} \to 0$, as indicated from Friedlander's approximations in eqn. (10.72). Though experimental observations are relatively scarce for wide ranges of the size ratio $\hat{K}$, they indicate the aerodynamic influences have to be included for accurate estimates of η. The extension of the collision theory of small droplets in a gravity field to include the aerodynamics is treated in Section 10.3.

The Role of Turbulence in a Coagulating Cloud

Turbulence can affect coagulation by two different mechanisms. Suppose particles are suspended in a turbulent medium whose characteristic length scale of small eddies is larger than a characteristic particle size R_i. In the first case, since spatial non-homogeneities are brought about by the turbulent flow, different velocities of neighboring particles must appear, causing collisions by a mechanism analogous to the mechanism of laminar shearing flow. In addition, each particle has a relative motion differing from that of the turbulent air because its inertia will not be the same as an equivalent mass of air. Since the inertia of particles depends on their size, the first mechanism can give rise to collision between both equal and unequal sized particles, while the second mechanism results only in encounters between unequal sized particles. Levich[54] and Saffman and Turner[362] both have made estimates of the coagulation rate caused by these two mechanisms. As in the remarks in Chapter 3, their results apply to cases where the particle radii are smaller by an order of magnitude than the size of the smallest eddies. That is, these theories estimate the effect of turbulent motion on the scales determined largely by the viscosity of the fluid, and the dissipation rate of turbulent energy (i.e., the small scale turbulence discussed, for example, by Kolmogoroff[363] and Batchelor,[183] Although Levich used the concept of a turbulent diffusivity with Smoluchowski's method to obtain these results, Saffman and Turner applied a technique similar to kinetic theory, using probability theory to reach essentially the same conclusions.

Assuming that the collision efficiency, though not known precisely, is the order of unity for particles, it is possible to calculate the collision rate by turbulent motion purely on a kinematic basis. The calculation proceeds by considering the flux of particle entering a sphere of influence of one particle, to be absorbed on that particle.

Saffman and Turner have commented that it is useful to bear in mind the qualitative picture of the local shearing motions that bring the particles together. For two points close together in a turbulent gas, the relative motion of the (fluid) particles is that of uniform strain. Taking an origin of coordinate fixed to the center of one particle, and the coordinate axes along the directions of the principal rates of strain, the streamlines of relative motion in one of the coordinate planes curve towards the reference particle as indicated in Fig. 10.6.

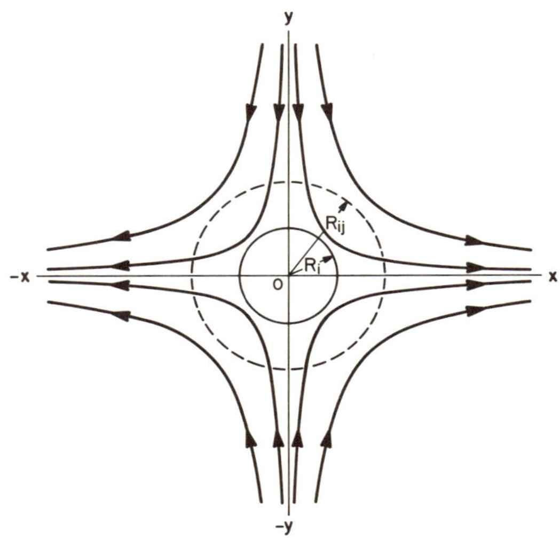

FIG. 10.6. Geometry of collision in a field of uniformly straining fluid (From Saffman and Turner[362]).

The particles that encounter the reference particle move with the gas along the streamlines in the Stokes regime because we disregard distortion of the flow field by the particles except very near the spheres. Then the collision rate of particles on the fixed body refers to the flux of gas moving inwards across the sphere of radius R_{ij} multiplied by the number concentration of the other particles.

To demonstrate the applicability of probabilistic arguments to the continuum model, let us follow the arguments of Saffman and Turner. In this theory, both the effect of particle motion with the air and relative to the air are considered along with the relative motion of droplets resulting from gravity. This will provide additional information about how these three different effects may be added together in accounting for their simultaneous influence on coagulation.

The rate of collision between droplets of different size R_i and R_j, whose concentrations are n_i and n_j, may be given in terms of a probability distribution $\tilde{P}(\mathbf{v}_r)$ of relative velocity $\mathbf{v}_r$ between droplets just before collision. In other words, $\tilde{P}(\mathbf{v}_r)\, d\mathbf{v}_r$ denotes the probability that the relative velocity $\mathbf{v}_r$ between two particles just before encounter lies in the range $\mathbf{v}_r$ and $\mathbf{v}_r + d\mathbf{v}_r$. The use of $\tilde{P}(\mathbf{v}_r)$ alone in determining the collision rate assumes that the relative velocity is a function only of the particle separation. That is, $\mathbf{v}_r$ is a function only of the distance along the line of centers between the particles, $\mathbf{r}$. When the separation of particles is small compared with the characteristic length of small eddies, $\tilde{P}(\mathbf{v}_r)$ may be deduced from the equations of motion of a particle, provided the distortion of flow by the particle is disregarded, and $\eta \approx 1$.

As before, collision between i and j will occur when the centers of droplets are separated by a distance

$$R_{ij} \leqslant R_i + R_j.$$

In a time interval Δt, smaller than the time scale of small eddies, the particles will move in straight paths during Δt. Assuming that particle i is at rest, then the j species has a velocity $\mathbf{v}_r$ just before collision. Hence, in the interval Δt, a collision will take place if the center of the jth species lies in the volume $\pi R_{ij}^2 v_r \Delta t$, where $v_r = |\mathbf{v}_r|$. If we integrate over velocity space, and multiply the result by the concentration of j particles, the number of collisions in the interval Δt is obtained. Taking account of the relative motion between particles by multiplying by n_i, we find the collision rate density is

$$L_{ij}^{(T)} = \pi R_{ij}^2 n_i n_j \int \mathbf{v}_r \tilde{P}(\mathbf{v}_r)\, d\mathbf{v}_r. \tag{10.73}$$

This expression is analogous to eqn. (10.20), and the arguments for its formulation are the same.

Accounting for the relative motion by only multiplying by n_i in eqn. (10.73) assumes that the average velocity between the two colliding particles is statistically independent of the relative velocity. Otherwise, the local concentrations of particles would depend on $\mathbf{v}_r$. Roughly speaking, the mean velocity of particles in a turbulent medium depends primarily on the large, energy-containing eddies while their relative velocity is controlled by the small eddies. It is known that the large and small eddies are statistically independent for conditions of large Reynolds number. In the atmosphere, the Reynolds number characteristic of the turbulent air motion is generally very large, $\sim 10^5$, so that this criterion should be met at least under geophysical conditions.

The problem now remains to evaluate the function $\tilde{P}(\mathbf{v}_r)$, which involves (a) the influence of spatial variations in air velocity (shear); giving collisions from particle motion with the air, (b) acceleration of air, giving collisions from particle motion relative to air, and (c) gravitational settling. $\tilde{P}(\mathbf{v}_r)$ can be derived formally by accounting for these three effects, but the collision rate cannot be evaluated in closed form. Instead, to obtain $\tilde{P}(\mathbf{v}_r)$, Saffman and Turner employed an interesting mathematical trick. By using a well known probability function which would yield the same variance as "exact" function $\tilde{P}(\mathbf{v}_r)$ the integrals in eqn. (10.73) can be obtained approximately since integrated statistical properties are relatively insensitive to the exact form of the probability distribution functions. The justification for such a technique is given *a posteriori*, in that for limits of either motion with air, or of gravity alone, there is no significant error in estimate of these effects.

The variance of $\mathbf{v}_r$, var $\mathbf{v}_r$, may be evaluated directly for the statistical properties of turbulence. To calculate var $\mathbf{v}_r$, consider two droplets in an encounter with velocities $\mathbf{q}_i$ and $\mathbf{q}_j$. The velocities $\mathbf{q}_{iG}$ and $\mathbf{q}_{jG}$ denote the undisturbed air velocities around the particles. Thus the relative velocities between air and the particle are $\mathbf{c}_i = \mathbf{q}_i - \mathbf{q}_{iG}$ and $\mathbf{c}_j = (\mathbf{q}_j - \mathbf{q}_{jG}$. Noting that

$$\mathbf{v}_r = \mathbf{q}_j - \mathbf{q}_i,$$

the variance of $\mathbf{v}_r$ is

$$\text{var}\ \mathbf{v}_r = \text{var}\,(\mathbf{q}_j - q_i) = \text{var}\,(\mathbf{c}_j - \mathbf{c}_i) + \text{var}\,(\mathbf{q}_{jG} - \mathbf{q}_{iG}) \tag{10.74}$$

provided that $\mathbf{q}_{jG} - \mathbf{q}_{iG}$ does not depend statistically on $\mathbf{c}_j$ and $\mathbf{c}_i$. The last term on the right is given by

$$\left|\text{var}\,(\mathbf{q}_{jG} - \mathbf{q}_{iG})\right| = \left|R_{ij}^2 \overline{\left(\frac{\partial \mathbf{q}}{\partial x}\right)^2}\right| = 5R_{ij}^2 \overline{\left(\frac{\partial q}{\partial x}\right)^2} = \frac{1}{3}\left(\frac{\epsilon_t}{\nu_G}\right) R_{ij}^2, \tag{10.75}$$

using the theory of local isotropy in small eddies (e.g. Batchelor[364]). In this expression ϵ_t denotes the rate of dissipation of turbulent energy.

If the equation of motion for particles is applied, the first term on the right of eqn. (10.75) may be estimated. That is, we assume that the particle acceleration is governed by

$$\frac{\mathrm{d}\mathbf{v}_i}{\mathrm{d}t} = \mathscr{A}_i(\mathbf{v}_i - \mathbf{q}) + \frac{\rho_G}{\rho_i}\frac{\mathrm{d}\mathbf{q}}{\mathrm{d}t} + \left(\frac{\rho_G}{\rho_i} - 1\right)\mathscr{g}, \tag{10.76}$$

where the particle density is $\rho_i = \rho_j$ and $\mathscr{A}_i$ is the inverse of the particle relaxation time, equal to $9\mu_G/2R_i^2\rho_i$.

When $\mathscr{A}_i^{-1}$ is much less than the time scale of smallest eddies, given by Batchelor[183] as $(\nu_G/\epsilon_t)^{\frac{1}{2}}$, the acceleration, $\mathrm{d}v_i/\mathrm{d}t$ of the ith particle is much smaller than the other terms in eqn. (10.76), and we find upon squaring and averaging eqn. (10.76), that

$$\mathbf{c}_i^2 = \mathscr{A}_i^{-2}\left(1 - \frac{\rho_G}{\rho_i}\right)^2 \left[\overline{\left(\frac{\mathrm{d}\mathbf{q}}{\mathrm{d}t}\right)^2} + \mathscr{g}^2\right]. \tag{10.77}$$

The relative motion of the particles resulting from $\mathbf{q}_{iG}$ and $\mathbf{q}_{jG}$ already has been accounted for. Then considering $\mathbf{c}_i$ and $\mathbf{c}_j$, it is reasonable to assume that the velocities of the gas are the same near two spheres that are close together. Under this condition the correlation between $\mathbf{c}_i$ and $\mathbf{c}_j$ is unity ($\overline{c_i c_j} = 1$), and

$$\text{var}\,(\mathbf{c}_j - \mathbf{c}_i) = (\mathscr{A}_j^{-1} - \mathscr{A}_i^{-1})^2 \left(1 - \frac{\rho_G}{\rho_i}\right)^2 \overline{\left(\frac{\mathrm{d}\mathbf{q}}{\mathrm{d}t}\right)^2} + \mathscr{g}^2. \tag{10.78}$$

For conditions of local isotropy, $\overline{(\mathrm{d}|\mathbf{q}|^2/\mathrm{d}t)} \approx \overline{3(Dq/Dt)^2}$, where in this case the operator D/Dt follows a fluid element instead of $\mathrm{d}/\mathrm{d}t$, following a particle (droplet). Using this approximation, we can write the variance of relative particle velocity $\mathbf{v}_r$ as

$$\text{var}\,(\mathbf{v}_r) = 3\left(1 - \frac{\rho_G}{\rho_i}\right)^2 (\mathscr{A}_j^{-1} - \mathscr{A}_i^{-1})^2 \overline{\left(\frac{Dq}{Dt}\right)^2} + \left(1 - \frac{\rho_G}{\rho_i}\right)^2 \times (\mathscr{A}_j^{-1} - \mathscr{A}_i^{-1})^2\, \mathscr{g}^2 + \frac{1}{3}\left(\frac{\epsilon_t}{\nu_G}\right) R_{ij}^2. \tag{10.79}$$

Now suppose we assume that $\tilde{P}(\mathbf{v}_r)$ can be replaced by a normal distribution function

$$\tilde{P}(\mathbf{v}_r) = (\xi/\pi)^{\frac{1}{2}} \exp(-\xi v_r) \tag{10.80}$$

such that the variance of the normal distribution is the same as that for $\tilde{P}(\mathbf{v}_r)$. That is, $(3/2\,\xi)$ equals eqn. (10.79).

With the integration in eqn. (10.73) the collision rate density with eqn. (10.80) becomes

$$L_{ij}^{(T)} = \pi R_{ij}^2 n_i n_j \int \mathbf{v}_r \tilde{P}(\mathbf{v}_r)\,\mathrm{d}\mathbf{v}_r = 2(2\pi)^{\frac{1}{2}} R_{ij}^2 n_i n_j \times \left[\left(1 - \frac{\rho_G}{\rho_i}\right)^2 (\mathscr{A}_j^{-1} - \mathscr{A}_i^{-1})^2 \overline{\left(\frac{Dq}{Dt}\right)^2} + \frac{1}{3}\left(1 - \frac{\rho_G}{\rho_i}\right)^2 (\mathscr{A}_j^{-1} - \mathscr{A}_i^{-1})^2 \mathscr{g}^2 + \frac{1}{9} R_{ij}^2 \frac{\epsilon_t}{\nu_G}\right]^{\frac{1}{2}}. \tag{10.81}$$

There may be several different cases where the first two terms on the right-hand side of (10.81) may vanish. This will be true if (a) the particles move with air, (b) $\mathscr{A}_i^{-1} = \mathscr{A}_j^{-1}$, (c) $\mathscr{A}_i^{-1}$ are zero, (d) $i = j$, or (e) if the particles are of uniform density. Under any one of these conditions,

$$L_{ij}^{(T)} \approx 1.67 \quad R_{ij}^3 n_i n_j (\epsilon_t/\nu_G)^{\frac{1}{2}}. \tag{10.82}$$

This relation is essentially the expression accounting for the influence of turbulence on coagulation when particles move with the gas. Equation (10.82) is the same as obtained by Saffman and Turner by a simpler model[362] and by Levich[54] using Smoluchowski's arguments and the turbulent (eddy) diffusivity.

When there is no turbulence and we realize that $\rho_G/\rho_i \ll 1$ in gases, eqn. (10.81) reduces to

$$L_{ij}^{(g)} \approx 2.9 R_{ij}^2 n_i n_j \mathscr{g} \left(1 - \frac{\rho_G}{\rho_i}\right)(\mathscr{A}_j^{-1} - \mathscr{A}_i^{-1}), \tag{10.83}$$

which contains a form of $b_{ij}^{(g)}$ equivalent to eqn. (10.63) except for the constant, 2.9. However, the ratio of the difference, $\pi/2.9$, is quite small for practical purposes.

The effect of the motion of particles relative to the air can be estimated by applying Batchelor's[364] relation:

$$\overline{\left(\frac{dq}{dt}\right)^2} \approx 1.3 \nu_G^{-\frac{1}{2}} \epsilon_t^{\frac{3}{2}}, \tag{10.84}$$

for large Reynolds numbers. Substitution of eqn. (10.84) into the first term on the right of eqn. (10.81) yields the relation (neglecting the last two terms):

$$L_{ij}^{(T)} \approx 5.7 R_{ij}^2 n_i n_j (\mathscr{A}_i^{-1} - \mathscr{A}_j^{-1}) \ \epsilon_t^{\frac{3}{4}} \nu_G^{-\frac{1}{4}}, \tag{10.85}$$

which is a form similar to that found by Levich[54] on dimensional grounds. Note that this mechanism cannot yield collision with particles of the same size, while the first mechanism written, eqn. (10.82) can.

Saffman and Turner[362] note that eqn. (10.85) is a reasonable approximation for conditions where there is a relative velocity resulting only from particle motion relative to air. This expression replaces a constant relative velocity arising from the differential fall of different sized particles under gravity by a random relative velocity with same mean square value.

If the gravitational velocity is zero and motion relative to the gas is zero, eqn. (10.80) yields a relation that is inconsistent with the assumption of isotropy in the small scale eddies. For example, if the separation of two droplets is parallel to the x axis, isotropy indicates that $\overline{v_{rx}^2} = \frac{1}{2}\overline{v_{ry}^2} = \frac{1}{2}\overline{v_{rz}^2}$. But eqn. (10.80) yields $\overline{v_{rx}^2} = \overline{v_{ry}^2} = \overline{v_{rz}^2}$. Despite these anomalies, the relation obtained for the collision rate using eqn. (10.82) gives rise to results obtained by other arguments.

Equation (10.81) can be used now to evaluate the relative importance of the two mechanisms for collisions by turbulence. This relation is restricted to conditions in which $\mathscr{A}_i \ll (\nu_G/\epsilon_t)^{-\frac{1}{2}}$, and $R_i \ll (\nu_G^3/\epsilon_t)^{\frac{1}{4}}$. These limitations are satisfied for particles up to $\sim 20\ \mu$ in radius even for strong turbulence in which $\epsilon_t \sim 1000\ \text{cm}^2\ \text{sec}^{-3}$.

The ratio of the first term to the third in (10.81) is

$$11.7\left(1 - \frac{\rho_G}{\rho_i}\right)^2 \frac{(\nu_G \epsilon_t)^{\frac{1}{2}}}{R_{ij}^2}(\mathscr{A}_i^{-1} - \mathscr{A}_j^{-1})^2 = 8.2 \times 10^6 \epsilon_t^{\frac{1}{2}} (R_i - R_j)^2, \tag{10.86}$$

for particles of unit density in air. This ratio becomes about unity or the effects are the same magnitude, when $R_i - R_j \approx 3\,\mu$ if $\epsilon_t = 5$ cm^{2}sec^{-3}, and when $R_i - R_j \approx 1\,\mu$ for $\epsilon_t = 1000$ cm^{2}sec^{-3}. This implies qualitatively that, except for very small particles, $R_i < 1\,\mu$, for which continuum theory will not strictly apply anyway even at atmospheric pressure, collisions in not too polydispersed clouds will be controlled primarily by motion relative to the air.

Comparison between the second and first terms in eqn. (10.81) indicates that the motion of particles relative to air resulting from gravity and turbulence become the same magnitude if $g \approx 2\nu_G^{-\frac{1}{4}}\epsilon_t^{\frac{3}{4}}$, corresponding to $\epsilon_t \approx 2100$ cm^{2}sec^{-3}. This value of ϵ is particularly large for atmospheric conditions, even in vigorous development of cumulus clouds. Thus, the action of gravity is expected to dominate coagulation resulting from particles moving relative to air even in comparatively intense turbulent air.

For particles of a micron radius and smaller, Brownian motion should influence the rate of coagulation in a turbulent system. Taking, for example, the ratio of the collision rate for Brownian motion to the turbulent shearing mechanism for equal sized particles,

$$\frac{L^{(s)}}{L^{(c)}} = \frac{13.6R_i^3N^2(\epsilon_t/\nu_G)^{\frac{1}{2}}}{8\pi D_i R_i N^2} = \frac{1.7R_i^2(\epsilon_t/\nu_G)^{\frac{1}{2}}}{\pi D_i}, \tag{10.87}$$

or, in the absence of gravitational effects, turbulent shear exceeds Brownian motion in controlling coagulation if

$$R_i > \left[\frac{\pi D_i}{1.7(\epsilon_t/\nu_G)^{\frac{1}{2}}}\right]^{\frac{1}{2}}. \tag{10.88}$$

The regimes of control of coagulation mechanisms also can be visualized by examining the curves in Fig. 10.7. Here, the curves representing collision parameters b_{ij} between particles of R_j, given on the abscissa, with 1 μ radius particles is shown for the cases of Brownian motion, gravitational coagulation, and coagulation resulting from turbulence by mechanism 1, eqn. (10.82). Below $R_j \approx 1\,\mu$, Brownian motion tends to control coagulation of particles, while for $\epsilon_t \approx 5$ cm^{2}sec^{-3}, turbulence and gravity play the predominant role in cases where $R_j > 1\,\mu$.

10.3. Influences of Aerodynamic and Electrical Forces Arising from Interaction between Particles

When large numbers of particles travel together, their individual motion may be modified somewhat by aerodynamic forces or electrical forces induced between the particles. The aerodynamic interactions are negligible in the free molecule regime, and, as noted in Section 10.2A, the electrical forces can be taken into account in a fairly general way in this case. However, when $Kn_i < 10$, any particle movement relative to the suspending fluid will give rise to currents, which in turn will induce aerodynamic interaction between the particles. The aerodynamic forces and electrical forces induced by the particles themselves become especially important in collisional processes not controlled by Brownian agitation.

Most of the effort to elucidate the contribution of these effects on the behavior of clouds has been confined to the continuum regime. To illustrate the importance of the interaction between particles in a viscous medium, let us take the case of two spheres in translational motion induced by an external field of force. The analysis is restricted to the Stokes regime where $Re_i \ll 1$, and $Kn_i \to 0$.

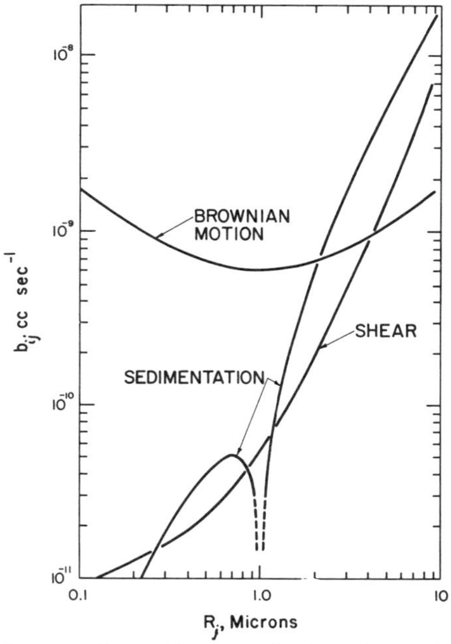

FIG. 10.7. Comparison between three collision mechanisms for particle of one micron radius as a function of the radius of the second interacting particle. Spheres are suspended in a turbulent medium with a dissipation rate $\epsilon_t = 5\ \text{cm}^2\ \text{sec}^{-3}$. (From Friedlander[361]).

A. Aerodynamic Forces

In general, the magnitude of the aerodynamic forces acting on neighboring particles depends on (a) their shape and size, (b) the distances between them, (c) the orientation relative to one another and relative to the external force field, and (d) their speeds and rotation with respect to the fluid. For a set of translational and rotational velocities of particles, the parameters of primary interest are the aerodynamic forces and the torques exerted by the local action of the fluid on the bodies. Having determined these induced forces for a given cloud, it is possible to solve the inverse problem for the motion of the particles from knowledge of external forces.

To determine the viscous forces acting on the particles, the equations of fluid motion given by Stokes, eqns. (3.1–3.5), are applied. The spheres are assumed to be rigid, and travelling through an unbounded medium at rest at infinity. The boundary condition for the fluid around individual particles is $\mathbf{q} = \mathbf{v}_{si}$ on particle i, and the condition at infinity is $\mathbf{q} \to 0$ as $r \to \infty$.

Even approximate solutions for a set of N particles becomes quite complicated for this system of equations and boundary conditions. Detailed discussions of this subject is beyond the scope of this monograph, but a rather comprehensive account may be found in Happel and Brenner's recent book.[41]

Solutions in the Stokes Regime

If we consider a pair of particles i and j falling together at distances far apart an approximate solution for the force acting on either body may be found. The geometry of the two interacting spheres is drawn in Fig. 10.8. The coordinate system is oriented in such a way

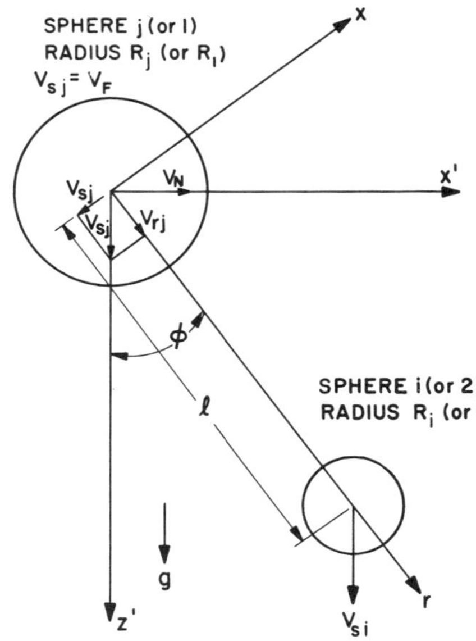

FIG. 10.8. Two interacting spheres falling by the action of gravity through a continuum.

that the r-axis falls along the line of centers between the bodies. The distance between the particles is given as l, and the induced aerodynamic forces are $\mathbf{F}_i$ and $\mathbf{F}_j$. The particle velocities are given by $\mathbf{v}_{si}$ and $\mathbf{v}_{sj}$. For this case, Happel and Brenner[41] give the result within the approximation that both spheres are governed by equations for steady flow: †

$$\frac{\mathbf{F}_i}{6\pi\mu_G R_i} = -\mathbf{i}\,\frac{q_{ix} - (6\pi R_j q_{jx})(8\pi l)}{1 - (36\pi^2 R_i R_j)(8\pi l)^2} - \mathbf{k}\,\frac{q_{ir} - (6\pi R_j q_{jr})(4\pi l)}{1 - (36\pi^2 R_i R_j)(4\pi l)^2}, \tag{10.89}$$

where the unit vectors $\mathbf{i}$ and $\mathbf{k}$ refer to the coordinate directions x and r.

The force on particle j, $\mathbf{F}_j$, is exactly the same as that given in (10.89), except the subscripts i and j are interchanged.

We note that the external force $\mathbf{F}_e$ is generally given, so that the particle velocities are determined upon solving simultaneously a set of the four equations corresponding to the vector components in (10.89).

In the case of equal sized particles ($\mathbf{q}_i = \mathbf{q}_j = \mathbf{q}$), we may rewrite the expression (10.89) in a simpler form:

$$-\frac{\mathbf{F}}{6\pi\mu_G R_i} = \mathbf{i}\,\frac{q_x}{1 + (3/4)(\mathrm{R}_i/l)} + \mathbf{k}\,\frac{q_r}{1 + (3/2)(R_i/l)}. \tag{10.90}$$

Thus the forces exerted on either particle is the same, and their movement is parallel with the same velocity. The particles may drift sidewise, but they will always retain the same spacing.

For equal-sized spheres, Happel and Brenner have indicated that the velocity component in the direction of the external force $q_F(=v_s)$ and the component normal to it, $q_{x'}$, are determined from eqn. (10.90). If the angle ϕ is the angle between the direction of the force field, and the line of centers,

† When the spheres are close together, refinement of the equations of motion must include unsteady terms (Ref. 41, p. 244).

$$q_{z'} = v_{si} = -\frac{F}{6\pi\mu_G R_i}\left(1 + \frac{3}{4}\frac{R_i}{l}\sin^2\phi\right), \tag{10.91a}$$

$$q_{ix'} = q_{x'} = -\frac{F}{6\pi\mu_G R_i}\frac{3R_i}{4}\sin\phi\cos\phi. \tag{10.91b}$$

Then a drift of the pair of spheres together in direction normal to the direction of the force field develops only if $\phi \neq 0$, or $90°$.

It is possible to demonstrate from eqn. (10.89) that a larger sphere will overtake a smaller particle under any conditions. Consider only the component of motion in the z' direction. Noting that $\mathbf{F}_i$ and $\mathbf{F}_j$ are proportional to the particle volume,[41]

$$\frac{v_{si}}{v_{sj}} = \frac{\hat{K}^3 + (3/2)(R_i/l)}{1 + (3/2)\hat{K}^2(R_i/l)}. \tag{10.92}$$

The particle will collide if $l = R_i + R_j$, wherein

$$\frac{v_{si}}{v_{sj}} = \frac{\hat{K}^3 + \hat{K}^2 + 3/2}{1 + 1/\hat{K} + (3/2)\hat{K}^2}. \tag{10.93}$$

If $R_i = R_j$, $v_{si}/v_{sj} = 1$ as expected. Where $R_j \gg R_i$, $v_{si}/v_j \to (3/2)\hat{K}$, indicating that the smaller sphere will always move slower than the larger one.

It is interesting to remark that the result, eqn. (10.89), indicates that the resistance of an individual particle is always lowered in the presence of other particles. Assuming that the effects of larger numbers of spheres is additive, this suggests that a cloud of aerosols will travel through a stagnant medium at a faster rate than a single particle alone.

Equation (10.89) may be generalized for particle interaction in the Stokes regime. Happel and Brenner[41] write the following forms:

(no rotation)
$$-\frac{\mathbf{F}_i}{6\pi\mu_G R_i} = \mathbf{i}T_1 + \mathbf{k}T_2. \tag{10.94}$$

(rotational freedom)
$$-\frac{\mathbf{F}_i}{6\pi\mu_G R_i} = \mathbf{i}T_1 + \mathbf{k}T_3. \tag{10.95}$$

The parameters T have been determined by the method of reflections[41] and are listed in Table 10.3 as a function of the dimensionless separation distance $(l/2R_i)$.

The values of T given by the method of reflections are approximations. To confirm their validity, these results have been checked with an exact solution of the equations of motion for equal sized spheres moving parallel to their line of centers (Stimson and Jeffery[365]). The approximate results reported by Happel and Brenner equal those estimated from the exact solution to less than 0.2 per cent.

Experimental data of Bart[41] and others for the case of particle fall by gravity in parallel orientations to the line of centers, and perpendicular to the line of centers are in extremely close agreement with the theoretical predictions using Stokes' approximation, provided the value of $Re_i < 0{\cdot}25$.

Inertial Effects

Above Reynolds numbers of about one-quarter, inertial effects in the fluid motion have been observed[41] and equal particles no longer tend to maintain a fixed position relative

TABLE 10.3. FUNCTIONS FOR THE PARAMETERS OF AERODYNAMIC FORCES FOR EQUAL SIZED SPHERES AS DETERMINED BY THE METHOD OF REFLECTIONS (FROM HAPPEL AND BRENNER[41])

Length along line of centers/diameter	T_1	T_2	T_3
1.0	0.6466	0.7158	0.6937
2.0	0.7423	0.8386	0.8366
3.0	0.8047	0.8875	0.8871
4.0	0.8441	0.9136	0.9298
5.0	0.8706	0.9299	0.9298
6.0	0.8895	0.9409	0.9409
7.0	0.9036	0.9490	0.9490
8.0	0.9145	0.9551	0.9551
9.0	0.9232	0.9599	0.9599
10.0	0.9304	0.9638	0.9638

to one another.† The Oseen approximations appear to offer a point of departure for deducing the effects of inertial forces in the aerodynamic interaction between particles. Some years ago, Oseen himself[13] studied the effect of inertial perturbations on creeping motion of the fluid around two spheres of equal size. His analysis resulted from substitution of the Oseen velocity fields into the equations governing fluid motion near the two bodies following each other along their line of centers. Oseen's analysis indicated that the force exerted by the leading sphere on the surrounding fluid was slightly larger than the following sphere. The force acting on the fluid by the following particle was the same as estimated for Stokes flow. Thus the leading sphere is slowed up to a greater extent than the following particle, and the distance between them will tend to decrease by the influence of inertial forces in the fluid.

If the spheres fall side by side with orientation of the force field along the x-axis, the force exerted on each sphere by the fluid will be the same, but not necessarily directed entirely along the direction of fall. In this particular orientation, the force exerted on the bodies along their line of centers r will be one corresponding to repulsion.

Happel and Brenner[41] have remarked that even though Oseen's equations are linear, they cannot be used to treat the case of particle interaction in arbitrary spatial orientation by simple vector addition. This is because the velocity $\mathbf{v}_{si}$ appearing in the approximation $\rho_G \mathbf{v}_{si} \cdot \nabla \mathbf{q}$ is different for different orientation of the line of centers with respect to the direction of fall.

Though Oseen's original derivation relates to two spheres of arbitrary size, his analysis for the case appears incorrect for this case because no inertial correction develops if either particle radius is set at zero. In addition, Oseen's equations do not approach the Oseen flow relation for drag on two isolated spheres if the separation is large. Happel and Brenner note that this comes about apparently from a failure to match boundary conditions for each particle separately when the spheres are at distances apart, as well as at infinity in the fluid.

† The recent experiments of Pruppacher and Steinberger[419] suggest that inertial effects are significant at Reynolds number much lower than one-quarter.

Generalization to Include Electrical Forces

So far we have discussed only aerodynamic interaction between particles in the absence of Brownian motion. The equations developed for the forces may be generalized to include electrical forces induced between the falling spheres. The expression for the electrical forces have been incorporated recently into the relations for aerodynamic interaction by Krasnogorskaya[366] for particles falling by the action of gravity through a stagnant medium. Her equations are summarized below in dimensionless form in the coordinate system x', z' shown in Fig. 10.8. The length scale is the radius of the larger particle R_j $(=R_1)$. The time scale is the ratio $R_1/2v_{s1}$, and the velocity scale is $\rho_G R_1/9\mu_G$.

$$\hat{I}\frac{dq_{1z'}}{dt} = 1 - \frac{\hat{a}_1 z' - \hat{b}_1 x'}{r_1} + \hat{c}_1 F_{1z'}. \tag{10.96a}$$

Particle 1:

$$\hat{I}\frac{dq_{1x'}}{dt} = -\frac{\hat{a}_1 x' + \hat{b}_1 z'}{r_1} + \hat{c}_1 F_{1x'} \tag{10.96b}$$

Particle 2:

$$\hat{K}^{-1}\hat{I}\frac{dq_{2z'}}{dt} = \hat{K}^{-2} - \frac{\hat{a}_2 z - \hat{b}_2 x}{r_1} + \hat{c}_2 F_{2z'} \tag{10.97a}$$

$$\hat{K}^{-1}\hat{I}\frac{dq_{2x'}}{dt} = -\frac{\hat{a}_2 x' - \hat{b}_2 z'}{r_1} + \hat{c}_2 F_{2x'}. \tag{10.97a}$$

where

$$r_1 = \frac{r}{R_1} = \frac{1}{R_1}\sqrt{(x'^2 + z'^2)}.$$

Here it is assumed that the terminal velocity v_{s1} is the fluid velocity as $r \to \infty$ when the coordinate system is placed on the larger particle (particle 1). the last term in the equation denotes the electrical force. The coefficient $\hat{c}_1$ is given by $1/6\pi\mu_G R_1 v_{s1}$. The second term on the right-hand side of eqns. (10.96a) and (10.97a), and the first terms on the right of (10.96b and 10.97b) specify forces on the spheres associated with aerodynamic interaction. These are analogous to the coefficients T_1, T_2, and T_3 in eqns. (10.94) and (10.95). Appropriate relations for the coefficients $\hat{a}_1$, $\hat{a}_2$, $\hat{b}_1$, and $\hat{b}_2$ in terms of $\mathbf{q}_1$, $\mathbf{q}_2$, and r_1 have been derived by Hocking.[368] His forms are probably correct to second or third order, but have been shown[371] to give only qualitatively correct relations for the aerodynamic forces.

The velocity, length and time scales are made dimensionless in terms of the parameter $\hat{I} = 2R_1\rho_1 v_{s1}/9\mu_G$, where v_{s1} is the terminal settling velocity, $2R_1^2\rho_1 g/9\mu_G$. This velocity corresponds to the value of the fluid velocity at infinity if the coordinate system is placed on the larger particle (particle 1).

The last terms in eqns. (10.96) and (10.97) denote the forces acting on the two particles resulting from electrical effects. These components have been specified by Davis[367] from a general solution of Laplace's equation for the electrostatic potential function of two conducting spheres in a uniform electrical field. In the coordinate system shown in Fig. 10.8, assume that the electric field vector $\mathscr{E}_0$ is oriented in the same direction as the gravity vector $\mathbf{q}$. Then the components of the electrical force are given by:

$$F_{iz'} = F_x \sin\phi + F_r \cos\phi,$$

$$F_{ix'} = F_x \cos\phi + F_r \sin\phi.$$

The components of force in the x and r coordinate system have been given by Davis.[367] For example, the forces acting on the smaller sphere are:

$$\left.\begin{aligned} F_{2r} = \Big[& \epsilon R_2^2 \mathscr{E}_0(\hat{F}_1 \cos^2\phi + \hat{F}_2 \sin^2\phi) + \mathscr{E}_0 \cos\phi(\hat{F}_3 Q_1 + \hat{F}_4 Q_2) \\ & + \frac{1}{\epsilon R_2^2}(\hat{F}_5 Q_1^2 + \hat{F}_6 Q_1 Q_2 + \hat{F}_7 Q_2^2)\Big] + \mathscr{E}_0 Q_2 \cos\phi. \end{aligned}\right\} \quad (10.98)$$

$$F_{2x} = [\epsilon R_2 \mathscr{E}_0^2 \hat{F}_8 \sin 2\phi + \mathscr{E}_0 \sin\phi(\hat{F}_9 Q_1 + \hat{F}_{10} Q_2)] + \mathscr{E}_0 Q_2 \sin\phi. \quad (10.99)$$

The coefficients $\hat{F}_1, \hat{F}_2, \ldots, \hat{F}_{10}$ are complicated series expressions dependent on the separation distance s, between the spheres, $s = l - (R_1 + R_2)$. These coefficients have been calculated by Davis[367] and are listed in Table 10.4 for various combinations of particle sizes.

B. Collision Efficiencies Accounting for Induced Forces

The coagulation of particles by gravitational settling appears to be a controlling mechanism for growth of relatively large droplets found in low concentrations in a mildly turbulent medium like the atmosphere. The formation of rain from clouds of water droplets must

TABLE 10.4. COEFFICIENTS FOR ELECTROSTATIC FORCES ACTING ON TWO CHARGED CONDUCTING SPHERES IN A UNIFORM ELECTRIC FIELD (FROM DAVIS[367])

s/R_2	$\hat{F}_1$	$\hat{F}_2$	$\hat{F}_3$	$\hat{F}_4$	$\hat{F}_5$
($R_1 = R_2$)					
10.0	0.00030	−0.00014	−0.00116	0.00116	8.1×10^{-6}
1.0	0.0927	−0.0330	−0.0951	0.0951	0.0101
0.1	1.400	−0.0945	−0.9741	0.9741	0.1533
0.01	9.554	−0.1032	−5.994	5.994	0.9263
0.001	59.49	−0.1041	−36.39	36.39	5.552
($R_1 = 2R_2$)					
10.0	0.00169	−0.00084	−0.00093	0.00730	5.4×10^{-6}
1.0	0.2758	−0.0759	−0.0576	0.3063	0.00227
0.1	3.136	−0.1436	−0.5370	2.128	0.0224
0.01	20.59	−0.1505	−3.356	11.97	0.1364
0.001	127.70	−0.1511	−20.59	71.46	0.8299
($R_1 = 5R_2$)					
10.0	0.0122	−0.00554	−0.00056	0.0611	1.9×10^{-6}
1.0	0.6422	−0.0927	−0.0191	0.8537	0.00014
0.1	5.310	−0.1033	−0.1436	3.617	0.00097
0.01	34.63	−0.1023	−0.9203	18.17	0.00611
0.001	213.0	−0.1022	−5.641	105.0	0.0373
($R_1 = 10R_2$)					
10.0	0.0376	−0.0137	−0.00033	0.2162	4.9×10^{-7}
1.0	0.7689	−0.0579	−0.00546	1.322	9.6×10^{-6}
0.1	6.127	−0.0483	−0.0411	4.336	6.9×10^{-5}
0.01	41.38	−0.0465	−0.2757	20.60	0.00046
0.0001	254.8	−0.0464	−1.696	116.9	0.00282

TABLE 10.4. (*cont.*)

s/R_2	$\hat{F}_6$	$\hat{F}_7$	$\hat{F}_7$	$\hat{F}_9$	$\hat{F}_{10}$
$(R_1 = R_2)$					
10.0	−0.00694	8.1×10^{-6}	−0.00014	0.00058	−0.00058
1.0	−0.1146	0.0101	−0.0391	0.0407	−0.0407
0.1	−0.4532	0.1533	−0.2248	0.1736	−0.1736
0.01	−2.005	0.9263	−0.3756	0.2707	−0.2707
0.001	−11.26	5.552	−0.4652	0.3260	−0.3260
$(R_1 = 2R_2)$					
10.0	−0.00592	4.5×10^{-5}	−0.00084	0.00046	−0.00364
1.0	−0.0662	0.0255	−0.1033	0.0203	−0.1330
0.1	−0.2322	0.2977	−0.4048	0.0706	−0.3717
0.01	−1.029	1.685	−0.6202	0.1058	−0.5066
0.001	−5.816	9.946	−0.7478	0.1264	−0.5793
$(R_1 = 5R_2)$					
10.0	−0.00391	0.00028	−0.00581	0.00026	−0.0305
1.0	−0.0220	0.0494	−0.1915	0.00525	−0.3765
0.1	−0.0603	0.4249	−0.4606	0.0124	−0.6470
0.01	−0.2532	2.232	−0.6369	0.0171	−0.7555
0.001	−1.402	12.80	−0.7431	0.0199	−0.8098
$(R_1 = 10R_2)$					
10.0	−0.00227	0.00073	−0.0163	0.00013	−0.1080
1.0	−0.00749	0.0603	−0.1930	0.00129	−0.5893
0.1	−0.0177	0.4586	−0.3535	0.00236	−0.7965
0.01	−0.0718	2.337	−0.4648	0.00310	−0.8657
0.001	−0.3924	13.22	−0.5338	0.00356	−0.8990

involve coagulation or coalescence of droplets as well as condensation to explain the rates of growth of very large particles. Therefore, the question of the efficiency of collection of smaller particles by larger ones is of central importance to cloud physics. This conclusion has led scientists to seek more accurate relations for η than given by the kinematic results in the previous section.

Aerodynamic Forces

Hocking[(368)] first attempted to find the influence in the Stokes' regime of the falling particles. His analysis sought the critical trajectories for spheres approaching one another initially at their terminal settling velocities. Particles were separated horizontally by distances up to $x'/R_1 \approx 50$. Critical trajectories were those that just allowed the spheres to touch during their passage around one another. Like other later workers, such calculations were carried out by solving eqns. (10.96) and (10.97), using approximations for the terms accounting for aerodynamic interaction. Hocking's calculations excluded electrical forces. However, his results were later extended to incorporate these effects by Krasnogoskaya.[(366)]

The results of Hocking's computations are shown by the solid lines in Fig. 10.9. These curves indicate that the collision efficiency depends on the large particle j in contrast to the kinematic result, eqn. (10.71), and has a maximum at $\hat{K} \approx 0.65$–0.75. The rapid drop off of the $\sqrt{\eta}$ curves suggests that no collisions take place between particles of radius 19 μ or less. In addition, the limit $\eta \to 0$ as $\hat{K} \to 0$, as expected from Friedlander's equations, eqn. (10.72), is not predicted correctly in Hocking's calculations. We note that $\sqrt{\eta}$ can be greater than unity under some conditions, and it begins to approach the geometrical efficiency for very large particles j as calculated by Findeisen.[369] These different results indicate what markedly different conclusions can be suggested by various simplifying assumptions for the capture process.

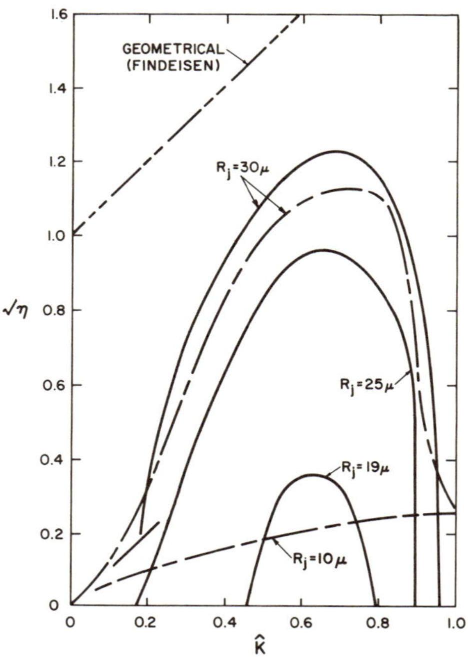

FIG. 10.9. Comparison between theoretical collection efficiencies. ——— Hocking, – – – — Geometrical, and— – — Davis and Sartor.

It is unlikely that Hocking's method for calculation accounts satisfactorily for the interference of particles on fluid motion when the particles approach each other. By including terms in $\hat{a}_1$ and $\hat{b}_1$ only up to the seventh power in R_1/l and R_2/l, Happel and Brenner[41] have noted that Hocking was able to obtain good agreement with other results for the resistance on two equal-sized spheres moving perpendicular to their line of centers and in contact. However, with two equal-sized spheres falling along their line of centers and touching, Hocking's method disagrees with Stimson and Jeffrey's[365] exact result for the drag on the spheres by a factor of 2.5. Hocking indicated that better accuracy could be attained with higher order terms. However, it is important to note that the steady state form of the equations for creeping motion will not yield correct results when the particles are very near each other. Better computations accounting at least for local acceleration of fluid should be employed for more accurate results.

The recent calculations of Davis and Sartor[371] are of interest in connection with Hocking's estimates. Davis[370] calculated collision efficiencies on the basis of Stimson and Jeffery's[365] analysis of Stokes' flow, while Sartor extended Hocking's power series expansions to include higher order terms up to 40. Davis and Sartor's independent calculations for collisions of particles in the Stokes regime of flow agree with each other, and are shown for two typical cases in Fig. 10.9. The principal disagreement between these new results and those of Hocking lie in the extreme range of $\hat{K}$ and in the estimate of the collisional "cut off" for $R_j \approx 19\ \mu$.

Davis and Sartor have found that finite collision efficiencies are predicted below $\hat{K} \approx 0.2$ with an asymptotic behavior for $\sqrt{\eta}$ as $\hat{K} \to 0$ in agreement with that predicted by eqn. (10.72) (i.e., $\sqrt{\eta} \sim \hat{K}$, as $\hat{K} \to 0$). For $\hat{K} \to 1$, Davis and Sartor have found a finite value for $\sqrt{\eta}$ of about 0.25. Particles of radius less than 19 μ are found to have a small but finite collision efficiency in disagreement with Hocking's earlier results. A typical curve for $R_j = 10\ \mu$ based on Sartor's and Davis' calculations is drawn in Fig. 10.9.

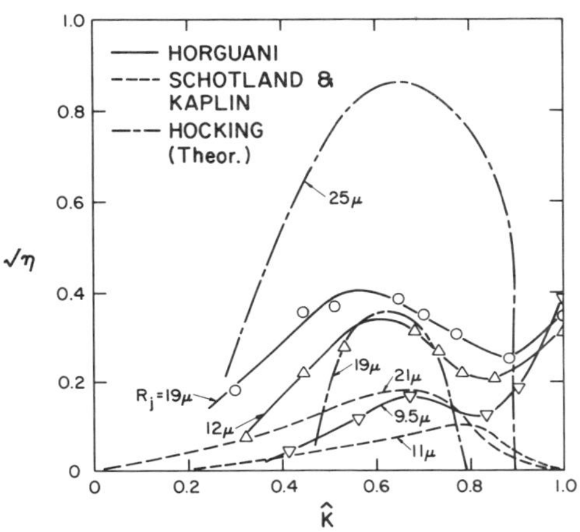

FIG. 10.10. Comparison between experimentally determined collection efficiencies and the theory of Hocking.

When falling spheres become sufficiently large for their Reynolds number to be greater than about 0.1, inertial effects must contribute more strongly to particle collisions. The first attempt to account for non-linear contributions to aerodynamic forces induced between falling spheres was published by Pearcey and Hill.[372] By superposition of Oseen's results for single spheres, a flow pattern near two approaching particles was determined, and critical trajectories could be estimated. Pearcey and Hill's results differ markedly from other results in the limit of very low Re, such as those of Hocking, and Sartor and Davis, so that they are now believed to be incorrect. It is currently believed that collision efficiencies for $Re > 0.1$, as reported by Shafrir and Neiburger[376] are at least good approximations for a theory to account for inertial effects. Shafrir and Neiburger's approximate calculations were based on the estimation of the drag force on each sphere using only the disturbance of the fluid flow field due to the other sphere. Extrapolation of their results to low Reynolds number substantially agree with Hocking's calculations, so that the Shafrir and Neiburger efficiencies are most often used for collision calculations up to Reynolds number ≈ 0.5.

Experimental methods for evaluating the collision efficiency largely have followed two paths, the first attempts to construct hydrodynamic analogies of collision processes in a gas by studying the interaction between small spheres and larger, heavier spheres in a very viscous, liquid medium. The second technique has attempted to simulate directly the behavior of large water droplets as they fall through smaller droplets suspended in air. Neither technique has provided completely accurate tests of the theoretical models, and the experimental results for η as a function of $\hat{K}$ vary considerably between different workers for the same conditions of Re_i and Stokes' number. Several of these experiments appearing in the literature up to the 1950's are discussed in detail by Fuchs[(2)] and will not be discussed further.

A newer set of results has been reported by Horguani.[(373)] These will be used here to indicate the typical conclusions to be reached from the hydrodynamic analogies. Horguani's experiments involved photographic observations of smooth steel spheres falling in glycerine. A number of trajectories of two interacting balls were estimated from photographs, and values of $\sqrt{\eta}$ were calculated from these results.

In general, complete similarity between the results of Horguani's experiment and aerosol behavior would require geometrical similarity plus similarity in the Reynolds number and the Stokes number. Because the pattern of flow around spheres for $Re_i < 1$ remains the same, then similarity between the hydrodynamic model and aerosol behavior should be maintained if the Stokes number, *Stk*, and the ratio of radii of colliding spheres remains the same between model and prototype.† Using the principles of similarity, Horguani has calculated curves for the collision efficiency for micron sized particles based on his model experiments. His results for an equivalent size range R_j of about 5 μ to 19 μ are shown in Fig. 10.10. Also drawn for comparison are curves of Hocking. Interestingly enough, as $\hat{K} \to 0$, the curves tend to show $\sqrt{\eta}$ approaching zero as expected. However, there is a minimum in the curves for $\sqrt{\eta}$ at $\hat{K} \approx 0.9$. As $\hat{K} \to 1$, rises to a second maximum roughly equivalent to the limit estimated by Davis and Sartor.[(371)] Curves very similar to these for 11 μ and 20 μ droplet collision in air have been reported earlier in similar experiments by Schotland and Kaplin.[(374)] Their curves for $\sqrt{\eta}$ lie about a factor of 3 lower than Horguani's, however, in the range of $\hat{K} \approx 0.6$. In the region of $\sqrt{\eta} < 0.8$, the experimental curves give larger values of $\sqrt{\eta}$ than predicted by Hocking on the basis of Stokes' flow. Therefore, these aerodynamic collision theories cannot be regarded as verified quantitatively even in these recent experimental results. We note, however, that departures from the creeping flow regime have been observed for $Re_i > 0.25$. Particle Reynolds numbers in Horguani's analogy experiments exceeded this value. Thus, these experiments, like most others in the literature, do not rigorously apply to the theory based on Stokes' hydrodynamics. It is of interest, however, to note that Shafrir and Neiburger's[(376)] should apply better to the model experiments. Here, again, there is only rough qualitative agreement between theory and the experiments.

The secondary maximum may be the result of a so-called wake capture effect resulting from small asymmetries of flow in the lee of particles of $Re_i > 0.25$. A consequent increase in $\sqrt{\eta}$ may result from the perturbations in the fluid flow around the body. Horguani observed a tendency for approximately equal sized particles to come together as rearward sphere entered the lee of the forward sphere. This effect is not accounted for in the theoretical calculations based on Stokes' flow.

† There remains some question whether or not it is actually possible to model the aerodynamics of collision by a hydrodynamic analogy. It is certainly true that for steady flow of single spheres, Re_i and *Stk* are the only relevant parameters. However, in the collision process, it should be necessary to dynamically model local acceleration of fluid as particles approach each other. This appears feasible only if the ratio (ρ_G/ρ_i) is the same between the model and prototype in addition to the other two parameters.

Typical of newer results using the second method of allowing droplets to fall through a quiescent gas are those of Woods and Mason.[375] They observed the collisions between water droplets of radius 30–55 μ and saline droplets of radius 1–12 μ. Though the Reynolds number for the larger droplets exceeded unity, their results substantially agreed with theoretical estimates of η except below $\hat{K} \approx 0.1$, where they found consistently higher values of η than predicted by Hocking's theory.[368] In a few measurements of large droplets of 17–19 μ radius capturing smaller ones of 10–14 μ radius, Woods and Mason found an upper limit of $\sqrt{\eta} \approx 0.3$. This agrees quite well with Davis' and Sartor's estimate of $\sqrt{\eta}$ for $\hat{K} \approx 0.6$, in contrast to Hocking's prediction that $\sqrt{\eta} \approx 0$ in this range of particle size.

In the results discussed here, only the case of fall through a quiescent gas has been studied. The more difficult, but important practical question of the influence of particle accelerations by turbulence, combined with aerodynamic interaction between particles, is much more complicated. It is possible that Sartor's and Davis'[371] results and those of Shafrir and Neiburger[376] could be used to improve the theory of Saffman and Turner.[362] At this stage, however, such improvements are probably not worthwhile, at least for application to the physics of natural clouds. The scale of turbulence containing most of the energy in atmospheric convection is so much larger than the scales that would contribute to causing differences in relative velocities between droplets that the quiescent gas model seems to be a sufficiently good first approximation in such cases. Certainly the values of η evaluated by Hocking or Pearcey and Hill cannot be used with any reliability to improve the theory of Saffman and Turner.[362] Therefore, in the absence of better information, one still has to consider $\eta \approx 1$ for examining the mechanism of coagulation by the influence of turbulence.

The slippage of flow between particles that are in the range of distance between centers less than or the order of the mean free path of the suspending gas has not been examined in any detail until recently.[370] Since the interference of aerodynamic forces on particle collision is most important at small interaction distance, the slippage effect has to be taken into account. In fact, it may be of fundamental significance in the collision process because particles can never actually collide in a viscous medium within the continuum approximation.

Electrical Forces

In addition to the effect of aerodynamic forces developing between colliding droplets, electrical interaction may influence strongly the efficiency of collision of falling particles. Experiments of Telford *et al.*,[377] for example, have indicated that bipolar charging with opposite sign of ~50 μ radius droplets to 10^{-4} e.s.u. above natural ionization levels can increase the coagulation rate substantially, while unipolar charging effectively stopped coagulation. Theoretical arguments of Pauthenier and Cochet[378] and later Levin,[379] neglecting aerodynamic interaction and droplet inertia, have indicated that coagulation of falling droplets of ~10 μ radius or less may be affected substantially by charging or by electrical induction forces with charges greater than 4×10^{-4} e.s.u. on one colliding droplet. More recently, Sartor[380] has estimated the electrical fields necessary to induce collision of ~20 μ droplets considering an arbitrary initial distance between particles. His calculation made use of modifications in Stokes' equations to include the effect of electrical forces. Davis,[435] analyses have indicated that the electrical forces contribute most significantly to collisions when particles are close to each other. Thus the parameter $Q_i Q_j/\mathbf{r}$ has most relevance to this problem.

Perhaps the most extensive calculations of electrical forces and the influence on collisions reported so far stem from Krasnogorskaya's work.[366, 381] She has computed numerically a multitude of trajectories of two colliding particles taking into account both aerodynamic interaction and electrical effects. From these trajectories, estimates were made of the collision efficiency of Krasnogorskaya's work involved the application of eqns. (10.96–10.97) with Hocking's functions for aerodynamic interaction. In most of her computations, the terms accounting for electrical forces approximated Davis' exact solutions given by eqns. (10.98) and (10.99). As in Davis' expressions, here approximate relation for F_{iz}, and F_{ix}, included a term for Coulombic interaction between charged particles, a part to account for the influence of an imposed, homogeneous electric field, and an approximation for the induced (image) forces resulting from dipole–dipole interaction.

Comparison between the electrical forces evaluated from her approximate expression. and those from eqns. (10.98) and (10.99) indicated that for (dimensionless) interparticle distances less than $r_1 \approx 2$ marked differences existed between the forces estimated by the approximate theory and Davis' exact calculations. For ranges $r_i > 2$, the two methods revealed very nearly the same curves for **F**. Though certain differences in the calculated collision efficiency were found using the different forms for the electrical force, Krasnogorskaya's approximation evidently does not overestimate the role of the electrical effects. Qualitatively the two analytical models showed the same variation with electrical parameters.

As we have seen previously, the influence of an electrical field in particle collisions will depend on the sizes of the particles, the signs and magnitude of net charge on the sphere, and their mutual location. In addition to polarizing the charge on droplets, an electrical field will cause a change in relative velocity of a charged particle if its size is small enough. Therefore, depending on a complicated combination of parameters of the interacting particles, the electric field can either speed up or slow down the coagulation process.

The results of Krasnogorskaya's[381] extensive computations indicated that electrical fields with voltages below 100 v cm^{-1} and weak bipolar charging of 10^{-9}–10^{-7} e.s.u. per droplet in absence of any external field affect the collision efficiency of droplets very little for particles of 10–25 μ in radius. However, larger fields of ~ 1000 v cm^{-1} can have a substantial effect on collision of uncharged particles in this size range. The lower the relative velocity between the spheres, the larger is the influence of electrical forces. Thus, electrical interaction aids collision significantly if the particles are similar in size. These computations indicated that, in the range of size 5–25 μ, electrical fields larger than 100 v cm^{-1}, and charges exceeding 10^{-7} e.s.u./droplet could cause agglomeration of spheres in the regimes where the collision efficiency based on Hocking's estimate of aerodynamic interaction is negligible.†

Krasnogorskaya[366] concluded from her computations that for electrical fields characteristic of cumulus clouds of average thickness in the atmosphere, say the order of tens of volts per centimeter, and weak charging up to 10^{-7} e.s.u./particle have a rather small effect on droplet collision, where $R_i > 10$ μ. This means that like charged particles of this size will not be inhibited from collisions, and will act as if they were effectively neutrally charged. However, as the size of cloud droplets decreases, the influence of electrical forces increases.

At present, the theoretical knowledge of collisional mechanism exceeds laboratory evidence to verify the theoretical work. The verification of theories and the detailed

† In view of the known inaccuracies in Hocking's calculations, Krasnogorskayas' results are probably only correct for particles with radii less than about 20 μ.

classification of importance of the various collisional mechanisms in geophysical problems must await new experimental studies, especially in the areas of turbulent interaction and electrical – aerodynamic interaction.

10.4. The Evolution of the Size Distribution of a Coagulating Aerosol

Many of the properties of dispersed systems such as the optical density depend on the concentration of suspended particles in various ranges of size. The size distribution, or the size spectrum of an aerosol thus plays a vital role in the phenomenology of the behavior of aerosols because of the complexities and the variety in processes contributing to shaping the distribution over wide ranges in size. We can readily visualize that the size spectrum of natural aerosol found in a volume of air, for example, will develop in accordance with the introduction of a multitude of sources, like trash burners or volcanoes, with collisional processes, with sweepout by rainfall, and by sedimentation from the air volume.

Despite the influences of widely varied physical processes that contribute to shaping aerosol size spectra, certain regularities have been observed in these distributions. Frequently, measurements of size spectra reveal that they are normally single peaked, and often monotonic, and frequently fit a logarithmic-normal distribution, or an exponential form over a range of particle size. In attempting to understand the complex mechanisms that tend to control the evolution of aerosol distributions, it is useful to consider the behavior of idealized clouds in which only one or two processes such as collision by Brownian motion govern the spectral shape. By examination of the highly simplified systems it may be possible to find insights into the processes controlling the development of the distribution of natural aerosols. Perhaps the most extensively studied "cloud" is that of a collection of rigid spheres in Brownian motion inside an enclosed vessel, without the influence of gravity. Other models of particulate clouds have been investigated, including ones formed by condensation in supersaturated vapor (Chapter 9).

Let us review first the general problem of mathematically specifying the rate equations for the developing spectrum governed only by collisional processes. We note that the coagulation equations arise as a special form of a general expression for the rate of change in number concentration of cloud particles as given in Chapter 9. These rate equations were given in discrete form as a set of ordinary differential equations specifying the changes in particle numbers of a particular size, k, as for example in eqn. (10.1).

In dealing with the set of equations for the discrete spectrum, one readily finds that the book-keeping on individual particles becomes untractable very quickly if the cloud exceeds more than 100–200 different sizes. Therefore, the rate equations for spectral development are often rewritten in continuous form. That is, the crucial assumption is made that the distribution of particles can be assumed continuous in its broad scale structure in the same way one can pass to continuous variables as a limiting process in ordinary calculus.

The continuous spectrum function may be defined in terms of linear scale dependence, i.e. particle radius $\tilde{R}$, or in terms of surface area ($\sim \tilde{R}^2$), or particle volume. For purposes of this discussion, let us choose to define the spectral function in terms of volume $\tilde{v}$. Then, the spectral density function $\equiv n(\tilde{v}, t)$. Using this definition, we note that the number of particles for unit volume in the size range $\tilde{v} + \mathrm{d}\tilde{v}$ over a time interval $t + \mathrm{d}t$ is just given by

$$\mathrm{d}N = n(\tilde{v}, t)\,\mathrm{d}\tilde{v}\,\mathrm{d}t. \tag{10.100}$$

For any given time, the concentration of particles of the kth class in the discrete spectrum was given as n_k. Equation (10.100) is approximately

$$\Delta N = n \Delta \tilde{v}. \tag{10.101}$$

If $\Delta \tilde{v} = v_1$, the unit particle volume, then $\Delta N = n_k$.

Using the definition of the continuous spectrum function $n(\tilde{v}, t)$, eqn. (10.1) may be rewritten as an integro-differentional equation. To derive this kinetic expression for the continuous spectrum, we seek an expression for the rate of change of particles in the volume size range $\tilde{v} + d\tilde{v}$. For the discrete collision parameter b_{ij}, the corresponding coefficient for the continuous spectrûm is $b(\tilde{m}, \tilde{n})$, where $\tilde{m}$ and $\tilde{n}$ denote the volumes of two colliding particles of different size. Then the total number of collisions between $\tilde{m} + d\tilde{m}$ and $\tilde{n} + d\tilde{n}$ to give particles of volume $\tilde{v}$ in the interval dt reads

$$b(\tilde{m}, \tilde{n})\, n(\tilde{m}, t)\, n(\tilde{n}, t)\, d\tilde{m}\, d\tilde{n}\, dt.$$

The net rate of change of particles of range in volumetric size $\tilde{v} + d\tilde{v}$ is the difference between the gains by collisions of all particles of volume $\tilde{m}$ and $\tilde{n}$ to give volume $\tilde{v}$, and the loss by collisions of particles of $\tilde{v}$ class with all other size ranges:

$$\frac{\partial(n(\tilde{v}, t)\, d\tilde{v})}{\partial t} = \tfrac{1}{2}\int_0^{\tilde{v}}\int_0^{\tilde{m}} b(\tilde{m}, \tilde{n})\, n(\tilde{m}, t)\, n(\tilde{n}, t)\, d\tilde{m}\, d\tilde{n} - n(\tilde{v}, t)\, d\tilde{v}\int_0^{\infty} b(\tilde{m}, \tilde{v})\, n(\tilde{m}, t)\, d\tilde{m}. \tag{10.102}$$

Making use of the equation for conservation of mass during collision, $\tilde{v} = \tilde{m} + \tilde{n}$, the first term on the right-hand side can be transformed into a more tractable form

$$\tfrac{1}{2}\int_0^{\tilde{v}}\int_0^{\tilde{m}} b(\tilde{m}, \tilde{n})\, n(\tilde{m}, t)\, n(\tilde{n}, t)\, d\tilde{m}\, d\tilde{n} = \frac{d\tilde{v}}{2}\int_0^{\tilde{v}} b(\tilde{m}, \tilde{v} - \tilde{m})\, n(\tilde{m}, t)\, n(\tilde{v} - \tilde{m}, t)\, d\tilde{m}.$$

Substituting this relation in eqn. (10.102) the rate expression prescribing the changes in the continuous spectrum function becomes

$$\frac{\partial n}{\partial t} = \frac{1}{2}\int_0^{\tilde{v}} b(\tilde{m}, \tilde{n})\, n(\tilde{m}, t)\, n(\tilde{v} - \tilde{m}, t)\, d\tilde{m} - n(\tilde{v}, t)\int_0^{\infty} b(\tilde{m}, \tilde{v})\, n(\tilde{m}, t)\, d\tilde{m}. \tag{10.103}$$

This is entirely equivalent to the expression

$$\frac{dn_k}{dt} = \frac{1}{2}\sum_{\substack{j=1\\ i=k-j}}^{i=k-1} b_{ij} n_i n_j - n_k \sum_{i=1}^{\infty} b_{ik} n_i \tag{10.1}$$

For the discrete case, the complete spectrum is determined by an interlocking set of ordinary non-linear differential equations, while the continuous spectrum is determined by a single non-linear integro-differential equation.

The model for the continuous spectrum might be considered less realistic physically than a model using a discrete formulation. However, there is arbitrariness in the discrete model in the division into size classes based on a "single" sphere. The continuous approach has proved to be more useful analyzing the behavior of coagulating sols because in most cases, measured size spectra are observed to be distributed over a range of size so wide as to make the discrete model unmanageable. Equations (10.1) and (10.103) apply only to the case of a uniformly distributed non-sedimenting cloud confined in an enclosure. These expressions can be modified to include sources and sinks for particles[382] and loss by sedimentation.[361]

A. Solutions for the Coagulation Equations

Because of the complexities in the form for the rate equations for coagulation, no general solutions have been found with arbitrary collision parameter b, either for the discrete spectrum, or for the continuous distribution. However, there are a number of special cases which have been analyzed in some detail.

Some Analytical Solutions

The earliest and perhaps the only exact analytical solution to the set of equations describing the discrete spectrum as given by eqn. (10.1) was found by Smoluchowski.[282] In this case, an initially monodisperse sol is considered in which the collision parameter b_{ij} is a constant b_0. The solution to eqn. (10.1) under these conditions is

$$\frac{n_k(t)}{N_0} = \frac{(2b_0tN_0)^{k-1}}{(2 + 2b_0tN_0)^{k+1}}, \tag{10.104}$$

where N_0 is the initial concentration of particles present in a unit volume.

Frequently, the decay with time of the total concentration of particles is of interest, especially as a possible method for verifying the theory by experiment. The rate of change in the total number of particles per unit volume $N(t)$ may be obtained easily by summing n_k over all values of k:

$$\frac{dN(t)}{dt} = \frac{d\sum_{k=1}^{\infty} n_k}{dt} = \tfrac{1}{2}\sum_{k=1}^{\infty}\sum_{\substack{j=1\\ i=k-j}}^{i=k-1} b_{ij}n_in_j - \sum_{k=1}^{\infty} n_k \sum_{j=1}^{\infty} b_{kj}n_j = -\tfrac{1}{2}\sum_{i=j}^{\infty}\sum_{j=1}^{\infty} b_{ij}n_in_j. \tag{10.105}$$

Similarly, in the continuous spectrum,

$$\frac{dN(t)}{dt} = -\frac{1}{2}\int_0^{\infty}\int_0^{\infty} b(\tilde{m}, \tilde{v})\, n(\tilde{m}, t)\, n(\tilde{v}, t)\, d\tilde{m}\, d\tilde{v}. \tag{10.106}$$

For Smoluchowski's discrete case, where $b_{ij} = \text{const}$, $N = N_0$ initially, and

$$\frac{dN}{dt} = -b_0N^2, \tag{10.107}$$

or

$$\frac{N}{N_0} = \frac{1}{1 + 2b_0tN_0}. \tag{10.108}$$

A similar expression also is found for the integrals in eqn. (10.106).

In much of the early literature on coagulation of sols, eqn. (10.108) was rewritten as

$$\frac{1}{N(t)} - \frac{1}{N_0} = 2b_0t, \tag{10.109}$$

indicating that, for a constant value of the collisional parameter, a plot of $1/N$ with t should be a straight line where slope is $2b_0$.

Experimental studies of aerosol coagulation reviewed over a period of many years by Fuchs,[2] Green and Lane,[4] and Whytlaw-Gray and Patterson[271] have noted that the change in N^{-1} for many aerosol systems presumed to be coagulating only by Brownian motion is linear with time. This observation has been widely used to evaluate b_0, in attempting to learn the ranges of applicability of Smoluchowski's theory for Brownian motion.

Without correcting for the effect of Knudsen number on collision rates, Smoluchowski's theory would predict the constant

$$b_0 = 4kT/3\pi\mu_G. \tag{10.110}$$

We realize, however, by considering eqns. (10.1) and (10.103) that the parameter b is nearly always a function of the volume of colliding particles in aerosol clouds. The only exception might be the earliest stages of coagulation of a monodispersed aerosol. Thus the relation like eqn. (10.109) should not hold in general. However, a linear plot as predicted by eqn. (10.109) often fits the observations satisfactorily. The actual values of the constant b_0 usually are larger than estimated by eqn. (10.110) because of electrical interaction, spurious air currents, the increase by Kn_i, and the correction for polydispersivity. Fuchs[2] has found, for example, in the last case that a linear relation like eqn. (10.109) should appear if $Kn_i < 0.10$ even for a polydispersed aerosol. Fuchs suggests from averaging combinations of particle size and values of b_{ij} that the effect of polydispersivity should not increase b_0 by more than about 10 per cent. Gillespie[383] also estimated that the increase in coagulation rate in polydispersed sols should not exceed the order of 5 per cent. These results give an explanation why the highly simplified theory for coagulation by Brownian motion fits experimental observation fairly satisfactorily despite the poor applicability of the theoretical assumptions.

The examination of the decay with time of particle concentration N to verify the applicability of the theory of collisions by Brownian motion is not adequate. This method is rather insensitive to the actual details of the statistics of interaction given in the theoretical formulation. One more sensitive test of the coagulation theory would center attention on the changes of concentration of species of a particular size range, say $n_k(t)$ or $n(\tilde{v}, t)$ rather than $N(t)$. Another more sensitive test would consist of an examination of the collision process between two beams of particles of essentially the same size, in analogy to molecular beam experiments. Though aerosol beams have been invented,[384, 385] no studies of collision processes in these streams have been reported as yet.

The change of N^{-1} with time exceeds $2b_0$ by much more than 10 per cent in many cases and must be explained by other effects. Until recently even the total particle charging in most experiments has not been monitored, let alone the distribution of charge on particles of a certain size. We have found on theoretical grounds, for example, that bipolar charging, among other effects, can substantially increase coagulation rates over the rates computed from Brownian motion alone. However, the exact details of the interaction of charged particles has not been thoroughly studied experimentally as yet.

Because of its wide practical utility, the integro-differential equation for the behavior of the continuous spectrum has been studied analytically in much more detail than the discrete case. Recently, for example, Martynov and Bakanov[386] have demonstrated that it is possible to find general solutions to eqn. (10.103) in the form of power series by the method of series decomposition. Except for special forms of $b(\tilde{m}, \tilde{n})$, this technique suffers, however, from the fact that the power series required to specify $n(\tilde{v}, t)$ needs as many terms as the number of classes of particles in the size distribution, which continually becomes more diffuse in time. Therefore, this method seems to be practical, in general, only for the initial coagulation of a rather narrow size spectrum.

Analytical solutions to eqn. (10.103) for the cases of $b(\tilde{m}, \tilde{n}) = \text{const.} = b_0$, $b(\tilde{m}, \tilde{n}) = b_1(\tilde{m} + \tilde{n})$, and $b(\tilde{m}, \tilde{n}) = b_2(\tilde{m}\tilde{n})$ have been found for arbitrary initial conditions by Martynov and Bakanov,[386] Melzak,[387] and Scott.[388] Melzak and Scott's analysis made use of the theory of Laplace transformations for solving the kinetic equation, eqn. (10.103), while Martynov and Bakanov's results stem from their method of series decomposition discussed above.

A sample of the solutions obtained by these workers is listed in Table 10.5. For uniformity, these equations are presented in dimensionless form using the parameters:

distribution function $$\tilde{\phi}(\tilde{x}, t) = \left(\frac{N_0}{\tilde{v}_0}\right)^{-1} n(\tilde{v}, t), \tag{10.111}$$

size $$\tilde{x} = \tilde{v}/\tilde{v}_0, \tag{10.111b}$$

where $$\phi/N_0 = (1/N_0) \int \tilde{v} n \, \mathrm{d}\tilde{v} = \tilde{v}_0. \tag{10.111c}$$

[TAB]LE 10.5. GENERAL SOLUTIONS FOR THE KINETIC EQUATIONS OF COAGULATION FOR THE CASES WHERE THE COLLISION PARAMETER b EQUALS b_0, A CONSTANT; $b_1\ (\tilde{m} + \tilde{n})$ AND $b_2\ ((\tilde{m}\tilde{\eta})$

[E]quation	Reference	Collision parameter	Time variable	Solution
[(1]0.112)	(386)	Arbitrary $b(\tilde{m}, \tilde{n})$	$\tau = 1 - \dfrac{N}{N_0}$	$\tilde{\phi}(\tilde{x}, \tau) = \sum_{n=0}^{\infty} \dfrac{1}{n!} \tilde{\phi}_n(\tilde{x})\, \tau^n$, where $\tilde{\phi}_n(\tilde{x}) = \dfrac{\partial^{n-1}}{\partial \tau^{n-1}} \left[\dfrac{\int_0^{\tilde{x}} b(\tilde{y}, (\tilde{x} - \tilde{y}))\, \tilde{\phi}(y, \tau)\, \tilde{\phi}(\tilde{x} - \tilde{y}, \tau)\, \mathrm{d}\tilde{y} - 2 \int_0^{\infty} \dots}{\int_0^{\infty} \int_0^{\infty} b(\tilde{y}, \tilde{x})\, \tilde{\phi}(\tilde{y}, \tau)\, \tilde{\phi}(x, \tau)\, \mathrm{d}\tilde{y}\, \mathrm{d}\tilde{x}} \right]_{\tau = 0}$
[(1]0.113)	(388)	$b = b_0$	$\tau = \dfrac{2}{b_0 N_0 t + 2}$	$\tilde{\phi}(\tilde{x}, \tau) = \tau^2 \sum_{k=0}^{\infty} (1 - \tau)^k \dfrac{1}{2\pi i} \int_c [f(p, 0)]^{k+1} \mathrm{e}^{p\tilde{x}}\, \mathrm{d}p$
[(1]0.114)	(388)	$b = b_1(\tilde{m} + \tilde{n})$	$\tau = 1 - \mathrm{e}^{b_1 \tilde{\phi} t}$	$\tilde{\phi}(\tilde{x}, \tau) = \mathrm{e}^{-\tilde{x}\tau} \sum_{k=0}^{\infty} \dfrac{(\tilde{x}\tau)^k}{(k+1)!} \dfrac{1}{2\pi i} \int_c [f(p, 0)]^{k+1} \mathrm{e}^{p\tilde{x}}\, \mathrm{d}p$
[(1]0.115)	(388)	$b = b_2(\tilde{m}\tilde{n})$	$\tau = \dfrac{b_2}{2} \tilde{\phi} \tilde{v}_0 t$	$\tilde{\phi}(\tilde{x}, \tau) = \dfrac{\mathrm{e}^{-2\tilde{x}(\tau - 1)}}{\tilde{x}} \sum_{k=0}^{\infty} \dfrac{[2\tilde{x}(-1)]^k}{(k+1)!} \dfrac{1}{2\pi i} \int_c \left[-\dfrac{\partial f(p, 0)}{\partial p} \right]^{k+1} \mathrm{e}^{p\tilde{x}}\, \mathrm{d}p$

* The Laplace transformation is defined as:

$$f(p, \tau) = \int_0^{\infty} \mathrm{e}^{-p\tilde{x}}\, \tilde{\phi}(\tilde{x}, \tau)\, \mathrm{d}\tilde{x}$$

$\tilde{\phi}(\tilde{x}, \tau) = \int_0^{\infty} \mathrm{e}^{-p\tilde{x}} f(p, \tau)\, \mathrm{d}p$, where the integral is taken over the contour C parallel to the imaginary axis to the right of all singularities in $f(p, \tau)$.

The corresponding forms for the change in total concentration of particles N for the solutions listed in Table 10.5 are tabulated in Table 10.6.

We note in general that the solutions for the distribution function are quite complicated, but they contain exponential forms, and appear to evolve monotonically if the initial distribution is monotonic. That is, an initial shape of $\tilde{\phi}$ corresponding, for example, to a log-normal distribution will tend to maintain a shape containing a single peak, without developing any lumps, or secondary maxima.

TABLE 10.6. SOLUTIONS FOR THE CHANGE OF TOTAL PARTICLE CONCENTRATION WITH TIME FOR DIFFERENT COLLISION MECHANISMS

Reference	Collision parameter	Solution for $N(t)$
(282), (386) (387)	$b = b_0$	$\frac{1}{N} - \frac{1}{N_0} = 2b_0 t$
(386), (388)	$b = b_1(\tilde{m} + \tilde{n})$	$\frac{N}{N_0} = e^{-b_1 \phi t}$
(386), (388)	$b = b_2(\tilde{m}\tilde{n})$	$\frac{N}{N_0} = 1 - \frac{b_2}{2} \tilde{\phi} \tilde{v}_0 t$

The decrease in total number of particles with time only behaves in the classical manner of Smoluchowski when the collision parameter is constant. Other classes of solution display quite different curves for N as a function of time. It is interesting to note that the decay rate is independent of the initial distribution in all cases. This was first elucidated apparently by Martynov and Bakanov.[386]

To illustrate some specific forms of distribution functions, a sampling of particular solutions to (10.103) is listed in Table 10.7. Examples of analytical solutions corresponding to the three different forms of the collisional parameter are included.

Again these solutions display similar properties to the general cases in Table 10.5. The distribution functions have rather complicated forms, but uniformly display mixtures of exponential functions as multiples to a power series expression.

The asymptotic behavior at very long times of the solution for $b = c_1(\tilde{m} + \tilde{n})$ is of interest. Golovin[389] has shown that eqn. (10.120) in limiting long times reduces to

$$\underset{\lim t \to \infty}{n(\tilde{v}, t)} = \left(\frac{N_0}{2}\right)\left(1 - \frac{N_0}{N}\right) \left(\frac{\tilde{v}_0}{\pi}\right) \tilde{v}^{-\frac{3}{2}}. \tag{10.122}$$

This result is of interest because it indicates how a size distribution may evolve into a power law form from initially an exponential shape.

Sample curves corresponding to the evolution of the spectrum functions with time for different collision parameters are shown in Fig. 10.11*A* and *B*. These indicate how the spectra tend to spread by collision processes in the dimensionless coordinates of eqn. (10.111). The peak in the distribution for $b = b_1(\tilde{m} + \tilde{n})$ tends to drift towards smaller values of $\tilde{x}$, and relatively little spreading of $\tilde{\phi}$ develops with time. However, for $b = b_0$, a constant, the peak remains essentially at the same value of $\tilde{x}$, but a significantly larger spreading, or growth of larger particles is observed in time.

BLE 10.7. SOLUTIONS TO THE KINETIC EQUATIONS OF COAGULATION FOR DIFFERENT MECHANISMS OF COLLISION AND DIFFERENT INITIAL CONDITIONS

quation	Reference	Collision parameter	Initial condition	Solution
0.116)	(387)	$b = \text{const.} = b_0$	$n(\tilde{v}, 0) = N_0\delta(\tilde{v} - 1)$	$n(\tilde{v}, t) = \dfrac{b_0 N_0}{b_0(b_0N_0/2 + 1)^2} \sum_{j=1}^{\infty} \left(\dfrac{b_0N_0}{b_0N_0 + 2}\right)^j \delta(\tilde{v} - (j + 1))$
0.117)	(386)	$b = b_0$	$n(\tilde{v}, 0) = A_1\tilde{v}^j e^{-\alpha_2\tilde{v}}$ $\alpha_2 = j + 1/\tilde{v}_0,$ $A_1 = \text{const.}$	$n(\tilde{v}, t) = \dfrac{2e^{-\alpha_2\tilde{v}} X_1}{b_0 t(X_2 t + 1)(K + 1)} \sum_{j=1}^{K+1} \xi^j \exp(\xi^j X_1 \tilde{v})$ $X_2 = \dfrac{b_0 A_1 K!}{2\alpha_2^{K+1}}, X_1 = \left[\dfrac{b_0 K!\, tA_1}{2(X_2 t + 1)}\right]^{1/K+1}$ $\xi = e^{2\pi i/K+1}$
0.118)	(388)	$b = b_1(\tilde{m} + \tilde{n})$	$n(\tilde{v}, 0) = N_0\delta(\tilde{v} - 1)$	$n(\tilde{v}, t) = \dfrac{N_0}{\tilde{v}_0}[\exp[-\tilde{v}/\tilde{v}_0(1 - N/N_0)\,]$ $\sum_{k=0}^{\infty} \dfrac{\dfrac{\tilde{v}}{\tilde{v}_0}\left(1 - \dfrac{N}{N_0}\right)^k}{(k + 1)!} \delta(\tilde{v} - (k + 1))$
0.119)	(389)	$b = b_1(\tilde{m} + \tilde{n})$	$n(\tilde{v}, 0) = (N_0/\tilde{v}_0)e^{-\tilde{v}/\tilde{v}_0}$	$n(\tilde{v}, t) = \dfrac{N \exp[-\tilde{v}/\tilde{v}_0(2 + N/N_0)]}{(\tilde{v}/\tilde{v}_0)(1 - N/N_0)^{\frac{1}{2}}} I_1 [2(\tilde{v}/\tilde{v}_0)(1 - N/N_0)^{\frac{1}{2}}]$ where I_1 = modified Bessel function of the first kind.
0.120)	(388)	$b = b_2(\tilde{m}\tilde{n})$	$n(\tilde{v}, 0) = N_0\delta(\tilde{v} - 1)$	$n(\tilde{v}, t) = \dfrac{N_0}{\tilde{v}_0} \exp\left[-\dfrac{\tilde{v}}{\tilde{v}_0}\left(\dfrac{b_2}{2}\phi\tilde{v}_0 t - 1\right)\right]$ $\times \sum_{k=0}^{\infty} \dfrac{\left[\dfrac{2\tilde{v}}{\tilde{v}_0}\left(\dfrac{b_2}{2}\phi\tilde{v}_0 t - 1\right)\right]^k \delta(\tilde{v} - (k + 1))}{(k + 1)!}$
0.121)	(388)	$b = b_2(\tilde{m}\ \tilde{n})$	$n(\tilde{v}, 0) = \left(\dfrac{N_0}{\tilde{v}_0}\right) e^{-\tilde{v}/\tilde{v}_0}$	$n(\tilde{v}, t) \approx \dfrac{2N_0}{\tilde{v}_0}\left(\dfrac{\tilde{v}}{v_0}\right)^{-\frac{5}{2}} [6\pi]^{\frac{1}{2}} [2b_2N_0 t\tilde{v}_0^2]^{-\frac{5}{6}}$ $\times \exp\left\{-\dfrac{\tilde{v}}{\tilde{v}_0}(b_2N_0\tilde{v}_0^2 t + 1) + \dfrac{3\tilde{v}}{\tilde{v}_0}(b_2N_0\tilde{v}_0^2 t)^3 [\tfrac{1}{2}]^{\frac{2}{3}}\right\}$ $\times \left\{1 - \dfrac{9}{72(\tilde{v}/\tilde{v}_0)(b_2N_0\tilde{v}_0^2 t)^{\frac{1}{3}}(2)^{\frac{1}{3}}}\right\}$

The complete analytical solutions reported so far represent rather unrealistic models physically. We have observed that the case where the collision parameter is constant may have some relevance to the initial stages of coagulation by Brownian motion of a monodisperse sol. The collisional parameter $b = b_1(\tilde{m} + \tilde{n})$ may have some bearing on the behaviour of particle growth by shearing action in the fluid, or by turbulence. These examples both yielded b proportional to particle volume. However, because of the complexities in $b(\tilde{m}, \tilde{n})$, much of the information about size spectra of aerosols for more realistic collisional mechanisms has to be derived from numerical integrations of the kinetic equations.

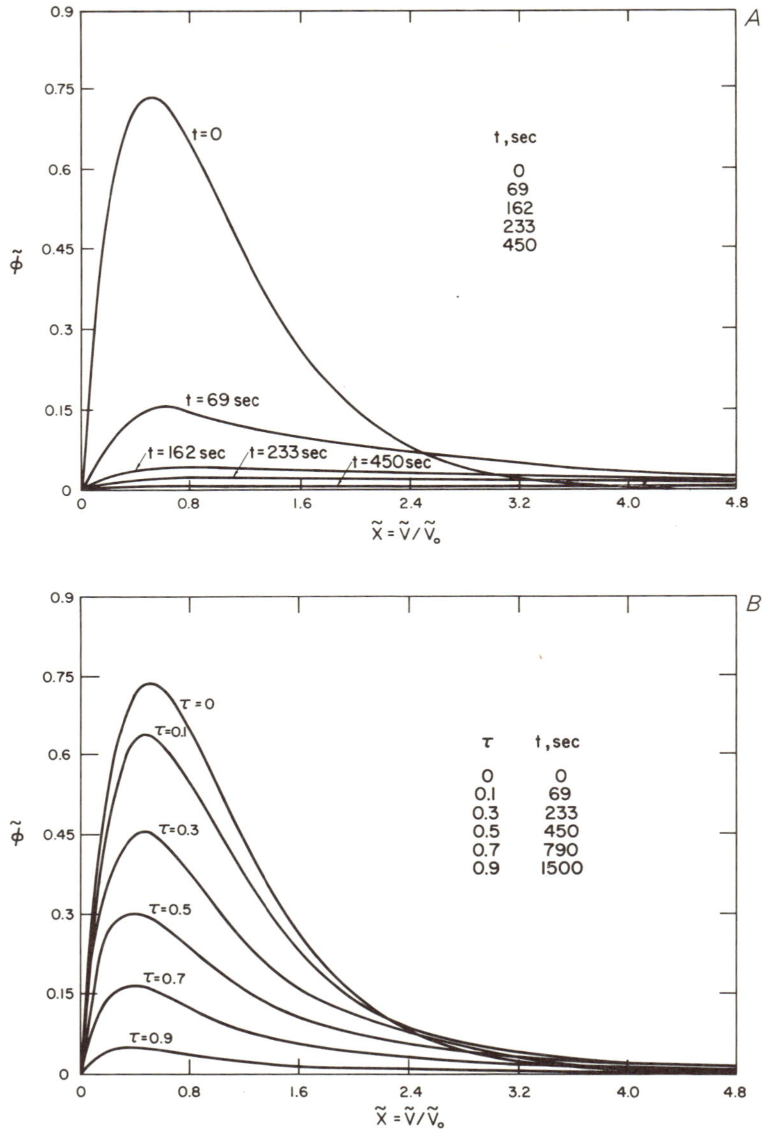

FIG. 10.11. Evaluation of size distribution functions for two different hypothetical collision processes. *A*. Collision parameter b = const., *B*. Collision parameter $b = b_1(\tilde{m} + \tilde{n})$. (From Scott[388]).

There are difficulties in obtaining numerical solutions for the coagulation equation because of its non-linearities. It has been found that integration of equations like eqn. (10.103) require great lengths of computer time even for the largest machines. Methods have been developed to take into account the behavior of a highly polydispersed sol by using a logarithmic scale for particle size.[390] However, the bookkeeping required to carry out integrations of eqn. (10.103) in time is still formidable.

Recently Thompson[391] has found that a generalized function

$$s(\tilde{v}, \gamma) = e^{-\gamma\tilde{v}} - 1 \tag{10.123}$$

yields upon multiplying both sides of eqn. (10.103), and integrating from zero to infinity

$$2\frac{\partial}{\partial t}\int_0^\infty s(\tilde{v}, \gamma)\, n(\tilde{v})\, d\tilde{v} = \int^\infty\int^\infty s(\tilde{v}, \gamma)\, s(\tilde{m}, \gamma)\, n(\tilde{v})\, n(\tilde{m})\, b(\tilde{v}, \tilde{m})\, d\tilde{v}\, d\tilde{m}. \tag{10.124}$$

One then can define a new dependent variable, which is an integral transform of the continuous droplet size spectrum:

$$\mathcal{N}(\gamma, t) = \int_0^\infty s(\tilde{v}, \gamma)\, n(\tilde{v}, t)\, d\tilde{v}. \tag{10.125}$$

Then eqn. (10.124) is a partial differential equation in $\mathcal{N}$ from which the special solution for the collision parameter $b = b_0, b = b_1(\tilde{m} + \tilde{n})\ b = b_2(\tilde{m}\tilde{n})$ may be deduced in a straightforward way. In taking more general expressions for b, Thompson[391] noted that eqn. (10.124) represents essentially the expression for the time derivative of the transform $\mathcal{N}$ as a sum of products of integrals, each of which is related to *derivatives* of the transform.

Even though the transformed coagulation equation remains non-linear (except in special cases of b) and has not been solved in closed form, it can be integrated numerically by finite-difference techniques. By including the most time-consuming numerical operation of integral calculation into the definition of the dependent variable $\mathcal{N}$, it is possible to reduce the multiplications per time stage involved in the numerical integration from 10^6 to 10^3–10^4. This substantial reduction factor may allow much more complex models of cloud evolution to be studied in present computers, than have been developed till now. Further explorations of Thompson's scheme may reveal numerical methods for treating condensation and coagulation simultaneously in a turbulent cloud. Such a procedure will be a useful alternative to other analytical methods being developed independently, for example, by Levin and Sedunov.[392]

With these considerations in mind about the limitations of numerical computations for solutions to the coagulation equation, let us outline briefly the results of such studies.

Some Numerical Solutions

Limited numerical experiments using eqn. (10.103) have been carried out for particular initial distributions, and for the case of Brownian motion of relatively large particles as early as 1958 by Zebel.[393] This work illustrated the evolution of size distributions, which revealed properties quite similar to those found in the analytical solutions, shown in Fig. 10.11. Some calculations for the evolution of cloud droplet population by coalescence have been carried out by Bartlett[436] using Hocking's collision efficiencies. More recently, Berry[390] has begun extensive numerical studies of various collision parameters associated especially with growth of droplets in clouds of sizes greater than about 20 μ. Twomey[394] also has undertaken some numerical experiments in this size range of particles. Using the model of gravitational sweep out, he has reported that the statistical interaction between droplets of different sizes can lead to growth of a meteorologically significant number of very large droplets in clouds at a rate of ten times that of the average growth of particles. This agrees substantially with a cruder analytical model reported earlier by Telford.[395]

A few computations have been made using the discrete form of the coagulation equation. For the case of Smoluchowski's theory, Fair and Gemmel[(396)] have reported a few limited results of integrations of eqn. (10.1). A somewhat more comprehensive numerical study has been given by Hidy,[(397)] who examined the properties of numerical solutions to eqn. (10.1) with Smoluchowski's relation for b_{ij} as corrected for Kn_i by the Stokes–Cunningham relation. These integrations were carried out mainly for initially monodisperse sols, but a few samples of polydispersed cases were also considered. The results of Hidy's study verified Todes'[(398)] idea that bumps and vacancies in the size distribution were filled in by the coagulation process in such a way as to smooth out the size spectrum as time increased. The effect of increasing heterogeneities in an initially monodisperse sol tended to effectively increase the rate of coagulation over that estimated for the limiting case of $b_0 = 4kT/3\pi\mu_G$, and tended to spread the size spectrum more than Smoluchowski's classical solution, eqn. (10.104) would predict. This can be seen easily in the data shown in Table 10.8. Here it is convenient to use the dimensionless time

$$\tilde{\tau} = 2\pi D_i R_i N_0 t \tag{10.126}$$

and the dimensionless concentration of species k

$$\nu_k = n_k/N_0. \tag{10.127}$$

Although there are relatively small differences between Smoluchowski's result for $N(t)$ and the numerical results listed in Table 10.8, the changes in the relative concentrations of individual classes of particles differed substantially from the classical solution for constant collision parameter b. This provides further evidence that truly sensitive tests of the theory for coagulation must lie in measurement of changes in the size spectra, and not in the changes of the total number of particles per unit volume.

A rather surprising conclusion was obtained from these numerical studies, however. It was found that Smoluchowski's results matched the results of the numerical computation most closely in the region where $(d\nu_k/d\tilde{\tau})_k$ is maximum, corresponding to a time $\tilde{\tau}_m = (k-1)/8$. Some typical values of the dimensionless concentration ν_{km} (for $\tilde{\tau}_m$) have been tabulated in Table 10.9.

TABLE 10.9. COMPARISON BETWEEN THE MAXIMUM VALUES OF THE DIMENSIONLESS CONCENTRATION OF PARTICLES ν_k FROM NUMERICAL INTEGRATIONS OF EQN. (10.1), AND THOSE ESTIMATED FROM EQN. (10.104), BASED ON b = const. $= 4kt/3\pi\mu_G$ (FROM HIDY [397])

k	$\tilde{\tau}_m$	Smoluchowski (ν_{km})	Numerical (ν_{km})
1	0	1.000	1.000
2	0.125	0.148	0.145
4	0.375	0.0346	0.0343
6	0.625	0.0152	0.0151
8	0.875	0.00850	0.00846
10	1.125	0.00543	0.00540
20	2.375	0.00135	0.00135
30	3.625	0.000602	0.000600
40	4.875	0.000338	0.000338
50	6.125	0.000217	0.000216

TABLE 10.8. SOLUTION TO THE DISCRETE FORM OF THE KINETIC EQUATIONS FOR COAGULATION IN THE CASE WHERE $Kn_i = 0$, AND THE INITIAL DISTRIBUTION IS MONODISPERSED. THE RESULTS OF THE NUMERICAL INTEGRATION ARE COMPARED WITH SMOLUCHOWSKI'S SOLUTION FOR n_k BASED ON EQN. (10.104), AND FOR THE TOTAL CONCENTRATION OF PARTICLES, $N(t)$ (FROM HIDY[397])

Time $\tilde{\tau}$	Smoluchowski				Numerical			
	ν_1	ν_{10}	ν_{100}	$\Sigma\nu_k$	ν_1	ν_{10}	ν_{100}	$\Sigma\nu_k$
0.00	1.000	0.000	0.000	1.000	1.000	0.000	0.000	1.000
0.20	0.309	2.09×10^{-4}	4.20×10^{-36}	0.555	0.307	2.28×10^{-4}	5.68×10^{-40}	0.553
0.40	0.148	1.87×10^{-3}	1.97×10^{-22}	0.385	0.146	1.96×10^{-3}	2.25×10^{-22}	0.381
0.60	0.0865	3.67×10^{-3}	0.15×10^{-17}	0.294	0.0843	3.86×10^{-3}	2.08×10^{-16}	0.290
0.80	0.0567	4.90×10^{-3}	1.15×10^{-13}	0.238	0.0545	4.95×10^{-3}	2.39×10^{-13}	0.234
1.00	0.0400	5.37×10^{-3}	10.02×10^{-11}	0.200	0.0379	5.37×10^{-3}	1.88×10^{-11}	0.196
2.00	0.0123	4.28×10^{-3}	1.06×10^{-7}	0.111	0.0108	4.19×10^{-3}	1.46×10^{-7}	0.107
3.00	5.92×10^{-3}	2.88×10^{-3}	2.14×10^{-6}	0.0769	4.81×10^{-3}	2.80×10^{-3}	2.60×10^{-6}	0.0738
4.00	3.46×10^{-3}	2.00×10^{-3}	8.56×10^{-6}	0.0588	2.61×10^{-3}	1.94×10^{-3}	9.76×10^{-6}	0.0562
5.00	2.27×10^{-3}	1.46×10^{-3}	1.81×10^{-5}	0.0476	1.58×10^{-3}	1.40×10^{-3}	1.98×10^{-5}	0.0453
6.00	1.60×10^{-3}	1.11×10^{-3}	2.81×10^{-5}	0.0400	1.04×10^{-3}	1.06×10^{-3}	3.00×10^{-5}	0.0380
7.00	1.19×10^{-3}	8.67×10^{-4}	3.69×10^{-5}	0.0345	7.26×10^{-4}	8.24×10^{-4}	3.85×10^{-5}	0.0327
8.00	9.18×10^{-4}	6.96×10^{-4}	4.36×10^{-5}	0.0303	5.25×10^{-4}	6.57×10^{-4}	4.50×10^{-5}	0.0287
9.00	7.30×10^{-4}	5.71×10^{-4}	4.85×10^{-5}	0.0270	3.92×10^{-4}	5.36×10^{-4}	4.95×10^{-5}	0.0255
10.00	5.95×10^{-4}	4.76×10^{-4}	5.17×10^{-5}	0.0244	3.00×10^{-4}	4.44×10^{-4}	5.22×10^{-5}	0.0230

TABLE 10.10. TABULATION OF VALUES OF ν_k VS. $\tilde{\tau}$ FOR TWO DIFFERENT VALUES OF THE KNUDEN RATIO, λ_G/R_1 (FROM HIDY[397]).

Time $\tilde{\tau}$	$\lambda_G/R_1 = 0.10$				$\lambda_G/R_1 = 1.00$			
	ν_1	ν_{10}	ν_{100}	$\Sigma\nu_k$	ν_1	ν_{10}	ν_{100}	$\Sigma\nu_k$
0·00	1.000	0.000	0.000	1.000	1.000	0.000	0.000	1.000
0.20	0.275	3.33×10^{-4}	1.75×10^{-34}	0.526	0.119	2.08×10^{-3}	4.41×10^{-22}	0.357
0.40	0.126	2.38×10^{-3}	1.09×10^{-20}	0.356	0.0419	5.11×10^{-3}	3.17×10^{-14}	0.223
0·60	0.0710	4.27×10^{-2}	4.27×10^{-15}	0.269	0.0203	5.72×10^{-3}	5.25×10^{-11}	0.165
0.80	0.0452	5.21×10^{-3}	7.15×10^{-12}	0.216	0.0116	5.36×10^{-3}	2.64×10^{-9}	0.131
1.00	0.0311	5.46×10^{-3}	6.51×10^{-11}	0.181	7.28×10^{-3}	4.76×10^{-3}	2.95×10^{-3}	0.109
2.00	8.53×10^{-3}	3.97×10^{-3}	2.53×10^{-7}	0.0993	1.43×10^{-3}	2.46×10^{-3}	3.72×10^{-6}	0.0614
3.00	3.67×10^{-3}	2.59×10^{-3}	3.56×10^{-6}	0.0685	4.82×10^{-4}	1.43×10^{-3}	1.65×10^{-5}	0.0434
4.00	1.95×10^{-3}	1.77×10^{-3}	1.19×10^{-5}	0.0523	2.06×10^{-4}	9.19×10^{-4}	3.13×10^{-5}	0.0338

If the Stokes–Cunningham correction is incorporated into b_{ij} to account approximately for transitional values of Kn_i up to about 1, one finds that the rate of coagulation increases with Kn_i, and the rate of spreading of the size spectrum increases with Kn_i for a given value of dimensionless time $\tilde{\tau}$, as indicated by the data in Table 10.10. Later calculations of Hidy indicated that collisional mechanisms for the simple shearing models, eqns. (10.60) or (10.82), the collision model for uncharged free molecule aerosols, eqn. (10.25), and the model for Brownian motion using Smoluchowski's relation modified by the Stokes–Cunningham correction, eqn. (10.8), all produced a more polydisperse distribution than the classical diffusion model given by Smoluchowski. The spreading effect is shown typically for $\tilde{\tau} \approx 0.3$ in Fig. 10.12.

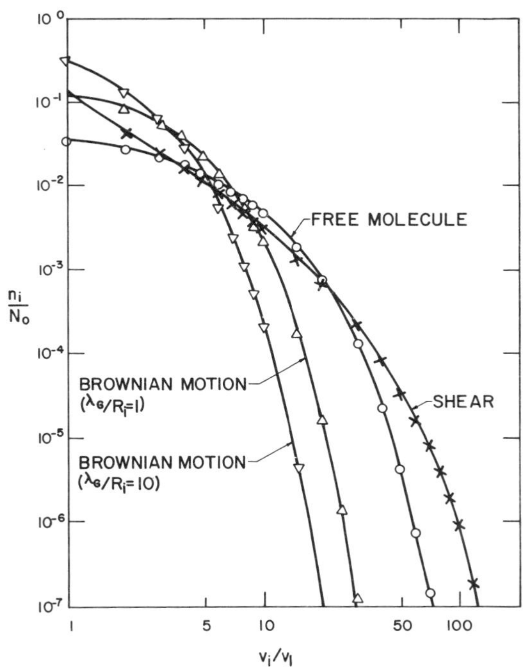

FIG. 10.12. Comparison between the development of the size spectrum of an initially monodisperse aerosol for different collision models. Curves are plotted for dimensionless times $\tilde{\tau} \approx 0.3$. The cases of Brownian motion were calculated using Smoluchowski's theory with the Stokes–Cunningham correction. The case for shear coagulation was calculated for $G_g = 10$ sec^{-1}.

B. Asymptotic Behavior of the Size Spectrum

The prediction of the behavior of aerosols after long periods of time is of practical interest, so that theoretical knowledge of the shapes of asymptotic distributions is of considerable value. It is known that particle size distributions of aerosols tend towards certain asymptotic properties after long periods of time which are independent of initial distributions. We determined from Table 10.3 that $N(t)$ as $t \to \infty$ often can be independent of the initial conditions, but the asymptotic form may depend on the collisional mechanism. The shapes of asymptotic distributions sometimes resemble closely logarithmic normal distributions.

Others like eqn. (10.122) display a power law form. For the case of the constant collisional parameter b, Schumann[(399)] found some years ago, for example, that an initially monodisperse distribution of particles whose size is greater than $\sim 10^{-5}$ cm approached a form $n(\tilde{v}, t) = N^2(t)/\phi \exp(-N(t)\tilde{v}/\phi)$ as $t \to \infty$. He concluded that this form might be the asymptotic solution for any initial distribution. Similarly, Todes[(398)] found that for coagulation of particles of 10^{-5} cm radius or larger, the size distribution becomes exceedingly diffuse over long time periods, so that individual peculiarities in the initial size spectrum are smoothed out, and the spectrum tends to approach an asymptotic form independent of initial conditions. Todes deduced that for $b(\tilde{m}, \tilde{n})$ being a homogeneous function of its arguments, eqn. (10.122) may be transformed into an ordinary integro-differential equation by the use of the parameter $n(\tilde{v}, t) = (\hat{M}/t^2)\,\phi(z)$, where $z = \hat{Q}\tilde{v}/t$, and $\hat{M}$ and $\hat{Q}$ are constants. However, Friedlander and Wang[(400)] have noted that this transformation does not appear to behave properly when $t \to 0$ because the initial distribution having the form $n(\tilde{v}, t) = \hat{M}/t^2)\,\phi(\hat{Q}\tilde{v}/t)$ tends to zero at $\tilde{v} = 0$.

Self-preserving Distributions

In a recent series of reports, Friedlander[(341, 352, 361, 401)] has developed a similarity theory for the uniformity in shapes of observed size spectra. Recognizing that the rate equations for coagulation are very similar to the equations for the spectrum of turbulent energy in a fluid, Friedlander employed dimensional arguments similar to those of Kolmogoroff[(363)] to represent a class of asymptotic size distributions. The similarity transformation proposed by Friedlander and colleagues is

$$n(\tilde{v}, t) = \tilde{g}(t)\,\tilde{\psi}(\tilde{v}/\tilde{v}^+) = \tilde{g}(t)\,\tilde{\psi}(\tilde{\eta}), \tag{10.128}$$

where $g(t)$ and $\tilde{v}^+$ are functions of time. This is the so-called *self-preserving* hypothesis. Substitution of this relation into eqn. (10.103) yields an ordinary integro-differential equation with $\tilde{\eta}$ as the independent variable.[(341)] This demonstrates that eqn. (10.128) is a particular solution to the coagulation equation, but not necessarily the only solution. The reduced spectrum function $\tilde{\psi}(\tilde{\eta})$ is termed "self-preserving" since it depends only on the reduced size parameter $\tilde{v}/\tilde{v}^+$, and not explicitly on time. The choice of the form for $\tilde{\psi}(\tilde{\eta})$ and $\tilde{\eta}$ of Friedlander is quite similar to those of Todes.[(398)]

The nature of the self-preserving solution of Friedlander can be determined by introducing the following non-dimensional variables into eqn. (10.103):

$$n(\tilde{v}, t) = \frac{[N(t)]^2}{\phi}\tilde{\psi}(\tilde{\eta}, \tilde{\gamma}), \tag{10.129a†}$$

$$\tilde{n} = \frac{N(t)\,\tilde{v}}{\phi}, \quad \tilde{\gamma} = \left[\frac{N(t)}{N_0}\right]^2 \tag{10.129b†}$$

† We note that these dimensionless variables are related to those in (10.111) as follows:

$$\tilde{\psi}(\tilde{\eta}, \tilde{v}) = \frac{\phi\tilde{\phi}(\tilde{v}, \tau)}{(1-\tau)}, \qquad \tilde{\eta} = \frac{N_0(1-\tau)\tilde{v}}{\phi}, \qquad \tilde{v} = (1-\tau)^2,$$

where

$$\tau = 1 - N/N_0.$$

The transformation given by Friedlander and Wang[400] now involves no special assumptions about the form of $n(\tilde{v}, t)$ and $\tilde{\psi}(\tilde{\eta}, \tilde{\gamma})$ remains perfectly general. If $b(\tilde{m}, \tilde{n})$ is a homogeneous function of its arguments, eqn. (10.103) reads with eqn. (10.129):

$$\left[\int_0^\infty \int_1^\infty b(\tilde{\eta}, \tilde{\eta}')\, \tilde{\psi}(\tilde{\eta}, \tilde{\gamma})\, \tilde{\psi}(\tilde{\eta}', \tilde{\gamma})\, d\tilde{\eta}\, d\tilde{\eta}'\right] \left[\tilde{\eta} \frac{\partial \tilde{\psi}(\tilde{\eta}, \tilde{\gamma})}{\partial \tilde{\eta}} + 2\tilde{\psi}(\eta, \gamma) + 2\tilde{\gamma} \frac{\partial \tilde{\psi}(\tilde{\eta}, \tilde{\gamma})}{\partial \tilde{\gamma}}\right.$$

$$\left. + \int_0^{\tilde{\eta}} b(\tilde{\eta}, \tilde{\eta} - \tilde{\eta}')\, \tilde{\psi}(\tilde{\eta}, \tilde{\gamma})\, \tilde{\psi}(\tilde{\eta} - \tilde{\eta}', \tilde{\gamma})\, d\tilde{\eta}' - \int_0^\infty b(\tilde{\eta}, \tilde{\eta}')\, \tilde{\psi}(\tilde{\eta}, \tilde{\gamma})\, \tilde{\psi}(\tilde{\eta}', \tilde{\gamma})\, d\tilde{\eta}'\right] = 0. \tag{10.130}$$

The following integral constraints also are used:

$$\int_0^\infty \tilde{\psi}(\tilde{\eta}, \tilde{\gamma})\, d\tilde{\eta} = 1, \tag{10.131}$$

$$\int_0^\infty \tilde{\psi} \tilde{\eta}\, d\tilde{\eta} = 1, \tag{10.132}$$

which correspond to the expressions for conservation of number of the species and for conservation of their mass (assumed all of the same density).

The only part of the transformed eqn. (10.130) where $\tilde{\gamma}$ explicitly enters is the term $2\tilde{\gamma}(\partial\tilde{\psi}/\partial\tilde{\gamma})$. If this term vanishes, or approaches a universal function of only $\tilde{\eta}$ in the limit $t \to \infty$, the solution to eqn. (10.130) will be self-preserving, and independent of the initial distribution. For

$$\tilde{\gamma} \frac{\partial \tilde{\psi}(\tilde{\eta}, \tilde{\gamma})}{\partial \tilde{\gamma}} \to 0, \quad t \to \infty.$$

To see how the self-preserving transformation yields information about the distribution function, consider the simplest case where $b(\tilde{m}, \tilde{n}) = \text{const}$. Equation (10.128) reduces to

$$2\tilde{\gamma} \frac{\partial \tilde{\psi}}{\partial \tilde{\gamma}} + \tilde{\eta} \frac{\partial \tilde{\psi}}{\partial \tilde{\eta}} + \int_0^{\eta} \tilde{\psi}(\tilde{\eta}, \tilde{\gamma})\, \tilde{\psi}(\tilde{\eta} - \tilde{\eta}', \tilde{\gamma})\, d\tilde{\eta}' = 0. \tag{10.133}$$

The solution to this relation, as found by Laplace transforms[400] for the initial distribution of class

$$\tilde{G}(\tilde{\eta}) = \frac{\tilde{p}^p}{\Gamma(\tilde{p})} \tilde{\eta}^{p-1} e^{-p} \quad (p \text{ positive integer}), \tag{10.134}$$

has the form

$$\tilde{\psi}(\tilde{\eta}, \tilde{\gamma}) = \sum_{k=1}^{p} \frac{\exp\left[(2\pi k i/p) + S_k \tilde{\eta}\right]}{(1 - \sqrt{\tilde{\gamma}})^{(p-1)/p}}, \tag{10.135}$$

where $\tilde{S}_k = p/\sqrt{\tilde{\gamma}}[(1 - \sqrt{\tilde{\gamma}})^{1/p} \exp(2\pi k i/p) - 1]$, i being the imaginary unit. This solution, of course, corresponds to those found by Melzak[387] and Scott.[388]

The limiting form of eqn. (10.135) as $t \to \infty$ may be obtained by letting $\tilde{\gamma} \to 0$, giving

$$\lim_{\tilde{\gamma} \to 0} \tilde{\psi}(\tilde{\eta}, \tilde{\gamma}) = e^{-\tilde{\eta}}. \tag{10.136}$$

This same conclusion was reached independently in examining the numerical results of Hidy and Lilly,[402] and it is equivalent to Schumann's[399] solution.

Wang and Friedlander[403] have shown later that this asymptotic form, eqn. (10.136), holds true in the case of constant b for any initial distribution. This demonstrates that the kinetic equations for coagulation have a universal limiting spectrum which is explicitly independent of time provided the collisional parameter has a limiting simple form.

Limits of the Self-preservation Theory

The collision parameters for systems in which $b(\tilde{m}, \tilde{n}) \neq$ const. are generally too complicated to determine easily analytical solutions of the self-preserving class. We can see, however, from typical examples of solutions for special cases of $b(\tilde{m}, \tilde{n})$ that in general the solutions found for $b_1(\tilde{m} + \tilde{n})$ and $b_2(\tilde{m}\tilde{n})$ are not self-preserving in the limit $t \to \infty$. As an example, this is easily verified by rewriting eqn. (10.122) in the similarity variables.

In the case of Brownian motion there exists evidence that self preserving distributions develop for Smoluchowski's form of the collisional parameter where $Kn_i \to 0$. In the observations of thermal coagulation of hydrosols when Kn_i effectively equals zero, Swift and Friedlander[341] found strong indications that self-preserving distributions will develop. Furthermore, numerical studies[397] have also shed some light on the nature of the self-preserving spectra generated during thermal coagulation. The solutions given, for example in Table 10.8, can be transformed into the similarity form using the following variables:

$$\tilde{\eta}_i = \frac{v_i N}{\phi_i} = \frac{i v_1 N}{\phi_i}, \tag{10.137a}$$

$$\tilde{\psi}(\tilde{\eta}_i) = \frac{n_i \phi_i}{N^2 v_1}, \tag{10.137b}$$

where

$$\phi_i = \sum_{i=1}^{s} v_i n_i. \tag{10.137c}$$

This relation is equivalent to the integral in eqn. (10.111c).

For the case of coagulation by Brownian motion in the limit $Kn_i \to 0$, the numerical results showed an asymptotic behavior after a period of time. Regardless of whether or not the initial distribution was monodisperse or polydisperse, the numerical experiments indicated that the distribution $\tilde{\psi}(\tilde{\eta}_i)$ would have the asymptotic form drawn in Fig. 10.13. These calculations showed that a self-preserving size spectrum develops in a dimensionless time $\tilde{\tau} = 2\pi D_1 R_1 N_0 t \approx 3$. However, under conditions where the initial distribution given by $n_1/N_0 = 0.819$, $n_5/N_0 = 0.173$, $n_{10}/N_0 = 6.74 \times 10^{-3}$, $n_{15}/N_0 = 7.57 \times 10^{-4}$, with all other $n_{i0} = 0$, the tendency towards self-preservation over the entire size distribution took much longer. In this case of a polydisperse sol, the same self-preserving form as found for the initially monodisperse case evolved, but appeared in the high end of the size spectrum first. After $\tilde{\tau} \approx 5$, $\tilde{\psi}(\tilde{\eta}_i)$ was attained for the range of $\tilde{\eta}_i > 0.15$. As time increases, there was a tendency for this size distribution to approach the same form of $\tilde{\psi}$ in lower and lower ranges of $\tilde{\eta}_i$.

Similarly, in cases where λ_G/R_1 was greater than zero, a tendency towards apparent self-preservation was observed in the calculated size spectra. The data from numerical integration of eqn. (10.1) with the addition of the Stokes–Cunningham correction in b_{ij}, given as eqn. (10.10) displayed a spectral shape for $\lambda_G/R_i \neq 0$ rather similar to that in Fig. 10.13. For the case where $\lambda_G/R_1 = 0.100$, the shape of $\tilde{\psi}$ remained virtually unaffected in contrast to the case where $Kn_i \to 0$. But by the time λ_G/R_1 reached unity, the shape of $\tilde{\psi}$ is changed, though it appeared to approach an asymptotic form. Down to $\tilde{\eta}_i \approx 0.25$ $\tilde{\psi}$ remained approximately the same for all cases studied, but in the lower end of the size spectrum below $\tilde{\eta}_i = 0.28$ the curve for $\tilde{\psi}(\lambda_G/R_1 = 1.00)$ was pressed downward.

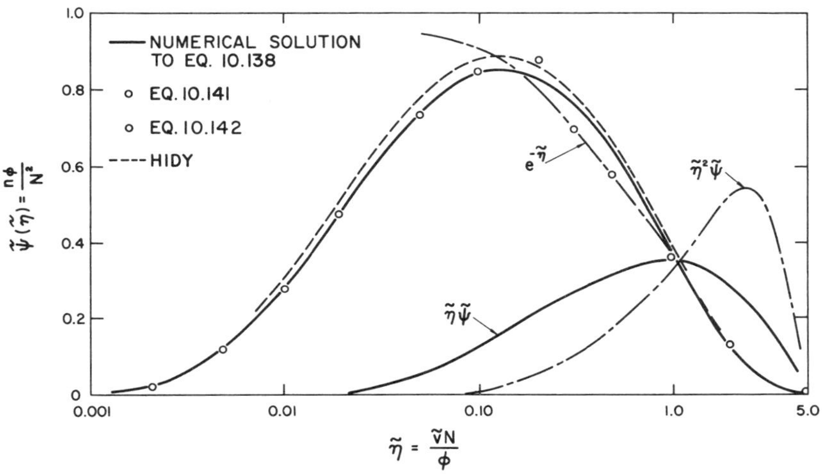

FIG. 10.13. Self preserving size spectrum for aerosol coagulating by Brownian motion. Smoluchowski's theory is assumed, where $\lambda_G/R_i = 0$.

All of the curves displaying an asymptotic similarity form resemble a logarithmic normal distribution. The case where $Kn_i = 0$, is an especially close approximation to this well recognized form.

Encouraged by the conclusions of the numerical experiments,(397) Friedlander and Wang(400) examined the kinetic equations of thermal coagulation in their dimensionless form assuming that $b \neq$ const. but their results apply to uncharged spheres governed by Smoluchowski's collision parameter for $Kn_i \to 0$, eqn. (10.8). Noting that $\tilde{\psi}(\tilde{\eta}, \tilde{\gamma})$ is a function of $\tilde{\eta}$ only, the simplified coagulation equation in transformed variables, become

$$(1 + \tilde{\alpha}\tilde{\beta})\tilde{\eta}\frac{d\tilde{\psi}(\tilde{\eta})}{d\tilde{\eta}} + (2\tilde{\alpha}\tilde{\beta} - \tilde{\beta}\tilde{\eta}^{\frac{1}{3}} - \tilde{\alpha}\tilde{\eta}^{-\frac{1}{3}})\,\tilde{\psi}(\tilde{\eta}) + \int_0^{\tilde{\eta}} \tilde{\psi}(\tilde{\eta})\,\tilde{\psi}(\tilde{\eta} - \tilde{\eta}')\left[1 + \left(\frac{\tilde{\eta} - \tilde{\eta}'}{\tilde{\eta}'}\right)^{\frac{1}{3}} d\tilde{\eta}\right] = 0. \tag{10.138}$$

The two constants in this relation are given by

$$\tilde{\alpha} = \int_0^{\infty} \tilde{\eta}^{\frac{1}{3}}\,\tilde{\psi}(\tilde{\eta})\,d\tilde{\eta}, \tag{10.139}$$

$$\tilde{\beta} = \int_0^\infty \tilde{\eta}^{-\frac{1}{2}} \tilde{\psi}(\tilde{\eta}) \, d\tilde{\eta}. \tag{10.140}$$

Friedlander and Wang solved eqn. (10.138) numerically and obtained the expression shown graphically in Fig. 10.13. Their values for $\tilde{\psi}$ as a function of $\tilde{\eta}$ are listed in Table 10.11. Shown for comparison is the distribution function from Hidy's[397] calculations for the discrete spectrum.

TABLE 10.11. TABULATION OF VALUES OF $\tilde{\psi}$ AND $\tilde{\eta}$ FOR THE SELF-PRESERVING SOLUTION TO THE COAGULATION EQUATION, WHERE $Kn_i = 0$ (AFTER FRIEDLANDER AND WANG[400])

$\tilde{\eta}$	$\tilde{\psi}(\tilde{\eta})$
0.0010	0.0030
0.0050	0.1088
0.0100	0.2645
0.0500	0.7307
0.100	0.8327
0.500	0.6368
1.00	0.3884
5.00	0.0069

Two approximate analytical solutions for the lower and upper extremes are also given as:[400]

lower end:
$$\tilde{\psi}(\tilde{\eta}) = \frac{0.5086}{\tilde{\eta}^{1.06}} \exp(1.758\,\tilde{\eta}^{\frac{1}{2}} - 1.275\,\tilde{\eta}^{-\frac{1}{2}}) \tag{10.141}$$

and

upper end:
$$\tilde{\psi}(\tilde{\eta}) = 0.915 \exp(-0.95\,\tilde{\eta}). \tag{10.142}$$

These approximate expressions are shown in Fig. 10.13. Equation (10.141) fits the numerical solution well up to $\tilde{\eta} \approx 0.01$, while eqn. (10.142) fits the upper extreme down to $\tilde{\eta} \approx 1.0$ satisfactorily.

Friedlander and Wang's results are approximately the same as the values from the numerical experiments. The greatest deviation between the two computations lies in the region where $\tilde{\psi}$ is maximum. It is likely that the differences lie in the inherent cumulative errors of numerical computation over long integration times in the particular scheme used by Hidy,[397] or in the inherent errors of Wang's numerical methods.

In view of the numerical results combined with those of Friedlander and Wang, the upper end of the self-preserving spectrum $\tilde{\psi}$ is of particular interest. Equation (10.142) is essentially the same as the result obtained for the case of constant $b(\tilde{m}, \tilde{n})$, eqn. (10.136). From the comparison of Smoluchowski's results with the numerical computation,[397] as shown in Table 10.9, it is clear that Smoluchowski's solution should be the same as $\tilde{\psi}$ in the region where $(dv_k/d\tilde{\tau})_k$ is maximum. Since Smoluchowski's solution, given by eqn. (10.104), is proportional to $\tilde{\eta}_k^2\tilde{\psi}$, it may be shown[402] that the maximization condition is satisfied best in the range where

$$\left[\frac{\partial(\tilde{\eta}_k^2\tilde{\psi})}{\partial\tilde{\eta}_k}\right] \approx 0. \tag{10.143}$$

It was found that the function $e^{-\tilde{\eta}_k}$ (or $e^{-\tilde{\eta}}$) corresponding to the solution of the coagulation equations for $b(\tilde{m}, \tilde{n}) = \text{const.}$ fits best in the region where eqn. (10.143) applies. That is, Smoluchowski's solution is best in the range $1 \leqslant \tilde{\eta} \leqslant 4$ (see also Fig. 10.13).

It is also of interest that eqn. (10.104) is a satisfactory approximation for both $\tilde{\psi}$, $Kn_i \to 0$, and $\tilde{\psi}$, $Kn_i = 1$, from $1 \leqslant \tilde{\eta}_k \leqslant 8$.[402] Frieldlander and Wang's solution for the upper end of the self-preserving spectrum comes closest to the numerical estimates of $\tilde{\psi}$ over the widest range of the dimensionless volume $\tilde{\eta}$.

Extensions of the work on the theory of self-preserving size distribution has yielded further information about the properties of these idealized forms. For instance, Wang and Friedlander[403] have found that these spectra develop for conditions of simultaneous Brownian motion and shearing flow. The shape of the resulting similarity spectrum is greatly influenced by the assumed collisional mechanism. Suppose we assume that the collisional mechanisms for Brownian motion and shear are additive:

$$b(\tilde{m}, \tilde{n}) = \frac{2}{3}\frac{kT}{\mu_G}(\tilde{m}^{\frac{1}{3}} + \tilde{n}^{\frac{1}{3}})\left(\frac{1}{\tilde{m}^{\frac{1}{3}}} + \frac{1}{\tilde{n}^{\frac{1}{3}}}\right) + \frac{G_q}{\pi}(\tilde{m}^{\frac{1}{3}} + \tilde{n}^{\frac{1}{3}})^3; \tag{10.144}$$

where G_q equals the rate of shear in the fluid. A similarity form is achieved in eqn. (10.102) with $b(\tilde{m}, \tilde{n})$ given by eqn. (10.144), and with the parameter:

$$\tilde{\mathscr{P}} = \frac{3\mu_G G_q \phi}{2\pi k T N} \tag{10.145}$$

expressing the ratios of the magnitude of shear vs. Brownian coagulation.

The similarity spectra for these cases are shown in Fig. 10.14.

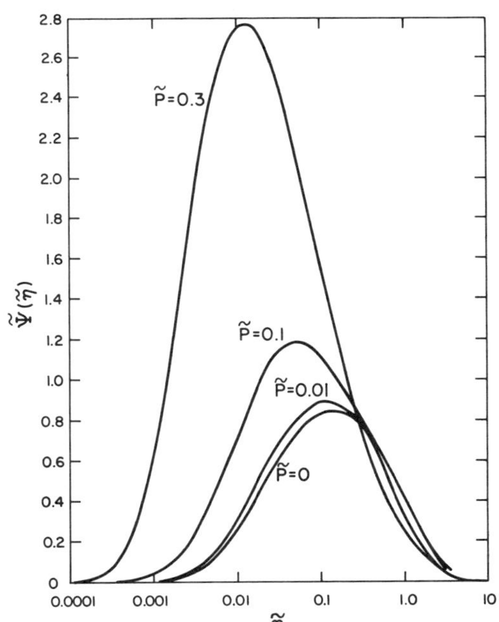

FIG. 10.14. Similarity spectra for collision of aerosols modeled by a linear combination of the mechanism for Brownian motion and shear (From Wang and Friedlander[403]).

Note the strong changes in $\tilde{\psi}(\tilde{\eta})$ for different values of $\tilde{\mathscr{P}}$. Collision only by Brownian motion occurs when $\tilde{\mathscr{P}} = 0$.

Wang and Friedlander's[403] analysis also has shown that Hidy's[397] conclusion that self-preserving spectra exist for conditions where the collision rate is given for finite Knudsen number is not entirely correct. If Smoluchowski's collision parameter, the first term on the right of eqn. (10.144), is corrected for Kn_i with the Stokes–Cunningham

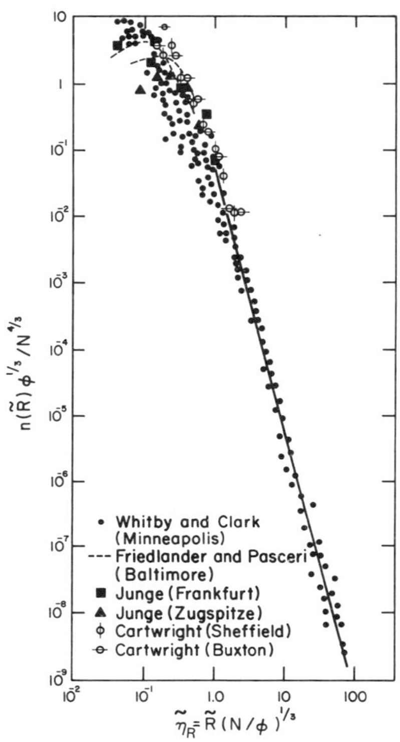

FIG. 10.15. Self-preserving size distributions for atmospheric aerosols sampled near the earth's surface (After Clark and Whitby[404]).

correction, the self-preserving form cannot develop because the term $\tilde{\gamma}\partial\tilde{\psi}/\partial\tilde{\gamma}$ does not approach zero with infinite time in these cases. However, Wang determined that classes of solutions to eqn. (10.102) actually exist that are asymptotically very slowly varying with time. These "quasi-self-preserving" distributions depend on the parameter

$$\tilde{\Lambda} = 1.257\,\lambda_G\left(\frac{4\pi N}{3\phi}\right)^{\frac{1}{3}}. \tag{10.146}$$

For two different values of $\tilde{\Lambda}$, corresponding approximately to the numerical solutions for $\lambda_G/R_1 = 0.100$, and $\lambda_G/R_1 = 1.00$ given by Hidy.[397] Wang has found numerical solutions to eqn. (10.102) written in similarity variables. His estimates of $\tilde{\psi}$ show satisfactory agreements with the numerical solutions to (10.1). Had Hidy been able to continue his integrations much farther in time he should have found that all of his apparent similarity curves would eventually approach the asymptotic form for $Kn_i = 0$.

Experimental Evidence of Self-preservation

At present, there exists no direct verification of the theory of self-preserving distributions in aerosols. Swift and Friedlander[341] have found that this theory applies to hydrosols coagulating by Brownian motion and by simultaneous Brownian motion and shear, where $b(\tilde{m}, \tilde{n})$ was assumed to have the form of eqn. (10.144). Their results, however, depended on cumulative distributions rather than $\tilde{\psi}(\tilde{\eta})$. It is well known that cumulative distributions of the form

$$\tilde{\chi} = \int_{\tilde{\eta}}^{\infty} \tilde{\psi}\, d\tilde{\eta}$$

are much less sensitive to collisional mechanisms than $\tilde{\psi}$ itself.

As mentioned above, Friedlander[352, 401] has found that several spectra of atmospheric aerosols may be placed in a similarity form if it is assumed that the spectral shape is controlled only by Brownian motion and sedimentation. This analysis differs from the self-preserving theory, and relies on the assumption of a quasi-stationary state in the atmospheric aerosol. Recently, Clark and Whitby[404] have reported data showing that the similarity variable in eqn. (10.129) may be used to correlate the distribution of natural aerosols taken at ground level over an exceedingly wide range of $\tilde{\psi}_R$ and $\tilde{\eta}_R$.† Their similarity correlation is drawn in Fig. 10.15. Clark and Whitby determined by their measurements that the similarity spectrum appears to take shape in atmospheric aerosols in matters of hours despite a wide variety of available, and intermittent sources for particulate matter near his sampling station. For comparison, the distributions of other workers are included in Fig. 10.15. The distributions all appear to fit a "universal" form over a range of $\tilde{\eta}$ which is not that predicted by the theory on p. 353–354.

A rather peculiar feature is displayed by the spectra for atmospheric aerosols. If the sampling is taken near the Earth's surface, Junge's[405] power law rule that $\tilde{\psi}_R \propto \tilde{\eta}_R^{-4}$ commonly develops. If the atmospheric aerosol is controlled by coagulation resulting from Brownian agitation, one would expect a closer resemblance to the exponential form $e^{-\tilde{\eta}}$ in the high range of $\tilde{\eta}_R$. Friedlander and Pasceri[406] have compared the similarity spectrum based on the similarity curve of Hidy for $Kn_i = 0$ (i.e. Fig. 10.14) with some very careful measurements of aerosol size distribution taken at ground level in Baltimore, Maryland. This comparison is shown in Fig. 10.16. The ψ_R curve of Hidy is too narrow and falls off too steeply to fit the experimental observations. The reasons for the strong tendency towards the similarity shape shown in Clark and Whitby's data and those of Friedlander and Pasceri remain unresolved. It would be very helpful to have asymptotic size distributions produced experimentally by the coagulation process alone in a laboratory vessel to see if it is possible to realize the similarity spectrum in its near log-normal form, given by Friedlander and Wang,[400] and by Hidy.[397]

Although it does not appear possible to explain the shape of the size spectrum by coagulation alone in a cloud maintaining constant mass of particles the power law form,

$$\tilde{\psi}_R(\tilde{\eta}) = 0.05\, \phi \tilde{\eta}_R^{-4}, \tag{10.147}$$

† Note here that Clark and Whitby's results are given in terms of similarity variables written for the particle radius $\tilde{R}$, In other words, $\tilde{\psi}_R = (36\pi)^{1/3} \tilde{\eta}^{2/3} \tilde{\psi}$, and $\tilde{\eta}_R = (3/4\pi)^{1/3} \tilde{\eta}^{1/3}$.

can be achieved at least qualitatively by introducing sources or sinks of particles. This has been demonstrated, for example by the calculations of Mockros *et al.*[382] Liu and Whitby[407] have noted also that solutions of the class of eqn. (10.147) are steady state solutions to eqn. (10.102). Though of theoretical interest, steady state solutions where $\partial n(\tilde{v}, t)/\partial t = 0$ in a closed system cannot be realistic physical models. The only simple way a steady state can be maintained in eqn. (10.102) is by means of introducing a source of

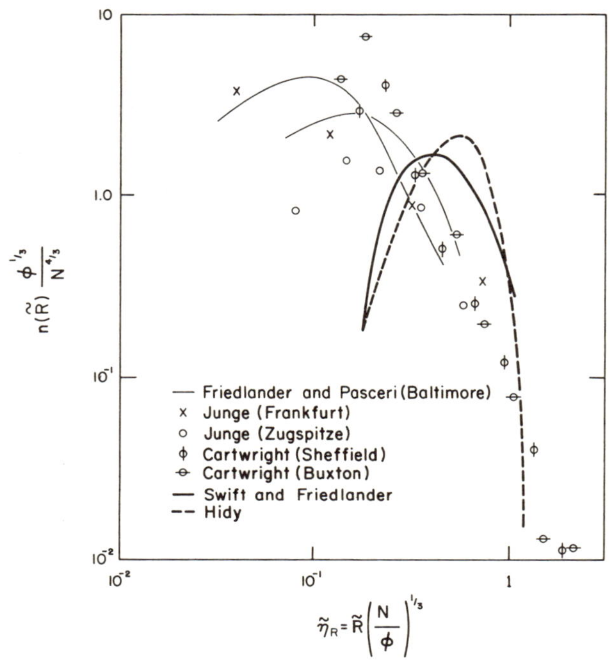

FIG. 10.16. Comparison between the theoretical curve for size distributions of self-preserving form based on Smoluchowski's theory, and observed size distributions for aerosol samples in Baltimore, Md. (From Friedlander and Pasceri[406]).

small particles to just balance a removal of large particles by sedimentation or some other mechanism.† The mechanism wherein particles are removed from the upper end of the spectrum by sedimentation has been examined by Friedlander.[401] In the quasi-steady case, loss by sedimentation may be compensated for by a steady migration of small particles up the spectrum by Brownian coagulation. For such a restricted process shaping the spectrum, Friedlander[401] found power law forms for $n(\tilde{R})$ that differed somewhat from eqn. (10.147). Thus, the apparent tendency to achieve a similarity spectrum of a shape proportional to $\tilde{R}^{-4}$, on the average, remains unexplained at this time.‡

† To account for these effects, other terms would have to be added to eqn. (10.102).

‡ In connection with this question, see also the recent comments of Junge[445], and Friedlander and Hidy.[313]

REFERENCES

1. NORGREN, C. T. and GOLDIN, D. S., *AIAA Journal* **3**, 1943 (1965).
2. FUCHS, N. A., *Mechanics of Aerosols*, Pergamon Press, N.Y. (1964).
3. PODZIMEK, J., "The importance of aerosols in meteorology," in *Aerosols: Physical Chemistry and Applications*, ed. K. Spurny, Gordon & Breach Inc., N.Y. (1965).
4. GREEN, H. and LANE, W., *Particulate Clouds: Dusts, Smokes and Mists*, 2nd edn., Van Nostrand Co. Inc., Princeton, N.J. (1964).
5. CHANDRASEKHAR, S., *Rev. Mod. Phys.* **15**, 61 (1943).
6. HILL, T. S. *Introduction to Statistical Thermodynamics*, Addison-Wesley Inc., Reading, Mass. (1960).
7. BALESCU, R., *Statistical Mechanics of Charged Particles*, Interscience, N.Y. (1963).
8. CHAPMAN, S. and COWLING, T., *The Mathematical Theory of Non-uniform Gases*, Cambridge (1960).
9. LÖBNER, A., *Staub* **25**, 9 (1965).
10. LOEB, L., *Static Electrification*, Springer-Verlag, Berlin (1958).
11. SEVERNE, G., Thesis, University of Brussels, Brussels (1963).
12. STREET, R. E., in *Rarified Gas Dynamics*, ed. F. Devienne, Pergamon, N.Y. p. 276 (1960).
13. OSEEN, C., *Neure Methoden und Ergebnisse in der Hydrodynamik*, Akademische Verlag, Leipzig (1927).
14. PROUDMAN, I. and PEARSON, J., *J. Fluid Mech.* **2**, 237 (1957).
15. EISNER, F., *3rd Intl. Congr. Appl. Mech.*, Stockholm (1930).
16. JEANS, J., *The Dynamical Theory of Gases*, Dover, N.Y. (1954).
17. HIRSCHFELDER, J., CURTISS, J. and BIRD, R., *Molecular Theory of Gases and Liquids*, Wiley, N.Y. (1951).
18. PRIGOGINE, I., *Non-equilibrium Statistical Mechanics*, Interscience, N.Y. (1963).
19. BOGOLIUBOV, N., *J. Phys. USSR* **10**, 265 (1946).
20. BORN, M. and GREEN, H., *Proc. Roy. Soc. (London)*, A **188**, 10 (1946).
21. KIRKWOOD, J., *J. Chem. Phys.* **14**, 180 (1946).
22. YVON, J., *La Theorie Statistique des Fluides et l'Equation d'Etat*, Hermann et Cie, Paris (1935).
23. GREEN, H., *The Molecular Theory of Fluids*, Interscience, N.Y. (1952).
24. BOLTZMANN, L., *Wien Ber.* **58**, 517 (1868).
25. GRAD, H., *Comm. Pure Appl. Math.* **5**, 257 (1952).
26. BHATNAGAR, P., GROSS, E. and KROOK, M., *Phys. Rev.* **94**, 511 (1954).
27. LEES, L., GALCIT Report No. 51 (1959).
28. BROCK, J. R., *J. Colloid Interface Sci.* **23**, 448 (1967).
29. JEFFREYS, H., *Cartesian Tensors*, Cambridge (1931).
30. WILLIS, D., *Phys. Fluids* **8**, 2522 (1966).
31. DEFAY, I., PRIGOGINE, I., BELLEMANS, A. and EVERETT, D., *Surface Tension and Adsorption*, Wiley (1966).
32. BIGG, E., *Proc. Phys. Soc. (London)* B **66**, 688 (1953).
33. WADE, W. and HACKERMAN, N., *Adv. in Chemistry*, **43**, 222 (1964).
34. PLATT, J., *Astrophys. J.* **123**, 486 (1956).
35. HARBURY, L., *J. Phys. Chem.* **50**, 190 (1946).
36. LENNARD-JONES, J., *Zeit. Krist.* **75**, 215 (1930).
37. NAIDITCH, S., AFCRL Report 64-689 (1964).
38. JAGER, H. and KIESCHKE, H., *Nature* **217**, 934 (1968).
39. CORN, M. and REITZ, R., *Science* **159**, 1350 (1968).
40. BLACKMAN, M., LISGARTEN, N. and SKINNER, L., *Nature* **217**, 1245 (1968).
41. HAPPEL, J. and BRENNER, H., *Low Reynolds Number Hydrodynamics*, Prentice-Hall, Englewood Cliffs, N.J. (1965).
42. VAN DYKE, M., *Perturbation Methods in Fluid Mechanics*, Academic Press, N.Y. (1964).
43. PRANDTL, L., "The mechanics of viscous fluids", in *Aerodynamic Theory*, Vol. 2, ed. W. Durand, J. Springer (1935).
44. LAMB, H., *Hydrodynamics*, 6th edn., Cambridge (1932).
45. GOLDSTEIN, S. (Ed.), *Modern Developments in Fluid Dynamics*, Vol. 1, Oxford (1938).
46. MAXWORTHY, T., *J. Fluid Mech.* **23**, 369 (1965).

47. RYBCZNSKI, W., *Bull. Acad. Sci. Cracovie* A **40** (1911).
48. HADAMARD, J., *Compt. Rend. Acad. Sci. (Paris)* **152**, 1735 (1911).
49. SCHILLER, L., *Handbuch Experimentalphys. Leipzig* **4** (II), 339 (1932).
50. MILLIKAN, R., *Phys. Rev.* **21**, 217 (1923); also **22**, 1 (1924).
51. KNUDSEN, M. and WEBER, S., *Ann. Physik* **36**, 982 (1911).
52. DAVIES, C. N., *Proc. Phys. Soc. (London)* **15**, 61 (1943).
53. BASSET, A., *A Treatise on Hydrodynamics,* Dover, N.Y. (1961).
54. LEVICH, V., *Physicochemical Hydrodynamics,* Prentice-Hall, Englewood Cliffs, N.J. (1962).
55. HINZE, J., *Turbulence,* McGraw-Hill, N.Y. (1959).
56. KUCHANOV, S. and LEVICH, V., *Soviet Phys. Doklady* **12**, 549 (1967).
57. CORRSIN, S. and LUMLEY, J., *Appl. Sci. Res.* **6A**, 114 (1956).
58. TCHEN, C., *Mean Value and Correlation Problems Connected with the Motion of Small Particles in Turbulent Fluid,* The Hague (1947).
59. BOGGIO, T., *Rendi Conti.* **16**, 613, 730 (1907).
60. PRANDTL, L. and TIETJENS, O., *Fundamentals of Hydro- and Aeromechanics,* Dover, N.Y. (1957).
61. ABBOTT, R., *Phys. Rev.* **12**, 381 (1918).
62. THOMAS, J., *J. Air Pollution Contr. Assc.* **8**, 32 (1958).
63. BRIDGMAN, P., *Dimensional Analysis,* Yale, New Haven, Conn. (1931).
64. LANGMUIR, I., *J. Meteor.* **5**, 175 (1948).
65. LEVIN, L., *Dokl. Akad. Nauk SSSR* **91**, 1329 (1953).
66. YUREV, I., *Soviet Physics Doklady* **11**, 844 (1967).
67. SCHLICHTING, H., *Boundary Layer Theory,* 4th edn., McGraw-Hill, N.Y. (1960).
68. DAVIES, C. N. and AYLWARD, M., *Proc. Phys. Soc. (London)* B **64**, 889 (1951).
69. RANZ, W. and WONG, J., *Ind. Eng. Chem.* **44**, 1371 (1952).
70. MAY, K., *J. Scient. Instr.* **22**, 187 (1947).
71. DORSCH, R., SAPER, P. and KADOW, C., NACA TN 3587 (1955).
72. WALTON, W. and WOOLCOCK, A., *Aerodynamic Capture of Particles,* Pergamon, N.Y. (1960).
73. LANGMUIR, I. and BLODGETT, K., U.S. Army TR 5418 (1946).
74. DAVIES, C. N., *Proc. Inst. Mech. Engr. (London)* **1B**, 185 (1952).
75. BRUN, R., LEWIS, W., PERKINS, P. and SERAFINI, J., NACA Rep. No. 1215 (1955).
76. DUKHIN, S. and DERJAGUIN, B., *Koll. Zh.* **20**, 326 (1958).
77. KRAEMER, H. and JOHNSTONE, H., *Ind. Eng. Chem.* **47**, 2726 (1955).
78. NATANSON, G., *Dokl. Akad. Nauk SSSR* **112**, 696 (1957).
79. ZEBEL, G., *J. Colloid Sci.* **20**, 522 (1965).
80. ZEBEL, G., *J. Colloid Interface Sci.* **27**, 294 (1968).
81. DEGROOT, S. and MAZUR, P., *Non-Equilibrium Thermodynamics,* North Holland, Amsterdam (1962).
82. BIRD, R., in *Adv. in Chem. Engng.* **1**, Academic Press, N.Y. (1956).
83. MERK, H., *Appl. Sci. Res.* A **8**, 73 (1958).
84. PRIGOGINE, I. and DEFAY, R., *Chemical Thermodynamics,* Longmans, Green, N.Y. (1954).
85. DERJAGUIN, B. and YALAMOV, Y., *J. Colloid Sci.* **20**, 555 (1965).
86. DERJAGUIN, B., YALAMOV, Y. and STOROZHILOVA, A., *J. Colloid Interface Sci.* **22**, 117 (1966).
87. CARSLAW, H. and JAEGER, S., *Conduction of Heat in Solids,* 2nd edn., Oxford, London (1959).
88. SCHRAGE, R., *A Theoretical Study of Interphase Mass Transfer,* Columbia University Press, N.Y. (1953).
89. MAXWELL, J., *Collected Scientific Papers,* Cambridge (1890).
90. FUCHS, N. A., *Evaporation and Droplet Growth in Gaseous Media,* Pergamon, N.Y. (1959).
91. PHILIP, J., *J. Atmos. Sci.* **22**, 196 (1965).
92. CABLE, M. and EVANS, D., *J. Appl. Phys.* **38**, 2899 (1967).
93. LIGHTHILL, M., *Proc. Roy. Soc. (London)* A **202**, 369 (1950).
94. FUCHS, N. A., *Zh. Exp.-Theor. Phys.* **4**, 7 (1934).
95. KRÖNIG, R. and BRUIJSTEN, J., *J. Appl. Sci. Res.* A **2**, 439 (1951).
96. BREIMAN, L., Norman Bridge Laboratory, Report No. 2F-2, Calif. Inst. Tech., Pasadena, Calif. (1952).
97. FRISCH, H., *J. Chem. Phys.* **22**, 123 (1954).
98. FRIEDLANDER, S. K., *AIChE Journal* **3**, 43 (1957).
99. ACRIVOS, A. and TAYLOR, T., *Phys. Fluids* **5**, 387 (1962).
100. MORGAN, G. and WARNER, W., *J. Aero. Sci.* **23**, 937 (1956).
101. HASEN, W., *Rev. Sci. Instr.* **13**, 247 (1942).
102. LUND, L. and RIVERS, J., *J. Chem. Phys.* **45**, 4612 (1966).
103. SCHUSTER, B., Phd. Dissertation, New Mexico State University, Las Crucas, N.M. (1967).
104. KINZER, G. and GUNN, R., *J. Meteor.* **81**, 71 (1951).
105. STEFAN, J., *Wien Ber.* **83**, 943 (1881).
106. DERJAGUIN, B. and DUKHIN, S., *Dokl. Akad. Nauk SSSR* **106**, 851 (1956).
107. DERJAGUIN, B. and DUKHIN, S., *Dokl. Akad. Nauk SSSR* **111**, 613 (1956).

108. DERJAGUIN, B., BAKANOV, S., DUKHIN, S. and BATOVA, G., in *Surface Forces,* ed. B. Derjaguin, Consultants Bureau, N.Y. (1964).
109. FACY, L., *Compt. Rend. Acad. Sci. (Paris)* **246**, 102, 3161 (1958).
110. DUKHIN, S. and DERJAGUIN, B., *Dokl. Akad. Nauk SSSR* **112**, 407 (1957).
111. FRESNEL, A., *Ann de Chimie et Physique* **29**, 107 (1825).
112. SMOLUCHOWSKI, M. v., *Phil. Mag.* **21**, 11 (1911).
113. WACHMANN, H., *ARS Journal* **32**, 2 (1962).
114. PAUL, B., *ARS Journal* **32**, 1321 (1962).
115. BRUNDIN, C. (Ed.), *Rarified Gas Dynamics,* Academic Press, N.Y. (1967).
116. BROCK, J. R., *J. Phys. Chem.* **68**, 2857 (1962).
117. KENNARD, E., *Kinetic Theory of Gases,* McGraw-Hill, N.Y., ch. 8 (1938).
118. GENTRY, J. and BROCK, J. R., *J. Colloid Interface Sci.* **27**, 691 (1968).
119. BROCK, J. R., *J. Phys. Chem.* **68**, 2863 (1962).
120. WALDMANN, L., *Z. Naturforsch.* **14A**, 589 (1959).
121. BAKANOV, S. and DERJAGUIN, B., *Disc. Faraday Soc.* **30**, 130 (1960).
122. HIDY, G. and BROCK, J. R., *J. Geophys. Res.* **72**, 455 (1967).
123. RICE, S. and GRAY, P., *Statistical Mechanics of Simple Fluids,* Interscience, N.Y. (1965).
124. LEBOWITZ, J. and RUBIN, E., *Phys. Rev.* **131**, 2381 (1963).
125. MONCHICK, L. and REISS, H., *J. Chem. Phys.* **22**, 831 (1954).
126. BIRKS, J. and BRADLEY, R., *Proc. Roy. Soc. (London)* A **198**, 239 (1949).
127. SHERESHEVSKY, J. and STECKLER, S., *J. Chem. Phys.* **4**, 108 (1936).
128. BROCK, J. R., *J. Colloid Interface Sci.* **24**, 344 (1967).
129. TAKAO, K., in *Rarified Gas Dynamics,* Vol. 2, ed. J. Laurmann, Academic Press, N.Y. (1963).
130. HARBOUR, P., Ph.D. Dissertation, Cambridge (1963).
131. SHERMANN, F., in *Rarified Gas Dynamics,* Vol. 2, ed. J. Laurmann, Academic Press, N.Y. (1963).
132. LEES, L., *J. Soc. Ind. Appl. Math.* **13**, 278 (1965).
133. LORD, R. and HARBOUR, P., *AIAA Journal* **6**, 244 (1968).
134. SIROVICH, L., *Phys. Fluids* **5**, 908 (1962).
135. HAMEL, B., *Phys. Fluids* **8**, 418 (1964).
136. BROCK, J. R., *Phys. Fluids* **9**, 1602 (1966).
137. WILLIS, D., University of California Aero. Sci. Report No. AS-65-16 (1965).
138. STACY, L., *Phys. Rev.* **21**, 239 (1923).
139. MATTAUCH, J., *Z. Phys.* **32**, 439 (1925).
140. SCHMITT, K., *Z. Naturforsch.* **14A**, 870 (1959).
141. LIU, V., PANG, S. and JEW, H., *Phys. Fluids* **8**, 788 (1965).
142. JACOBSEN, S. and BROCK, J. R., *J. Colloid Sci.* **20**, 544 (1965).
143. SCHADT, C. and CADLE, R., *J. Phys. Chem.* **65**, 1689 (1961).
144. ROSENBLATT, P. and LAMER, V., *Phys. Rev.* **70**, 385 (1946).
145. WALDMANN, L., in *Rarified Gas Dynamics,* ed. L. Talbot, Academic Press, N.Y. (1960).
146. WALDMANN, L. and SCHMITT, K., in *Aerosol Science,* ed. C. N. Davies, Academic Press, N.Y. (1967).
147. BROCK, J. R., *J. Colloid Interface Sci.* **27**, 95 (1968).
148. BROCK, J. R., *J. Phys. Chem.* **72**, 747 (1968).
149. SPRINGER, G. and TSAI, S., *Phys. Fluids* **8**, 1561 (1965).
150. BURNETT, D., *Proc. London Math Soc.* **40**, 382 (1935).
151. GRAD, H., *Comm. Pure Appl. Math.* **2**, 331 (1949).
152. SCHAAF, S. and CHAMBRE, P., in *Fundamentals of Gas Dynamics,* ed. H. Emmons, Princeton Univ. Press, Princeton (1958).
153. PAYNE, H., *J. Chem. Phys.* **21**, 2127 (1953).
154. BROCK, J. R., *J. Catalysis* **2**, 248 (1963).
155. DERJAGUIN, B. and YALAMOV, Y., *J. Colloid Interface Sci.* **22**, 195 (1966).
156. DERJAGUIN, B. and BAKANOV, S., *Dokl. Akad. Nauk SSSR* **147**, 139 (1962).
157. WALDMANN, L., *Z. Naturforschung* **22A**, 1269 (1967).
158. EPSTEIN, P., *Z. Physik.* **54**, 537 (1929).
159. BROCK, J. R., *J. Colloid Sci.* **17**, 768 (1962).
160. EINSTEIN, A., *Investigations on the Theory of Brownian Motion,* Dover, N.Y. (1956).
161. WANG, M. C. and UHLENBECK, G., *Rev. Mod. Phys.* **17**, 323 (1945).
162. GOLDSTEIN, S., *Quart. J. Mech. Appl. Math.* **4**, 129 (1951).
163. MICHELSON, I., Amer. Inst. Aerological Res., TR No. 193, Pasadena, Calif. (1954).
164. FRIEDLANDER, S. K., *Unpubl. Remarks on Diffusion by Brownian Motion,* Johns Hopkins Univ. (1961).
165. SNEDDON, I., *Elements of Partial Differential Equations,* McGraw-Hill, N.Y. (1957).
166. CRANK, J., *Mathematics of Diffusion,* Oxford (1959).
167. CARSLAW, H. and JAEGER, J., *Proc. London Math. Soc.* **46**, 361 (1940).

168. Sparrow, E. M. and Gregg, J., *J. Heat Transfer* **82**, 294 (1960).
169. Friedlander, S. K. and Pasceri, R., *J. Atmos. Sci.* **22**, 571 (1965).
170. Jakob, M., *Heat Transfer*, Vol. 1, Wiley, N.Y. (1949).
171. Gormley, P. and Kennedy, M., *Proc. Roy. Inst. Acad.* **52A**, 163 (1949).
172. Davis, R., *Int. J. Air Water Pollution* **8**, 177 (1964).
173. Frössling, N., NACA TM 1432 (1958).
174. Friedlander, S. K. and Pasceri, R., *Aerosol Filtration by Fibrous Filters*, Dept. Chem. Engng. Report. Johns Hopkins Univ., Baltimore (1962); also Friedlander, S. K. in *Biochemical and Biological Engineering Sci.*, **1**, N. Blakebrough ed. Academic Press, N.Y. (1967).
175. Friedlander, S. K., *J. Colloid Interface Sci.* **23**, 157 (1967).
176. Stechkina, I. and Fuchs, N. A., *Ann. Occup. Hyg.* **9**, 59 (1966).
177. Hochrainer, D., Hidy, G. and Zebel, G., *J. Colloid Interface Sci.* **30**, 553 (1969).
178. Langmuir, I., OSRD Report No. 865 (1942).
179. Spurny, K. and Pich, J., *Int. J. Air Water Pollution* **8**, 193 (1964).
180. Pich, J., in *Aerosol Science*, ed. C. N. Davies, Academic Press, N.Y. (1966).
181. Brinkman, H., *Appl. Sci. Res.*, A1, 27 (1947).
182. Spielman, L. and Goren, S., *Environ. Sci. Tech.* **2**, 279 (1968).
183. Batchelor, G., *Homogeneous Turbulence*, Cambridge (1955).
184. Taylor, G. I., *Proc. London Math. Soc.* A **20**, 196 (1921).
185. Kampe de Feriet, J., *Ann. Soc. Sci. de Brux.* **59**, 145 (1939).
186. Batchelor, G., *Aust. J. Sci. Res.* **2**, 437 (1949).
187. Lumley, J., in *Mechanics of Turbulence*, Gordon & Breach, N.Y. (1964).
188. Corrsin, S., in *Mechanics of Turbulence*, Gordon & Breach, N.Y. (1964).
189. Friedlander, S. K., *AIChE Journal* **3**, 381 (1957).
190. Longwell, J. P. and Weiss, M., *Ind. Eng. Chem.* **45**, 667 (1953).
191. Hidy, G. and Keller, K., *Diffusion and Dispersion of Particles in Turbulent Flow*, unpubl. report, Dept. of Chem. Engng., Columbia Univ., N.Y. (1957).
192. Kennedy, D., Ph.D. Dissertation, Johns Hopkins Univ., Baltimore (1965).
193. Goldschmidt, V., Ph.D. Dissertation, Syracuse Univ., Syracuse, N.Y. (1964).
194. Davies, C. N., in *Aerosol Science*, ed. C. N. Davies, Academic Press, N.Y. (1966).
195. Sinclair, D. and LaMer, V., *Chem. Rev.* **44**, 245 (1949).
196. Kasiosis, F. and Fish, B., *J. Colloid Sci.* **17**, 155 (1962).
197. Monk, G., *J. Appl. Phys.* **23**, 288 (1952).
198. Lane, W. and Green, H., in *Surveys in Mechanics*, ed. G. Batchelor and R. Davies, Cambridge (1956).
199. Rayleigh, Lord, *Proc. London Math. Soc.* **10**, 4 (1878); *Proc. Roy. Soc. (London)*, **24**, 71 (1879).
200. Lapple, C., in *Advances in Chemical Engineering*, **8**, Academic Press, N.Y. (1969).
201. Schultz, R. and L. Bronson, in *2nd Symposium on Adv. Propulsion Concepts*, Boston (1959).
202. Zeleny, J., *Proc. Cambridge Phil. Soc.* **18**, 71 (1916).
203. Drozin, V., *J. Colloid Sci.* **10**, 158 (1955).
204. Peskin, R. and Raco, R., *AIAA Journal* **2**, 781 (1964).
205. Nayyar, N. and Murty, G., *Proc. Phys. Soc. (London)* **75**, 369 (1960).
206. Walton, W. and Prewett, W., *Proc. Phys. Soc. (London)* B **62**, 341 (1949).
207. Straus, R., Ph.D. dissertation, London Univ. (1950).
208. Dunskii, V. and Nikitin, N., *Inzhenerno-Fizcheskii Zh.* **9**, 54 (1965).
209. Volynskii, M., *Dokl. Akad. Nauk SSSR* **62**, 301 (1948).
210. Bukhman, S., *Vestik Akad Nauk Kazakhskoi SSR*, No. 11 (1954).
211. Merrington, A. and Richardson, E., *Proc. Phys. Soc. (London)* **59**, 1 (1947).
212. Basanaev, M. E. Teverovskii and Tregubova, E., *Dokl. Akad. Nauk SSSR* **66**, 821 (1949).
213. Taylor, G. I., *Quart. J. Mech. Appl. Math.* **3**, 129 (1950).
214. Watson, E., *Proc. Inst. Mech. Engr. (London)* **158**, 187 (1948).
215. McCubbin, T., *J. Acoust. Soc. Amer.* **25**, 1013 (1953).
216. Green, H., in *Flow Properties of Disperse Systems*, ed. J. Hermans, North Holland, Amsterdam (1953).
217. Bisa, K., Dirough, K. and Esche, R., *Siemans Zeit.* **28**, 341 (1954).
218. Lang, R., *J. Acoust. Soc. Amer.* **34**, 6 (1962).
219. Peskin, R. and Raco, R., *J. Acoust. Soc. Amer.* **35**, 1378 (1963).
220. Crawford, A., *J. Acoust. Soc. Amer.* **27**, 176 (1955).
221. Castleman, R., *J. Res. Natl. Bur. Stds. (Wash.)* **6**, 369 (1931).
222. Kelvin, Lord, *Math. and Physical Papers*, Vol. 4, Cambridge (1910).
223. Miles, J., *J. Fluid Mech.* **6**, 583 (1959).
224. Miles, J., *J. Fluid Mech.* **13**, 433 (1962).
225. Goren, S., *J. Colloid Sci.* **19**, 81 (1964).
226. Fraser, R., Eisenklam, E. and Dombrowski, N., *(British) Chem. Engng.* **2**, 496 (1957).

227. LANE, W., *Ind. Eng. Chem.* **43**, 1312 (1951).
228. LITTAYE, G., *Compte Rend. Acad. Sci. (Paris)* **217**, 99, 340 (1943).
229. HINZE, J., *Appl. Sci. Res.* A **1**, 273 (1948).
230. DAUTREBANDE, L., *Microaerosols*, Academic Press, N.Y. (1962).
231. COLLISON, W., *Inhalation Therapy Technique*, Heinemann, London (1935).
232. PREINING, O., SHEESLEY, D. and DJORDJERIC, N., *J. Colloidal Interface Sci.* **23**, 458 (1967).
233. TAYLOR, G. I., U.K. Ministry of Supply Report, London (1949).
234. DUNSKII, V., *Zhur. Tekh. Fiz.* **26**, 1262 (1956).
235. LIU, B. Y., in Particle Tech. Lab. Report No. 112, Univ. of Minn., Minneapolis, Minn. (1967).
236. NUKIYAMA, S. and TANASAWA, Y., *Trans. Soc. Mech. Engr. (Japan)* **5**, 62, 68 (1939).
237. BITRON, M., *Ind. Eng. Chem.* **47**, 23 (1955).
238. HRUBECKY, H., *J. Appl. Phys.* **29**, 572 (1958).
239. GREEN, H., *Trans. Faraday Soc.* **32**, 1091 (1936).
240. BOWDEN, F. and TABOR, D., *Friction and the Lubrication of Solids*, Oxford, London (1950).
241. BRADLEY, R., *Phil. Mag.* **13**, 853 (1932).
242. HAMAKER, H., *Physica* **4**, 1058 (1937).
243. VOLD, M., *J. Colloid Sci.* **9**, 451 (1954).
244. CORN, M., in *Aerosol Science*, ed. C. N. Davies, Academic Press, N.Y. (1966).
245. CASIMIR, H. and POLDER, D., *Phys. Rev.* **73**, 360 (1948).
246. BLACK, W., DEJOUGH, J., OVERBEEK, J. T. G. and SPARNAAY, M., *Trans. Faraday Soc.* **56**, 1597 (1960).
247. DERJAGUIN, B., ABRIKOSOVA, I. and LIFSCHITZ, E., *Quart. Rev. Chem. Soc.* **10**, 295 (1956).
248. KITCHENER, J. and PROSSER, A., *Proc. Roy. Soc. (London)* A **242**, 403 (1957).
249. LIFSCHITZ, E., *Dokl. Akad. Nauk SSSR* **97**, 643 (1954); also *Sov. Phys. JETP* **2**, 73 (1956).
250. BRADLEY, R., *Trans. Faraday Soc.* **32**, 1088 (1936).
251. OVERBEEK, J. T. G. and SPARNAAY, M., *Disc. Faraday Soc.* **18**, 12 (1954).
252. BAILEY, A. and COURTNEY-PRATT, J., *Proc. Roy. Soc. (London)* A **227**, 500 (1955).
253. PROSSER, A. and KITCHENER, J., *Nature* **178**, 1339 (1956).
254. HOWE, P., BENTON, D. and PUDDINGTON, I., *Can. J. Chem.* **33**, 1375 (1955).
255. BEISHER, D., *Kolloid Zeit.* **89**, 215 (1939).
256. DERJAGUIN, B. and SMILGA, V., *J. Appl. Phys.* **38**, 4609 (1967).
257. DUKHOSKOI, E., KRAGELSKII, I. and SILIN, A., *Soviet Phys. Doklady* **12**, 730 (1967).
258. MACFARLANE, J. and TABOR, D., *Proc. 7th Congr. Appl. Mech.* 335 (1948).
259. CORN, M. and STEIN, F., in *Surface Contamination*, ed. B. Fish, Pergamon, N.Y. (1967).
260. BILLINGS, C., SILVERMAN, L., DENNIS, R. and LEVENBAUM, L., *J. Air Pollution Contr. Assc.* **10**, 318 (1960).
261. GERRARD, J., *Brit. J. Appl. Phys.* **14**, 186 (1963).
262. BAGNOLD, R., *Phys. of Blown Sand and Desert Dunes*, Metheun, London (1941).
263. LÖFFLER, F., *Staub (Engl.)* **26** (7), 10 (1967).
264. ZENZ, F., *Ind. Eng. Chem. (Fundamentals)* **3**, 65 (1964).
265. JORDAN, D., *Brit. J. Appl. Phys.* **3**, 5194 (1954).
266. DAVIDSON, J. and HARRISON, D., *Fluidized Particles*, Cambridge (1963).
267. ANDERSON, A., *Kolloid Zeit.* **86**, 70 (1939).
268. WILSON, C. T. R., *Phil. Trans. Roy. Soc. (London)* A **189**, 265 (1897); A **193**, 289 (1900).
269. WEGENER, P. and MACK, L., in *Adv. in Appl. Mech.*, Vol. 5, Academic Press, N.Y. (1958).
270. WEGENER, P. and POURING, A., *Phys. Fluids* **7**, 352 (1964).
271. WHYTLAW-GRAY, R. and PATTERSON, H., *Smoke*, E. Arnold, London (1932).
272. POWELL, W. and OSWALD, L., Lawrence Radiation Laboratory Report UCRL-3590, Berkeley, Calif. (1956).
273. AMELIN, A., *Kolloid Zh.* **10**, 168 (1948).
274. HIDY, G. and FRIEDLANDER, S. K., *AIChE Journal* **10**, 115 (1964).
275. AMELIN, A., *Theory of Fog Condensation*, D. Davey, N.Y. (1966).
276. STODOLA, A., *Steam and Gas Turbines*, McGraw-Hill, N.Y. (1927).
277. REISS, H., *Methods of Thermodynamics*, Blaisdell Publ. Co., N.Y., 1965.
278. FRENKEL, J., *Kinetic Theory of Liquids*, Oxford, London (1946).
279. ABRAHAM, F., *J. Atmos. Sci.* **25**, 76 (1968).
280. FRIEDLANDER, S. K., *Phys. Fluids* **3**, 693 (1960).
281. SMOLUCHOWSKI, M.v., *Z. Physik. Chem.* **92**, 129 (1917).
282. SMOLUCHOWSKI, M.v., *Z. Physik.* **17**, 557 (1916).
283. ANDRES, R. P., *Ind. Eng. Chem.* **57**, 24 (1965).
284. PARLANGE, J., *J. Chem. Phys.* **48**, 776 (1968).
285. STOGRYN, D. and HIRSCHFELDER, J., *J. Chem. Phys.* **31**, 1531 (1959).
286. REED, S., Jr., *J. Chem. Phys.* **20**, 208 (1958).
287. HIRTH, J. and POUND, G., *Condensation and Evaporation*, Pergamon, N.Y. (1963).
288. REISS, H., *J. Chem. Phys.* **20**, 1216 (1952).

289. COURTNEY, W., *J. Chem. Phys.* **35**, 2249 (1961).
290. KUHRT, F., *Z. Physik* **131**, 185 (1952).
291. DUNNING, W., *Prog. Astron. Aeron.* **15**, 739 (1964).
292. LOTHE, J. and POUND, G., *J. Chem. Phys.* **36**, 2080 (1962).
293. LOTHE, J. and POUND, G., *J. Chem. Phys.* **45**, 630 (1966).
294. REISS, H. and KATZ, J., *J. Chem. Phys.* **46**, 2496 (1967).
295. ABRAHAM, F. and POUND, G., *J. Chem. Phys.* **48**, 732 (1968).
296. REISS, H., KATZ, J. and COHEN, E., *J. Chem. Phys.* **48**, 5553 (1968).
297. VOLMER, M. and WEBER, A., *Z. Phys. Chem.* **119**, 277 (1925).
298. COURTNEY, W. and CLARKE, W., TM-1250, Texaco Experimental, Inc. (1961).
299. BECKER, R. and DÖRING, W., *Ann. Phys.* (5), **24**, 719 (1935).
300. KATZ, J., SALTSBURG, H. and REISS, H., *J. Colloid Interface Sci.* **21**, 560 (1966).
301. ANDRES, R. and BOUDART, M., *J. Chem. Phys.* **42**, 2057 (1965).
302. FLETCHER, N. H., *The Physics of Rainclouds*, Cambridge (1966).
303. FLOOD, H., *Z. Phys. Chem.* A **170**, 286 (1934).
304. HIDY, G., *J. Appl. Meteor.* **2**, 545 (1963).
305. KATZ, J. and OSTERMIER, B., *J. Chem. Phys.* **47**, 478 (1967).
306. VOLMER, M. and FLOOD, H., *Z. Phys. Chem.* A **170**, 2713 (1934).
307. POWELL, C., *Proc. Roy. Soc. (London)* A **119**, 553 (1928).
308. DUNNING, W., *Disc. Faraday Soc.* **30**, 9 (1961).
309. FRANCK, J. and H. HERTZ, *Z. Physik* **143**, 559 (1956).
310. MASON, B., *Disc. Faraday Soc.* **30**, 20 (1961).
311. ALLARD, E. and KASSNER, J., Jr., *J. Chem. Phys.* **42**, 140 (1965).
312. KASSNER, J., Jr. and SCHMITT, R., *J. Chem. Phys.* **44**, 4166 (1967).
313. FRIEDLANDER, S. K., and HIDY, G., in *Proc. of 7th Intl. Conf. on Condensation and Ice Nuclei*, Prague, Czechoslovakia (1969), pp. 21–25.
314. OSWATITSCH, K., *Z. Angew Math. Mechanik* **22**, 1 (1942).
315. BINNIE, A. and WOODS, M., *Proc. Inst. Mech. Engr. (London)* **138**, 29 (1938).
316. HIGUCHI, W. and O'KONSKI, C., *J. Colloid Sci.* **15**, 14 (1960).
317. RAMSEY, N., *Molecular Beams*, Oxford, London (1956).
318. BECKER, E., BIER, K. and HENKES, W., *Z. Physik* **146**, 333 (1956).
319. LECKENBY, R., ROBBINS, E. and FRENALION, P. *Proc. Roy. Soc. (London)* A **280**, 409 (1964).
320. VOLMER, M., *Kinetik der Phasenbildung*, Steinkopff, Dresden (1939).
321. DEBYE, P., *Polar Molecules*, Dover, N.Y. (1945).
322. HENNIKER, S., *Rev. Mod. Phys.* **21**, 322 (1949).
323. PRZIBAM, K., *Sitz. Akad Wiss. Wien.* **115**, 81 (1906).
324. LABY, T., *Phil. Trans. Roy. Soc. (London)* A **208**, 445 (1908).
325. ANDREN, L., *Ann. Physik (Leipzig)* **52**, 1 (1917).
326. SCHARRER, L., *Ann. Physik* **35**, 619 (1939).
327. SANDER, A. and DAMKÖHLER, G., *Naturwiss.* **31**, 460 (1943).
328. TWOMEY, S., *J. Chem. Phys.* **30**, 941 (1959).
329. KOHLER, H., *Geofys. Publ.* **2**, No. 3 and No. 6 (1921).
330. WRIGHT, H., *Proc. Phys. Soc.* **48**, 675 (1936).
331. JUNGE, C., *Ann. Metor. (Beiheft)* (1952).
332. TURNBULL, D. and VONNEGUT, B., *Ind. Eng. Chem.* **44**, 1292 (1952).
333. MASON, B. J., *The Physics of Clouds*, Oxford, London (1957).
334. REISS, H., *J. Chem. Phys.* **19**, 482 (1951).
335. FRISCH, H. and COLLINS, F., *J. Chem. Phys.* **21**, 2158 (1953).
336. ZUNG, J., *J. Chem. Phys.* **46**, 2064 (1967).
337. HIDY, G., D. Eng. Dissertation, Johns Hopkins University (1962).
338. FUCHS, N. A., *Z. Physik* **89**, 736 (1934).
339. DEBYE, P., *Trans. Electrochem. Soc.* **82**, 265 (1942).
340. KRAMERS, H., *Physica* **7**, 290 (1940).
341. SWIFT, D. and FRIEDLANDER, S., *J. Colloid Sci.* **19**, 621 (1964).
342. ZEBEL, G., *Kolloid Z.* **157**, 37 (1958).
343. GILLESPIE, T., *Proc. Roy. Soc. (London)* A **216**, 569 (1953).
344. DEVIR (WEINSTOCK), S., *J. Colloid Interface Sci.* **23**, 80 (1967).
345. MASON, E., *J. Chem. Phys.* **27**, 782 (1957).
346. KIHARA, T., *Rev. Mod. Phys.* **25**, 844 (1953).
347. MASON, E. and CHAPMAN, S., *J. Chem. Phys.* **36**, 627 (1962).
348. BROCK, J. R. and HIDY, G., *J. Appl. Phys.* **36**, 1857 (1965).
349. FUCHS, N. A. and SUTUGIN, A., *J. Colloid Sci.* **20**, 442 (1965).

350. Overbeek, J. T. G., in *Colloid Science*, Vol. 1, ed. H. Kruyt, Elsevier, N.Y. (1952).
351. Junge, C., Changnon, C. and Manson, J., *J. Meteor.* **18**, 81 (1961).
352. Friedlander, S. K., *J. Meteor.* **18**, 753 (1961).
353. Fuchs, N. A., *Z. Phys. Chem.* **171A**, 199 (1934).
354. Astakhov, A., *Dokl. Akad. Nauk SSSR* **161**, 1114 (1965).
355. Hidy, G. and Brock, J. R., *J. Colloid Sci.* **20**, 477 (1965).
356. Junge, C., *Air Chemistry and Radioactivity*, Academic Press, N.Y., ch. 2 (1963).
357. Patterson, H. and Cawood, W., *Proc. Roy. Soc. (London)* A **136**, 358 (1932).
358. Cawood, W. and Whytlaw-Gray, R., *Trans. Faraday Soc.* **32**, 1059 (1936).
359. Quon, J., *Int. J. Air Water Pollution* **8**, 355 (1964).
360. Fuchs, N. A., *Dokl. Akad. Nauk SSSR* **81**, 1053 (1951).
361. Friedlander, S. K., in *Aerosols; Physical Chemistry and Applications*, ed. K. Spurny, Gordon & Breach, Inc., N.Y. (1965).
362. Saffman, P. and Turner, J. S., *J. Fluid Mech.* **1**, 16 (1956).
363. Kolmogoroff, A. N., *Dokl. Akad. Nauk SSSR* **30**, 301 (1941).
364. Batchelor, G. K., *Proc. Cambridge Phil Soc.* **47**, 359 (1951).
365. Stimson, M. and Jeffery, G., *Proc. Roy. Soc. (London)* A **111**, 110 (1926).
366. Krasnogoskaya, N., *Izvestiya Akad. Sci. Nauk SSSR, Atmospheric and Oceanic Physics* **1**, 237 (1965).
367. Davis, M. H., *Quart. J. Mech. Appl. Math.* **17**, 499 (1964).
368. Hocking, L. M., *Quart. J. Roy. Meteor. Soc.* **85**, 44 (1959).
369. Findeisen, W., *Meteor. Z.* **56**, 356 (1939).
370. Davis, M. H., *J. Geophys. Res.* **71**, 3101 (1965).
371. Davis, M. H. and Sartor, J. D., *Nature* **215**, 1371 (1967).
372. Pearcey, T. and Hill, G., *Quart. J. Roy. Meteor. Soc.* **83**, 77, 555 (1957).
373. Horguani, V., *Izvestiya Akad. Sci. Nauk SSSR, Atmospheric and Oceanic Physics* **1**, 122 (1965).
374. Schotland, R. and Kaplan, E., in *Artificial Stimulation of Rain*, Pergamon, N.Y. (1957).
375. Woods, J. and Mason, B. J., *Quart. J. Roy. Meteor. Soc.* **90**, 373 (1964).
376. Shafrir, V. and Neiburger, M., *J. Geophys. Res.* **68**, 4141 (1963).
377. Telford, J., Thorndike, N. and Bowen, E., *Quart. J. Roy. Meteor. Soc.* **81**, 241 (1955).
378. Pauthenier, M. and Cochet, R., *Rev. Gener. Electricitie* **62**, 255 (1953).
379. Levin, L., *Dokl. Akad. Nauk SSSR* **94**, 467 (1954).
380. Sartor, J. D., *J. Geophys. Res.* **66**, 3 (1961).
381. Krasnogorskaya, N., *Izvestiya Akad. Sci. Nauk SSSR, Atmospheric and Oceanic Physics* **1**, 200 (1967).
382. Mockros, L., Quon, J. and Hjelmfelt, T., Jr., *J. Colloid Interface Sci.* **23**, 90 (1967).
383. Gillespie, T., *J. Colloid Sci.* **18**, 562 (1963).
384. Murphy, W. and Sears, G., *J. Appl. Phys.* **35**, 1986 (1964).
385. Israel, G. and Friedlander, S. K., *J. Colloid Interface Sci.* **24**, 330 (1967).
386. Martynov, G. and Bakanov, S., in *Surface Forces*, ed. B. Derjaguin, Consultants Bureau, N.Y. (1963).
387. Melzak, Z., *Quart. J. Appl. Math.* **11**, 231 (1953).
388. Scott, W., *J. Atmos. Sci.* **25**, 54 (1968).
389. Golovin, A., *Soviet Phys. Doklady* **8**, 191 (1963).
390. Berry, E., *J. Atmos. Sci.* **24**, 688 (1967).
391. Thompson, P. D., NCAR MS No. 68-66, National Center for Atmospheric Research, Boulder, Colo. (1968).
392. Levin, L. and Sedunov, Y., *Izvestiya Akad. Sci. Nauk SSSR, Atmospheric and Oceanic Physics* **3**, 242 (1967).
393. Zebel, G., *Kolloid Z.* **156**, 102 (1958).
394. Twomey, S., *J. Atmos. Sci.* **21**, 553 (1964).
395. Telford, J., *J. Meteor.* **12**, 436 (1955).
396. Fair, G. and Gemmel, R., *J. Colloid Sci.* **19**, 360 (1964).
397. Hidy, G., *J. Colloid Sci.* **20**, 123 (1965).
398. Todes, O., *Problems in Kinetics and Catalysis*, Vol. VII, Izd-vo An SSSR (1949).
399. Schumann, T., *Quart. J. Roy. Meteor. Soc.* **66**, 195 (1940).
400. Friedlander, S. K. and Wang, C. S., *J. Colloid Interface Sci.* **22**, 126 (1966).
401. Friedlander, S. K., *J. Meteor.* **17**, 479 (1960).
402. Hidy, G. and Lilly, D., *J. Colloid Sci.* **20**, 867 (1965).
403. Wang, C. S. and Friedlander, S. K., *J. Colloid Interface Sci.* **24**, 170 (1967).
404. Clark, W. and Whitby, K., *J. Atmos. Sci.* **24**, 677 (1967).
405. Junge, C., in *Artificial Stimulation of Rain*, Pergamon, N.Y. (1957).
406. Friedlander, S. K. and Pasceri, R., *J. Atmos. Sci.* **23**, 577 (1964).
407. Liu, B. Y. and Whitby, K., *J. Colloid Interface Sci.* **26**, 161 (1968).
408. Liebster, H. and Schiller, L., *Phys. Z.* **25**, 670 (1924).
409. Schmiedel, J., *Phys. Z.* **29**, 593 (1928).

410. CHEN, C., *Chem. Rev.* **55**, 595 (1955).
411. WONG, J. and JOHNSTONE, H., Tech. Report No. 11, Eng. Explt. Station, Univ. of Illinois, Urbana, Illinois (1953).
412. HUMPHREY, A. and GADEN, A., *Ind. Eng. Chem.* **47**, 924 (1955).
413. RAMSKILL, E. and ANDERSON, W., *J. Colloid Sci.* **6**, 416 (1951).
414. SADOFF, H. and ALMLOF, J., *Ind. Eng. Chem.* **48**, 2199 (1956).
415. STERN, S., ZELLER, H. and SCHEKMAN, A., *J. Colloid Sci.* **15**, 546 (1960).
416. THOMAS, D. and LAPPLE, C., *AIChE Journal* **7**, 203 (1961).
417. THOMAS, J. and YODER, R., *AMA Arch. Industr. Health* **13**, 545, 550 (1956).
418. WONG, J., RANZ, W. and JOHNSTONE, H., *J. Appl. Phys.* **27**, 161 (1956).
419. PRUPPACHER, H. and STEINBERGER, E., *J. Appl. Phys.* **39**, 4129 (1968).
420. TABOR, D. and WINTERTON, R., *Nature* **219**, 1120 (1968).
421. HAGERTY, W. and SHEA, J., *J. Appl. Mech.* **22**, 509 (1955).
422. DOMBROWSKI, N. and JOHNS, W., *Chem. Eng. Sci.* **18**, 203 (1963).
423. HASSON, D. and MIZRAHI, J., *Trans. Inst. Chem. Eng. (London)* **38**, 415 (1961).
424. HINZE, J. and MILBORN, H., *J. Appl. Mech.* **17**, 145 (1950).
425. BOSHOFF, W., *Proc. Inst. Mech. Engr. (London)* A **166**, 443 (1952).
426. RYLEY, D., *British J. Appl. Phys.* **10**, 180 (1959).
427. PATTISON, J. and ALDRIDGE, J., *Engineer,* **203**, 514 (1957).
428. FRIEDMAN, S., GLUCKERT, F. and MARSHALL, W., *Chem. Eng. Prog.* **48**, 181 (1952).
429. CARSTENS, J., GRAYSON, M., KASSNER, J., RIVERS, J. and LUND, L., *J. Recherches Atmos.* **2**, 95 (1966).
430. EMSLIE, A., BONNER, F. and PECK, L., *J. Appl. Phys.* **29**, 858 (1958).
431. FRASER, R., DOMBROWSKI and ROUTLEY, J., *Chem. Eng. Sci.* **18**, 315, 323 (1963).
432. HOMANN, K., *Combust Flame* **11**, 265 (1967); also HOMANN, K. and WAGNER, H., *Symp. Combustion* **11**, 371 (1967).
433. FRASER, R., DOMBROWSKI, N. and ROUTLEY, J., *Chem. Eng. Sci.* **18**, 339 (1963).
434. ROSIN, P. and RAMMLER, E., *J. Inst. Fuel* **7**, 29 (1933).
435. DAVIS, M. H., "The effect of electric charges and fields on the collisions of very small drops," in *Proc. of the Intl. Conf. on Cloud Phys,* Tokyo, Japan, p. 118 (1965).
436. BARTLETT, J., *Quart. J. Roy. Meteor. Soc.* **92**, 93 (1966).
437. REISS, H., *Ind. Eng. Chem.* **44**, 1284 (1952).
438. REISS, H., *J. Chem. Phys.* **18**, 840 (1950).
439. DOYLE, G., *J. Chem. Phys.* **35**, 795 (1961).
440. KASSNER, J., Jr., CARSTENS, J. and ALLEN, L., *J. Atmos. Sci.* **25**, 919 (1968).
441. HOMANN, K. H. and WAGNER, H., *Ber. Bunsengesellschaft* **69**, 20 (1965).
442. PICH, J., *Staub.* **22**, 15 (1962).
443. KATZ, J. L., *J. Chem. Phys.* **52** (1970).
444. JAEGER, H., WILSON, E., HILL, P. and RUSSELL, K., *J. Chem. Phys.* **51**, 5380 (1969).
445. JUNGE, C., *J. Atmos. Sci.* **26**, 603 (1969).
446. DAWSON, D. B., WILLSON, E., HILL, P., and RUSSELL, K., *J. Chem. Phys.* **51**, 5389 (1969).

AUTHOR INDEX

(Only the name of the first author of two or more collaborators is indexed. The numbers in italics relate to reference numbers. The numbers in ordinary type are page numbers.)

SUBJECT INDEX